# Rapidly Solidified Alloys and Their Mechanical and Magnetic Properties

## MATERIALS RESEARCH SOCIETY SYMPOSIA PROCEEDINGS

ISSN 0272 - 9172

Volume 1—Laser and Electron-Beam Solid Interactions and Materials Processing, J. F. Gibbons, L. D. Hess, T. W. Sigmon, 1981

Volume 2—Defects in Semiconductors, J. Narayan, T. Y. Tan, 1981

Volume 3—Nuclear and Electron Resonance Spectroscopies Applied to Materials Science, E. N. Kaufmann, G. K. Shenoy, 1981

Volume 4—Laser and Electron-Beam Interactions with Solids, B. R. Appleton, G. K. Celler, 1982

Volume 5—Grain Boundaries in Semiconductors, H. J. Leamy, G. E. Pike, C. H. Seager, 1982

Volume 6—Scientific Basis for Nuclear Waste Management, S. V. Topp, 1982

Volume 7—Metastable Materials Formation by Ion Implantation, S. T. Picraux, W. J. Choyke, 1982

Volume 8—Rapidly Solidified Amorphous and Crystalline Alloys, B. H. Kear, B. C. Giessen, M. Cohen, 1982

Volume 9—Materials Processing in the Reduced Gravity Environment of Space, G. E. Rindone, 1982

Volume 10—Thin Films and Interfaces, P. S. Ho, K.-N. Tu, 1982

Volume 11—Scientific Basis for Nuclear Waste Management V, W. Lutze, 1982

Volume 12—In Situ Composites IV, F. D. Lemkey, H. E. Cline, M. McLean, 1982

Volume 13—Laser Solid Interactions and Transient Thermal Processing of Materials, J. Narayan, W. L. Brown, R. A. Lemons, 1983

Volume 14—Defects in Semiconductors II, S. Mahajan, J. W. Corbett, 1983

Volume 15—Scientific Basis for Nuclear Waste Management VI, D. G. Brookins, 1983

Volume 16—Nuclear Radiation Detector Materials, E. E. Haller, H. W. Kraner, W. A. Higinbotham, 1983

Volume 17—Laser Diagnostics and Photochemical Processing for Semiconductor Devices, R. M. Osgood, S. R. J. Brueck, H. R. Schlossberg, 1983

Volume 18—Interfaces and Contacts, R. Ludeke, K. Rose, 1983

Volume 19—Alloy Phase Diagrams, L. H. Bennett, T. B. Massalski, B. C. Giessen, 1983

Volume 20—Intercalated Graphite, M. S. Dresselhaus, G. Dresselhaus, J. E. Fischer, M. J. Moran, 1983

Volume 21—Phase Transformations in Solids, T. Tsakalakos, 1984

Volume 22—High Pressure in Science and Technology, C. Homan, R. K. MacCrone, E. Whalley, 1984

Volume 23—Energy Beam-Solid Interactions and Transient Thermal Processing, J. C. C. Fan, N. M. Johnson, 1984

Volume 24—Defect Properties and Processing of High-Technology Nonmetallic Materials, J. H. Crawford, Jr., Y. Chen, W. A. Sibley, 1984

## MATERIALS RESEARCH SOCIETY SYMPOSIA PROCEEDINGS

Volume 25—Thin Films and Interfaces II, J. E. E. Baglin, D. R. Campbell, W. K. Chu, 1984

Volume 26—Scientific Basis for Nuclear Waste Management VII, G. L. McVay, 1984

Volume 27—Ion Implantation and Ion Beam Processing of Materials, G. K. Hubler, O. W. Holland, C. R. Clayton, C. W. White, 1984

Volume 28—Rapidly Solidified Metastable Materials, B. H. Kear, B. C. Giessen, 1984

Volume 29—Laser-Controlled Chemical Processing of Surfaces, A. W. Johnson, D. J. Ehrlich, H. R. Schlossberg, 1984.

Volume 30—Plasma Processsing and Synthesis of Materials, J. Szekely, D. Apelian, 1984

Volume 31—Electron Microscopy of Materials, W. Krakow, D. Smith, L. W. Hobbs, 1984

Volume 32—Better Ceramics Through Chemistry, C. J. Brinker, D. E. Clark, D. R. Ulrich, 1984

Volume 33—Comparison of Thin Film Transistor and SOI Technologies, H. W. Lam, M. J. Thompson, 1984

Volume 34—Physical Metallurgy of Cast Iron, H. Fredriksson, M. Hillerts, 1985

Volume 35—Energy Beam-Solid Interactions and Transient Thermal Processing/1984, D. K. Biegelsen, G. Rozgonyi, C. Shank, 1985

Volume 36—Impurity Diffusion and Gettering in Silicon, R. B. Fair, C. W. Pearce, J. Washburn, 1985

Volume 37—Layered Structures, Epitaxy and Interfaces, J. M. Gibson, L. R. Dawson, 1985

Volume 38—Plasma Synthesis and Etching of Electronic Materials, R. P. H. Chang, B. Abeles, 1985

Volume 39—High-Temperature Ordered Intermetallic Alloys, C. C. Koch, C. T. Liu, N. S. Stoloff, 1985

Volume 40—Electronic Packaging Materials Science, E. A. Giess, K.-N. Tu, D. R. Uhlmann, 1985

Volume 41—Advanced Photon and Particle Techniques for the Characterization of Defects in Solids, J. B. Roberto, R. W. Carpenter, M. C. Wittels, 1985

Volume 42—Very High Strength Cement-Based Materials, J. F. Young, 1985.

Volume 43—Coal Combustion and Conversion Wastes: Characterization, Utilization, and Disposal, G. J. McCarthy, R. J. Lauf, 1985

Volume 44—Scientific Basis for Nuclear Waste Management VIII, C. M. Jantzen, J. A. Stone, R. C. Ewing, 1985

Volume 45—Ion Beam Processes in Advanced Electronic Materials and Device Technology, F. H. Eisen, T. W. Sigmon, B. R. Appleton, 1985

Volume 46—Microscopic Identification of Electronic Defects in Semiconductors, N. M. Johnson, S. G. Bishop, G. D. Watkins, 1985

# MATERIALS RESEARCH SOCIETY SYMPOSIA PROCEEDINGS

Volume 47—Thin Films: The Relationship of Structure to Properties, C. R. Aita, K. S. SreeHarsha, 1985

Volume 48—Applied Material Characterization, W. Katz, P. Williams, 1985

Volume 49—Materials Issues in Applications of Amorphous Silicon Technology, D. Adler, A. Madan, M. J. Thompson, 1985

Volume 50—Scientific Basis for Nuclear Waste Management IX, L. O. Werme, 1985

Volume 51—Beam-Solid Interactions and Phase Transformations, H. Kurz, G. L. Olson, J. M. Poate, 1986

Volume 52—Rapid Thermal Processing, T. O. Sedgwick, T. E. Siedel, B. Y. Tsaur, 1986

Volume 53—Semiconductor-on-Insulator and Thin Film Transistor Technology, A. Chiang. M. W. Geis, L. Pfeiffer, 1986

Volume 54—Interfaces and Phenomena, R. H. Nemanich, P. S. Ho, S. S. Lau, 1986

Volume 55—Biomedical Materials, M. F. Nichols, J. M. Williams, W. Zingg, 1986

Volume 56—Layered Structures and Epitaxy, M. Gibson, G. C. Osbourn, R. M. Tromp, 1986

Volume 57—Phase Transitions in Condensed Systems—Experiments and Theory, G. S. Cargill III, F. Spaepen, K. N. Tu, 1986

Volume 58—Rapidly Solidified Alloys and Their Mechanical and Magnetic Properties, B. C. Giessen, D. E. Polk, A. I. Taub, 1986

Volume 59—Oxygen, Carbon, Hydrogen, and Nitrogen in Crystalline Silicon, J. W. Corbett, J. C. Mikkelsen, Jr., S. J. Pearton, S. J. Pennycook, 1986

Volume 60—Defect Properties and Processing of High-Technology Nonmetallic Materials, Y. Chen, W. D. Kingery, R. J. Stokes, 1986

Volume 61—Defects in Glasses, Frank L. Galeener, David L. Griscom, Marvin J. Weber, 1986

Volume 62—Materials Problem Solving with the Transmission Electron Microscope, L. W. Hobbs, K. H. Westmacott, D. B. Williams, 1986

Volume 63—Computer-Based Microscopic Description of the Structure and Properties of Materials, J. Broughton, W. Krakow, S. T. Pantelides, 1986

Volume 64—Cement-Based Composites: Strain Rate Effects on Fracture, S. Mindess, S. P. Shah, 1986

Volume 65—Fly Ash and Coal Conversion By-Products: Characterization, Utilization and Disposal III, G. J. McCarthy , D. M. Roy, 1986

Volume 66—Frontiers in Materials Education, G. L. Liedl, L. W. Hobbs, 1986

MATERIALS RESEARCH SOCIETY SYMPOSIA PROCEEDINGS VOLUME 58

# Rapidly Solidified Alloys and Their Mechanical and Magnetic Properties

Symposium held December 2-4, 1985, Boston, Massachusetts, USA

EDITORS:

B. C. Giessen
Barnett Institute, Northeastern University, Boston, MA, U.S.A.

D. E. Polk
Office of Naval Research, Arlington, VA, U.S.A.

A. I. Taub
G.E. Corporate Research and Development, Schenectady, NY, U.S.A.

MRS MATERIALS RESEARCH SOCIETY
Pittsburgh, Pennsylvania

Published by:

Materials Research Society
9800 McKnight Road, Suite 327
Pittsburgh, Pennsylvania 15237
telephone (412) 367-3003

Library of Congress Cataloging in Publication Data

Rapidly solidified alloys and their mechanical and magnetic properties.

(Materials Research Society symposia proceedings, ISSN 0272-9172 ; v. 58)
Includes bibliographies and index.
1. Alloys—Rapid solidification processing—Congresses. 2. Alloys—Mechanical properties—Congresses. 3. Alloys—Magnetic properties—Congresses. I. Giessen, B. C., 1932- II. Polk, D. E. (Donald E.) III. Taub, A. I. (Alan I.)
IV. Materials Research Society. V. Series.
TS247.R367 1986 620.1′6 86-12711
ISBN 0-931837-23-5

Manufactured in the United States of America

Manufactured by Publishers Choice Book Mfg. Co.
Mars, Pennsylvania 16046

# Contents



*Invited Paper

*Invited Paper

## III. CRYSTALLINE MAGNETIC MATERIALS



## PART IV. QUASICRYSTALS



*Invited Paper

## PART VA: MICROSTRUCTURE AND PROPERTIES OF CRYSTALLINE RSP ALLOYS-LIGHT METALS: ALUMINUM, MAGNESIUM, TITANIUM



*Invited Paper

*Invited Paper

# Preface

The papers in this volume comprise 70 contributions on rapidly solidified alloys invited or submitted for oral or poster presentation at the 1985 Fall Meeting of the Materials Research Society, held in Boston December 2-7, 1985.

Since the initiation of the MRS Symposia Proceedings series, this book is the third to be devoted to rapid solidification processing (RSP) and the alloys prepared or treated by methods falling within this category. The two previous volumes are Rapidly Solidified Amorphous and Crystalline Alloys, MRS Symposia Proceedings Vol. 8 (1982), edited by B.H. Kear, B.C. Giessen, and M. Cohen and Rapidly Solidified Metastable Materials, MRS Symposia Proceedings Vol. 28 (1984), edited by B.H. Kear and B.C. Giessen. With the present volume, the Materials Research Society continues its effort to promote the exchange of information in this field. A related symposium is planned for next year's meeting, to be arranged by W. Johnson, L. Tanner and M. Tenhover.

While the emphasis in this symposium was on the mechanical and magnetic properties of rapidly solidified alloys, the papers cover a much wider spectrum of topics including new processing techniques and amorphous alloy formation and phase transformation. Recent developments in the field of rapid solidification were reflected in the large number of papers in two new areas: iron-rare earth permanent magnets and quasicrystals.

The papers are arranged into five sections: Processing, Amorphous Alloys, Crystalline Magnetic Materials, Quasicrystals, and Microstructure and Properties of Crystalline RSP Alloys. A subject index is provided at the back of the book.

The editors and participants are indebted to the following companies for financial support of the symposium:

Allied Corporation
General Electric Company
Martin-Marietta Laboratories
Pratt and Whitney Aircraft

For assistance and session chairmanship, we wish to thank Dr. M.J. Blackburn (Pratt and Whitney), Dr. L.A. Davis (Allied Corporation), Prof. T. Mizoguchi (Gakushuin University), Prof. B.L. Mordike (Technical University Clausthal), Dr. J.R. Pickens (Martin-Marietta) and Dr. R.J. Reynik (NSF). Special thanks to Mrs. Mary Dean (General Electric) who assisted in every aspect of editing this volume.

Bill C. Giessen
Northeastern University
Boston, MA

Donald E. Polk
Office of Naval Research
Arlington, VA

Alan I. Taub
General Electric
Schenectady, NY

# PART I

# Processing

# RECENT DEVELOPMENTS IN LASER SURFACE TREATMENT

B.L. MORDIKE AND H.W. BERGMANN
Institut für Werkstoffkunde und Werkstofftechnik, Technische Universität, Agricolastr. 2, 3392 Clausthal-Zellerfeld, FRG

## ABSTRACT

Some unrelated examples are given where progress has been made in laser surface treatment in the last few years. These are large scale transformation hardening, reproducible surface melting of cast iron, surface alloying with a reaction with the surrounding atmosphere, fundamental studies of solidification and evaporation cleaning.

## INTRODUCTION

Developments in laser technology and in manipulating techniques as well as computer technology have resulted in systems being available for a wide range of surface treatments. The highly localized heat source enables non equilibrium conditions to be achieved. Required microstructures can be produced, enabling tailor making of components with desired surface properties. The types of surface treatment can be subdivided into the classes surface modification, solid state (eg martensitic transformation) and surface melting as shown in Table I.

TABLE I: Types of Surface Heat Treatment

| Treatments without melting | Treatments with melting |
|---|---|
| Rapid homogenization | Simple melting (finer structure, supersaturation) |
| Transformation hardening (Martensitic transformation) | Refinement by evaporation of impurities |
| | Surface alloying |
| | Injecting particles into surface (Alloy hard metals) |
| Surface annealing | Cladding (thick hard metal coatings) |

Each of the applications relies on the specific features of laser treatment ie rapid heating and homogenization and subsequent self quenching. In transformation hardening the rapid quench produces a martensitic structure in the surface which improves wear and fatigue behaviour. In other applications the desired properties are achieved by homogenization of the surface layer by producing a single phase or finely divided microstructure (eg corrosion) or by surface softening (eg increasing ductility for thread rolling) or by changing the composition of the surface or by dispersion/precipitation hardening (eg internal oxidation or nitriding).

Some applications of lasers are discussed in the following with examples of where new developments have recently taken place either in the handling enabling better reproducibility and higher quality of existing techniques or in new applications.

## SOLID STATE TRANSFORMATIONS

Laser hardening has been used successfully to harden a variety of steels [1,2,3], see Table II.

TABLE II: Materials and components hardened by lasers in commercial processes

| Component | Material |
|---|---|
| Steering Gear Housing | Ferritic malleable iron |
| Camshafts and Camfollowers | Ferritic, pearlitic and bainitic SG iron |
| Mineral Sieves | Bainitic SG iron |
| Motor Blocks and Cylinders | Cast iron |
| Motor Shaft Splines | AISI 1050 steel |
| Cutting Blades | AISI 1050 steel |
| Firing Zone Cutout Cams | AISI 4340 steel |
| Electric Razor Cutter Combs | 0.7 % C steel |

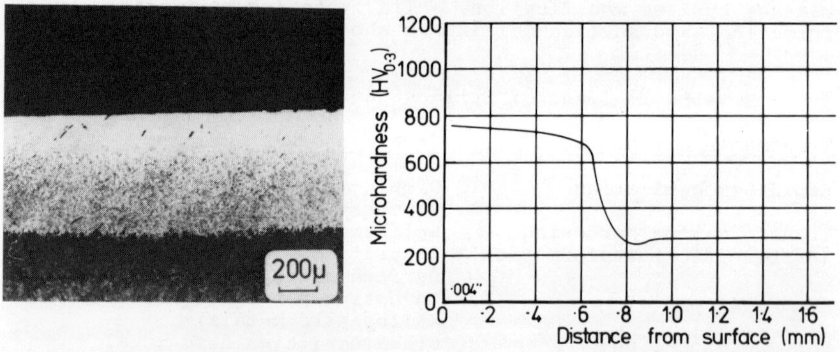


Fig. 1: Typical micrograph with corresponding hardness profile, En8 (1.0503, AISI 1043)

A typical micrograph is shown in Fig. 1, together with the corresponding hardness profile. The effort in recent times has been devoted to harden reproducibly large surfaces without overlapping effects which can lead to tensile residual stresses and possibly cracking. One method of reducing the cracking problem is to increase the effective track front so that instead of a narrow track a wide strip is treated. This can be achieved by beam shaping eg via beam integrators, wobbling mirrors or more elegantly with slit shaped beams, as is available in the Heraeus laser. A typical laser system is shown in Fig. 2. The table is capable of supporting components up to 100 kg in weight.

A problem of laser hardening in production has been the variable absorption by the workpiece. To overcome this colloidal graphite coating has been used thus enabling a very high degree of reproducibility as demonstrated by Dausinger [4] who hardened and tested 40.000 set screws. A further development has been the treatment of very large components, see Fig. 3. Such components could not be treated previously and inherently suitable materials had to be used. Laser hardening is not required over the whole surface - a supporting hardened surface is produced. This reduces total wear.

Fig. 2: 5 kW Heraeus Laser and Manipulator
Computer control of five translations/rotations possible

Fig. 3: Laser hardened diesel cylinders
750 mm id., MAN München photograph

Another form of solid state transformation is homogenization and retention of the homogenized phase on rapid cooling. An example of this is the production of a homogeneous surface on Zircalloy thereby preventing selective corrosion by the dispersed phase. The laser treatment produces very thin layers without distortion unlike other methods. The hardness and strength in depth is retained.

The advances in the field of transformation hardening have been therefore in the reproducibility of the treatment, size of components treated and range of applications.

## LASER MELTING

Laser melting of metallic surfaces enables a rapid and directional solidification of the liquid phase [5]. Depending on the composition and melting conditions different solidification morphologies are obtained [6]. A progressively finer and finer structure and ultimately amorphous solidification is observed with increasing quench rate. Two examples of this, iron-boron and iron-carbon-silicon alloys are given to demonstrate the principle.

### Iron-Boron Alloys

It is now possible to devise a solidification map showing the structure produced as a function of composition and solidification rate for a constant temperature gradient (Fig. 4,5).

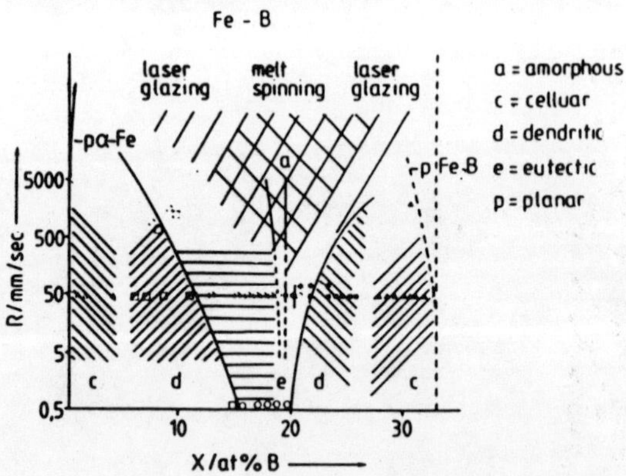


Fig. 4: Solidification map

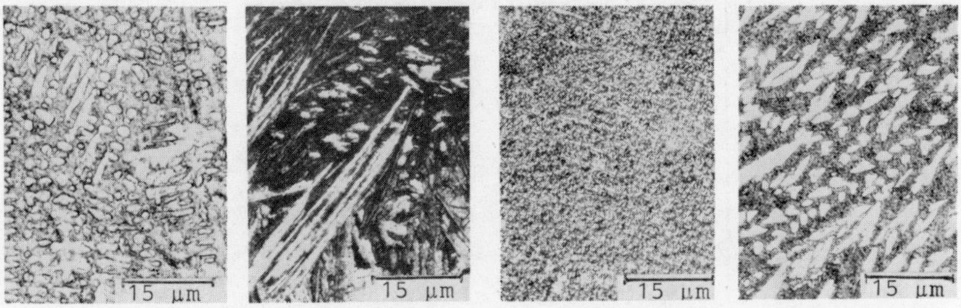


Fig. 5: Microstructure from which Fig. 4 was constructed
a) dendritic b) acicular eutectic
c) $Fe_{23}B_6$ d) $Fe_2B$ dendrites + eutectic

On the iron rich side the limited solubility of boron in iron is hardly increased by rapid quenching. At low boron contents the structure is cellular; on increasing the amount of boron a transition to first dendritic, then feathery and acicular eutectics is observed. Within a composition range of 19-20,5% B $Fe_{23}B_6$ appears. Still higher boron contents result in a reappearance of dendritic and cellular structures, this time the primary phase is $Fe_2B$. The effect of increasing the quenching rate is to displace the boundary between the various structures. At sufficiently high rates amorphous solidification is observed. In the transition region metastable $Fe_3B$ is observed.

Iron-Carbon-Silicon Alloys

If the composition is maintained constant and the solidification velocity and thermal gradient varied, different microstructures can be produced. Fig. 6 shows the variation of solidification mode with temperature gradient and solidification velocity for the alloy Fe3.5%C2%Si. It can be seen that a change in the G/R-ratio leads to macrostructural changes whereas changes in the quenching rate modify the degree of refinement.

Improvements to laser and manipulative devices has enabled a better control of the microstructures obtained and hence properties.

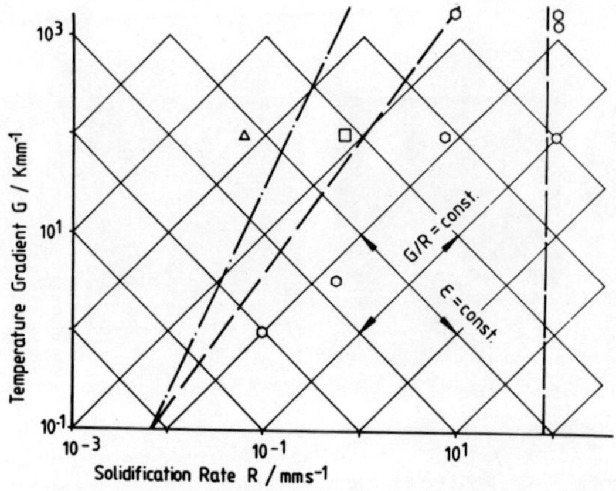


Fig. 6: Variation of the solidification mode
□ cellular — ○ planar, single phase
O dendritic — △ planar, two phase

Surface Melting of Cast Iron

Surface melting of iron base alloys has been carried out successively in the past. It was shown that non metallic inclusions in the surface layer of steels could be rendered harmless by surface melting when they reprecipitated in an extremely fine form or even evaporated. The rolling wear fatigue properties could thus be improved dramatically [6].

Surface remelting of cast iron components for the automobile industry (Fig. 7,8) has now been optimized to a high degree and better service properties have been observed. Laser treatment is distortion free and can produce shiny, crack and pore free surfaces with a roughness of typically $\pm$ 5 µm. This high quality was only possible by specifying the nature, composition and microstructures of the cast iron, for example the quantity, size and morphology of graphite. Elements which affect the hardenability must be known to enable a proper choice of the laser parameters. Irons containing only traces of Cr, Mo, Ni etc. can be remelted crack free without preheating while with 1 to 2% of the alloying elements they must be preheated to 300 - 400°C. Some elements from the moulding sand are also found in the surface regions of a casting and can cause pores or affect the plasma formation.

Control of these parameters has lead to laser treated components with a life four times longer than hitherto with TIG, chill cast iron or $ZrO_2$ plates [7].

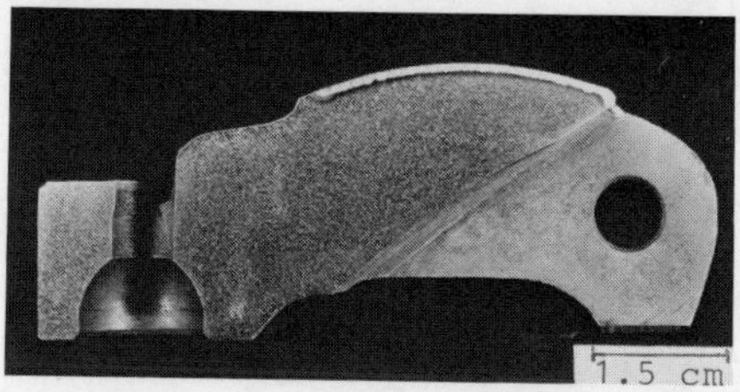


Fig. 7: Camfollower

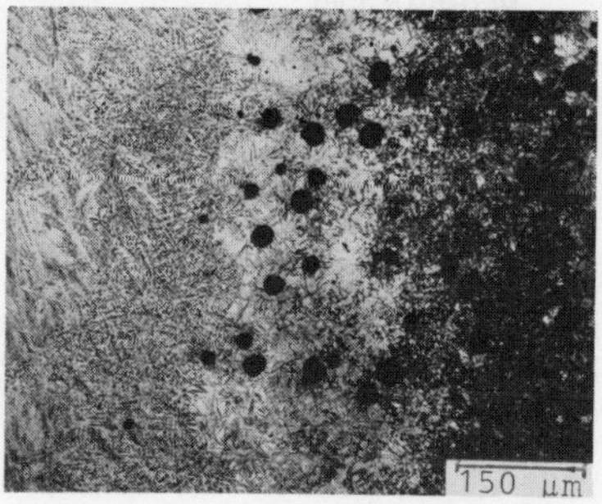


Fig. 8: Microstructure of above camfollower

## SURFACE MELTING WITH BASEOUS ALLOYING

### Internal Oxidation of Silver Base Contact Alloys

Silver base contact alloys are normally produced using powder metallurgical techniques. The requirement is an alloy contact which on the one side can be soldered but on the other contains oxide to prevent welding during switching. Such oxides make soldering difficult. Recent work has shown that it is possible to cause a surface to react with the surrounding atmosphere by heating it with a laser. This has been exploited for carburizing iron base alloys and carburizing and nitriding of titanium alloys. In this present application the laser is used to heat the surface in oxygen thus inducing internal oxidation.

Fig. 9 shows an X-ray microprobe distribution map for Sn in an AgSn alloy. The tin is homogeneously distributed in the matrix and in the surface the tin level is generally low but peaks are observed associated with the formation of tin oxide [8].

Internal oxidation is a very interesting extension of gas alloying with some promising technical applications.

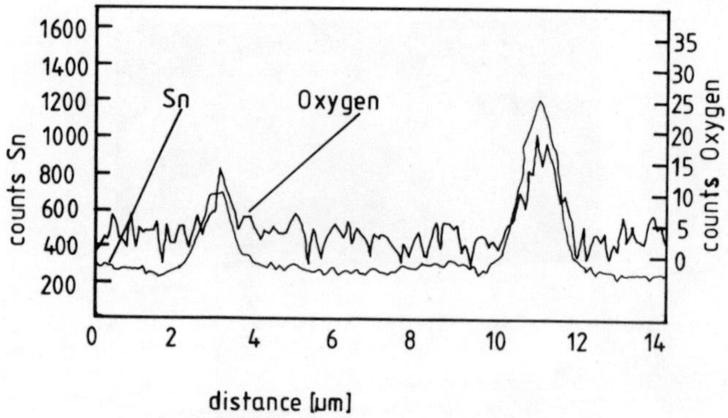


Fig. 9: An X-ray microprobe line scan across the surface

## Gas Alloying of Titanium

Reactive metals, e.g. titanium, zirconium, aluminium, magnesium form compounds when melted in a reactive atmosphere. Such compounds are usually very hard and hence it is possible by a laser surface treatment to produce hard surface layers. Fig. 10 shows a titanium specimen, surface melted in a nitrogen/argon mixture with a typical roughness of a few microns. The homogeneity of the layer is shown in Fig. 11. The hardness values obtained depend on the quantity of nitride and hence interaction time.

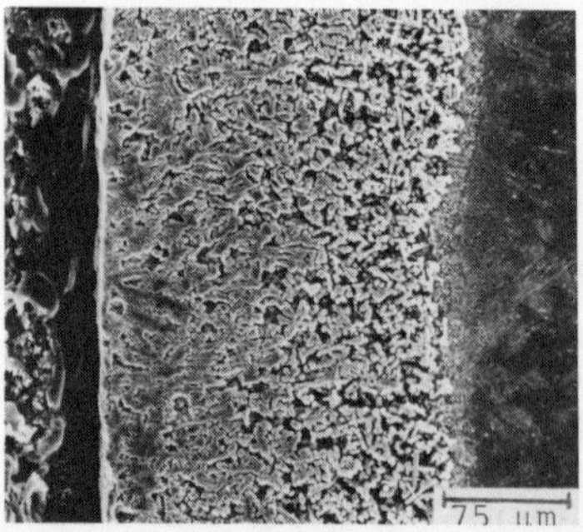


Fig. 10:
Surface melted titanium

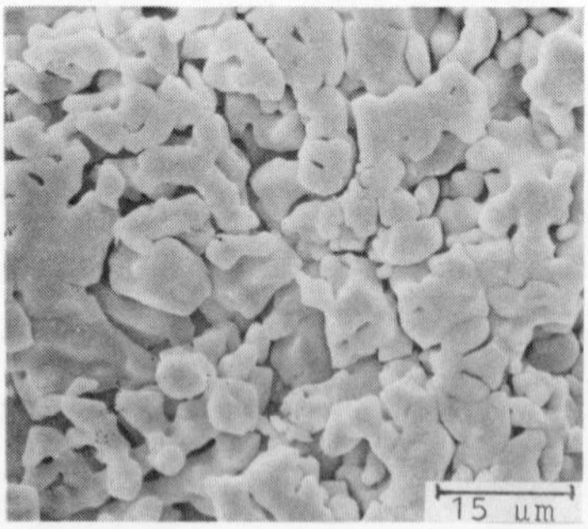


Fig. 11:
Homogeneity of the layer

Surface nitriding of titanium can be considered to have passed the development stage. There are numerous possible applications, where the wear, hardness or corrosion can be improved [9]. Nitriding of titanium is only one possible form of gas alloying - carburizing and carbo-nitriding of titanium is equally successful. Gas carburizing of iron-carbon alloys has also been optimised. Increasing the carbon content increases the amount of martensite and hence hardness in low carbon alloys. Cast irons can also be gas carburized to increase the amount of ledeburitics. Gas alloying is one aspect of laser surface treatment in which significant improvements have been made.

## METALLIC GLASSES

The production of amorphous surface layers by lasers has been widely reported. This is achieved by surface melting and rapidly cooling by a massive substrate alloys of the requisite composition. The production of amorphous surface layers usually has the aim of producing corrosion resistant layers for hostile environments. The high temperature corrosion resistance of Fe-Cr-B layers, for eg, is sufficiently high for applications in heat exchanger tubes for primary circuits in nuclear power stations.

Laser melting of melt spun tapes has been studied both for practical applications and for fundamental research. The direct applications of tapes is naturally limited and melt spun tapes may contain quenched in crystals on the surface. The rate of heating and cooling possible with a laser is such that an amorphous layer can be remelted and on self quenching again produce an amorphous layer without a melt boundary being discernible. In the case of thin tapes and long heating periods crystallization in the heat treated zone may occur. It is also possible to remove quenched in crystals by melting with the laser, see Fig. 12.

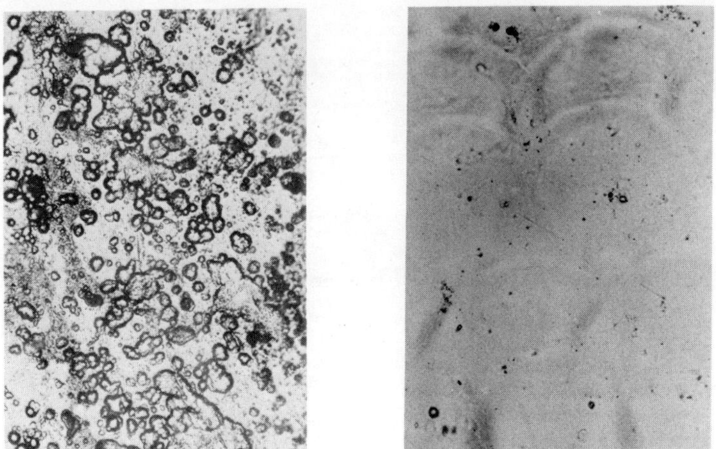


Fig. 12: Crystallization of a metallic glass layer, etched surface. a) as received b) after Nd:YAG laser melting

A fundamental question is whether laser amorphous surfaces are more stable than melt spun tapes. In order to answer this a melt spun tape was crystallized and subsequently laser surface melted. The TTT-curves of the two cases demonstrate no significant difference so long as short period annealing experiments are concerned, see Fig. 13. Long term experiments have not yet been evaluated.

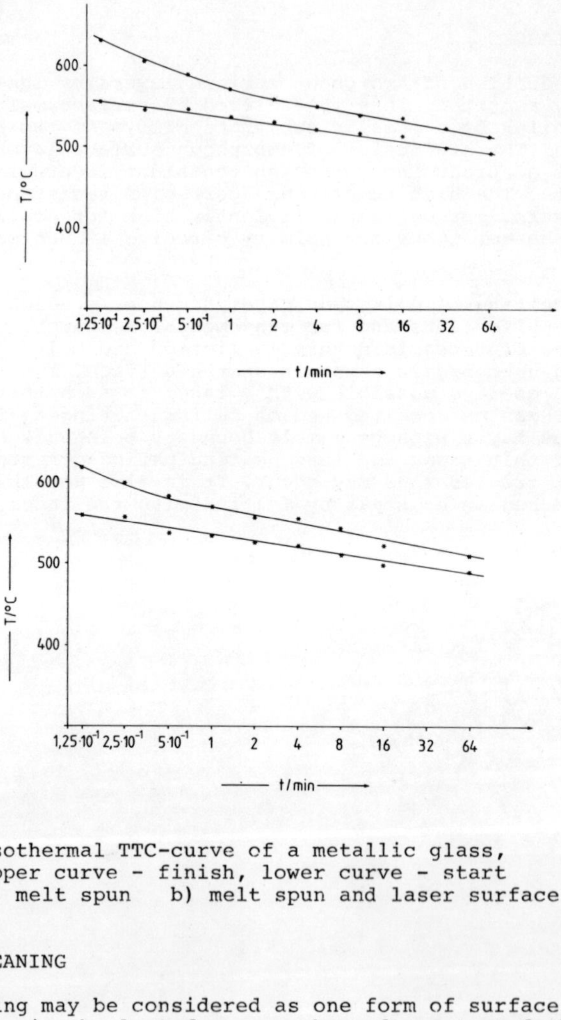


Fig. 13: Isothermal TTC-curve of a metallic glass, upper curve - finish, lower curve - start
a) melt spun b) melt spun and laser surface melted

## SURFACE CLEANING

Decoating may be considered as one form of surface cleaning. Developments in the last few years have demonstrated that the method is useful in removing deposits. Fig. 14 shows TiN coated WC-Co cutting tool chips, before and after laser cleaning. In decoating, the substrate need not be affected and can be used for other coating processes. Decoating can in addition be used in selective welding processes where the area to be pressure welded is selectively cleaned.

Fig. 14: TiN coated cutting tool chips decoated with Nd:YAG laser
a) SEM micrograph
b) X-ray distribution

## CONCLUSIONS

The examples demonstrate that significant advances are still being made in the application of laser surface treated materials. One of the major limitations, hitherto, has been the problem in handling. Developments in computer control have ensured that surfaces can now be treated reproducibly and reliably.

## ACKNOWLEDGEMENTS

The authors are grateful to the Volkswagen Foundation and the Bundesministerium für Forschung und Technologie.

## REFERENCES

1. B.L. Mordike and H.W. Bergmann, Proc. Mat. Res. Soc. Symp., "Rapidly Solidified Metastable Materials", Boston, 45 (1983)
2. B.L. Mordike and H.W. Bergmann, ibid, 29
3. H.W. Bergmann and T. Bell, Laser Surface Treatment, London 1985
4. F. Dausinger, Laser Surface Treatment, London 1985
5. H.W. Bergmann, Surface Engineering Vol.1 No 2, 137 (1985)
6. A. Bloyce, I. Hancock and H.W. Bergmann, 7th Int. Cong. and Int. Trade Fair with 2nd Int. Nd:YAG Laser Conf., München, 1985, to be published
7. H.W. Bergmann, Inf. Conf. on Foundry and Heat Treatment, Johannisburg, 1985 , to be published
8. L. Schultz and H.W. Bergmann, to be published
9. B.L. Mordike and H.W. Bergmann, this volume

# ANALYSIS OF RAPID SOLIDIFICATION PROCESS IN THE DOUBLE-ROLLER METHOD

F. KOGIKU, M. YUKUMOTO, K. SHIBUYA, M. OZAWA AND T. KAN
High Technology Research Laboratories, Kawasaki Steel Corporation
Kawasaki-cho 1, Chiba, 260, Japan

## ABSTRACT

High-silicon steel was rapidly solidified to thin strips by the double roller method. Two typical macrostructures were observed: one with an equiaxed zone and the other without. The formation of the equiaxed zone is caused by an excessive gap between the rollers. Heat transfer calculations and dendrite arm spacing measurements both suggested that the cooling rate is about $10^3$ to $10^4$ K/sec.

## INTRODUCTION

Production of thin sheet directly from molten metal using rapid solidification techniques has received increasing attention. These techniques offer not only cost advantages associated with streamlining of the sheet-making process, but also unique material properties brought about by the rapid solidification process itself.

The crystallized thin sheets produced by rapid solidification have the following unique features:
(1) Increased solid solubility
(2) Decreased segregation and precipitation
(3) Finer crystalline structure
(4) Manufacture of thin sheets from materials difficult to shape
(5) Simplification of the thin sheet making process

These features are advantageous for manufacturing high-silicon steel thin sheet, which is difficult to roll but superior in magnetic and mechanical properties [1]. Kawasaki Steel is experimentally producing such high-silicon steel thin sheets by the double roller method.

Examined below are the solidification characteristics of the rapidly solidified high-silicon steel strips directly cast by the double-roller method.

## MANUFACTURING CONDITIONS

Molten steel with silicon contents as high as 4.5-5.5 wt.% was introduced between two rollers with internal water-cooling systems. The rollers were 400-550mm in diameter and rotated at circumferential speeds of 3-12 m/sec. The rapidly solidified thin steel strip (RS strip) thus obtained measured 120-500 μm in thickness. An example is shown in Photo. 1.

Photo. 1 High-silicon steel rapidly solidified strip

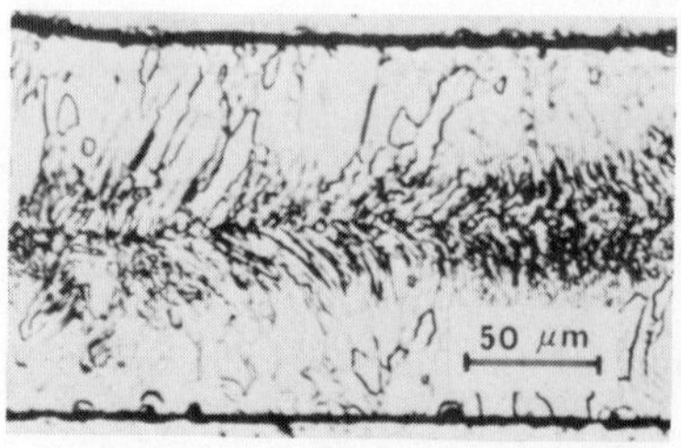


roller rotation direction
Photo. 2 Structure of RS strip
(without equiaxed zone)

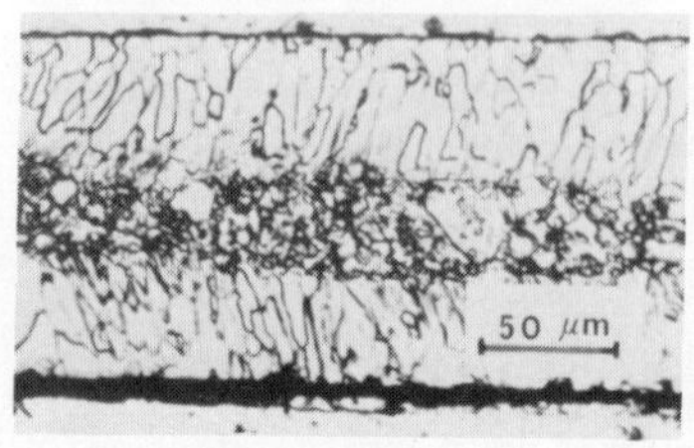


roller rotation direction
Photo. 3 Structure of RS strip
(with equiaxed zone)

## RESULTS AND DISCUSSION

### Macrostructure of Rapidly Solidified Strip

Photos. 2 and 3 show typical examples of the mactostructure of RS strip. In Photo. 2, fine columnar crystals have developed from both surfaces, and extend almost to the center thickness. By contrast, in Photo. 3, the fine columnar crystals do not reach the center of the strip thickness, and the central area consists of equiaxed crystals [2].

The equiaxed zone is considered to be formed by roller-misalignment or roller heat crown, which are schematically shown in Fig. 1. When the gap of the rollers is like the one shown in Fig. 1(B) or (C), the equiaxed zone is formed at the large gap part (a). Fig. 2 supports this idea. In Fig. 2, which is for heat crown case (the strip is 250mm in width), the thickness of the equiaxed zone varies along the RS strip's transverse direction, i.e., the zone thickness is low at the center, but high at the edge. From the results, the mechanism of the formation of the equiaxed zone is considered as follows: Initial chill crystals are formed in the molten meral, then fall and laterally spread into the large gap part without remelting and function as nuclei for equiaxed crystal formation.

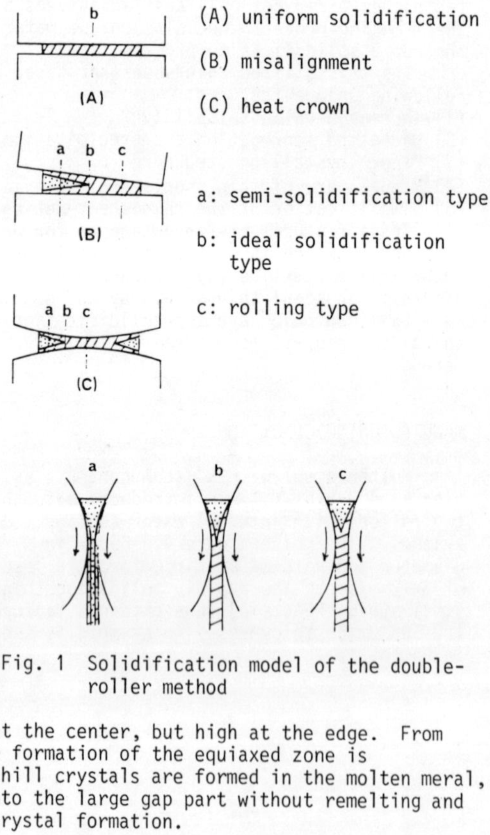


Fig. 1 Solidification model of the double-roller method

## Cooling Rate of Rapidly Solidified Strip

The primary dendrite arm spacing observed in the RS strip is shown in Photo. 4. The spacing increases from 1 to 7 μm along the thickness direction.

For the primary dendrite arm spacing in silicon steel and other alloy steels, Kishitake and Okamoto have obtained the following relationship [3]:

$$d_I = 110\ [mc\ (k - 1)]^{1/6}/v^{1/2} \quad . \ . \ . \ . \ . \ (1)$$

where $d_I$: Primary dendrite arm spacing (μm)
m : gradient of liquidus line (K/%)
c : concentration of solute (%)
k : equilibrium distribution coefficient
v : cooling rate (K/sec)

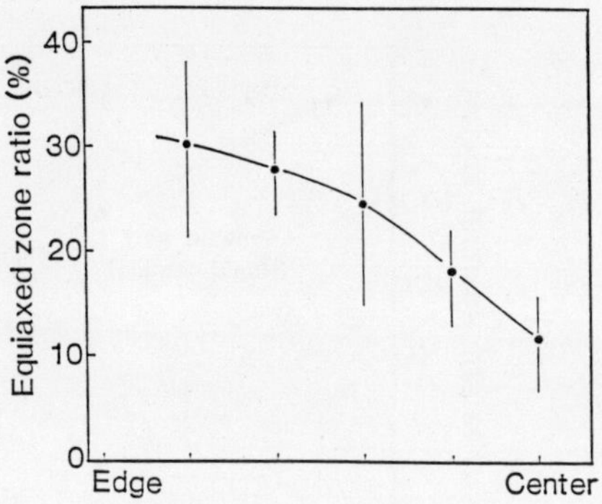


Fig. 2 Ratio variations of equiaxed zone along transverse direction of strip

For 4.5 to 5.5 wt% Si-steel, equation (1) becomes

$$d_I = 270\ v^{-1/2} \quad . \ . \ . \ . \ . \ . \ . \ . \ . \ . \ . \ . \ . \ . \ . \ . \ (2)$$

If it is assumed that equation (1) is applicable to the present rapid solidification process, the cooling rate becomes to be $10^3$ to $10^4$ K/sec.

The adequacy of this value has been independently examined in the present study by an experiment and a theoretical calculation. In the experiment, high-silicon steel was cast into a small copper mold to obtain the cooling curve. In the calculation, pseudo two-dimensional heat transfer models were adopted [4], and the heat transfer coefficient between the roller and steel was fixed at an estimated value of 0.80 cal/sec.cm$^2$.K.

The results of these examinations are shown in Fig. 3. The experimental and calculated values of the cooling rate show good agreement with that derived from equation (2). It is concluded that equation (2) is applicable to the rapid solidification by the double roller.

Surface

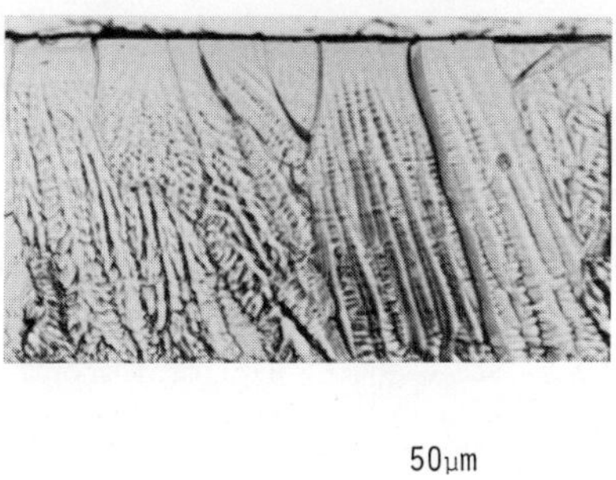

Center

50μm

Photo. 4 Dendritic structure of rapidly solidified strip

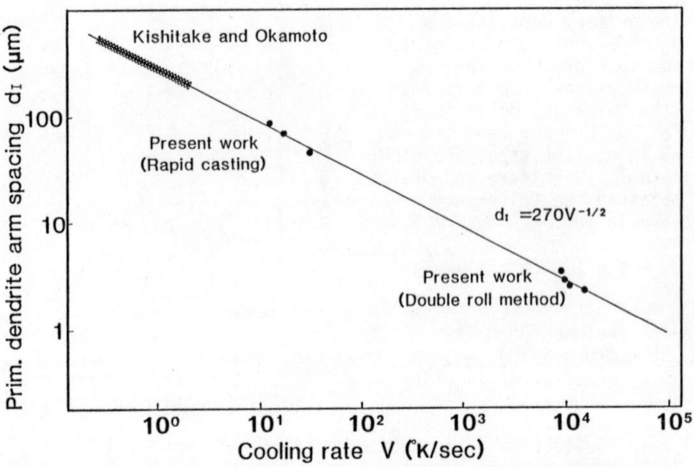


Fig. 3 Relation between cooling rate and primary dendrite arm spacing

## CONCLUSION

Rapid solidification by the double roller technique, with its unique advantages, is expected to be utilized in the future for the production of high-silicon steel strip, a material which is difficult to manufacture by conventional means but superior in magnetic and mechanical properties.

High-silicon steel strips produced in the present study by the double roller method have shown two typical macrostructures, one with an equiaxed zone, and the other without. This difference has been attributed to an excessive gap between rollers which leads to the formation of the equiaxed zone.

The cooling rate with double-roller quenching has been extimated to be $10^3$ to $10^4$ K/sec.

## REFERENCES


1. T. Kan, Y. Ito and H. Shimanaka, J. Magnetism and Magnetic Materials. 26, 127 (1982)
2. K. Shibuya, T. Kan, Y. Shimizu and Y. Ito, The 112th AIME Annual Meeting, (1983), No. A83-15
3. K. Kishitake and T. Okamoto, "Tetsu to Hagane" (Iron & Steel). 63. 425 (1977)
4. M. Yukumoto, K. Shibuya, T. Kan and Y. Ito, in Rapidly Quenched Metals, edited by S. Steeb and H. Warlimont (North-Holland Physics Publishing, Amsterdam, 1985), P.91

# INNOVATIVE MELT-SPINNING TECHNIQUE FOR FERROMAGNETIC AMORPHOUS RIBBONS

H. Y. Yu*, D. R. Huang**, P. C. Yao** and S. E. Hsu**
*Chung Cheng Institute of Technology, Tachi, Taiwan 335, R.O.C.
**Chung Shan Institute of Science and Technology, Lungtan, Taiwan 325, R.O.C.

## ABSTRACT

A new melt-spinning technique (field quenching) for ferromagnetic amorphous ribbon is suggested to produce better magnetic properties than conventional as-quenched amorphous ribbon. An external field is applied during the quenching either parallel or perpendicular to the casting direction. The magnetic moment and initial permeability are increased when the field direction is parallel to the casting direction. These effects are opposite when these two directions mentioned are perpendicular to each other.

## INTRODUCTION

The amorphous alloy or the glassy metal produced by continuous melt-quenching processes are recognized as important materials for industrial applications as well as for fundamental scientific research. Some of them show attractive engineering characteristics, such as soft magnetic properties, high strength or good corrosion resistance, which are equivalent to or superior to those of conventional crystalline materials. The occurrence of ferromagnetism in amorphous materials was predicted on a theoretical basis by Gubanov [1]. The first strongly ferromagnetic glassy metal obtained by liquid quenching was $Fe_{80}P_{12.5}C_{7.5}$ [2]. Recent studies of magnetic behavior of glassy metal have been centered on glassy ribbon because of their potential industrial applications [3].

The ferromagnetic glassy metal ribbons having general composition, transition metals ~20% metalloid, exhibit low coercive force Hc ~0.1-0.001 Oe. and high permeabilities comparable to soft magnetic alloys. A number of investigators have shown that significant improvement in magnetic properties of glassy metal can be obtained by field annealing [4~8]. However, there is a drawback effect of heat treatment, which causes embrittlement of iron-containing amorphous alloys [9]. In some cases, the alloys become so brittle after annealing that they can not be handled without breaking. Since the combination of excellent magnetic and mechanical properties is the most attractive feature of the amorphous alloys, the loss of ductility by annealing is a serious problem.

Our purpose in this paper is to report a new melt-spinning technique with the result that is believed to be a highlight and possibly can produce amorphous ribbons with better soft magnetic properties than conventional as quenched ribbons. The method is defined as "Field Quenching Technique."

## EXPERIMENT PROCEDURE

The castings were carried out in air on a rotating pure copper wheel with the diameter of 29.5cm and 2060 rpm rotating speed which gave a linear surface speed of 3187.2cm/s. The quartz crucibles had retangular slot openings of 0.35cm by 20cm. The crucible is held vertically and displaced about 3.5cm away from the vertical central line of the wheel, and the gap between crucible bottom surface and wheel surface was typically 1mm during casting. The ejection pressure was between 0.4 to 2.5kg/cm$^2$. The small variation in the casting conditions stated above did not change the products significantly. A magnetic field is applied to the melt during casting.

The magnetic field applied during casting are along two directions, "longitudinal" and "transverse." The longitudinal direction $\hat{L}$ means that the direction of magnetic field during casting is parallel to the casting direction ($\hat{R}$). The transverse direction ($\hat{T}$) means the field direction is perpendicular to both casting direction and ribbon surface normal. During present experiments, the longitudinal direction $\hat{L}$ is not exactly parallel to $\hat{R}$ but with an angle θ~6.8° due to the off center of the crucible (Fig. 1).

FIG. 1

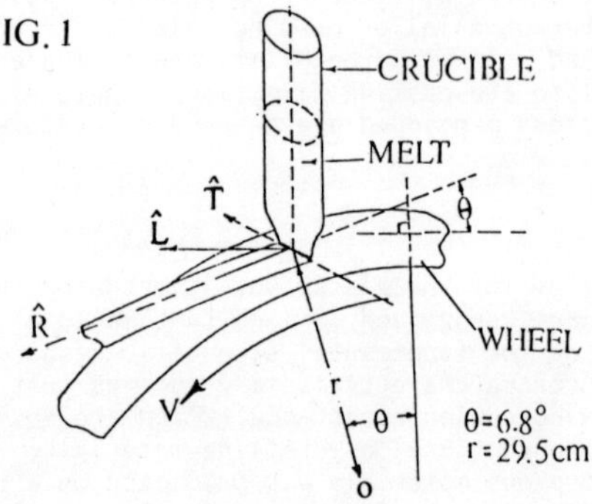


The composition of the alloys prepared is $Fe_{42}Ni_{38}Mo_4B_{16}$ with about 1wt% unidentified impurity and in the form of ribbon with ~40μm thick and ~19mm wide. The samples were tested by differential scanning calorimeter (DSC) and X-ray diffraction to make sure they were amorphous. The crystallization temperature is 430°C as obtained by DSC. All the samples were bent to be sure that they were ductile. Magnetisation at room temperature was determined in a vibrating sample magnetometer (EG & G Modle 155 VSM) to a maximum field of 9KOe. on specimen about 6mm long x 2.5mm wide with thickness about 40μm.

It has been shown in the centrifugal spinning process that the resultant domain structures appear to reflect the flow pattern such that the direction of magnetization is parallel to the direction of spin $\hat{R}$ in the center of the ribbon and perpendicular to $\hat{R}$ at the ribbon edges [3]. The VSM specimen were cut from the center of the ribbon in order to avoid this flow inhomogeneity as much as possible. The specimen was tested with its length direction ($\hat{Z}$) parallel to VSM vibrating axis and its surface normal ($\hat{X}$) perpendicular to both $\hat{Z}$ direction and VSM magnetic field ($\hat{H}v$). The various specimen orientations are shown in Table 1.

Table I The relation among field directions and rolling direction for different sample.

| Sample* | LN | LP | TN | TP | N | P |
|---|---|---|---|---|---|---|
| Direction Relation | $\hat{H}_f // \hat{R} \perp \hat{H}v$ | $\hat{H}_f // \hat{R} // \hat{H}v$ | $\hat{H}_f \perp \hat{R} \perp \hat{H}v$ | $\hat{H}_f \perp \hat{R} // \hat{H}v$ | $\hat{R} \perp \hat{H}v$ | $\hat{R} // \hat{H}v$ |

*The number labeled (eg. LN-50) is the applied field strength (in Oersted) during quenching.

RESULTS AND DISCUSSION

The low field (100 Oe.) initial magnetization curves obtained are shown in Fig. 2. The measurement have been repeated on several different specimens. The results showed that the magnetic moment and initial permeability are increased when the direction of applied field $H_f$ is parallel to the casting direction (curves a, b and c), while the transverse field has the deterioration effect, i.e. decreasing the moment and initial permeability (curves e and f). Curve d is the result for specimen N, P, TN-50 and TP-50 which means 1) low field initial magnetization shows no obvious casting-induced anisotropy when there is no magnetic field applied during quenching, 2) the sample with the transverse magnetic field applied has to be larger than 50 Oe. in order to show the negative influence(as shown by curves e and f), 3) the field during quenching parallel to the casting direction is more helpful than perpendicular to the casting direction.

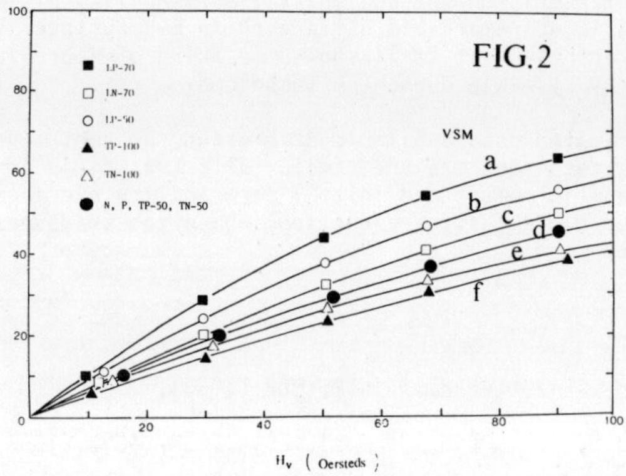


Further evidence of field quenching in FeNiMoB alloy is shown by the high field (9KOe.) initial magnetization curve of Fig. 3 and the B-H hysteresis of Fig. 4 which is measured by hysteresisgraph (Walker Sci. Inc. Model MH-20). Fig. 3 shows the induced anisotropy of as-casted sample without magnetic field applied during quenching(at high field)together with the effect of

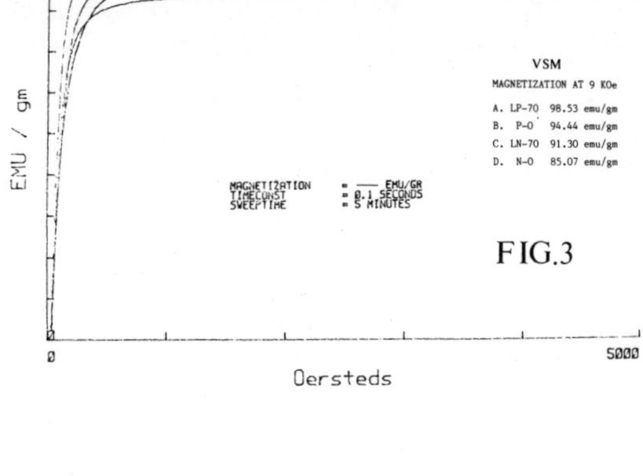

FIG.4

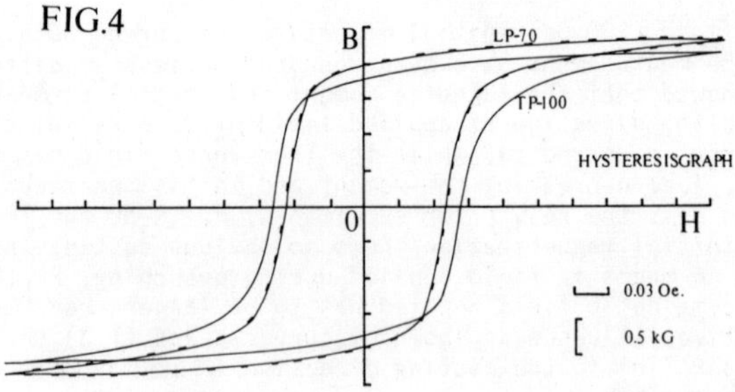


increasing the magnetic moment and initial permeability by longitudinal field applied. There is no remarkable difference in B-H hysteresis loop obtained so far, but nevertheless it still shows the trend of improving the soft magnetic property by field quenching technique.

From the limited data available at present, we tentatively conclude that field quenching technique has the similar effect as field annealing process [4~8]. The magnetic moment and initial permeability are increased when the field is applied along casting direction, while the transverse field has opposite effect.

REFERENCES

1. A. I. Gubanov, Fiz. tverd. Tela 2, 502 (1960).

2. P. Duwez and S. C. H. Lin, J. Appl. Phys., 38, 4096 (1967).

3. H. S. Chen, Rep. Prog. Phys., 43, 353 (1980).

4. T. Egami, P. J. Flandred and C. D. Graham, Jr., AIP Conf. Proc., 24, 697 (1974).

5. T. Egami, P. J. Flandred and C. D. Graham, Jr., Appl. Phys. Letters, 23, 128 (1975).

6. R. C. Sherwood, E. M. Gyorgy, H. S. Chen, S. D. Ferris, G. Noeman and H. J. Leamy, AIP Conf. Proc., 24, 745 (1974).

7. H. S. Chen, S. D. Ferris, E. M. Gyorgy, H. J. Leamy and R. C. Sherwood, Appl. Phys. Letters, 26. 405 (1975).

8. F. E. Luborsky, J. J. Becker and R. O. McCary, IEEE Trans. Magn., Mag-11, 6, 1644 (1975).

9. T. Masumoto and T. Egami, Mat. Sci. Engr., 48, 147 (1981).

# RAPIDLY SOLIDIFIED METAL FOILS BY MELT OVERFLOW

THOMAS GASPAR*, LLOYD E. HACKMAN*, YOGESHWAR SAHAI**, W.A.T. CLARK** AND JOHN WOOD***
* Ribbon Technology Corporation, Box 30758, Gahanna, OH 43230
** Ohio State University, 116 West 19th Avenue, Columbus, OH 43210
*** Open University, Walton Hall, Milton Keynes, UK MK7 6AA

## ABSTRACT

A single roller technique called melt overflow for direct casting metallic sheet, strip or foil is described. A simple heat transfer model is developed to predict foil thickness from process parameters. The model agrees with thickness of Ti-6Al-4V cast strip. Cooling rates as high as $10^5$ K/s are estimated from measurements of secondary dendrite arm spacings in 7075 aluminum alloys. Applications include magnesium-graphite composites.

## PROCESS DESCRIPTION

Ribtec's melt overflow rapid solidification process is shown schematically in Figure 1. Molten metal overflows a reservoir onto the surface of a rotating chill block. The melt stream is not extruded through an orifice like melt spinning techniques, rather, the melt pool overflows a reservoir and is channeled by a runner or tundish to contact the moving chill surface. Rapidly solidified metallic fiber, filaments, flakes, particulate, foil, sheet and strip can be cast depending on the surface geometry of the chill block.

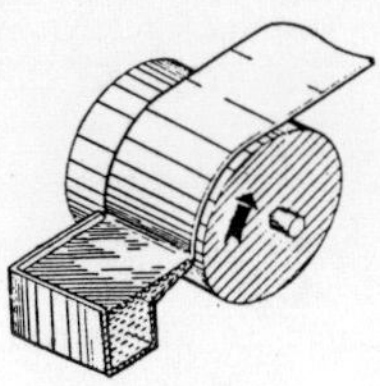

Fig. 1 Melt Overflow Process

Ribbon Technology Corporation operates a 200 pound steel equivalent air induction furnace and water-cooled chill block to cast metallic fiber, filaments, flakes, foils, sheet and strip by melt overflow. Typically, the chill blocks are 23 cm (9 in) to 36 cm (14 in) diameter and up to 30 cm (12 in) wide. Chill blocks have been fabricated from copper, brass or steel. Casting rates as high as 20 m/s (66 ft/s) are used to cast fine filaments and fiber. Foil, sheet and strip products are cast at rates up to 10 m/s (33 ft/s) depending on the thickness desired.

## HEAT TRANSFER MODEL

The analysis of heat transfer, when molten metal comes in contact with a metallic chill surface, is very complicated. Molten metals do not normally "weld to the substrate, so a thermal contact resistance exists at the solidified metal-mold interface. This resistance results in a large temperature drop at the interface. The thermal conductivity of the solidifying metal also contributes to the overall resistance to heat flow and results in a thermal gradient within the solidifying foil.

This model is based on the assumptions that the interface resistance predominates over the resistance offered by the metal;the temperature in the water cooled copper mold and at the casting surface is constant, To; the metal freezes as a plane at N=M where the temperature is at the freezing point of the metal, Tm; and the temperature profile within the solidifying

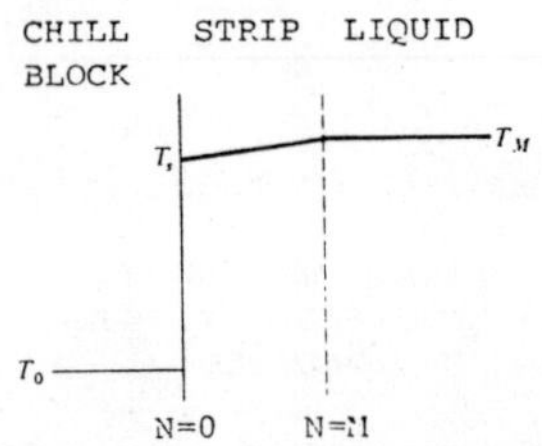


metal layer is a linear function. Figure 2 presents a schematic diagram of the temperature profile in the solidifying foil. The solidified metal temperature at the interface is Ts, which is higher than the roll temperature, To due to interface resistance.

* Fig. 2 Model temperature profile

Heat transfer due to conduction in the solidified layers is:

$$q=k\,\frac{Tm - Ts}{M} \tag{1}$$

Where k is the thermal conductivity of the metal and M is the thickness of the solidified layers.

Heat transfer at the metal - chill block interface is given by:

$$q=h\,(Ts - T_0) \tag{2}$$

Where h is the heat transfer coefficient. Combining equations (1) and (2) yields:

$$q = \frac{Tm - T_0}{\frac{1}{h} + \frac{M}{k}} \tag{3}$$

The total heat evolved due to latent and superheat is given by:

$$q = \rho H'_f \frac{dM}{dt} \tag{4}$$

Where $H'_f$ is effective latent heat: $H'_f = H_f + Cp_1\ \Delta Ts$

and, $H_f$ is the latent heat of fusion, $Cp_1$ is the heat capacity of the molten metal and $\Delta Ts$ is the superheat of the melt. The densities of liquid and solid metal at the melting temperature are assumed to be equal.

Equating equations (3) and (4), we get:

$$\rho H'_f \frac{dM}{dt} = \frac{Tm - T_0}{\frac{1}{h} + \frac{M}{k}} \tag{5}$$

Upon integrating with M=0 at time t=0 and M=M at time t=t;

$$M + \frac{M^2 h}{2k} = h\,\frac{(Tm - To)\,t}{\rho H'_f} \tag{6}$$

Equation (6) can be used to determine the thickness of solidified layer, M for any time t that the molten metal is in contact with the substrate.

## AS-CAST SHEET AND STRIP

A computer program was written to calculate the thickness of a solidified layer M according to equation (6). Values for the physical properties of titanium used in the model were: $k = 0.2$ W/cm K; $\rho = 4.51$ g/cm$^3$ ; $H_f = 428$ J/g. The heat transfer coefficient, h, used for all calculations was 10 W/cm$^2$ K and the length of contact 2.8 cm.

The feasibility of casting Ti-6Al-4V alloy foils by melt overflow was demonstrated using a vacuum melting system [1]. Each experiment produced approximately 450 gm (1 lb) of 2.5 cm (1 in) wide Ti-6Al-4V alloy strip. The thickness of the strip depends on the casting speed, as shown in Figure 3. The solid line in Figure 3 represents the strip thickness predicted the heat transfer model while the experimental values (measured with a micrometer) are plotted for ten different experiments. The microstructure of the rapidly solidified Ti-6Al-4V alloy foils reveals a fine acicular structure that is martensitic in appearance (Figure 4).

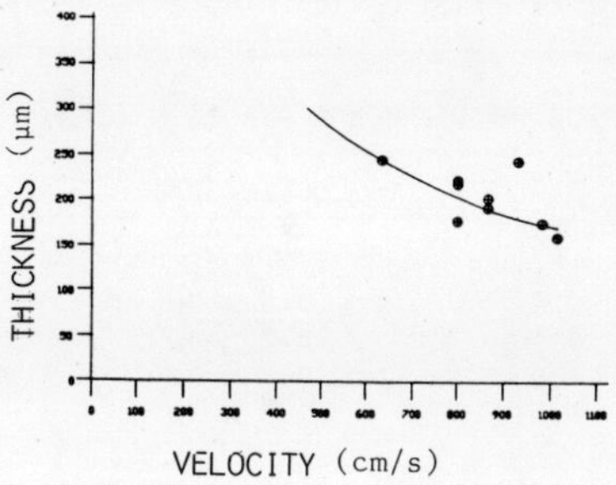


Fig. 3 Thickness of Ti-6Al-4V strip

Aluminum alloys A360 (9.5 Si-0.5 Mg) and 7075 (5.6 Zn-2.5 Mg-1.6 Cu-0.23 Cr) and magnesium alloys AM60 and AZ61 were cast into sheet up to 28 cm (11 in) wide and up to 1.5 mm (0.060 in) thick by melt overflow. Foils as thin as 0.075 mm (0.003 in) were also cast. The magnesium alloys were cast under type 230 flux in a steel crucible while the aluminum alloys were cast in air. The microstructures and cooling rates of the aluminum alloy foils strongly depend on thickness. Thin foils of 7075 alloy exhibit a fine dendritic structure (Figure 5). Cooling rates during melt overflow have been estimated based on measurements of secondary dendrite arm spacings for foils produced at different casting rates [2].

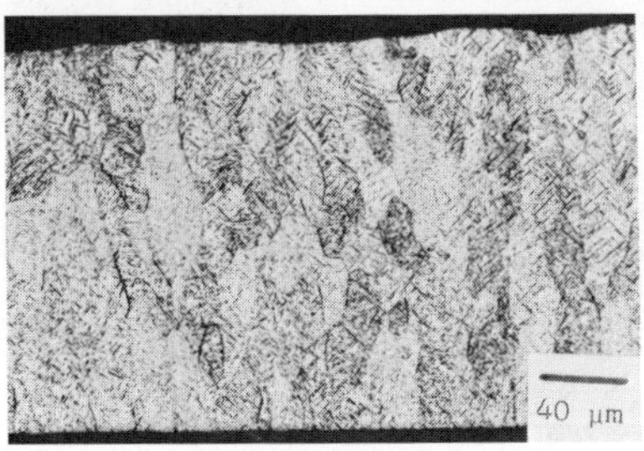


Fig. 4 Ti-6Al-4V strip

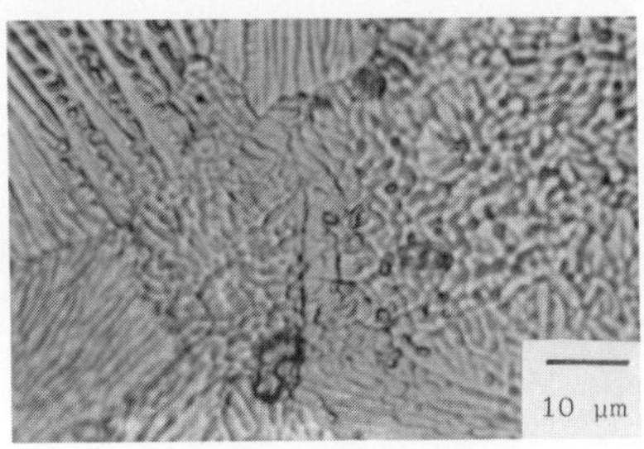


Fig. 5 7075 Aluminum Foil

Flemings [3] gives the following relationship between dendrite arm spacing, d, and the cooling rate U:

$$d = b\,U^{-n} \qquad (7)$$

were b and n are constants. Jones [4] documented values for for the constants b and n for various alloys. Values of b = 44 microns K/s and n = 0.36 were used in estimating cooling rates from the melt overflow 7075 aluminum alloy foils.

Table I gives the dendrite arm spacings and cooling rates for 7075 foils cast at different speeds. The foil thickness decreases as casting speed increases, all other factors being equal. Other process variables that affect thickness include, but are not limited to the length of contact between the melt and substrate and the properties of the liquid metal or alloy. Based on the data presented in Table I, it can be concluded that the melt overflow process is capable of cooling rates up to $10^5$ K/s, like other chill block casting techniques.

TABLE I. ESTIMATES OF COOLING RATES FROM DENDRITE ARM SPACINGS (DAS)

| SPEED (cm/s) | THICKNESS (mm) | DAS (microns) | COOLING RATE (K/s) |
|---|---|---|---|
| 400 | 0.085 | 2.0 | $5.4 \times 10^3$ |
| 500 | 0.087 | 1.5 | $1.2 \times 10^4$ |
| 500 | 0.080 | 1.0 | $3.7 \times 10^4$ |
| 600 | 0.059 | 0.5 | $2.5 \times 10^5$ |

Ribbon Technology Corporation's melt overflow process has cast 28 cm wide (11 in) magnesium alloy AM-60 foil. The direct cast foil can be used in graphite-magnesium composites at potentially lower cost than rolled magnesium alloy foils. Figure 7 shows a 2.5 cm (1 in) diameter rapidly solidified magnesium-graphite clad tube fabricated by the MCI Pultrusion process. A micrograph of the clad is shown in Figure 8. It is anticipated that these rapidly solidified magnesium-graphite composites will be used in aerospace structural applications.

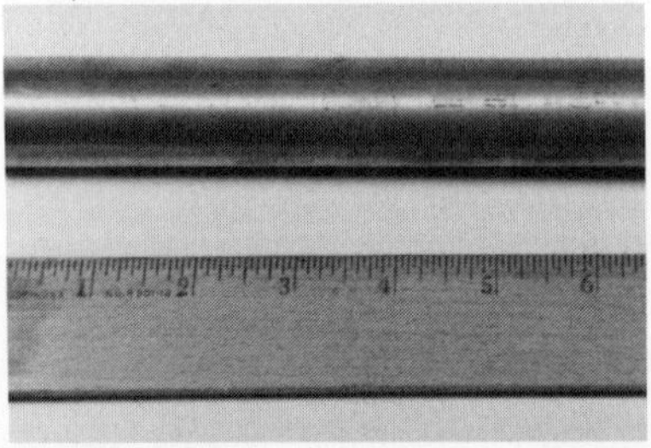

Fig. 7 Magnesium-graphite clad tube

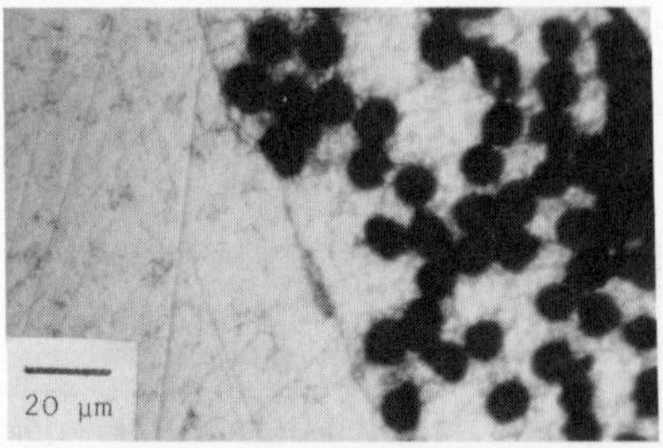


Fig. 8 Micrograph of clad tube

REFERENCES

1. Maringer, R. E. (private communication).

2. Gaspar, Thomas, Hackman, Lloyd, Sahai, Yogeshwar, Mozhi, T.A., and Clark, W.A.T., NASA Contract NAS1-17978, Final Report, June, 1985.

3. Flemings, M. C., Solidification Processing, (McGraw Hill, Inc. New York, New York) 1974.

4. Jones, H., from Treatise on Materials Science and Technology, 20 (edited by H. Herman, Academic Press, New York, New York).

# A STUDY OF AMORPHOUS ERBIUM-BASED ALLOYS FORMED BY NEAR-ISOTHERMAL COLD-ROLLING OF ELEMENTAL COMPOSITES

MICHAEL ATZMON*, KARL M. UNRUH**, CONSTANTIN POLITIS*** AND WILLIAM L. JOHNSON, W. M. Keck Laboratory of Engineering Materials, California Institute of Technology, Pasadena, California 91125.

* present address: Division of Applied Sciences, Harvard University, Cambridge, Massachusetts 02138.
** present address: Department of Physics, University of Delaware, Newark, Delaware 19711.
*** Kernforschungszentrum Karlsruhe, Institut fur Nukleare Festkorperphysik, Postfach 3640, D-7500 Karlsruhe, Fed. Rep. of Germany.

## ABSTRACT

We report the formation of single-phase amorphous Cu-Er and Ni-Er alloys in bulk form by cold-rolling of composites prepared from elemental foils. As for previously reported cases of metallic glass formation by solid-state reaction, the driving force for the reaction is the negative enthalpy of mixing of the constituent elements. It occurs during deformation close to room temperature. Amorphous $Cu_{72}Er_{28}$ was also produced by high-energy ball-milling of the elemental powders as well as by sputtering and liquid quenching. The alloys obtained were characterized by means of differential scanning calorimetry and x-ray diffraction. The crystallization behavior observed and the radial distribution functions obtained showed good agreement between the alloys prepared by different methods.

## INTRODUCTION

In recent years it has been shown that a solid-state reaction of elemental polycrystalline metals can result in the formation of an amorphous phase [1]. Thin layers of polycrystalline Au and La reacted at 80 °C to form amorphous Au-La alloys. The reaction was explained as driven by the large and negative enthalpy of mixing of the constituent elements. In a related experiment, Koch et al. [2] demonstrated that high-energy ball-milling of elemental Ni and Nb powders resulted in the formation of amorphous alloy powder. Further studies of amorphous alloys prepared by ball-milling and their comparison to rapidly quenched alloys have been reported [3-5].

We have recently reported the extension of the thin film diffusion experiment to three dimensions [6]. Amorphous Cu-Zr and Ni-Zr alloys were produced by solid-state reaction in elemental composites. The composites were prepared by mixing elemental foils or powders and cold-rolling them. Schultz [7] reported experiments based on the same principle.

In this study, Ni-Er and Cu-Er were selected for the following reason: 1) rare-earth alloys are of interest for magnetic applications, 2) amorphous rare-earth based alloys have been studied by many authors. Erbium was chosen as the rare-earth element because of its relatively low oxidation rate in air and its high ductility.

Based on reported calorimetric measurements of the enthalpy of dissolution of Er in liquid Cu [8] and the regular solution approximation, we estimate the enthalpy of formation of liquid CuEr from the liquid elements to be approximately 19 kJ/mol. We are not aware of calorimetric data for the Ni-Er system. However, considering the general proportionality between the number of intermetallic compounds and the enthalpy of mixing in a binary system [9] and the existence of several Ni-Er intermetallic compounds [10], we assume the enthalpy of mixing of Ni and Er to be negative and large.

We report the complete amorphization of Cu-Er and Ni-Er composites by cold-rolling near room temperature with no further thermal treatment. We also report the formation of amorphous $Cu_{72}Er_{28}$ by high-energy ball-milling as well as sputtering and liquid quenching. The radial distribution functions (RDF) and crystallization behavior were shown to be in close agreement for the alloys prepared by the different methods.

## EXPERIMENTAL PROCEDURE

Samples were prepared by sandwiching elemental foils and subsequently compacting and deforming them by cold-rolling. For details see Ref. 11. We shall denote by "one deformation pass" the following sequence: 1) folding a composite sample foil over so as to double its thickness, 2) rolling it in small successive steps to the original thickness. The nominal reduction after n passes is therefore $2^{-n}$. A total sample thickness of up to 0.4 cm was obtained. An estimate of the transient temperature during deformation was obtained by placing a thermocouple between two sample foils and rolling them using the usual procedure.

Samples were sputtered by magnetron sputtering, using argon (15 mTorr) as the sputter gas. Base pressures prior to sample deposition were in the $10^{-7}$ torr range. Total film thickness was 5-10 µm. Elemental powders were ball-milled in purified argon atmosphere sealed in a stainless steel container with copper gaskets. Liquid-quenched samples were prepared by the "piston and anvil" [12] method.

Thermal reaction of samples was performed in vacuum-sealed pyrex ampules gettered with Zr at 550 °C. For details see Ref. 11.This method produced an environment cleaner than in a vacuum furnace with continuous pumping at the $10^{-7}$ torr range.

The DSC measurements for the ball-milled sample were performed with a Perkin Elmer DSC 2 and for the other samples with a DuPont 1090 calorimeter. The baseline was measured separately and subtracted from the signal.

The RDF's were measured using a vertical Norelco diffractometer in step-scanning mode. The radiation employed was Mo-$K_\alpha$, monochromatized with a curved LiF crystal. The samples used were at least 60 µm thick (i.e. three times the absorption depth of Mo-$K_\alpha$ radiation in $Cu_{72}Er_{28}$), so the infinite thickness approximation was valid for the analysis. Sputtered samples were stacked and glued with diluted Duco Cement. Ball-milled powder was compressed into disks, so as to minimize shadow effects in the diffraction due to surface roughness. The data analysis was performed following Williams and Johnson [13].

## RESULTS AND DISCUSSION

Samples were deformed in 10-55 passes. Fig. 1 (bottom) shows the x-ray diffraction pattern for a Cu-Er composite deformed in 14 passes. The line intensities indicate preferential grain orientations due to the cold-rolling (see Ref. 6). We studied the Bragg peak broadening by plotting the line full width at half maximum ΔK (obtained by fitting a Lorentzian lineshape) as a function of the magnitude of the scattering vector K. According to Ref. 14, the line broadening due to strain is given by:

$$\Delta K = A\langle e^2\rangle^{0.5}K$$

where $\langle e^2\rangle^{0.5}$ is the rms strain and A is a geometrical constant approximately equal to one for a random distribution of dislocations. On the other hand line broadening due to finite grain size (Scherrer broadening) [15] is independent of K and given by

$$\Delta K = 0.9\cdot 2\pi/L.$$

One observes that the major contribution to the line broadening (Fig. 2) is from strains. Using A=1 as a first approximation, one obtains an rms strain of approximately 2 percent.

Samples deformed in 12-14 passes were thermally reacted at 70-170 °C for one to ten hours. Fig. 1 shows the results for a Cu-Er composite. The decrease of Bragg peak intensity and increase of a broad maximum indicate the formation of an amorphous phase. In all cases, elemental crystalline material is still present. Attempts to complete the reaction by increasing the time or temperature led to the formation of crystalline intermetallic compounds. For the case of Cu-Er the equilibrium compound CuEr formed. The lattice parameter obtained (3.438±0.005 Å) agrees with previously reported values [16]. In the Ni-Er system the orthorombic eqiulibrium compound NiZr formed above 130 °C.

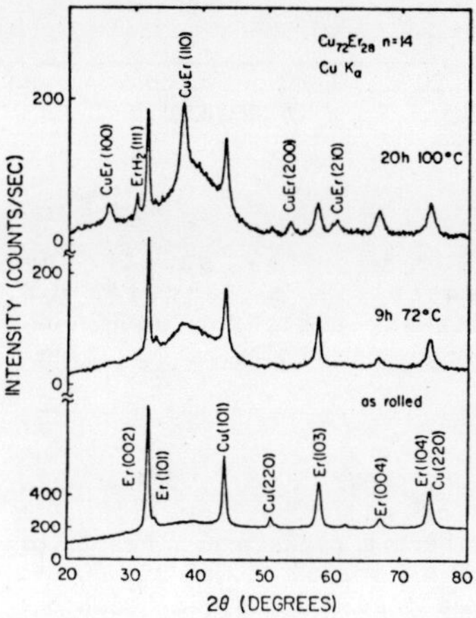


**Figure 1** X-ray scattering intensity as a function of scattering angle for as-rolled (bottom) and isothermally annealed (middle and top) $Cu_{72}Er_{28}$ for 14 deformation passes.

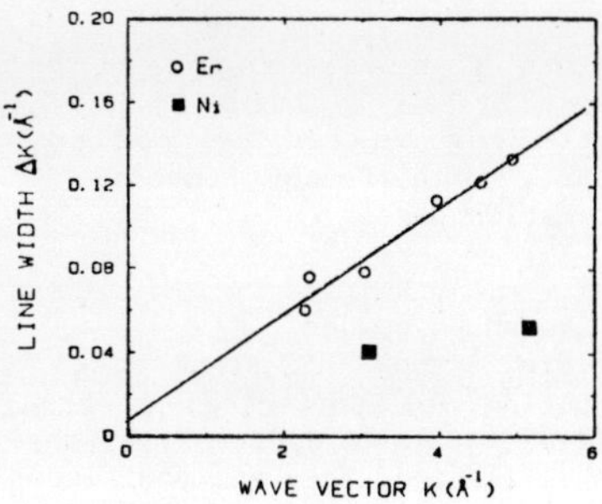


**Figure 2** Width of Bragg peaks ΔK as a function of magnitude of scattering vector K for Ni-Er composite deformed in 12 passes.

Complete amorphization was achieved for $Cu_{72}Er_{28}$ and $Ni_{55}Er_{45}$ by extensive deformation up to 55 passes (see Fig. 3). The sample temperature rose to 40-50 °C for a total time of the order of one minute, as indicated by thermocouple measurement as well as estimates based on the theoretical limit for the yield stress of a metal. For both Ni-Er and Cu-Er systems, several weak Bragg peaks indicating an FCC $ErH_2$ phase appeared during reaction. The lattice parameter of 5.125±0.005 Å is in agreement with the literature [17]. The use of higher purity erbium (less than 0.1 atomic percent hydrogen) resulted in less intense $ErH_2$ peaks. This implies that the hydride was formed from hydrogen present in the erbium as a dilute impurity. Since hydrogen mobility in erbium is higher than that of a transition metal in erbium, we can assume the hydrogen to be in equilibrium with the erbium during the metal-metal reaction. As the reaction proceeds, the hydrogen concentration in the erbium increases until it reaches a critical concentration and $ErH_2$ forms.

The ball-milled powders were amorphous with small traces of the original elemental metals left., as seen in the x-ray diffraction pattern (Fig. 4). A small peak also suggests the presence of crystalline $ErH_2$ in

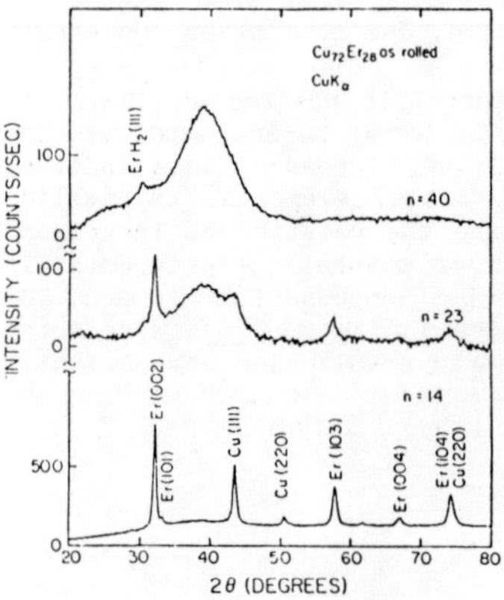


**Figure 3** X-ray scattering intensity as a function of scattering angle for rolled $Cu_{72}Er_{28}$ at different numbers of deformation passes n.

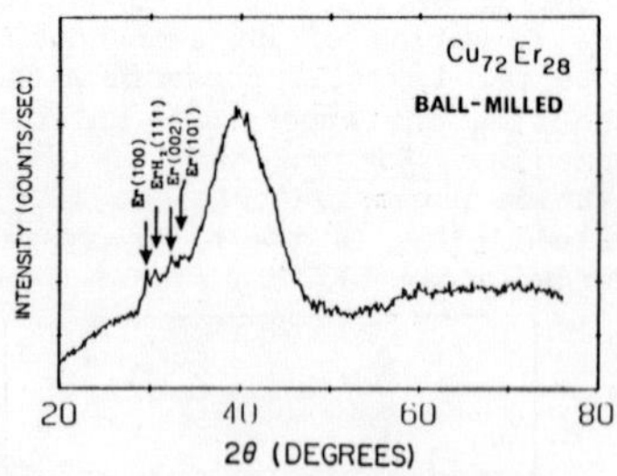


**Figure 4** X-ray scattering intensity as a function of scattering angle for amorphous $Cu_{72}Er_{28}$ prepared by ball-milling.

the amorphous powder.

Fig. 5 shows DSC scans at 40 °C/min for amorphous $Cu_{72}Er_{28}$ prepared by a) rolling, b) sputtering, c) liquid quenching and d) ball-milling. As verified by x-ray diffraction, the sharp exothermic peaks are a result of crystallization. They are positioned at a) 367 °C, b) 352 °C, c) 371 °C and d) 382 °C, correspondingly. The liquid-quenched sample contained about one percent crystalline material, as estimated from an x-ray diffraction pattern. The tail on the low-temperature side of its crystallization peak can be explained as heterogeneous nucleation catalyzed by the crystalline nuclei (compare Ref. 18). For the ball-milled sample, an exothermic peak appeared below 200 °C. Using x-ray diffraction and cycling the DSC scans, it was verified to be a result of the completion of the reaction of the elemental phases. After scanning to 200 °C, cooling down to 50 °C and scanning again, the exothermic peak below 200 °C was not present. The exothermic peaks indicate similar crystallization behavior for the amorphous alloys prepared by four different methods. The small differences in crystallization temperatures could be caused by variations in the composition from sample to sample. There are remarkable differences in the secondary crystallization events, possibly due to the influence of differences in number and types of heterogeneous nucleation sites.

Fig. 6 shows the RDF's for amorphous $Cu_{72}Er_{28}$ alloys prepared by rolling, ball-milling and sputtering. There is good agreement between the different samples, especially between the rolled and the ball-milled. These samples show a shoulder on the right side of the first peak. It is believed to be a result of the presence of $ErH_2$, since a complete subtraction of its Bragg peaks is not possible. The shoulder is centered at 3.63 A and the Er-Er distance in the hydride is 3.624 Å. The above explanation is supported by the fact that before subtracting the $ErH_2$ (111) peak (the only one visible), the shoulder was more intense. However, we cannot exclude the possibility of distinct Er-Er distances in the above alloys. One notices that the RDF for the sputtered alloy has sharper features than those for the two mechanically alloyed samples. We attribute this to shadow effects caused by surface roughness and leading to a reduction of the peak intensities of

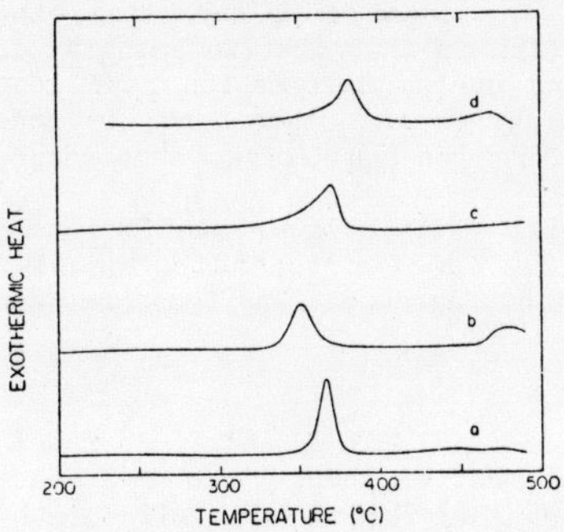


**Figure 5** DSC scans at 40 °C/min for amorphous $Cu_{72}Er_{28}$ prepared by: (a) rolling, (b) sputtering, (c) liquid quenching and (d) ball-milling. Samples c and d contain traces of crystalline material.

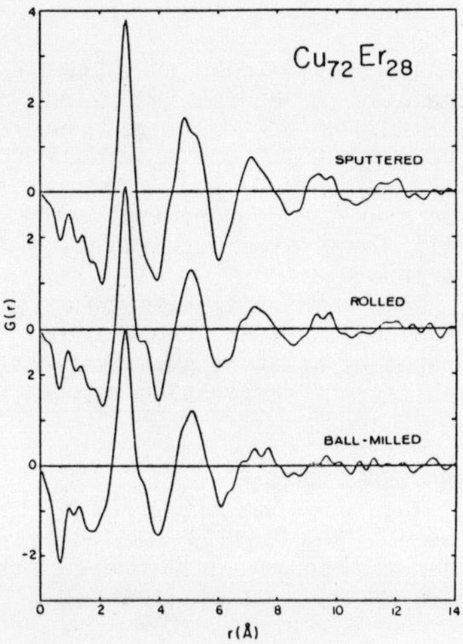


**Figure 6** Reduced radial distribution function for amorphous $Cu_{72}Er_{28}$ prepared by rolling, sputtering and ball-milling, respectively.

the diffraction function at low angles. This reduction of intensity weakens the peak intensities in the RDF.

In addition to the chemical driving force in the reaction of the cold-rolled composite, one has to consider the contribution of the microstructure, in particular strain and interfacial energy, to the balance of free energy. We use the value for the rms strain obtained from the x-ray line broadening to estimate the strain energy u in the crystalline composite as

$$u = E/2 \langle e^2 \rangle$$

where E is Young's modulus. A value of $u \approx 1$ kJ/mol is obtained. An upper limit to the interfacial energies can be obtained from tabulated values of energies of high-angle grain boundaries between grains of the same phase. For 50 Å thick layers, one obtains about 1 kJ/mol. Since the enthalpy of mixing ($\approx$ 19 kJ/mol) is larger than these values we conclude that the dominant driving force for the amorphous phase formation is chemical. An indirect confirmation of this claim was given by cold-rolling a Cu-Nb composite. In this system, the enthalpy of mixing is positive or very small [19]. After 40 deformation passes, x-ray diffraction showed no indication of an amorphous phase.

The reaction in the cold-rolled composite is believed to occur by diffusion across several atomic distances close to room temperature. We speculate that the strains during deformation assist the interdiffusion process. Due to the low deformation rates, heating effects are small, as confirmed by measurement and estimates. As a result, the reaction occurs at near-isothermal conditions. In contrast, the ball-milling process involves very high strain rates and is possibly very far from equilibrium. At present, it is not possible to exclude severe local heating and even melting for this method.

## SUMMARY

Amorphous Cu-Er and Ni-Er alloys were synthesized by two alternative methods of mechanical alloying: cold-rolling of elemental composites and ball-milling of elemental powders. As for the case of cold-rolling, the driving force for the amorphous phase formation is the negative enthalpy of mixing of the elements. The reaction occurs during deformation near room temperature. The amorphization by ball-milling, on the other hand, is more likely to involve local heating and therefore non-equilibrium temperature distributions.

The crystallization behavior and radial distribution functions were determined for the above alloys as well as for corresponding samples prepared by sputtering and rapid quenching. Good agreement was found between the different preperation methods.


## ACKNOWLEDGEMENTS

This work was supported by the US department of Energy through Project Agreement No. DE-AT03-81ER10870, contract DE-AM03-76SF-00767, and by the Caltech program in Advanced Technologies, supported by Aerojet General, General Motors GTE and TRW.

## REFERENCES


1. R. B. Schwarz and W. L. Johnson, Phys. Rev. Lett. **42**,242(1983).
2. C. C. Koch, O. B. Cavin, C. G. McKamey and J. O. Scarborough, Appl. Phys. Lett. **43**, 1017 (1983).
3. C. Politis, Proc. of the Intern. Conf. on the Materials and Mechanisms of superconductivity, Ames, Iowa (May 1985).
4. C. Politis and W. L. Johnson, accepted for publication in J. Appl. Phys. (1986).
5. R. B. Schwarz, R. R. Petrich and C. K. Saw, to be published.
6. M. Atzmon, J. D. Verhoeven, E. D. Gibson and W. L. Johnson, Appl. Phys. Lett. **45**, 1052(1984).
7. L. Schultz, in Amorphous Metals and Non-Equilibrium Processing, edited by M. von Allmen (Les editions de Physique, Les Ulis, 1984), p.135.
8. I. V. Nikolaenko, E. A. Beloborodova, G. I. Batalin, N. I. Frumina and V. S. Zhuravlev, Russ. J. Phys. Chem. **57**,1154, 1983. Translated from Zh. Fis. Khim. **57**, 1897 (1983).
9. A. R. Miedema, Phillips Tech. Rev. **36**,217 (1976).
10 W. G. Moffat, The Handbook of Binary Phase Diagrams (General Electric, Schenectedy, NY, 1984).
11 M. Atzmon, K. M. Unruh and W. L. Johnson, J. Appl. Phys. **58**, 3865 (1985).
12. P. Pietrokowsky, Rev. Sc. Instrum. **34**, 445 (1963).
13. A. R. Williams and W. L. Johnson, J. Non-Cryst. Solids **34**, 121 (1979).
14. J. Friedel, Dislocations (Pergamon, Oxford, 1964), p. 418.
15. A. Guinier, X-Ray Diffraction (Freeman, San Francisco, 1963), p. 124.
16. F. A. Shunk, Constitution of Binary Alloys, Second Suppl., (McGraw Hill, New York, 1969), p. 287.
17. Ref. 14, p. 311.
18. Y. D. Dong, G. Gregan and M. G. Scott, J. Non-Cryst. Solids **43**, 403 (1981).
19. A. K. Niessen, F. R. de Boer, R. Boom, P. F. de Chatel, W. C. M.Mattens and A. R. Miedema, Calphad **7**, 51 (1983).

# DEPOSITION OF TiN FILMS BY REACTIVE SPUTTERING WITH TARGETS FACING TYPE OF HIGH RATE SPUTTERING

M. Naoe* and Y. Hoshi**
* Tokyo Institute of Technology, Meguro-ku, Tokyo 152 Japan
** Tokyo Institute of Polytechnics, Atsugi-shi, Kanagawa-ken 243-02 Japan

ABSTRACT

TiN films 1μm thick were deposited on SKH55 alloy substrate at various substrate bias voltage by a reactive sputtering of Ti target in a mixture of Ar (1 mTorr) and $N_2$ (1 mTorr) with a Target Facing type of high rate sputtering apparatus. The color of the film changes in the order of yellow, golden, silver and golden as the substrate bias voltage increases from 0 to 200 V. The reflection ratio and Vicker's hardness of the film also changes with the bias voltage. Besides, the change of the reflection factor corresponds well to the change of the Vicker's hardness of the film. The films with silver color and maximum reflection ratio which are obtained at a rf bias voltage around 100 V, have the largest value of the Vicker's hardness about 3500. While, the films with golden color and small reflection factor, which are obtained at both rf bias voltage around 80 V and 150 V, have the lowest Vicker's hardness of about 1500.

## INTRODUCTION

It is well known that TiN is one of the useful coating materials to improve wear resistance and various coating techniques such as ion plating, sputtering etc. are used for the formation of the films.

We have developed a new high rate sputtering technique named Target Facing type of high rate sputtering (TF sputtering) method[1-3], and show the superiority of the sputtering technique compared to the other sputtering techniques. Therefore, we have applied the TF sputtering technique for the formation of TiN films by a reactive sputtering under the condition of various rf substrate bias voltages and investigate the film deposition process. In this paper, the special features in the preparation of TiN films by a reactive sputtering of Ti with dc TF sputtering and the effect of ion bombardment to the film surface during deposition on the properties of the films will be presented.

## EXPERIMENTAL

Figure1 shows the schematic diagram of a dc TF sputtering apparatus used in this work. Two Ti disks (99.9 % purity) same in size (10 cm in diameter) were used as the target and were arranged parallel facing their planes to be sputtered. The distance between the targets was fixed at 10 cm. A magnetic field of about 125 Oe at the midpoint between the targets was applied perpendicular to the target surface in order to confine the high energy $\gamma$-electrons and promote the ionization of gas in the space between the targets. As a result, high density plasma is easily produced in the space, which realizes high rate sputtering. In conventional diode sputtering system including planer magnetron type sputtering, not only the high energy $\gamma$-electrons but also the high energy particles produced from negative ions emitted from the targets bombards the substrate during sputtering. But the Target Facing type arrangement shown in Fig.1 can make these high energy particle bombardments almost negligible. As the consequence, the substrate can be maintained at low temperature. Moreover, the films deposited by this arrangement show excellent crystal orientation[4,5]. Besides, the substrate can be maintained at low temperature.

After the sputtering chamber was pumped down below $10^{-6}$ Torr, a mixture of $N_2$ and Ar (50% $N_2$-Ar) gas was introduced to the sputtering chamber. Then, sputtering was performed at a total gas pressure $P_{total}$ of 2 mTorr (Ar:1 mTorr and $N_2$:1 mTorr) and TiN films 1 μm thick were deposited on SKH55 alloy substrate under the condition of various rf substrate voltages. Typical deposition conditions are shown in Table I. The deposition rate for most of the films was about 1 μm/hour.

The crystal structure of the films was measured with X-ray diffractometry. Vicker's hardness of the film was measured with a micro Vicker's meter at a load of 25 g. Reflection factor of the films was measured with a spectrophotometer.

## RESULTS AND DISCUSSION

### (1) Reactive sputtering of Ti in Ar+$N_2$ mixture by means of dc TF sputtering

Figure 2 shows a typical dependence of discharge current Id and total gas pressure $P_{total}$ on applied target voltage Va at various $P_{total}$. These Id-Va characteristic curves are divided into three discharge regions. In the discharge region (1) where Id increases monotonically as Va increases, target surface is covered with TiN layer and the discharge characteristic coincides with that for TiN target sputtering in Ar+$N_2$ mixture. The films deposited in this region seems to be completely nitrified Ti films and have yellow color. While, further increase of Va leads a decrease in Id as shown in Fig.2 (discharge region (2)). In this region, small part of the target surface where sputtering rate is higher than the growth rate of nitride layer exposes metal Ti surface (see Fig.2). The area of this region increases as Va increases, which results in the increase of sputtering rate and decrease in Id, since sputtering yield of TiN is much larger than that of Ti and $\gamma$ coefficient of TiN is much larger than that of metal Ti. Besides, $P_{total}$

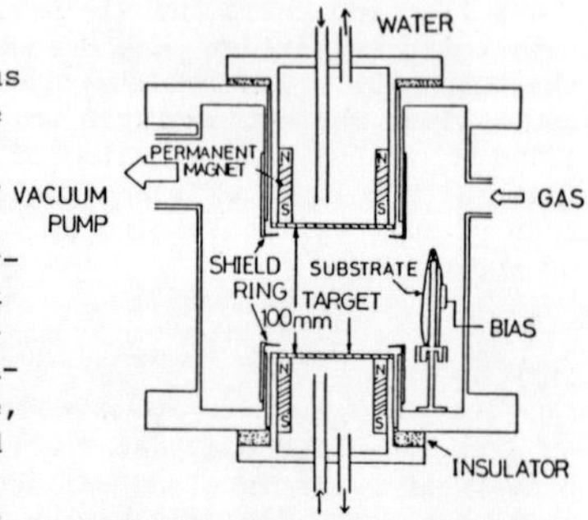

Fig.1 Target Facing type of sputtering apparatus.

Table I. Typical sputtering conditions.

| | |
|---|---|
| $V_{in}$ ( V ) | 600 - 1000 |
| $I_{in}$ ( A ) | 1 - 4 |
| $P_{in}$ ( W ) | 100 - 2500 |
| D.R. ( Å/min. ) | 20 - 2500 |
| $P_{Ar}$ ( mTorr ) | 1 |
| $P_{N_2}$ ( mTorr ) | 1 |
| $H_p$ ( Oe ) | 120 |
| Targets | 60 , 100 mm |
| Substrate | Stainless |
| d ( μm ) | 1 |
| $V_b$ ( V ) | 0 ~ -300 |

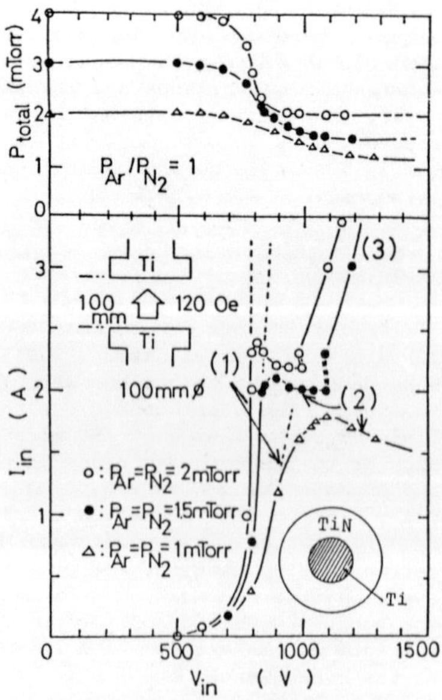

Fig.2 Discharge characteristics.

begin to decrease steeply to the value of partial Ar pressure, as Va increases. This decrease is caused by the consumption of nitrogen to form TiN films. Considering these phenomena, nitrogen content in the film deposited in this region is thought to decrease as Va increases. The color of the film changes in the order of yellow, golden and silver as Va increases.

The Id-Va characteristic curve of the discharge region (3) corresponds well to that observed in sputtering of metal Ti in pure argon gas. This result indicates that nitrogen gas is completely used up and there is no nitrogen in the space near the targets if sputtering is performed in the discharge region (3).

Most of the films were deposited without substrate heating at this region where the films with yellow and golden color were obtained. The films deposited without substrate bias have poor adhesion. Therefore, rf bias up to 300 V was applied to the substrate during deposition in order to improve the adhesion of the film.

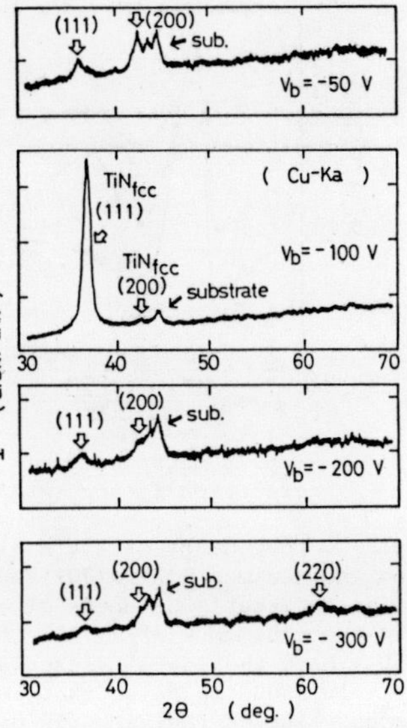


Fig.3 Typical X-ray diffraction patterns of bias sputtered films.

(2) Properties of the film

Figure 3 shows typical X-ray diffraction diagrams of the films deposited at various rf bias voltage. Reflection peaks from only fcc TiN are observed. The intensity of the (111) reflection peak increases as Vb increases from 0 to 100 V and the films deposited at Vb of 100 V show strong (111) orientation. Further increase in Vb leads the decrease in the (111) reflection peak and the (111) orientation disappears at Vb above 200 V.

Figure 4 shows the change of reflection factor R(λ)/R(706) of the films with wave length λ, where R(λ) and R(706) are the reflection factor at the wave length of λ and 706 nm, respectively. The color of the film changes in the order of yellow, golden, silver and golden as rf bias voltage increases from 0 to 300 V. It is evident from the figure that the films with golden color have quite a different λ dependence of the reflection factor than the films with silver color have.

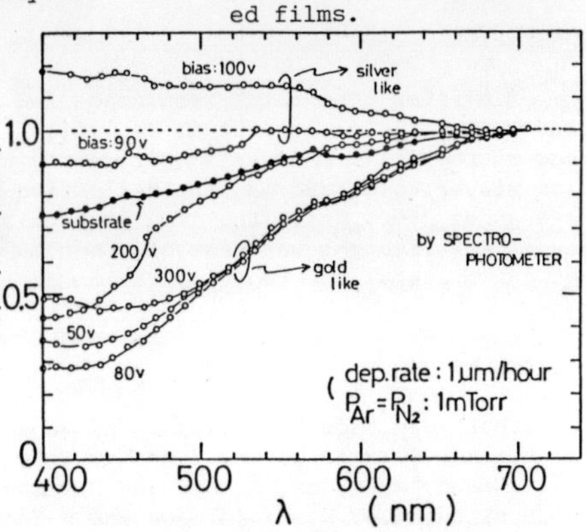


Fig.4 Wave length dependence of optical reflection factor : R(λ)/R(706).

Figure 5 shows the change of R(λ)/R(706) with rf bias voltage at λ of 400, 499 and 598 nm. R(λ)/R(706) takes a maximum value at rf bias voltage around 100 V. Corresponding to the change of R(λ)/R(706), the color of the film is also varied and the films with large R(λ)/R(706) have silver color. These changes may be caused by the change of the film composition, since the nitrogen content in the film is thought to be decreased as the bias voltage increases due to the preferential resputtering of nitrogen atoms.

Figure 6 shows the change of Vicker's hardness of the films with rf bias voltage. Compared with Fig.5, the change of the Vicker's hardness corresponds

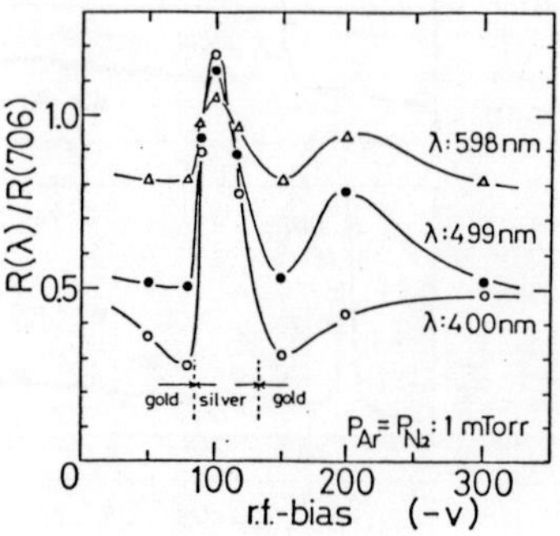


Fig.5 RF-bias dependence of optical reflection factor : R(λ)/R(706).

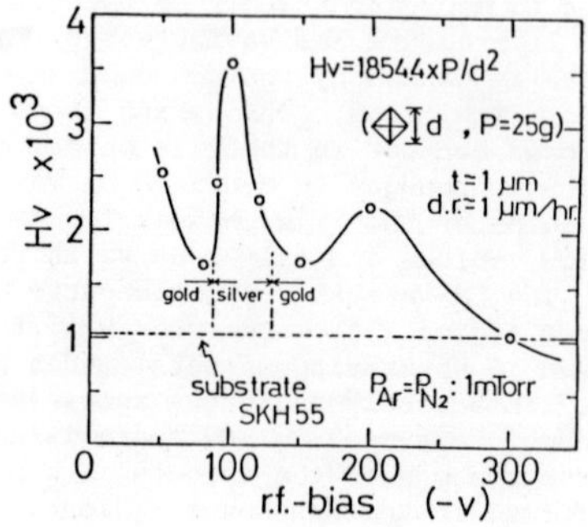


Fig.6 RF-bias dependence of Vicker's hardness.

well to the change of the R(λ)/R(706), that is, the films with silver color and the maximum R(λ)/R(706) obtained at a rf bias voltage around 100 V have largest Vicker's hardness of about 3500. While, the films with golden color and the minimum R(λ)/R(706) obtained at both rf bias voltage around 80 V and 150 V have the lowest Vicker's hardness of about 1500.

It shoud be noted that the films with maximum reflection factor have very large value of the Vicker's hardness above 3500 and seems to be useful for the films to improve wear resistance.

## CONCLUSION

TiN films 1 μm thick were deposited by a reactive sputtering of Ti with a dc TF sputtering apparatus. The color, reflection factor and Vicker's hardness of the films changes significantly with substrate bias voltage. The film with silver color and maximum reflection factor have very large Vicker's hardness above 3500. While the film with golden color and low reflection factor have small Vicker's hardness of about 1500. In conculsion, TF sputtering method is useful for the preparation of TiN films with large Vicker's hardness.

## REFERENCES

(1) Y.Hoshi, M.Naoe and S.Yamanaka; Japan.J.Appl.Phys.,16(1977)1715
(2) M.Naoe, S.Yamanaka and Y.Hoshi; IEEE Trans.Mag.,MAG-16(1980)646
(3) Y.Hoshi, M.Kojima, M.Naoe and S.Yamanaka; IEEE Trans.Mag.,MAG_18(1982)1433
(4) M.Matsuoka, Y.Hoshi, M.Naoe and S.Yamanaka; IEEE Trans.Mag.,MAG-18(1982) 1119
(5) M.Matsuoka, Y.Hoshi, M.Naoe and S.Yamanaka; Proc.Int'l Ion Engineering Congress, ISIAT'83 & IPAT'83 (Kyoto), (1983)1017

# EFFECT OF BOMBARDMENT OF HIGH ENERGY IONS ONTO GROWING SURFACE ON STRUCTURE AND PROPERTIES OF SPUTTERED MAGNETIC FILMS

N. Terada* and M. Naoe**

* Electrotechnical Laboratory, 1-1-4 Umezono, Sakura-mura, Ibaraki, Japan
**Tokyo Institute of Technology, 2-12-1 Oh-okayama, Meguro-ku, Tokyo, Japan

Magnetic thin films (Co-Ta, Co-Zr and pure iron films) have been deposited by means of dual ion beam sputtering and the effect of bombardment of high energy ions onto growing surface on their structure and properties has been investigated. The bombardment of argon ions with proper kinetic energy has the effects as follows;I. suppression of growth of crystallites, II. improvement of structural ordering in micro-scale and III. improvement of structural uniformity in macro-scale. Therefore, the films deposited with the proper bombardment are composed of fine crystallites with good atomic ordering. This causes an increase of $4\pi Ms$ and reduction of Hc of the films;Co-Zr:$4\pi Ms$=16 kG, Hc<0.6 Oe, Co-Ta:$4\pi Ms$=15 kG,Hc<0.3 Oe and iron films:$4\pi Ms$=21 kG,Hc=1 Oe.

## INTRODUCTION

In order to realize high recording density in magnetic recording system, thin films with high saturation magnetization $4\pi Ms$ and low Hc are required for magnetic layer of thin film type of head. Many kinds of magnetic films such as Co-based amorphous films[1-7] and iron based ones[8] have been produced by means of several kinds of sputtering techniques. However, crystal structure of these films depends on preparation conditions, especially on kinds of sputtering technique[2]. For example, in Co-based amorphous films, there is a significant difference in composition range where amorphous structure is obtained between the films deposited by means of conventional rf sputtering and those deposited by means of ion beam sputtering. Furthermore, their magnetic properties and chemical stabilities also depend on kind of sputtering technique. Therefore, it should be necessary to clear the origin and preparation condition dependence of them in order to obtain the soft magnetic films with good reproducibility. Bombardment of high energy particles onto growing film surface changes film structure and properties. It seems to be one of the important factors for the variations described above, since it always exists in sputtering process and degree of its stress depends mainly on the type of sputtering method. Dual ion beam sputtering is useful to investigate the effect of such bombardment, since flux and kinetic energy of bombarding ions can be controlled independently. In this study, the authors have attempted to deposit Co-Ta, Co-Zr and pure iron films with high $4\pi Ms$ and low Hc by using a dual ion beam sputtering apparatus, and to investigate the effect of the ion bombardment on their structure and magnetic properties.

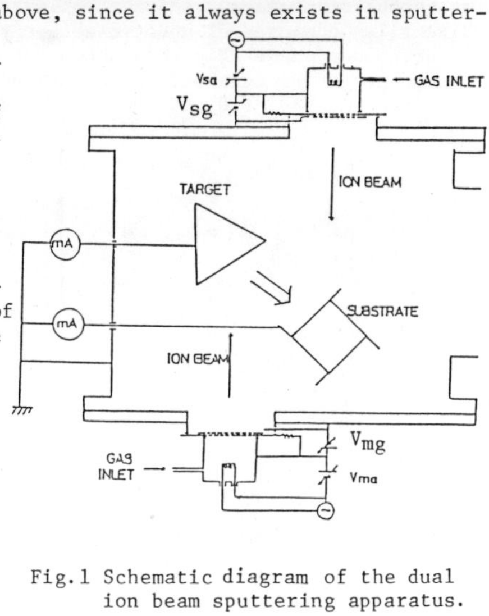


Fig.1 Schematic diagram of the dual ion beam sputtering apparatus.

## EXPERIMENTAL PROCEDURE

Figure 1 shows a schematic diagram of the dual ion beam sputtering apparatus used in this study. The kaufman type of ion source set on the lower side of deposition chamber and that set on the upper side are to sputter the target and to bombard the growing film surface, respectively. For Co-Ta and Co-Zr films, composit target composed of Co plate and Ta or Zr strips was used. For iron films, iron plate 99.99 % in purity was used as the target. Specimen films 2000~7000 Å thick were deposited glass slide substrate under the condition as follows: argon gas pressure of $3 \times 10^{-4}$ Torr, substrate temperature below 50°C and deposition rate about 20 Å/min. Accelerating voltage for bombarding argon ions Vsg was in the range up to 150 V. Incidence angle of the bombarding ions to the substrate and ion current density were fixed at 30° and 0.012 mA/cm$^2$, respectively. Structure of the films was characterized by means of transmission and scanning electron microscopy and x-ray diffractometry. $4\pi Ms$ and Hc were determined by using vibrating sample magnetometer in the field up to 10 kOe.

## RESULTS AND DISCUSSION

### I. Co-based Films

Figure 2 shows the variation of x-ray diffraction diagram of Co-Zr films with accelerating voltage for bombarding ions Vsg. The $Co_{97}Zr_3$ films deposited without ion bombardment (Vsg=0V) exhibit the diffraction peaks identified with (002) and (101) planes of hcp Co. The ones with 60 eV ion bombardment exhibit the weak peaks. On the other hand, the films with 120 eV ion bombardment does not exhibit any detectable peak. This result indicates that the ion bombardment onto grow ing surface prevents the films from forming even detectable crystallites. The change of film structure from detectable crystalline phase to amorphous or very fine crystallites results in a drastical increase of electrical resistivity ρ. Figure 3 shows the Zr content dependence of ρ of the films deposited various condition of Vsg. In the range of Zr content less than 10 at.%, the increase of Vsg results in a decrease of the lower limit of the range of Zr content where amorphous or fine crystallite phase is obtained. This result also indicates that the ion bombardment is an useful technique to synthesize films composed of undetectable crystallites or amorphous phase. Figure 4 shows the Zr content dependence of Hc of the films deposited at the condition of Vsg of 0, 60 and 120 V. For the

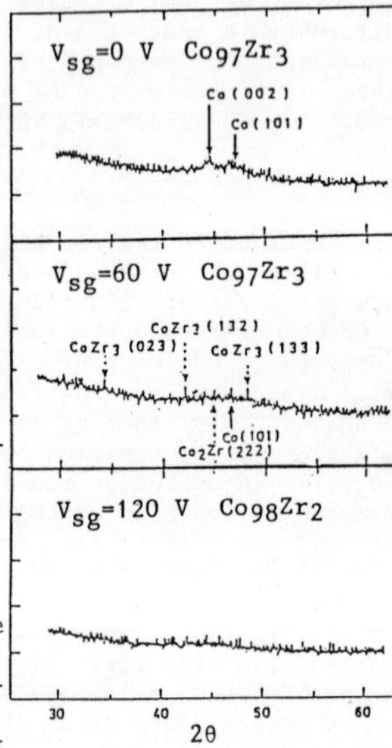


Fig.2 X-ray diffraction diagram of the Co-Zr films.

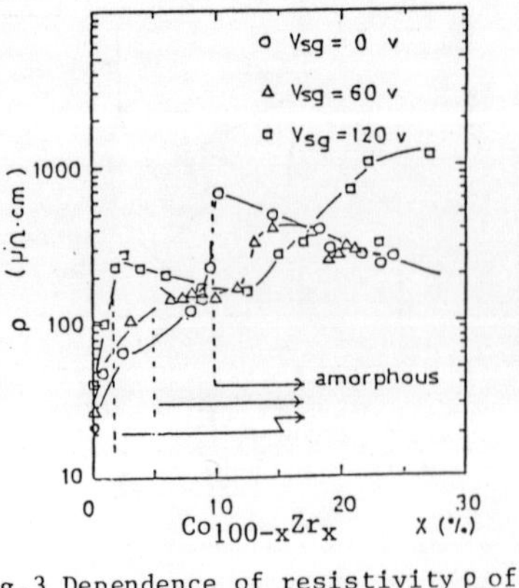


Fig.3 Dependence of resistivity ρ of the Co-Zr films on Zr content.

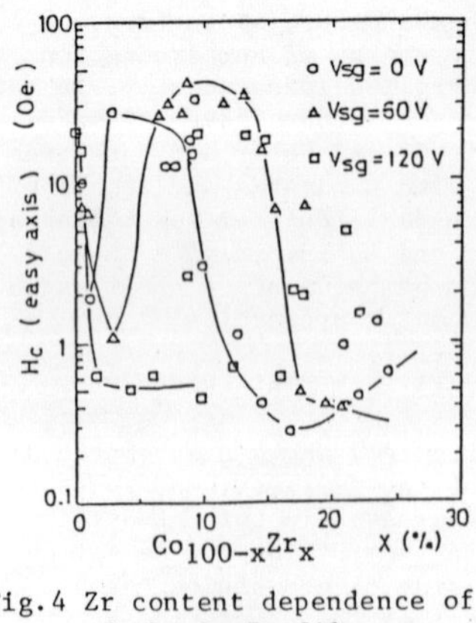


Fig.4 Zr content dependence of Hc of the Co-Zr films.

films with Zr content less than 10 at.%, Hc decreases as Vsg increases. Especially, on the condition of Vsg of 120 V, Hc decreases drastically as Zr content increases around 2 at.% and keeps low value below 0.6 Oe in the wide composition range. On the other hand, 4πMs of the Co-Zr films does not depend on the condition of ion bombardment (it decreases monotonically with an increase of Zr content). Therefore, in this study, the Co-Zr films of Co content of 98 at.% with high 4πMs about 16 kG and low Hc below 0.6 Oe are obtained by bombarding growing surface with 120 eV ion beam. These results concerning magnetic properties means that the bombardment of ions with proper kinetic energy decreases the density of pinning center for the motion of magnetic domain walls and it does not reduce magnetic ordering.

Structure, electrical resistivity and magnetic properties of Co-Ta films show similar dependence on the ion bombardment: the increase of ion accelerating voltage Vsg reduces the lower limit of Ta content where x-ray diffraction lines are to be undetectable, ρ increases steeply and Hc shows drastic decrease. By adjusting Vsg at 120 V, $Co_{94.3}Ta_{5.7}$ films with 4πMs of 15 kG and Hc below 0.3 Oe are obtained.

II. Iron Films

All iron films deposited in this study show no apparent peak in x-ray diffraction diagram. On the other hand, their magnetic properties and microstructure depend significantly on Vsg. Figure 5 shows the Vsg dependence of 4πMs and Hc. 4πMs of the films deposited without ion bombardment (Vsg=0 V) is rather low about 11 kG and an increase of Vsg up to 120 V results in the rise

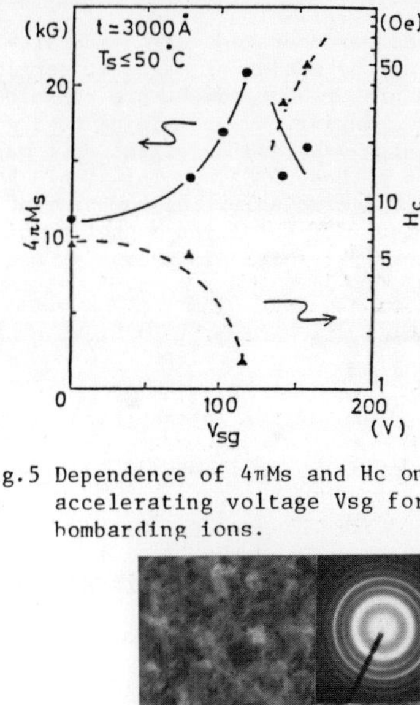


Fig.5 Dependence of 4πMs and Hc on accelerating voltage Vsg for bombarding ions.

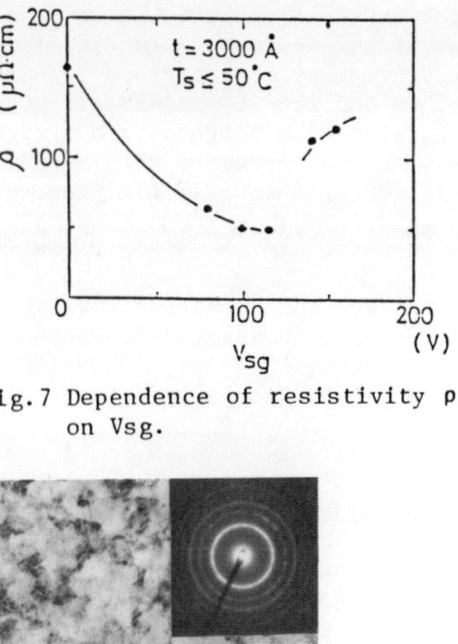


Fig.7 Dependence of resistivity ρ on Vsg.

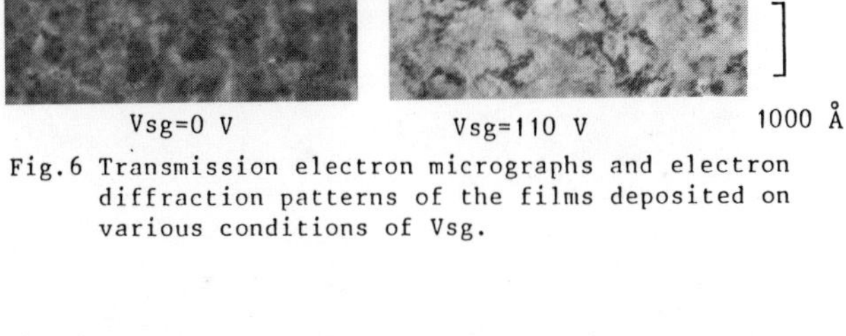


Fig.6 Transmission electron micrographs and electron diffraction patterns of the films deposited on various conditions of Vsg.

of it. (Maximum of $4\pi Ms$ about 21 kG is obtained on the condition of Vsg around 115 V.) Hc of the films shows the inverted dependence on Vsg. Figure 6 shows electron diffraction pattern and transmission electron micrograph of the films deposited on the condition of Vsg of 0 and 110 V. By bombarding growing surface, the width of diffraction rings becomes narrow and the area ratio of light part to dark one in the micrograph increases. Electrical resistivity $\rho$ also depends on Vsg. $\rho$ decreases with an increase of Vsg up to 120 V. High value of $\rho$ of the films deposited without ion bombardment results from electron scattering at grain boundaries and at disorders of atomic arrangement. $\rho$ of the films deposited with the proper bombardment results mainly from the former.

These results indicate that atomic and magnetic ordering of iron films are improved by the proper bombardment and that structural uniformity in macro-scale is also improved (at least, the bombardment reduces the density of defects which are effective for pinning the motion of magnetic domain walls).

## CONCLUSION

Magnetic thin films (Co-Ta, Co-Zr and pure iron films) have been deposited by means of dual ion beam sputtering and the effect of ion bombardment onto growing surface on their structure and properties has been investigated. The bombardment of argon ions with proper kinetic energy has three kinds of effects on film structure; I. suppression of growth of crystallites, II. improvement of structural ordering in micro-scale, III. improvement of structural uniformity in macro-scale(at least, the bombardment reduces the density of large defects which are effective for pinning the motion of magnetic demain walls). Therefore, films deposited with the proper bombardment are composed of fine crystallites which have good atomic ordering. By optimizing the degree of bombardment, magnetic films with high $4\pi Ms$ and excellent soft magnetic properties are obtained (Co-Zr;$4\pi Ms$=16 kG,Hc<0.6 Oe, Co-Ta;$4\pi Ms$=15 kG, Hc<0.3 Oe and iron films;$4\pi Ms$=21 kG,Hc=1 Oe). Accordingly, ion bombardment onto growing film surface is a useful technique for obtaining magnetic thin films with high $4\pi Ms$ and soft magnetic properties, which are useful as magnetic layer of thin film type recording head.

## REFERENCES


1. N. Heiman and N. Kazama;Phys. Rev., B17, 2215 (1978).
2. M. Naoe, Y. Hoshi and S. Yamanaka;Proceedings of the International Ion Engineering Congress, 1005 (1983).
3. M. Naoe, M. Kodaira, Y. Hoshi and S. Yamanaka;J. Appl. Phys., 53, 7846 (1983).
4. M. Hattori, Y. Maehata, S. Tsunashima and S. Uchiyama;Jpn. J. Magn. Soc, 6, 47 (1982).
5. Y. Shimada snd H. Kojima;Proceedings of Sendai Symposium on Perpendicular Magnetic Recording, 111 (1982).
6. M. Naoe, H. Kazama, Y. Hoshi and S. Yamanaka;IEEE Trans. Magn., MAG-19, 1958 (1983).
7. M. Naoe, N. Terada, Y. Hoshi and S. Yamanaka;IEEE Trans. Magn., MAG-20, 1311 (1984).
8 M. Naoe, M. Yamaga and N. Terada;IEEE Trans. Magn., MAG-21 (1985) to be published.

# TRANSITION FROM A PLANAR INTERFACE TO CELLULAR AND DENDRITIC STRUCTURES DURING RAPID SOLIDIFICATION PROCESSING

V. LAXMANAN, Department of Metallurgy and Materials Science, Case Western Reserve University, Cleveland, Ohio 44106. Concurrently, Visiting Scientist, NASA Lewis Research Center, Mail Stop 105-1, Cleveland, Ohio 44135.

## ABSTRACT

Microstructural transitions occurring during solidification processes may be described in terms of three dimensionless parameters, s, p and $\sigma_c$, which include among them various processing variables and relevant material constants. The dimensionless parameter, s, in particular offers a convenient description of both planar-cellular and cell-dendrite transitions. The relationships between $\sigma_c$, s and p predicted by various dendrite growth models have been summarized.

## INTRODUCTION

The transition from a planar to a cellular to a dendritic microstructure is observed quite frequently in many rapid solidification processes. Such a transition in microstructure is inevitably accompanied by a segregation pattern which in turn adversely affects mechanical, electrical, magnetic, chemical and electro-chemical properties of all rapidly solidified alloys. The main purpose of this paper is to describe a rather simple theoretical framework to characterize the above transitions. Specifically, the transition may be described by three dimensionless parameters, s, p and $\sigma_c$ which correlate microstructural features and processing conditions [1-3].

## 1. THE PLANAR-CELLULAR TRANSITION (PCT)

The conditions leading to the initial breakdown of a planar interface into a cellular structure are perhaps the best understood [4-6], although detailed quantitative predictions of the size and shape of the cells formed, and the spacing between the cells, have yet to be made. This continues to remain an area of active theoretical and experimental research [7-10]. At least two different approaches are available to describe PCT, the first based on the classical constitutional supercooling (CS) theory of Chalmers and co-workers [4] and the second based on morphological stability analysis [5, 6]. The simple CS criterion may be rewritten as follows, to yield an important dimensionless parameter, s, defined earlier by the author [12]:

$$s = \frac{D_L G_L}{R \Delta T_o} = \frac{ak}{k-1} = \frac{1}{2} \mathcal{G} k = 1 \qquad (1)$$

where $\mathcal{G} = 2 D_L G_L / k R \Delta T_o$ and $a = D_L G_L / m_L R C_o$ are dimensionless groups introduced by Trivedi and Flemings respectively [13,14]. Other symbols used here are explained in Appendix A and are the same as employed in Ref. 12.

All three parameters, s, a and $\mathcal{G}$ reflect the importance of the thermal gradient in determining the stability of the interface. Of the three, however, s is most conveniently used to describe the type of solidification microstructure obtained. Thus, when $G_L/R$ is very large, $s \to 1$. At the limit of stability of the planar interface s = 1 and $R = R_p = D_L G_L / \Delta T_o$. For $R > R_p$, the planar interface breaks down to form a cellular structure, at first, and eventually a dendritic structure when $R \gg R_p$ or $G_L/R \to 0$. It follows that $0 < s < 1$ during dendritic and cellular growth.

It is proposed that s be called the Chalmers number, since it was the work of Chalmers and co-workers (duly commemorated in a special issue of Materials Science and Engineering [15]) which lead to the modern understanding of the conditions leading to planar, cellular and dendritic growth in alloy systems.

## 2. CELL-DENDRITE TRANSITION (CDT)

The transition from a cellular to a dendritic microstructure or vice-versa is as yet not very well understood. Kurz and Fisher [16] were the first to propose that CDT occurs when $s = k$, where $k$ is the equilibrium partition ratio. Although recent experimental observations in succinonitrile-acetone (SCN-Ace) alloys seem to confirm this prediction [10, 17], it may be shown quite easily that relaxing certain assumptions and approximations made by Kurz and Fisher would lead to a very different result: $s \simeq k/2$ at CDT, as shown in Appendix B. In other words, their theoretically predicted transition velocity would differ by almost a factor of 2 from the experimentally observed value, see later in Fig. 2. Also, experimental results in other alloy systems do not seem to agree with this prediction. For example, cellular structures have been reported in SCN-salol for $s/k \ll 1$ [18].

Kirkaldy and co-workers [18, 19], on the other hand, have suggested that when $s \geq 1/e$ (where $e = 2.718$ ... is the natural logarithmic base) only "unstable" cellular states are possible. "Stable" cellular states exist only when $s \leq 1/e$, or $s \leq 0.368$. Cellular and dendritic growth can, therefore, only occur for $0.3\bar{6}8 \leq s \leq 0$.

Trivedi and co-workers [17] have suggested yet another criterion for CDT. According to these authors, the transition in microstructure occurs when the solute Peclet number $p = Rr_t/2D_L$ goes through a minimum, as the growth velocity is decreased, at constant $G_L$ (i.e. when $\partial p/\partial R = 0$ at constant $G_L$, $k$, $C_o$). Correspondingly, the primary arm spacings, $\lambda_p$, have been noted to go through a maximum. (Note that there is no firm theoretical justification for this criterion).

The mathematical complexity of the Trivedi analysis makes it difficult to examine, theoretically, the conditions leading to a minimum in $p$, or a maximum in $\lambda_p$. The simpler model for cellular and dendritic growth presented by the author [12] may, however, be used quite conveniently to determine the theoretical minimum in $p$ versus $R$, at constant $G_L$. This always yields $s = 1/2$ or $R = 2R_p$ when $\partial p/\partial R = 0$, regardless of the exact assumptions used to calculate the tip radius or Peclet number as discussed in Appendix B.† Thus, (according to this new model), the transition in morphology from cells to dendrites or vice versa, *if at all associated with the minimum in p, or maximum in* $\lambda_p$, must occur at a growth velocity slightly smaller than $eR_p$, just within the regime where a transition from "unstable" cells to stable cellular growth is predicted according to Kirkaldy.

Figure 1 summarizes this information in a graphical way. The ordinate here represents any measurable microstructural feature. Clearly, the regimes of "unstable" cellular, cellular and dendritic growth need more thorough investigation, particularly in binary alloys with widely different values of $k$. Solutes for which $k > 1/e$ or $k > 0.50$ would be of particular interest since for such a system the Kurz & Fisher transition velocity would fall well within the Kirkaldy regime of "unstable" cellular structures (also called 'chaotic' structures), or the new criterion for CDT, $s = 1/2$, suggested here.

---

†Hunt [20] also predicted a maximum in cellular or primary arm spacings at $s = 1/2$. It is also interesting to note that at the critical growth rate, $R_{cr}$, defined by the author, $s = 1/2$, provided $G_L = G_L^{cr}$, for which $R_{cr} = R_a$. The critical growth rate marks the beginning of a transition from a dendritic to a planar interface in the large growth rate regime [1, 12].

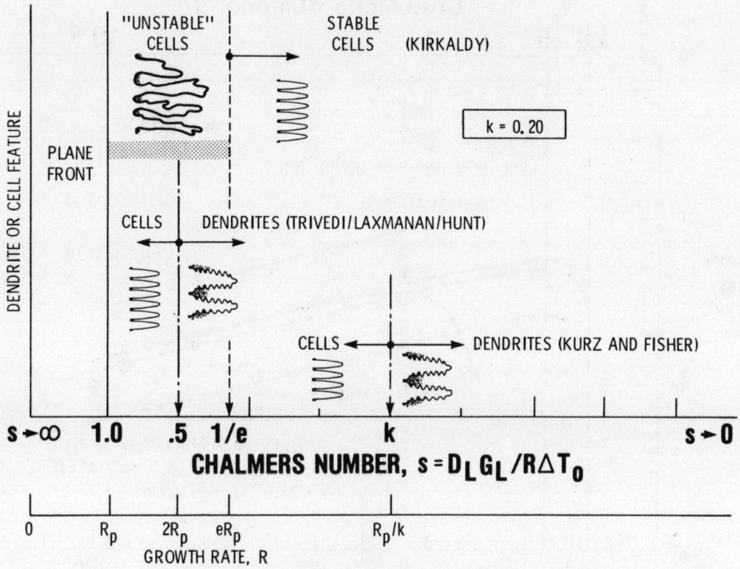


Figure 1. - Cell-dendrite transition summary.

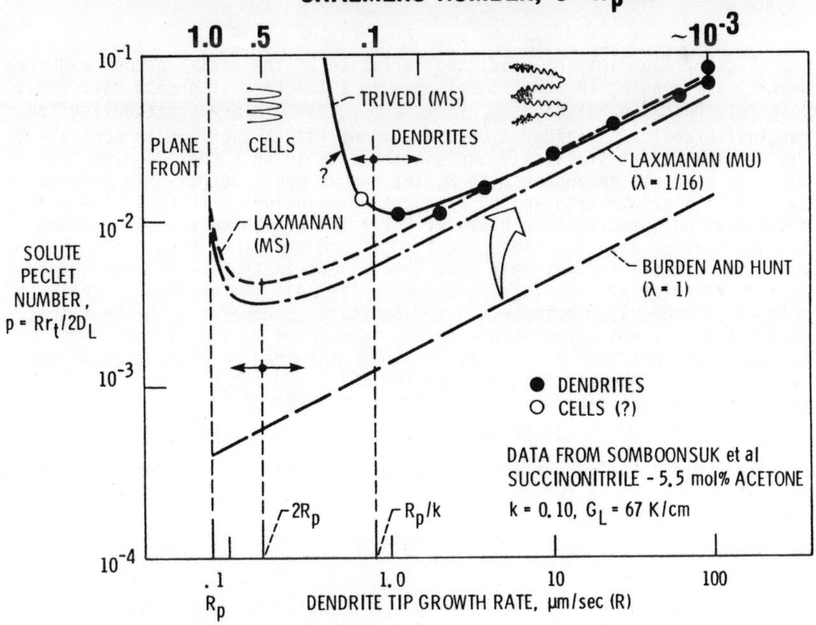


Figure 2. - Solute peclet number versus Chalmers number at cell-dendrite transition.

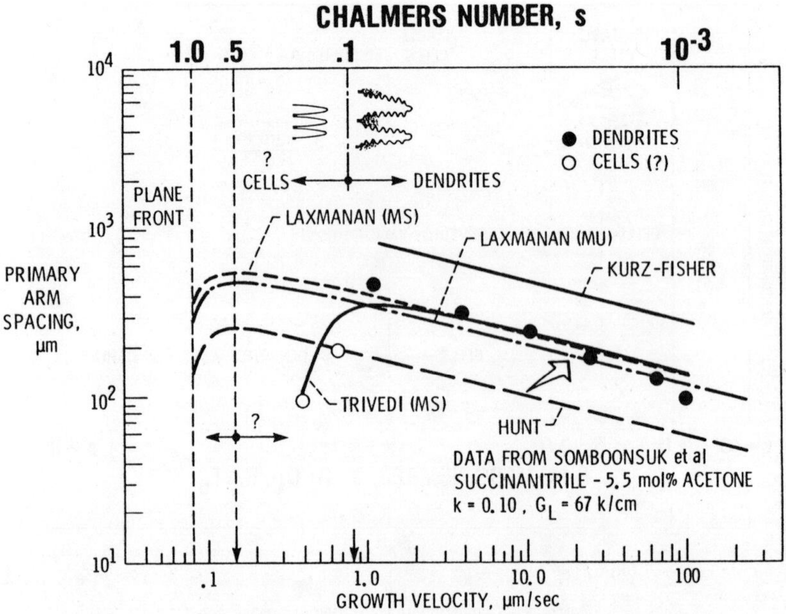


Figure 3. - Primary spacing versus Chalmers number at cell-dendrite transition.

Figs. 2 & 3 plot the predicted variation of the Peclet number and the primary arm spacing in SCN-5.5 mol% Ace as a function of growth rate for a constant thermal gradient of $G_L$ = 67 K/cm. Both minimum undercooling and marginal stability considerations, equations (3) through (6) in Appendix B, may be seen to satisfactorily explain the data in this regime. The Trivedi analysis predicts an increase in Peclet number and a decrease in primary spacing at a growth rate which "apparently" coincides with CDT. The present model does not predict this behavior until the growth velocity decreases to much smaller values. The present model, on the other hand, describes much more satisfactorily, than does the Trivedi analysis, the cell spacing data of Kirkaldy et al. This is shown in Fig. 4. The maximum in cell spacing is properly predicted. More important, however, is the fact that this maximum is not associated with any transition in microstructure. The structure was cellular in the entire growth rate regime being considered here. Clearly, the various criteria for CDT discussed here do not yet satisfactorily describe this transition in all respects.

## 3. THE SOLUTE PECLET NUMBER AND THE TIP STABILITY PARAMETER

The solute Peclet number, p, and the tip stability parameter, $\sigma_c$, are intimately related as follows:

$$\sigma_c = \frac{A}{p^2} = \frac{\ell_s \ell_c}{r_t^2} \qquad (2)$$

Here $A = \ell_c/\ell_s$ and $p = r_t/\ell_s$, with $\ell_c$ and $\ell_s$ being the capillary length and solute diffusion length respectively. Both $\sigma_c$ and p may also be used to

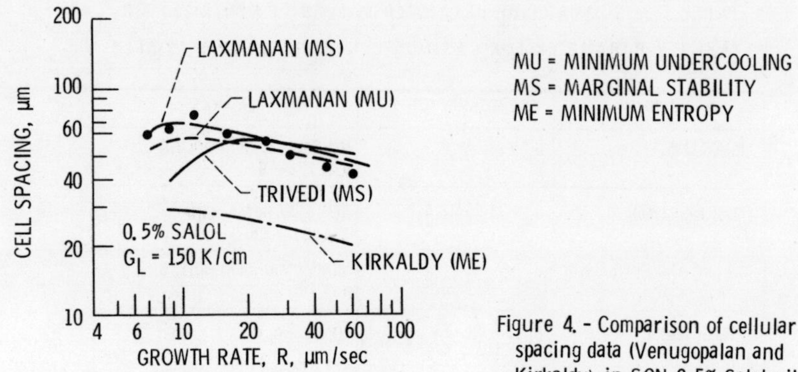


Figure 4. - Comparison of cellular spacing data (Venugopalan and Kirkaldy) in SCN-0.5% Salol with various growth models/criteria.

characterize microstructural transitions.

The tip stability parameter, $\sigma_c$, defined here is very similar to the stability parameter $\sigma$, first defined by Langer and Muller-Krumbhaar [21] to analyze dendritic growth in pure melts. As shown by the author [12], the parameter $\sigma_c$ is a very useful parameter to analyze dendritic growth in alloys. (The literature on this subject now discusses the product $Rr_t^2$ which is found to be a constant under the usual conditions of dendritic growth [13, 16, 22, 23]). Table 1 summarizes the predictions of various dendrite growth models in terms of the parameter $\sigma_c$.

Note that when the Peclet number is small all models predict $\sigma_c$ to be a function only of R and $G_L$. For fixed values of R and $G_L$, as in typical directional solidification experiments, $\sigma_c$ = constant. Only the value of this constant differs, depending on the details of the assumptions used to describe solute diffusional processes and other assumptions such as minimum undercooling or marginal stability used to determine the tip radius.

An important requirement of a satisfactory dendrite growth model is that the predicted dendrite tip radius must become infinite at <u>both</u> the CS limit of Chalmers and co-workers, as well as the absolute stability (AS) limit, $R_a$ of Mullins & Sekerka [5, 12]. It follows that a satisfactory cell or dendrite theory must yield $\sigma_c \to 0$ and $p \to \infty$, (consistent with a planar interface) *only* as $s \to 1$ in the low growth rate regime, *and* also $\sigma_c \to 0$ and $p \to \infty$ as $s \to 0$ in the large growth rate regime. These two regimes are also distinguished by the value of the dimensionless parameter $A = R/2k^2R_a$. Close to the CS limit $A \to 0$ whereas close to the AS limit $A \to 1/2k^2$.

From Table 1 it is clear that both the Trivedi, and, the Kurz & Fisher analyses do not satisfy this requirement, mainly because they do not apparently properly account for the influence of the temperature gradient. The function $N_g(p)$ derived by Trivedi has the limiting value of minus unity as p becomes large ($N_g(p) \to -1$ as $p \to \infty$; Trivedi has assumed $N_g(p) \to 0$) which is the reason for $\sigma_c$ being negative. The Kurz & Fisher result is equivalent to assuming negligibly small solute diffusion within

TABLE I. - SUMMARY OF PREDICTED VALUES OF DENTRITIC TIP STABILITY PARAMETER FOR VARIOUS DENDRITIC GROWTH MODELS

| MODEL | EXPRESSION FOR $\sigma_C$ | REMARKS |
|---|---|---|
| FISHER (1951) | $\sigma_C = \lambda = \frac{1}{2}$ | MINIMUM UNDERCOOLING $p \ll 1$, $s \to 0$ |
| FLEMINGS (1966) | $\sigma_C = \frac{(1-s)}{L} \simeq \frac{1}{L}$ | ALL p, WITH MARGINAL STABILITY CRITERION $\lambda = 0$ |
| BURDEN AND HUNT (1974) | $\sigma_C = \lambda = 1$ | MINIMUM UNDERCOOLING $p \ll 1/2$ |
| TRIVEDI (1980) | $\sigma_C \simeq \frac{1}{L}\left(1 - \frac{s}{k}\right) \simeq \frac{1}{L}$ | FOR $p \to 0$; $\Phi \ll 1$, $s \to 0$ <br> $\Phi = p\exp(p)E_1(p)$ |
| | $\sigma_C \simeq \frac{1}{Lk}\left(1 - \frac{s}{k}\right) \simeq \frac{1}{Lk}$ | FOR $p \to \infty$; $\Phi \to 1$, $s \to 0$ |
| KURZ AND FISHER (1981) | $\sigma_C \simeq \frac{1}{2\pi^2}\left(1 - \frac{s}{k}\right) \simeq \frac{1}{2\pi^2}$ | FOR $s < k$, "DENDRITIC REGIME" <br> $\lambda = 1/2$ |
| | $\sigma_C = 0(?)$ | FOR $s > k$, "CELLULAR REGIME" <br> $\Gamma = \ell_C k\,\Delta T_0 = 0$ |
| PRESENT WORK (1985) | $\sigma_C = \lambda(1 - s) \simeq \frac{1}{16}$ | FOR $p \ll 1/2\lambda(1 - k)$, MINIMUM UNDERCOOLING, $\lambda = 1/16$ |
| | $\sigma_C = \frac{(1-s)}{L} \simeq \frac{1}{L}$ | MARGINAL STABILITY, INDEPENDENT OF $\lambda$ |

the interdendritic or intercellular regions ($s \simeq 0$), hence the negative $\sigma_C$. Interdendritic diffusion is included in the model due to the present author and is reflected by the presence of the terms containing the Chalmers number, s, in the basic expression for solute buildup at the tips [12]. Table 1 indicates that this model yields the anticipated result, $\sigma_C \to 0$, as $s \to 1$.

Figure 5 plots the theoretically predicted Peclet number as a function of s, for various values of A, according to the simple model presented by the author. This plot may also be interpreted as depicting the variation of tip radius with increasing values of the thermal gradient at the interface (or tip), $G_L$, for various constant values of the growth rate, R. Note that the tip radius remains small, alternatively $p \ll 1/2\lambda(1-k)$, even very close to the CS limit. Also, the Peclet number or tip radius remains quite small even for very large growth velocities or $A > 1$.

This leads to two very important conclusions. First, the transition from an initially dendritic structure to a truly planar interface (infinite tip radius) occurs only at growth velocities extremely close ($s > 0.9999$) to the CS limit. There are no "unstable" cellular states as proposed by Kirkaldy. Second, and this is particularly relevant to rapid solidification processing, the transition from an initially dendritic structure to a truly planar interface with an infinite tip radius, at growth rates approaching the AS limit, can only occur in the presence of very steep temperature gradients. (Note the values predicted by curves with $A \geq 1$.) To obtain a truly planar interface, thermal gradients of the order of the critical thermal gradient, $G_L^{cr} = \Delta T_0/2\ell_C k^2$ are required, as discussed briefly in Refs. 1 & 12. The solid composition formed at the tip will approach the initial alloy composition (thus leading to a completely segregation-free structure)

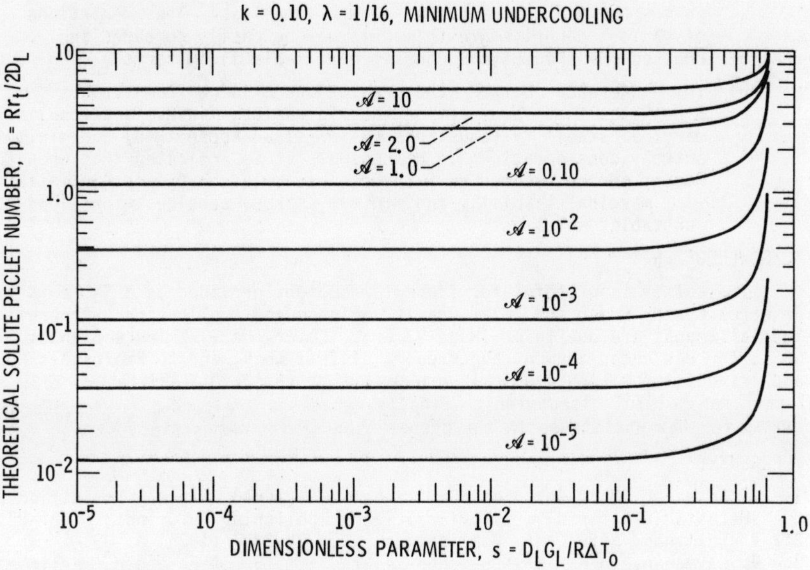


Figure 5. - Variation of tip radius (or p) at constant R (or $\mathscr{A}$) with increasing thermal gradient, $G_L$ (or s).

*only under these conditions.* Space considerations do not permit a more complete discussion of this very important point, which will be addressed in a future publication [24]. However, it must be stated here that marginally stability considerations do not yield an entirely satisfactory description of the transition from an initially dendritic structure to a planar interface at both the CS and the AS limits. Very close to these limits, the predicted tip radius becomes imaginary. (This has also been noted by Kurz, Giovanola and Trivedi in a recent publication [25]). Solid composition formed is, strictly speaking, less than the initial value. Indeed, the predictions obtained from marginal stability considerations resemble those of Kirkaldy et al (they used minimum entropy considerations to predict the tip radius) since only imaginary or "unstable" solutions are obtained.

## SUMMARY AND CONCLUSIONS

i) An important dimensionless parameter s has been identified which may be used to conveniently describe microstructural transitions. It is shown that minimum undercooling considerations require that $s \rightarrow 1$ at both very low growth rates and very large growth rates to yield a transition to a planar interface from an initially dendritic interface (Fig. 5).

ii) The various criteria proposed for cell-dendrite transition have been reviewed. It is shown that relaxing the various assumptions and approximations made by previous authors leads to a single criterion, viz s = 1/2 at CDT, if it is assumed that CDT coincides with the minimum in p or maximum in $\lambda_p$. However, this criterion, or others proposed previously, are not in agreement with presently available experimental data, Figs. 2 through 4.

iii) A dimensionless parameter $\sigma_c$, similar to the stability parameter $\sigma$

of Langer and Muller-Krumbhaar, is shown to be particularly useful in the analysis of dendritic growth in alloys. The predictions of various dendrite growth models may be easily compared and checked for theoretical consistency (Table 1) in terms $\sigma_C$.

iv) The differences between the predictions arising from various hypotheses used to determine the tip radius, minimum undercooling, marginal stability (equations (3) & (4) in Appendix B) and minimum entropy considerations is discussed. It is indicated that when the Peclet number becomes very large, near both the CS and AS limits, *both* marginal stability and minimum entropy considerations lead to "unstable" solutions.

ACKNOWLEDGEMENTS

The author is grateful for financial support provided by NASA Lewis Research Center under the "Microgravity Science and Applications" Program. Special thanks are due to Dr. Hugh R. Gray, Chief, Metal Science Branch, NASA, for his encouragement and support of this work, and to Professors Kurz and Trivedi for making available pre-prints of their work and to Dr. R.S. Israel for helpful discussions. Finally, thanks are also due to Mrs. Carol Thomas for her assistance in the preparation of the manuscript.

REFERENCES

1. V. Laxmanan, Mat. Res. Soc. Symp., Vol. 28 (1984), Rapidly Solidified Metastable Materials, Elsevier Science Publishing Co., Inc., p. 21-27.
2. V. Laxmanan, NASA Conf. Publication, NASA CP-2337 (1984) p. 33.
3. V. Laxmanan, in Rapidly Quenched Metals, Eds. S. Steeb and H. Warlimont, Elsevier Science Publishers, B.V. (1985) p. 807.
4. W. A. Tiller, K.A. Jackson, J.W. Rutter and B. Chalmers, Acta Metall, 1, 428 (1953).
5. W.W. Mullins and R.F. Sekerka, J. Appl. Phys., 35, 444, (1964).
6. J.W. Cahn, S.R. Coriell and W.J. Boettinger, in Laser and Electron Beam Processing of Materials, Eds. C.W.White and P.S. Peercy, Academic Press, New York (1980) p. 89.
7. L.H. Ungar and R.A. Brown, Phys. Rev. B 29(1984) 1367, 30 (1984) 3993.
8. S.R. Coriell, M.R. Cordes, W.J. Boettinger and R.F. Sekerka, J. Crystal Growth, 49, 13 (1980), see also in Ref. 2, p. 117.
9. R. Trivedi and K. Somboonsuk, Acta Metall, 33, 1061 (1985).
10. H. Esaka and W. Kurz, J. Crystal Growth, 72 (1985) p. 578.
11. S.R. Coriell and R.F. Sekerka, in Proc. 2nd Int. Conf. on Rapid Solidification Processing, Reston, Virginia (1980), p. 35. Eds. R. Mehrabian, B.H. Kear and M. Cohen, Claitors Publishing Division, Baton Rouge, LA 70821.
12. V. Laxmanan, Acta Metall 33, (1985) 1023, 1037, 1475.
13. R. Trivedi, J. Crystal Growth, 49 (1980) p. 219.
14. M.C. Flemings, in Solidification Processing, McGraw Hill (1974) p. 79.
15. Materials Science and Engineering, Special Issue on Solidification Microstructure, vol. 65, No. 1, July 1984, Eds. H. Jones and W. Kurz.
16. W. Kurz and D.J. Fisher, Acta Metall, 29, (1981) p. 11.
17. K. Somboonsuk, J.T. Mason and R. Trivedi, Metall. Trans. A, 15 (1984) 967, also R. Trivedi, metall. Trans. A, 15 (1984) 977.
18. J.S. Kirkaldy, Scripta Metall, 14 (1980) p. 739.
19. J.S. Kirkaldy, Metall. Trans A, 16A (1985) p. 1781.
20. J.D. Hunt, Solidification and Casting of Metals, The Metals Society Book 192, London, 1979, p. 3.
21. J.S. Langer and H. Muller-Krumbhaar, J. Crystal Growth, 42 (1977) p. 11, also Acta Metall, 26 (1978) 1681-1697.
22. R. Trivedi, J. Crystal Growth, 73 (1985) 289.
23. J. Lipton, M.E. Glicksman and W. Kurz, Mat. Sci. & Eng., vol. 65 (1984) p. 57.

24. V. Laxmanan, to be published in Proc. 8th Int. Conf. on Crystal Growth.
25. W. Kurz, B. Giovanola and R. Trivedi, Acta Metall (1986) in press.
26. V. Laxmanan, "Cellular and dendritic growth in a binary alloy melt: A marginal stability approach", J. Crystal Growth (in press).

## APPENDIX A

| | |
|---|---|
| $D_L$ | Diffusivity of solute in liquid |
| $G_L$ | Thermal gradient in liquid, at interface or tip |
| $R$ | Dendrite or cell tip growth rate |
| $\Delta T_o$ | Equilibrium freezing range of alloy = $m_L C_o(k-1)/k$ |
| $m_L$ | Slope of equilibrium liquidus line |
| $k$ | Equilibrium partition ratio |
| $r_t$ | Dendrite or cell tip radius |
| $\ell_c$ | Chemical capillary length = $(\gamma/\Delta S)/k\Delta T_o = \Gamma/k\Delta T_o$ |
| $\gamma$ | Surface tension of solid-liquid interface |
| $\Delta S$ | Entropy of fusion per unit volume |
| $\ell_s$ | Solute diffusion length = $2D_L/R$ |

## APPENDIX B

According to the present model, the tip radius is given by one of the following [12, 26]:

$$\sigma_c = \frac{\ell_s \ell_c}{r_t^2} = \frac{A(1-s)}{b} \tag{3}$$

or

$$\sigma_c = \frac{\lambda(1-s)}{b^2} \tag{4}$$

Here $b = [1-2p\lambda(1-k)]$. Equation (3) assumes marginal stability conditions for which $A = 1/L$, where L varies between 10 and 28, or $A = 1/2\pi^2$. Equation (4) assumes minimum undercooling at the tips. For this case, $\lambda = 1/16$.

Differentiating (3) and (4) wrt R, holding $G_L$ constant, it is readily shown that $\partial p/\partial R = 0$ when $s = 1/2$. The Kurz and Fisher model is a special case of the present model wherein $s = 0$ and $\lambda = 1/2$. Hence relaxing all approximations and assumptions made by these authors would also yield $s = 1/2$ at $\partial p/\partial R = 0$.

However, using the complete, cubic, expression for the tip radius given by Kurz and Fisher, equation (11) in Ref. 16, yields $s \simeq k/2$ when $\partial p/\partial R = 0$.

It may be noted here that equation (3) above is obtained by employing the stability criteria proposed in Refs. 13 and 16. These are only applicalbe for small Peclet numbers, see also Refs. 22, 24-26 for a more complete discussion. The modification of the stability criterion proposed by Trivedi [22] in the rapid solidification regime does <u>not</u> alter the fundamental difficulties associated with the behavior of the function $N_g(p)$ (derived earlier by Trivedi [13]) and hence the solute buildup at large p as given by equation (31) in Ref. 13. At large p, the solid composition remains less than $C_o$.

## APPENDIX C

The cell spacing or primary spacing is obtained from the following:

$$RG_L^2 \lambda_p^4 = 64\Gamma D_L k\Delta T_o \, [(1-s)/\lambda] \tag{5}$$

$$RG_L^2 \lambda_p^4 = 64\Gamma D_L k\Delta T_o \, [(1-s)L/b] \tag{6}$$

Equation (5) assumes minimum undercooling at tips and equation (6) marginal stability. Equation (5) will reduce to the classical Hunt result, equation (26) in Ref. 20, for $s \simeq 0$ and $\lambda = 1$. Equation (6) will yield the Trivedi result [17], for $s \simeq 0$ and $b \simeq 1$, i.e. for a small Peclet number and a small Chalmers number.

The general expression for cell or primary arm spacing in terms of $\sigma_c$ is (based on Hunt's approach as typified by the central result equation (13) in Ref. 20):

$$RG_L^2 \lambda_p^4 = 64\Gamma D_L k\Delta T_o \, [\frac{(1-s)^2}{b^2 \sigma_c}] \tag{7}$$

PART II A

# Amorphous Alloys—Formation

# ELECTRONIC STRUCTURE AND GLASS FORMABILITY IN METALLIC GLASSES

R. HASEGAWA* AND K. TANAKA**
* Nippon Amorphous Metals Co., Ltd, Kasumigaseki Bldg, Kasumigaseki, Chiyoda-ku, Tokyo 100 Japan and Allied-Signal Corporation, Morristown, NJ 07960***
**Nagoya Institute of Technology, Dept. of Metallurgical Eng., Showa-ku, Nagoya 466 Japan

## ABSTRACT

Recent developments in theory and experiment seem to enable us to relate the electronic structure to the formability of glassy states in transition metal-metalloid base alloys. Especially stressed is the role of the antibonding states in the glass formation.

## INTRODUCTION

The first attempt to relate electronic properties to glass stability in metallic glasses was made by Nagel and Tauc by using a nearly free electron model[1]. They have argued that a glassy alloy is most stable when its composition satisfies the condition that the wave number $q_p$ corresponding to the peak of the interference function is equal to the Fermi sphere diameter $2k_F$, at which composition the density of states $N(E_F)$ becomes a minimum. Due to lack of pertinent experimental evidence, this proposal was neither strongly rejected nor supported until around 1980. Using the results of x-ray (XPS), ultra-violet (UPS) and Auger (AES) spectra on a glassy Pd-Si alloy, Riley et al have derived partial Pd 4d, Si 3p and Si 3s densities of states (PDOS) and concluded that the conduction band of the glassy alloy is far from free-electron-like and that the s-p portion of the density of states near $E_F$ is a local maximum rather than a minimum [2]. This is consistent with the low temperature specific heat data obtained for glassy Pd-Si-Cu alloys by Mizutani et al [3]. The experimental situation involving glassy alloys containing Fe is less clear. For example, the electronic specific heat coefficient $\gamma$ increases rapidly as the concentration of boron is decreased in the glassy Fe-B system [4]. The increase may be associated with the Invar-like properties of these alloys and is likely magnetic in origin. The XPS data of the same alloy system by Matsuura et al [5] and the XPS and UPS measurements on Fe-B and Fe-P-B by Amamou and Krill [6] seem to be consistent with the PDOS calculations by Fujiwara [7]. These calculations indicate a maximum of the DOS at $E_F$ and furthermore a splitting of the bonding and antibonding states. The latter feature is closely related to the present investigation. The relatively high Fermi level density of states thus appears to be one of the general properties of transition metal base metallic glasses [8]. The free-electron model of Nagel and Tauc [1], however, should be most applicable to non-transition metal base metallic glasses. A recent work by Häussler et al using XPS, UPS and AES on glassy Au-Sn, Ag-Sn and Cu-Sn shows that the stability against crystallization is a maximum at a composition at which the DOS at $E_F$ is a minimum [9], being consistent with the model. Although the situation is not clear at this moment, the experiment and the theory developed thus far indicate that detailed studies of the PDOS are necessary to elucidate glass formability or stability in metallic glasses in terms of electronic structures. One such effort [10] made recently using soft x-ray spectroscopy (SXS) is reviewed and the results are used to discuss the subject matter of the present paper.

*** Permanent address

 

SXS DATA FOR GLASSY AND CRYSTALLINE ALLOYS

Unlike photoemission which maps the total DOS, SXS utilizes x-ray transitions subject to dipole selection rules, providing a technique to obtain the PDOS of an alloy. For example, K and L emission spectra probe p and d states of metalloids and transition metals respectively of a transition metal-metalloid base alloy. Figure 1 compares Si $K_\beta$ and B $K_\alpha$ emission spectra for rf sputtered Ni-Si and liquid-quenched Ni-B systems taken from Ref. 10. Using the $L_3$ emission spectra for Ni and the results of the PDOS calculation by Bisi and Calandra [11], features A, and B with B' (in Fig. 1a) have been identified as being associated with the p-d antibonding and bonding states respectively resulting from Si 3p-Ni 3d hybridization. Similarly, the features A and B in Fig. 1b have been attributed respectively to the p-d

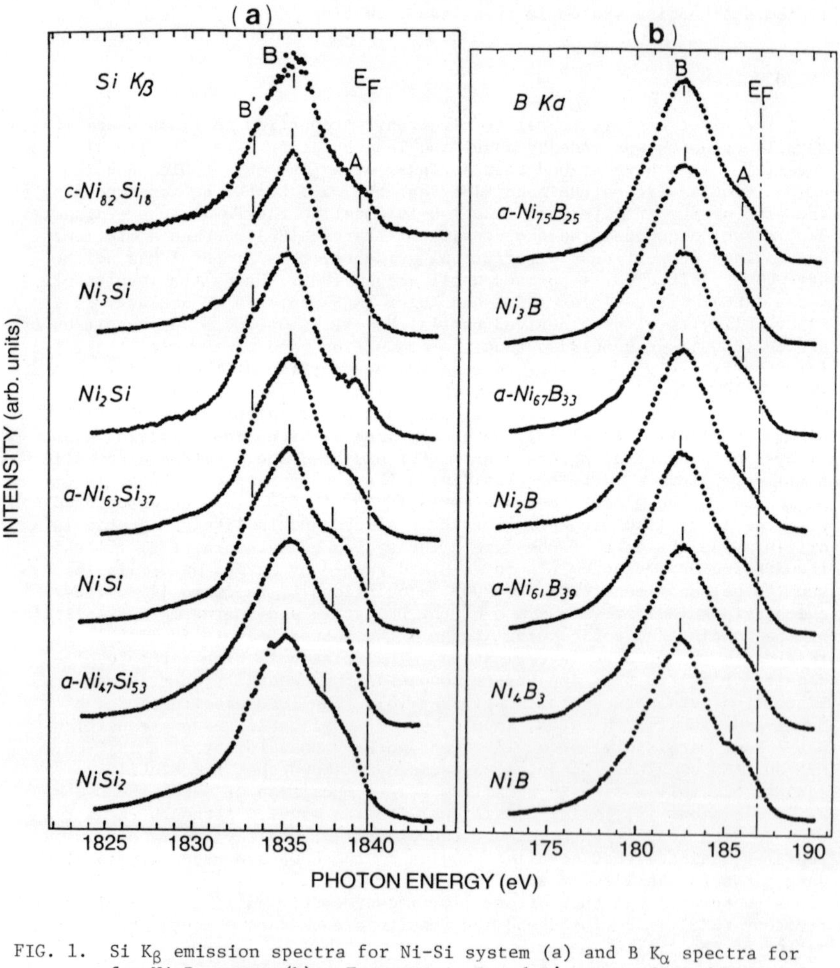


FIG. 1. Si $K_\beta$ emission spectra for Ni-Si system (a) and B $K_\alpha$ spectra for for Ni-B system (b). Features A, B and B' are explained in the text. Data were taken from Ref. 10.

antibonding and bonding states involving B 2p and Ni 3d states.

The above data may be summarized as follows: (i) The overall profile of the DOS for the Ni-Si system depends on the chemical composition, relatively irrespective of the atomic arrangement. The nature of the p-d antibonding states strongly depends on the chemical composition whereas that of the bonding states does not. The tail portion of the antibonding states near $E_F$ is negligible for low Si content for which the alloys tend to be crystalline. It tends to be pulled below $E_F$ with increasing Si content for which glassy states tend to form and extends further below $E_F$ for high Si content for which the alloys tend to be crystalline. (ii) For the Ni-B system, the overall profile of the DOS does not vary considerably with the composition, showing appreciable contribution of the antibonding states to the DOS at $E_F$. This is to be compared with the fact that glasses are easily formed by liquid quenching for a wide concentration range between about 18 at.% and about 42 at.% [12].

Based on these findings, following discussion can be made: The role of the antibonding states is important to the glass formability of the Ni-metalloid alloys. It appears that the glass formability is high when the antibonding states lie near $E_F$, resulting in a relatively high DOS near $E_F$. This is consistent with the conclusions of Refs. 7 and 8. In terms of the present reasoning, glass formability of $Ni_2B$ with a tetragonal local atomic arrangement should be low because the antibonding states appear to be pushed above $E_F$. This is indeed the case. As indicated by the recent NMR data on Ni-B system [13], glassy alloys with compositions near $Ni_2B$ have local atomic arrangements based on an admixture of $Ni_3B$ and $Ni_4B_3$ structures. These orthorhombic phases clearly exhibit contributions of the antibonding states to the DOS at $E_F$ as Fig. 1b shows. Although the overall profile of the DOS for a glassy metal and its crystalline counterpart is similar, differences in the local atomic arrangements appear to result in the enhancement, in the glassy metal, of the contribution of the antibonding states to the DOS at $E_F$. This enhancement amounts to an increase of the total electronic energy of about 1 kcal/mole [10]. This is of the order of magnitude of the energy released upon crystallization of metallic glasses in general.

## ELECTRONIC STRUCTURE AND GLASS FORMABILITY

It is clear from the above discussion that glass formability may be predicted from the electronic properties. Before such generalization is made, it would be instructive to examine the results obtained for Pd-Si system which is one of the most well studied among metallic glasses. Figure 2a shows the soft x-ray Si $K_\beta$ emission spectra for the Pd-Si system. The antibonding states, marked by A, contribute appreciably to the DOS at $E_F$ for the glassy alloys, but their position shifts to lower energies as the Si content increases, especially in crystalline $Pd_2Si$ and PdSi. This trend is consistent with the Ni-metalloid systems discussed above. The spectra of Fig. 2a are to be compared with those in the upper part of Fig. 2b which were taken for liquid $Pd_{100-x}Si_x$ [14]. The spectra in Fig. 2a fits quite well between the cases with x=5 and 56, following the trend described above. It appears that, irrespective of the atomic arrangement and structure, the antibonding states lying above $E_F$ for low Si content shift to lower energies with increasing Si content in the Pd-Si system. Glass formation for this system becomes easiest when the antibonding states are in the vicinity of $E_F$. When this arguement is applied to the Cu-Si system, it becomes evident that the glass formability of this system is relatively low. The lower portion of Fig. 2b gives the Si $K_\beta$ emission spectra for liquid Cu-Si [14], showing no such trend as observed in the glass forming systems.

The examples used above involve d-elements whose d states are mostly occupied. By alloying, the d states are pushed even further below $E_F$.

Nevertheless, the situation is less complex compared with the case in which the transition metal is Fe or Co. Fujiwara has calculated the PDOS for glassy Fe, Fe-P and Fe-B, which indicates that p-d antibonding states appear near $E_F$ in Fe-P and above $E_F$ in Fe-B and that the antibonding component is contained in the Fe 3d band near or above $E_F$ and its amount is much more in Fe-P than in Fe-B [7]. In terms of the present discussion, these results suggest that Fe-P is a better glass forming system than Fe-B. Obviously more work, especially experimental determination of the PDOS, is needed to further extend our understanding of the subject. What seems to be clear for the transition metal-metalloid base glassy metals, though, includes relatively high DOS at $E_F$, which is probably due to the antibonding states forming near $E_F$. Partial filling of the antibonding states tends to weaken the atomic bonding, providing a unfavorable condition for compound formation but a favorable one for glass formation.

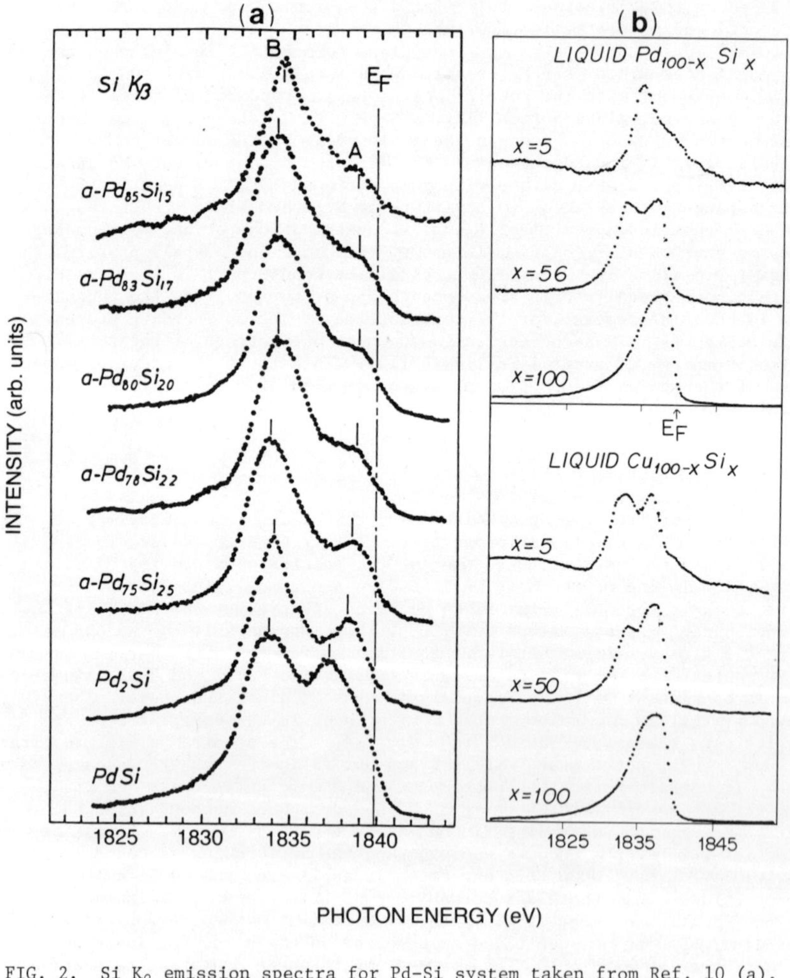


FIG. 2. Si $K_\beta$ emission spectra for Pd-Si system taken from Ref. 10 (a), and for liquid Pd-Si and Cu-Si taken from Ref. 14 (b).

REFERENCES


1. S. R. Nagel and J. Tauc, Phys. Rev. Lett. 35, 380 (1975).

2. J. D. Riley, L. Ley, J. Azoulay and K. Terakura, Phys. Rev. B 20, 776 (1979).

3. U. Mizutani, K. T. Hartwig, T. B. Massalski and R. W. Hopper, Phys. Rev. Lett. 41, 661 (1978).

4. M. Matsuura, U. Mizutani and Y. Yazawa, J. Phys. F: Metal Phys. 11, 393 (1981).

5. M. Matsuura, T. Nomoto, F. Itoh and K. Suzuki, Solid State Commun. 33, 895 (1980).

6. A. Amamou and G. Krill, Solid State Commun. 33, 1087 (1980).

7. T. Fujiwara, J. Phys. F: Metal Phys. 12, 661 (1982).

8. V. L. Moruzzi, P. Oelhafen and A. R. Williams, Phys. Rev. B 27, 7194 (1983).

9. P. Häussler, F. Baumann, J. Krieg, G. Indlekofer, P. Oelhafen and H.-J. Güntherodt, Phys. Rev. Lett. 51, 714 (1983).

10. K. Tanaka, T. Saito, K. Suzuki and R. Hasegawa, Phys. Rev. B 32, 6853 (1985).

11. O. Bisi and C. Calandra, J. Phys. C: Solid State Phys. 14, 5479 (1981).

12. J. E. Briggs and R. Hasegawa, US Patent No. 4 338 131.

13. P. Panissod, I. Bakonyi and R. Hasegawa, Phys. Rev. B 28, 2374 (1983).

14. C. F. Hague, J. Phys. (Paris) 41, C8-433 (1980).

# THERMAL STABILITY OF RAPIDLY QUENCHED TI-NI-AL AMORPHOUS ALLOYS

K. AOKI, K. HIRAGA AND T. MASUMOTO
The Research Institute for Iron, Steel and Other Metals,
Tohoku University, Sendai 980, Japan

## ABSTRACT

Ti-Ni-Al alloys were rapidly quenched from a molten state by the melt spinning method. Three kinds of metastable phases, namely, amorphous,nonequilibrium and quasicrystalline phases are formed in these alloys. The amorphous phase is formed in the range of 35 to 70 at% Ti and 0 to 25 at% Al. The nonequilibrium phases are formed in the composition range of 25 to 33 at% Ti. On the other hand, fine quasicrystalline phases are distributed in the amorphous matrix of the Ti-rich alloys. Crystallization temperatures and the hardness of the amorphous alloys were also examined.

## INTRODUCTION

In recent years, there has been an increased interest in the study of rapidly quenched metals and alloys. Among the products of rapid quenching, amorphous and nonequilibrium alloys occupy a prominent place because of the promise of industrial applications. The majority of amorphous and nonequilibrium alloys studied so far contain Fe, Ni, Co and/ or Cu. Rapid quenching has also been applied to Ti based alloys and amorphous phases have been obtained in some cases[1]. Polk et al. were able to produce an amorphous phase in the Ti-Ni-Si system [2]. On the other hand, the present authors have been successful in preparing TiNi ribbons, which exhibit shape memory effect, by rapid quenching[3].

The aim of the present work is to investigate how rapid quenching affects the structure of the Ti-Ni-Al alloys.

## EXPERIMENTAL

The Ti-Ni-Al alloys were arc melted from pure components. These alloys were then rapidly quenched using a single roller melt spinning technique in an argon atmosphere. The rotation speed of the copper roller( 200 mm in diameter) was 4000 rpm. Identification of the as-quenched phases were made by both X-ray diffraction analysis and transmission electron microscopy. Cu $K_{\alpha}$ radiation was applied in combination with an X-ray monochromator. Structural observations have been carried out by the JEM 200EX transmission electron microscope operating at 200 kV. Crystallization temperatures (Tx) of the amorphous alloys were measured in a differential scanning calorimeter(DSC) at a heating rate of 10 K/min. The Vickers hardness number(Hv) were measured by a Vickers microhardness tester with a 100g load.

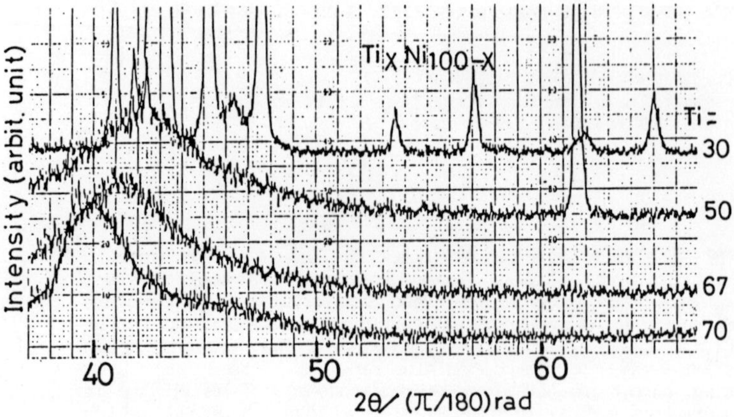


Fig.1 X-ray diffraction diagrams for the as-quenched Ti-Ni alloys

RESULTS AND DISCUSSION

Amorphous and Quasicrystalline Phases

Fig. 1 exemplifies the X-ray diffraction(XRD) diagrams for the binary Ti-Ni alloys. The $Ti_{70}Ni_{30}$ alloy is characterized by the diagram having no sharp diffraction peak but an extremely broad peak. On the other hand, sharp crystalline peaks are seen on the diagram for the as-quenched $Ti_{30}Ni_{70}$ alloy. The XRD pattern for the as-quenched $Ti_{50}Ni_{50}$ alloy is an intermediate of the above two, exhibiting a superposition of the broad amorphous peak and much sharper crystalline peaks of TiNi (CsCl structure).

Fig. 2 shows the formation range of the amorphous Ti-Ni-Al alloys. The numeral indicates the crystallization temperatures, Tx. In the present work, Tx was defined as the temperature corresponding to the intersection of the extrapolated base line and the steepest tangent to the exthothermic peak. Tx for the amorphous Ti-Ni-Al are in the range of 683 K to 784 K, which are lower than those of the Ti-Ni-Si alloys[2] by about 20-30 K at the same Ni and Ti content. That is, the contribution of Al to the thermal stability of the Ti-Ni amorphous alloys is lower in comparison with that of Si. Amorphous phases are formed in the composition range 35 to 70 at% Ti in the ternary system, but the formation range of an amorphous single phase is limitted to between 60 and 70 at% Ti. The latter result agrees well with the previous report[2]. The formation range of the amorphous Ti-Ni phase depends strongly on the qenching rate and extends prominently with the increase in the quenching rate. For example, the slowly quenched TiNi ribbons consist of CsCl type structures as already reported [3], while the rapidly quenched TiNi ribbons of about 0.01mm in thickness are amorphous. As seen in this figure, ternary amorphous phases are obtained in the alloys containing less than 25 at% Al. The amorphous alloys containing more than 15 at % Al are very brittle. The Vickers hardness number of ductile amorphous Ti-Ni-Al alloys are in the an order of 500.

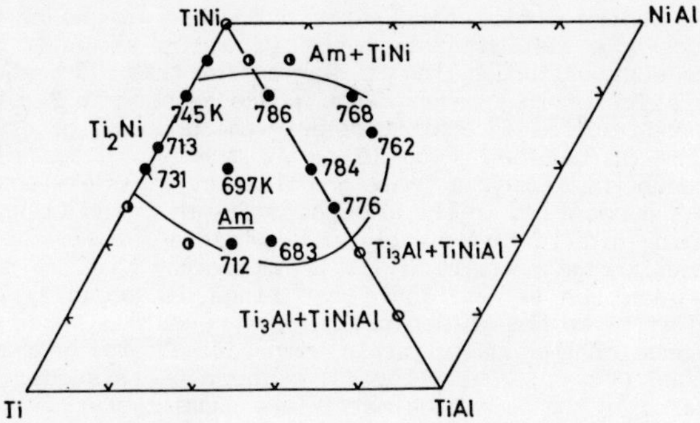


Fig.2 Compositions of Ti-Ni-Al alloys found to be amorphous(●), partly amorphous(◐) or crystalline(○) after rapid quenching

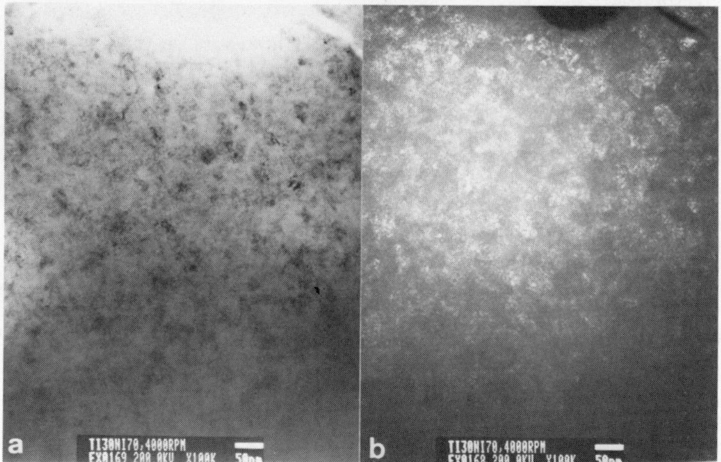


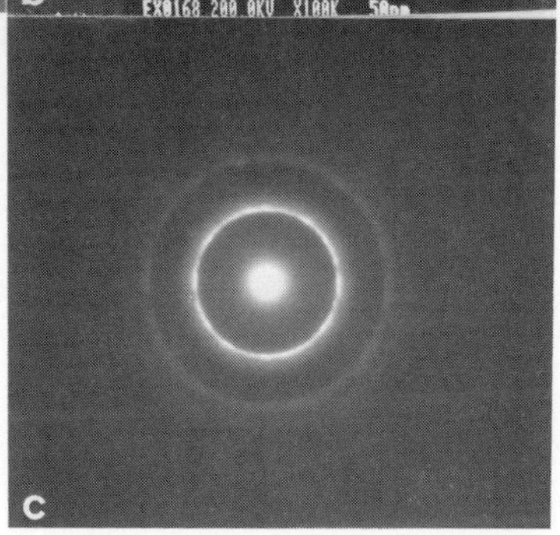


Fig. 3 (a)(b)(c) a bright- and dark electron micrographs and the corresponding diffraction pattern for the as-quenched $Ti_{70}Ni_{30}$ alloy

Fig. 3(a)(b)(c) show a bright- and a dark-field electron micrographs and the corresponding diffraction pattern for the as-quenched $Ti_{70}Ni_{30}$ alloy. Although the XRD diagram for this alloy shows a broad peak characteristics alloy. Although the XRD diagram for this alloy shows a broad peak charactristics of in an amorphous phase as seen in Fig.1, both the bright- and dark-field images exhibit clearly the presence of crystals with an average size of between 1 to 10 nm in the amorphous matrix. The boundaries between the amorphous phase and these crystals are obscure. These fine crystals are observed in the amorphous $Ti_{70}Ni_{30}$ and $Ti_{67}Ni_{33}$ alloys when the quenching rate is sufficiently high, while a slower quenching rate produces the equilibrium phase,$Ti_2Ni$. It is noteworthy that the ratio of the diameters between the second and first rings is about 1.6, which is approximately equal to the golden ratio. More recently, Zhang et al. have found the presence of the quasicrystals with icosahedral symmetry in the rapidly quenched $(Ti_{1-x}V_x)_2Ni$ alloy[4]. These facts suggest that the crystals scattered in the amorphous matrix are quasicrystalline phases with icosahedral symmetry as those observed in the rapidly quenched Al-14 at% Mn alloys [5]. Unfortunately , the crystals are too fine to obtain the five fold symmetry diffraction pattern in this alloy. The observation of the five fold symmetry diffraction pattern in the large crystals is the subject for a future work. Here, emphasized is the fact that the fine quasicrystalline phases are distributed in the amorphous matrix.

Nonequilibrium phases

The $Ti_{30}Ni_{70}$ alloys heat treated at 1073 K in vacuum or the as-arc melted samples could be identified as the equilibrium $Ni_3Ti$ phase , while the XRD pattern for the as-quenched $Ti_{30}Ni_{70}$ alloy could be indexed by neither NiTi nor $Ni_3Ti$ phase as seen in Fig. 1. This unidentified phase transforms to the equilibrium phase, $Ni_3Ti$ at about 900 K. Consequently, the as-queched $Ti_{30}Ni_{70}$ alloy is considered to be a nonequilibruim phase. This metastable phase is formed in the range of 25 to 33 at% Ti. Thus, three kinds of metastable phases are found out in the as-quenched condition.

## CONCLUSIONS

With the aid of rapid quenching, three kinds of metastable phases, amorphous, nonequilibrium and quasicrystalline phases are formed in the binary Ti-Ni system. The formation of the amorphous phase is observed in the range of 35 to 70 at% Ti, while that of the nonequilibruim phase is in the range of 25 to 33 at % Ti. On the other hand, the fine quasicrystalline particles are distributed thoroughout the amorphous matrix.

## REFERENCES

1. R. Ray, B. C. Giessen and N. J. Grant, Scripta Met. 2(1968)
2. D. E. Polk, A.Calk and B. C. Giessen, Acta Met. 26, 1097(1978)
3. Y. Schugo, K. Aoki, T. Masumoto and T. Honma, Bull. of the Research Institute of Mineral Dressing and Metallurgy, Tohoku University,38, 85(1982)
4. Z. Zhang, H. Q, Ye and K. H. Kuo, Phil. Mag. Lett. in the press
5. K. Shechtman, I. Blech, D. Gratias and J. W. Cahn, Phys. Rev. Lett. 53,1951(1984)

# THERMODYNAMICS AND KINETICS OF CRYSTAL-AMORPHOUS TRANSFORMATION IN $A_{1-x}B_x$ ALLOYS DURING REACTION WITH HYDROGEN

X. L. YEH, W. L. JOHNSON, J. Y. TANG* and C. R. SHI**
W. M. KECK Laboratory for Engneering Materials, California Institute of Technology, Pasadena, CA. 91125, U. S. A.
*Nuclear Science Department, Fudan University, Shanghai, China.
**Nuclear Research Institute, Lanzhou University, Lanzhou, China.

ABSTRACT

The kinetics of amorphous phase formation in polycrystalline $A_{1-x}B_x$ (A=Zr, Hf, B=Pd, Rh, .15<x<.25) alloys during reaction with hydrogen has been studied by x-ray diffraction, $^{+19}F$ nuclear reaction depth profiling, TEM and electron diffraction. The formation of the amorphous hydride phase is observed by TEM to begin at grain bounderies of the polycrystalline $Zr_{1-x}Rh_x$ much in the same manner that "melting" nucleates at grain boundaries. TEM micrographs further show that the phase boundary between the crystalline and amorphous phases remains sharp during the growth of the amorphous phase. Both x-ray diffraction and nuclear depth profiling studies suggest that the overall rate of transformation to the amorphous hydride phase is limited by the rate of hydrogen permeation through the sample surface.

Based on the present experiments and an analysis of the relevant free energy curves, we discuss the thermodynamic and kinetic aspects of this effect to explain why an amorphous phase is formed.

## INTRODUCTION

The key factor in producing amorphous alloys is to suppress the nucleation and growth of equilibrium crystalline phases. The conventional way to fulfill this requirement is by either rapidly quenching an alloy from the melt or depositing a metallic vapor onto a cold substrate. One drawback of these techniques is that one of the sample dimensions has to be small in order to achieve a high cooling rate. Recent experiments by Johnson et al. demonstrated the possibility of making bulk amorphous alloys via solid-state reactions [1,2]. The first example of such solid-state reactions was the amorphous phase formation from polycrystalline Zr-Rh alloys during hydriding [1]. The principle of this experiment is to raise the free energy of the crystalline phase above that of its amorphous counterpart through solid state diffusion at a low enough temperature so that crystallization to equilibrium is supressed, allowing solid-state melting to occur. The present research is an endeavor to understand the thermodynamic and kinetic aspects of this transformation. Based on the present experiments, we will discuss the thermodynamics and kinetics of the entire process. We also report the discovery of two other crystalline systems which transform to the amorphous state upon hydriding.

## EXPERIMENTAL PROCEDURES

$A_{1-x}B_x$ (A=Zr, Hf, B=Pd, Rh, .15<x<.25) ingots were processed into 30-50 $\mu$m thick ribbons by the melt-spinning technique and then examined by x-ray diffraction using Cu $K\alpha$ radiation. X-ray scans were made on both sides of the ribbons. Samples showing a single-phase solid solution having the fcc structure were selected for hydriding. The details of ingot preparation, rapid quenching and hydriding procedures are described elsewhere [1]. The TEM samples were prepared either by jet-thinning using 10% $HClO_4$ methanol solution or by ion-milling. TEM was done on a Philips EM420 electron microscope.

 

A resonant nuclear reaction using a tandem van de Graaff accelerator was used to examine the hydrogen concentration profile in amorphous $Zr_3Rh$. Groups of samples and a tourmaline stardard of known water concentration were placed in a sample chamber of the tandem accelerator at Cal Tech, and evacuated to a pressure of $10^{-6}$ torr. The nuclear reaction $^{19}F(^1H, \alpha\gamma)^{16}O$ was used, where $^{19}F$ nuclei at energies from 6-10 MeV react with protons. Oxygen in the form of $^{16}O$ is produced, along with alpha particles and gamma radiation. The detected intensity of gamma radiation is proportional to the concentration of hydrogen at various depths. The depth resolution of this technique is ±150Å at the surface and ±250Å at a depth of 1 $\mu$m.

RESULTS AND DISCUSSION

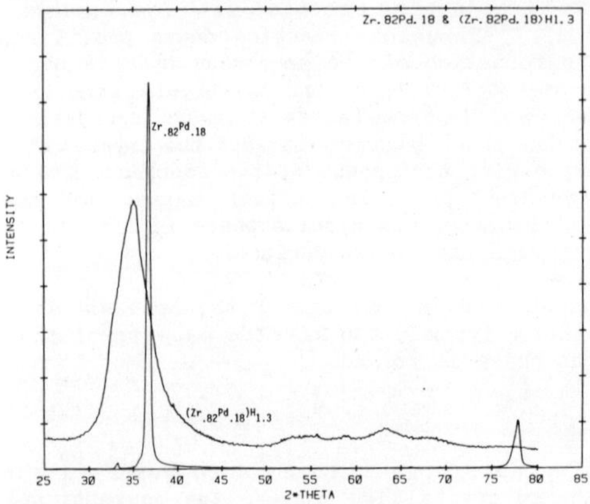


Fig.1a.

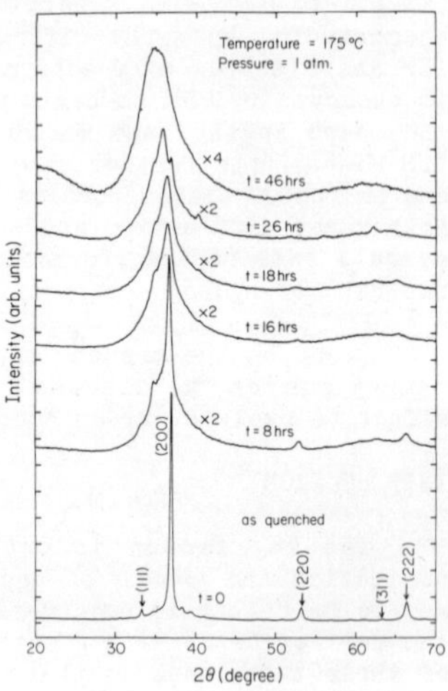


Fig.1b.

Fig.1. X-ray diffraction patterns for: (a) As-quenched $Zr_{.82}Pd_{.18}$ and $(Zr_{.82}Pd_{.18})H_{1.3}$ obtained by hydriding at 210°C; (b) $Zr_{.8}Rh_{.2}H_x$ at various hydriding times.

The crystal to amorphous transformation has been observed in fcc solution phases of the $Zr_{1-x}Rh_x$, $(.15<x<.25)$, $Zr_{1-x}Pd_x$, $(.17<x<.21)$, and $Hf_{1-x}Rh_x$, $(.19<x<.25)$ systems during reaction of as-quenched polycrystalline ribbons with hydrogen gas at temperatures below 220° C. Owing to space limitation, only two sets of the x-ray diffraction patterns (XRD) are displayed. Fig.1a shows the XRD of as-quenched polycrystalline $Zr_{.82}Pd_{.18}$ and amorphous $(Zr_{.82}Pd_{.18})H_{1.3}$ obtained by hydriding at 210°C. Fig.1b shows the XRD of $Zr_{.8}Rh_{.2}$ at various stages of reaction. We will focus our attention on the data obtained from $Zr_{.8}Rh_{.2}H_x$. One can see from Fig.1b that an amorphous band grows while the crystalline peaks disappear. There are no other intermediate phases developed throughout the reaction. One also notices that the crystalline Bragg peaks shift to lower angles, decrease in intensity and broaden as the reaction proceeds. This suggests that hydrogen is absorbed by the crystalline phase, and it becomes unstable relative to the amorphous phase as the hydrogen concentration increases, inducing a crystal to amorphous transformation.

Nuclear reaction hydrogen depth profile experiments were performed to study the distribution of hydrogen in the sample. Fig.2 shows the hydrogen

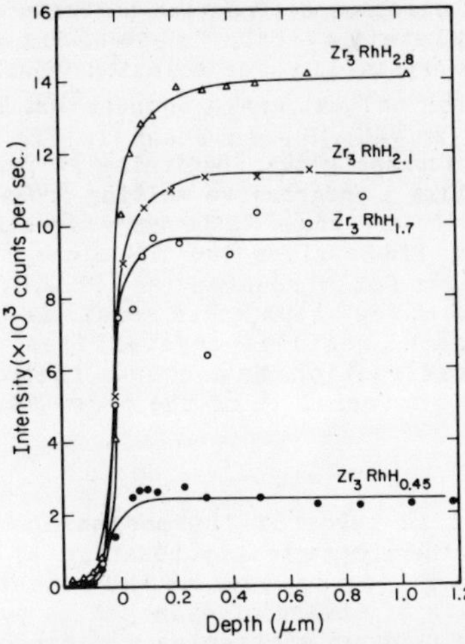


Fig.2. Hydrogen concentration profiles in amorphous $Zr_{.75}Rh_{.25}H_x$ during hydriding at 180 °C.

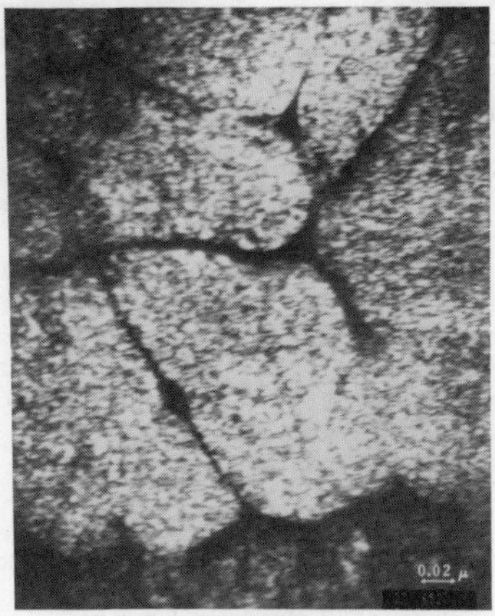


Fig.3a.

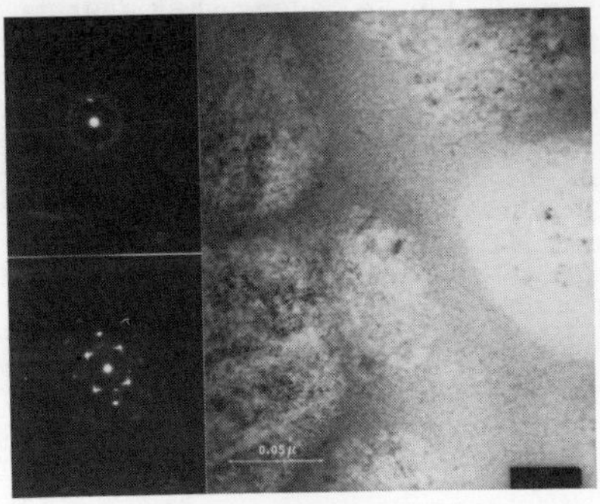

Fig.3b.

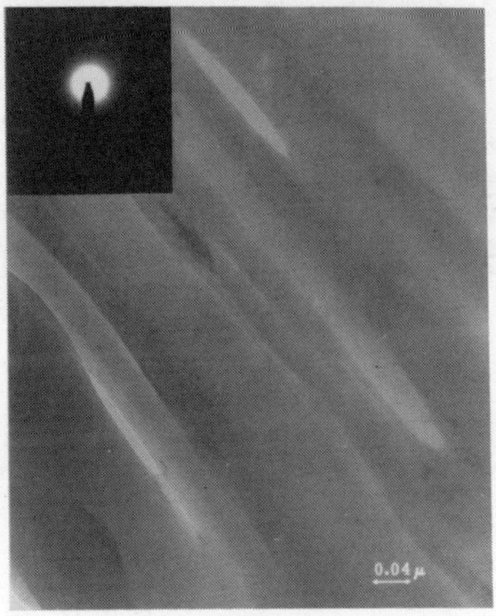


Fig.3c.

Fig.3. TEM micrographs and selected area electron diffraction patterns for $Zr_{.8}Rh_{.2}H_x$ during hydriding. a. As-quenched sample. b. Partially reacted sample with H/M = 0.5. c. Fully reacted sample with H/M = 1.0.

depth profile in hydrided amorphous $Zr_3Rh$ at various stages of reaction. The plateaus on the figure indicate that the hydrogen has a uniform distribution in the hydrided sample during each progressive hydriding step. This demonstrates that the time scale for hydrogen redistribution within the sample is much less than that of hydrogen permeation through the sample surface. In other words, this observation implies that a surface barrier to hydrogen permeation limits the reaction rate. By etching the sample in a dilute HF solution prior to hydriding, it is found that the zirconium oxide on the surface is removed and the rate of reaction increases dramatically, and also the reaction proceeds at much lower temperatures (e.g., at room temperature or even below). The surface barrier for hydrogen to enter the sample is apparently removed or reduced by this treatment, confirming that this barrier is rate limiting in untreated samples.

Fig.3 shows the TEM micrographs and electron diffraction patterns for as-quenched, partially reacted and completely reacted $Zr_{.8}Rh_{.2}$. The as-quenched sample was nearly a single-phase crystalline fcc solution (Fig.3a). Traces of amorphous phase were found by TEM only at grain boundaries. The TEM micrograph of partially hydrided $(Zr_{.8}Rh_{.2})H_{.5}$ is shown in Fig.2b. During hydrogen absorption, the amorphous phase grows, beginning at grain boundaries of the crystalline phase much like a progressive melting process. This is not surprising since grain boundaries are known to be easy sites for melting or vitrification in crystals [7]. Fig.3c gives the TEM micrograph and selected area diffraction pattern of a fully reacted $(Zr_{.8}Rh_{.2})H_{1.0}$. This sample is almost entirely amorphous. A few black spots remain in the micrograph and are believed to be unreacted residual crystallites. No crystallinity can be detected by x-ray diffraction. We conclude that the volume fraction of remaining crystalline material is of the order 1% or less.

We now discuss this transformation in terms of thermodynamics. A criterion to judge whether a reaction is thermodynamically possible is to see whether a decrease in Gibbs free energy is involved as the reaction proceeds. On the basis of calculated heats of mixing of hydrogen in pure metals [3] as well as NMR and inelastic neutron scattering experiments [4,5], hydrogen prefers residing in a tetrahedrally coordinated site having four Zr neighbors to one with three Zr and one Rh neighbor, since the former configuration has lower energy than the latter. A calculation by Samwer and Johnson suggests that amorphous $Zr_{1-x}A_x$ (where A represents late transition metals, $0<x<.25$) has more fourfold coordinated Zr tetrahedral holes than does an ordered fcc phase of this material [6]. We deduce that amorphous $Zr_{.8}Rh_{.2}H_{1.0}$ has lower free energy than fcc $Zr_{.8}Rh_{.2}H_{1.0}$. This is consistent with what we see in our experiments. Using the above information and the regular solution approximation, a schematic free energy diagram can be constructed as shown in Fig.4. Curve I and II in the figure denote the free energy of homogeneous fcc and amorphous phases of $Zr_{.8}Rh_{.2}H_x$, respectively, while curve III denotes the free energy of a two-phase mixture $(0.8ZrH_x+0.2Rh)$. The arrows on the figure indicate a possible path followed during our experiments, (i.e., an amorphous formation during hydriding). It is of obvious importance to establish these free energy diagrams on a quantitative basis. Calorimetric measurements to this are underway.

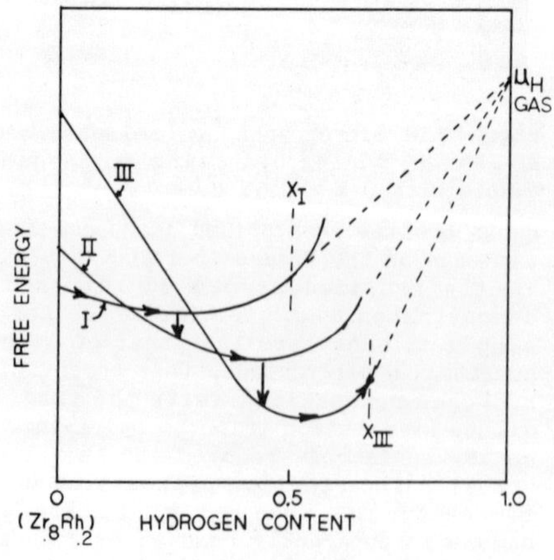


Fig.4. A schematic free energy diagram showing the free energy variation of fcc $Zr_{.8}Rh_{.2}H_x$ (curve I), amorphous $Zr_{.8}Rh_{.2}H_x$ (curve II), and two phase mixture $(.8ZrH_x+.2Rh)$ (curve III). Arrows indicate paths followed by various reactions.

The next question is why the final reaction product is not ($4ZrH_2$+Rh), which is known to be the equilibrium phase and thus to have a lower free energy than the amorphous phase [8]. Apparently, the reason is that the two types of metal atoms have such low atomic mobility at the reaction temperature (220°C), that such a phase separation is not possible, while hydrogen mobility is large enough to achieve long range diffusion. As shown previously, the overall reaction rate is controlled by a surface barrier. We thus have the following inequality:

$$\tau_{\text{hydrogen diffusion}} << \tau_{\text{reaction}} << \tau_{\text{metal diffusion}}$$

where $\tau$ is a characteristic time scale of various processes over the relevant distances. For instance, if distance is chosen as the size of a critical nucleus of $ZrH_2$ for the case of metal diffusion, then the above inequality implies that $ZrH_2$ can never be nucleated. The entire reaction takes place on a time scale very long compared to the characteristic hydrogen diffusion time. By measuring the sample tmeperature during hydriding, we found that the reaction is isothermal, even though the hydriding is exothermic in nature. This implies that the rate at which the heat is generated by the reaction is not large enough to raise the sample temperature. Thus, metal diffusion is not enhanced, and the formation of $ZrH_2$ is suppressed.

## CONCLUSION

Amorphous phase formation during hydriding has been observed in several as-quenched polycrystalline fcc phases of early transition metal-late transition metal systems. This effect can be described as a "solid state melting" process. The essential requirement is to create a situation where an amorphous phase has lower free energy than a crystalline phase of a material produced under constrained kinetic conditions which suppress the transformation of the crystalline phase to other crystalline phases or phase mixtures of still lower energy.

ACKNOWLEDGEMENT--- One of the authors (X. Y.) wishes to thank Standard Oil Company for financial support. We acknowledge the support of the U. S. Department of Energy (Project Agreement No. DE-AT03-81Er10870) under Contract No. DE-AA03-76SF00767.

## REFERENCES

1. X. L. Yeh, K. Samwer & W. L. Johnson, Appl. Phys. Lett. 42, 242 (1983).
2. R. Schwarz & W. L. Johnson, Phys. Rev. Lett. 51, 45 (1983).
3. C. D. Gelatt, Jr. et al, Phys. Rev. B 17, 1940 (1978).
4. A. Williams et al, J. Noncryst. Solids 61&62, 643 (1984).
5. R. C. Bowman et al, J. Noncryst. Solids 61&62, 649 (1984).
6. K. Samwer & W. L. Johnson, Phys. Rev. B 28, 2907 (1983).
7. J. W. Christian, The Theory of Transformations in Matals and Alloys, Part 1, 2nd ed., (Pergamon, Oxford Press), 448 (1981).
8. J. E. Wagner & R. C. Bowman, Jr., to be published in J. Appl. Phys., (1985).

# AMORPHOUS Cu-Zr AND Fe-Ti COMPOSITIONALLY MODULATED FILMS PRODUCED BY SPUTTERING FROM ELEMENTAL TARGETS

E. CHASON, H. KONDO AND T. MIZOGUCHI
Japan Research and Development Corporation, Gakushuin University Division, 1-5-1, Mejiro, Toshimaku, Tokyo 171, Japan

R.C. CAMMARATA,* AND F. SPAEPEN
Division of Applied Sciences, Harvard University, Cambridge, MA 02138, U.S.A.

B. WINDOW, J.B. DUNLOP AND R.K. DAY
CSIRO National Measurement Lab., Division of Applied Physics, P.O. Box 218, Lindfield, NSW 2070, Australia

## ABSTRACT

Amorphous compositionally modulated films of Cu-Zr and Fe-Ti have been prepared by sequential sputter deposition of the elemental metals. For Cu-Zr, a critical modulation amplitude has been found, below which the samples are entirely amorphous. The effective interdiffusion coefficient during deposition has been estimated, and found to be assisted by the free energy of mixing and the ion bombardment. X-ray studies of the modulation peaks, using grid scans, reveal a sharp mosaic, and asymmetrical peaks in some samples.

## INTRODUCTION

The first amorphous metallic compositionally modulated films were produced by Rosenblum et al. [1] for the purpose of studying interdiffusion upon subsequent annealing. Using d.c. sputtering on a movable substrate from two alloy targets, they prepared $(Pd_{85}Si_{15})_{61}/(Fe_{85}B_{15})_{39}$ films with modulation wavelengths around 3 nm. Greer et al. [2] and Cammarata and Greer [3] have reported results on similar materials. Rosenblum et al. [1] also succeeded in preparing fully amorphous modulated films of composition $(Au_7Pd_{80}Si_{13})_{70}/Fe_{30}$, with a wavelength of 2.1 nm, by using elemental Fe as one of the sputtering targets.

Lin et al. [4], and Lin and Spaepen [5] have reported the formation of fully amorphous modulated films of average composition $Ni_{50}Zr_{50}$ $Co_{40}Nb_{60}$, and $Ni_xNb_{1-x}$ $(40<x<79)$ by d.c. sputtering from two elemental targets, with modulation wavelengths of 2-3 nm. Modulated films of average composition $Ni_{50}Mo_{50}$ and $Co_{45}Mo_{55}$, prepared by the same method,were fully crystalline, however. Since these samples were prepared for the purpose of subsequent picosecond laser mixing and quenching, the conditions for amorphous phase formation were not systematically investigated.

Clemens and Buchholz [6] have prepared compositionally modulated films of average composition $Zr_{38}Ni_{62}$ over a large range of wavelengths by magnetron sputtering form elemental targets. They reported that the films were entirely amorphous if the individual layer thickness was less than four atomic planes. The main purpose of their work, however, was to study the crystalline-to-amorphous reaction upon annealing of the samples that were initially crystalline.

---

*Presently at the Department of Materials Science and Engineering, M.I.T., Cambridge, MA 02139, U.S.A.

In this paper, we report on the formation of amorphous compositionally modulated Cu-Zr and Fe-Ti films by sputtering from elemental targets. The amplitude of the resulting modulation and the structure of the films have been investigated systematically over a large range of wavelengths. The purpose was to study how the *in situ* interdiffusion process determines a range of wavelengths within which amorphous compositionally modulated films can be produced; the lower limit corresponds to the disappearance of the modulation, the upper limit to the appearance of crystallinity. At the same time, a comparison was made between three different deposition systems.

## SAMPLE PREPARATION

Samples were prepared in three different sputtering systems. Some of the Cu-Zr samples were produced at Harvard University in an Ar ion beam system described in Ref. 7. The remainder of the Cu-Zr samples were produced at Gakushuin University in a similar Ar ion beam system. In both systems, the ion guns were operated at a residual gas pressure of $2 \times 10^{-5}$ torr. In the Harvard system, the beam voltage was 1.5 kV, and average deposition rate was 0.1 nm/s. The substrate was water-cooled, although a few runs were also made with liquid nitrogen cooling. The total film thickness was about 800 nm. In the Gakushuin system the beam voltage was 1 kV, and the average deposition rate was 0.05 nm/s. The substrate was not cooled, but temperature monitoring during deposition showed that the temperature did not rise above 30°C. The total film thickness was about 1 μm.

The Fe-Ti samples were produced at CSIRO by magnetron sputtering. The average deposition rate was 0.039 nm/s, and the operating gas pressure $5 \times 10^{-3}$ torr. The substrate was not cooled, and the sample temperature was measured to have risen to 55°C during deposition. The total film thickness was about 280 nm.

## MEASUREMENTS

The Cu-Zr films were studied at Gakushuin University with an X-ray diffractometer with independent $\phi$ and $2\theta$ circles. The slit sizes were 0.166° for the incident beam, and 0.1° for the scattered beam. Soller slits limited the vertical divergence to 5°. Cr $K_\alpha$ radiation was used. The diffraction peaks resulting from the composition modulation were studied by a grid scan of reciprocal space, i.e., in both the direction of the modulation, $k_z$, and the direction perpendicular to it, $k_p$, as illustrated by Figure 1.

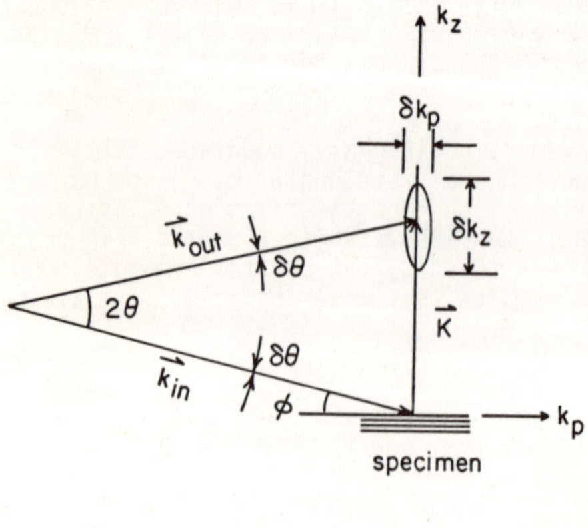


Figure 1. *Orientation of the sample with respect to the scattering vector* K. dθ *is the angular acceptance of the spectrometer slits. The resolution function in* k*-space is shown superimposed on the scattering vector.*

This type of scan is necessary because the resolution function along the $k_p$ direction is narrow compared to the $k_z$ direction at low scattering angles. It can easily be seen from Figure 1 that the widths of the resolution function are related to the angular acceptance of the slits by:

$$dk_z = \frac{4\pi}{\lambda} \cos\theta \, d\theta \tag{1}$$

$$dk_p = \frac{4\pi}{\lambda} \sin\theta \, d\theta$$

which makes $dk_p \ll dk_z$ at low angles $\theta$. This is not a problem if the sample mosaic is broad enough, but in these films it is often quite sharp: peak widths along $k_p$ were observed that were resolution-limited. Furthermore, peaks were often off the $k_z$ axis by as much as 0.05°. In these cases, a simple $\theta$-$2\theta$ scan would have given a peak intensity that would be as much as 50% in error.

The dominant source of error was the sample positioning, as a result of lateral inhomogeneity in the films. The sensitivity to positioning varied from sample to sample, between 20% and 1.5% per mm motion in the plane of the film. The stability of the X-ray system was checked periodically with a standard sample and was found to be within 2.5%.

The atomic scale structure of the Cu-Zr films was investigated by $\theta$-$2\theta$ scans at higher angles on the same system. For some samples, the results were checked by electron diffraction and dark field imaging in a Philips EM420 microscope at Harvard.

The Fe-Ti films were investigated at CSIRO by high angle $\theta$-$2\theta$ X-ray diffraction using Cu $K_\alpha$ radiation, and by Mössbauer spectroscopy.

## RESULTS

### 1. Cu-Zr

The samples with an average composition of $Cu_{50}Zr_{50}$, were deposited on glass slides. The Harvard samples were prepared with layer repeat lengths between 1.49 nm and 10.1 nm. The composition modulation could be observed for wavelengths greater than 1.4 nm. Attempts to produce a modulation at shorter wavelengths by cooling the substrate to liquid nitrogen temperature, by varying the beam voltage, the deposition rate, or the target angle were unsuccessful. The films were entirely amorphous for wavelengths less than 3.5 nm.

The Gakushuin samples were prepared with layer repeat lengths between 3.2 nm and 58.7 nm. The composition modulation could be observed for all wavelengths and the films were entirely amorphous for wavelengths below 2.5 nm.

Figure 2a shows a grid scan of the first modulation diffraction peak of a Harvard sample with 3.28 nm wavelength. Two peaks at different $k_z$ are observed. The tilt in the pattern relative to the $k_z$ and $k_p$ axes indicates a variation of the modulation wavelength at different parts of the mosaic. This may be a result of a variation of the deposition rate across the sample surface. Both these features are also seen in the grid scan of the second diffraction peak of this sample, shown in Figure 2b.

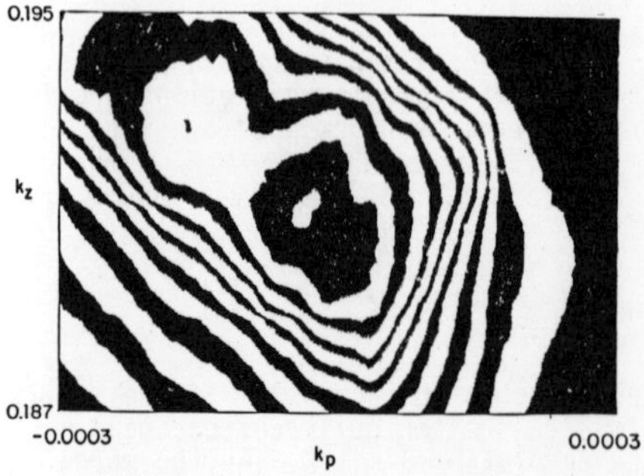


Figure 2(a). *Grid scan of the first modulation peak of a compositionally modulated amorphous* Cu-Zr *sample with wavelength of* 3.28 nm. *The range of the scans is indicated (in* Å$^{-1}$*). The peak intensity is* 31400 counts/s; *each contour corresponds to* 2000 counts/s.

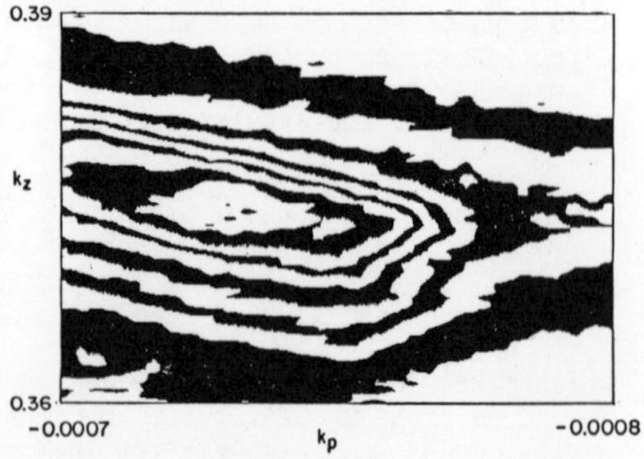


Figure 2(b). *Grid scan of the second modulation diffraction peak of the sample of Figure 2(a). The range of the scans is indicated (in* Å$^{-1}$*). The peak intensity is* 50 counts/s; *each contour corresponds to* 40 counts/s.

Figure 3a shows a grid scan of the first modulation diffraction peak of a Gakushuin sample with 4.72 nm wavelength. This peak is more symmetrical. Figure 3b shows a radial scan along a line from the origin of reciprocal space through the first and second diffraction peaks. This is necessary to determine the background intensity, which can be as high as 20% of the peak intensity.

Figure 4 shows a plot of the peak intensities of the first modulation peaks as a function of wavelengths, for both sets of samples. Note that the first crystallinity is observed if the *intensity*, and hence the modulation amplitude, exceeds a certain value. This indicates that a certain degree of mixing is required to stabilize the amorphous phase. The greater mixing in the Gakushuin system allows the amorphous phase to be formed at higher wavelengths, but also eliminates the modulation at the lower wavelengths.

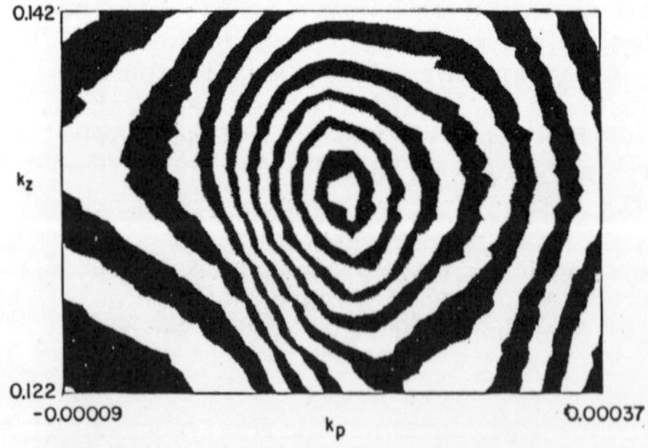


Figure 3(a). *Grid scan of the first modulation peak of a compositionally modulated amorphous* Cu-Zr *sample with wavelength of* 4.72 nm. *The range of the scans is indicated (in* Å$^{-1}$*). The peak intensity is* 10400 counts/s; *each contour corresponds to* 520 counts/s.

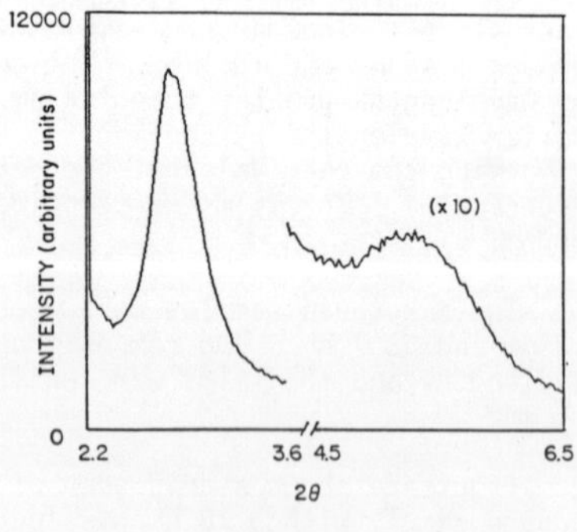


Figure 3(b). *Radial scan through the first and second modulation peaks of the sample of Figure 3(a).*

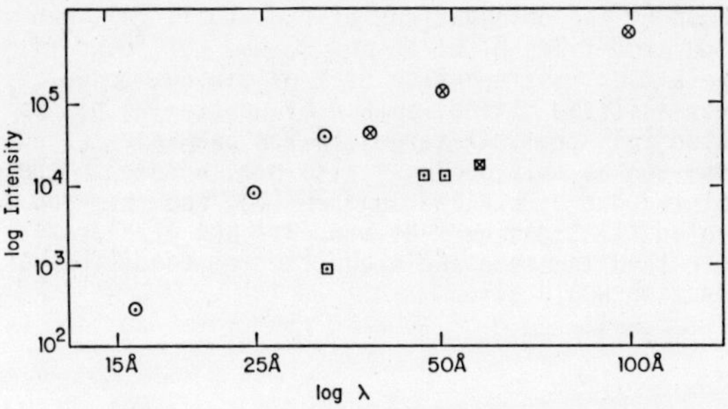


Figure 4. *Intensity of the first diffraction peak as a function of the modulation wavelength for all* Cu-Zr. *The circles and squares correspond to samples made at Harvard and Gakushuin, respectively. A cross inside a symbol indicates presence of crystallinity.*

2. Fe-Ti

The samples with an average composition of $Fe_{50}Ti_{50}$ were deposited on fused silica substrates for layer repeat lengths of 0.07 to 7.7 nm. The composition modulation could be observed for wavelengths of 1.6 nm and greater. The films were entirely amorphous for wavelengths less than 5.4 nm. At higher wavelengths, the hcp α-Ti, the bcc TiFe, and the hcp $TiFe_2$ phases could be identified.

Mössbauer spectroscopy was performed on samples deposited on a pure Aℓ foil. For wavelengths of less than 2 nm, the spectrum shows a doublet with a separation of approximately 0.71 mm/s. For wavelengths of 3.7 and 4.4 nm, which show an appreciable modulation amplitude, the doublet is smeared out. For the two largest wavelengths of 5.4 and 7.6 nm, which contain some crystals, there is separation into a ferromagnetic phase and a non-ferromagnetic one. The ferromagnetic phase has an internal field of approximately 300 kOe for $\lambda = 5.4$ nm, and 320 kOe for $\lambda = 7.6$ nm at room temperature. This certainly corresponds to the bcc FeTi phase. The non-ferromagnetic phase corresponds to amorphous Fe-Ti [8].

DISCUSSION

The effective interdiffusion during deposition, which determines both the minimum wavelength at which a modulation can be observed and that at which crystallinity begins to occur, can be estimated roughly from [9]:

$$Dt = \frac{\lambda^2}{8\pi^2} \ln\left(\frac{I_0}{I}\right) \qquad (3)$$

where D is the effective diffusion coefficient, t the mixing time (inversely proportional to the deposition rate), I the modulation peak intensity of the sample, and $I_0$ the same for an undiffused reference sample. If the sample with the longest wavelength is taken as the reference, the product Dt for all the Harvard samples is found to be $(1.3 \pm 0.2) \times 10^{-19}$ m$^2$, and for all the Gakushuin samples $(4.4 \pm 0.8) \times 10^{-19}$ m$^2$. Considering that in both cases the mixing time is no more than a few seconds, the effective diffusion coefficient is estimated to be on the order of $10^{-20}$ - $10^{-19}$ m$^2$/s. Given that the temperature of the samples never rose much above room temperature, this value is clearly many orders of magnitude greater than the chemical interdiffusivity in amorphous metals such as Cu-Zr or Fe-Ti at that temperature [9]. There are two important factors that enhance the mixing process. (i) The large negative heat of mixing of these systems. This is known to be an important factor in the formation of amorphous metals by solid state reaction during annealing of elemental composites [10]. That this process almost certainly also plays a role in these *in situ* reactions is shown by the observations of Lin et al. [4], who obtained *in situ* amorphous modulated films of Ni-Nb and Co-Nb, but found films of Ni-Mo and Co-Mo, which have a much less negative heat of mixing, to be entirely crystalline. (ii) Ion-assisted mixing, either by sputtering plasma ions, or by specularly reflected ions from the target in ion beam sputtering. That a non-equilibrium process such as this one must also play a role in these *in situ* reactions has been pointed out by Lin and Spaepen [5], who observed that *in situ* Ni-Nb amorphous modulated films can be made over a range of average compositions that is much wider than the metastable equilibrium conditions of the unassisted solid state reaction would allow.

## CONCLUSION

We have shown that it is possible to produce entirely amorphous compositionally modulated films of Cu-Zr and Fe-Ti by sequential deposition of the elemental metals. We have determined the conditions under which the modulation disappears, and at which crystallinity begins to occur. The latter seems to correlate with a maximum modulation amplitude, corresponding to a minimum amount of mixing. The effective interdiffusion coefficient during deposition of the Cu-Zr has been estimated; it was found to be enhanced by the chemical free energy decrease of the mixing reaction and by the ion bombardment. A systematic x-ray study of some of the samples by a grid scan of the modulation peaks revealed that the mosaic can be quite sharp. The observation of non-symmetrical peaks in some samples indicates possible inhomogeneity.

## ACKNOWLEDGEMENTS

The work at Gakushuin has been supported by the Japan Research and Development Corporation. The interest of Prof. T. Masumoto in promoting and supporting the work is gratefully acknowledged. The work at Harvard has been supported by the Office of Naval Research, under contracts N00014-83-K-0030 (development of the ion beam system), and N00014-85-K-0023 (study of amorphous metals).

## REFERENCES

1. M.P. Rosenblum, F. Spaepen and D. Turnbull, Appl. Phys. Lett. 37, 184 (1980).
2. A.L. Greer, C.J. Lin and F. Spaepen, Proc. 4th Int. Conf. on Rapidly Quenched Metals, ed. by T. Masumoto and K. Suzuki, Jap. Inst. Metals, Sendai, 1982, p. 567.
3. R.C. Cammarata and A.L. Greer, J. Non-Crystalline Solids 61/62, 889 (1984).
4. C.J. Lin, F. Spaepen and D. Turnbull, J. Non-Crystalline Solids 61/62, 767 (1984).
5. C.J. Lin and F. Spaepen, to appear in Acta Met.
6. B.M. Clemens and J.C. Buchholz, Mat. Res. Soc. Symp. Proc. 37, 559 (1985).
7. F. Spaepen, A.L. Greer, K.F. Kelton and J.L. Bell, Rev. Sci. Instr. 56, 1340 (1985).
8. S.H. Liou and C.L. Chien, J. Appl. Phys. 55, 1820 (1984).
9. A.L. Greer and F. Spaepen, in "Synthetic Modulated Structures," ed. by L. Chang and B.C. Giessen, Academic, N.Y. (1985), p. 419.
10. R.B. Schwartz and W.L. Johnson, Phys. Rev. Lett. 51, 415 (1983).

# DIFFERENCES BETWEEN SPUTTERING METHODS IN THE FORMATION OF AMORPHOUS MAGNETIC ALLOY FILMS

Y. Hoshi* and M. Naoe**
* Tokyo Institute of Polytechnics, Atsugi-shi, Kanagawa-ken 243-02 Japan
** Tokyo Institute of Technology, Meguro-ku, Tokyo 152 Japan

## ABSTRACT

Fe-Si, Co-Ta and Co-Zr amorphous films have been deposited by using various sputtering methods (conventional rf diode sputtering, rf triode sputtering, dc Targets Facing type sputtering (dc TF sputtering) and dual ion beam sputtering (DIB sputtering)). The lower limit of the Si and Ta content to form amorphous Fe-Si and Co-Ta films changes significantly with the sputtering method. These differences between the sputtering methods are mainly caused by the differences in the plasma potential which affects the amount of ion bombardment to the film surface during sputtering,and the minimum content of Si or Ta to obtain amorphous films decreases as the plasma potential increases. These results indicate that the ion bombardment suppresses the growth of crystallites and promotes the formation of the films with amorphous structures. This is confirmed by the deposition of Co-Ta and Co-Zr amorphous films under the condition of various amount of ion bombardment by using a DIB sputtering system.

## INTRODUCTION

Many kinds of amorphous magnetic thin films have been produced by means of various kinds of sputtering techniques[1-6]. But different sputtering methods have different deposition conditions for the formation of the amorphous films. However, the differences between the sputtering methods in the formation of the amorphous magnetic films are still unclear. In this work, at first we deposited Fe-Si and Co-Ta amorphous magnetic thin films by using various sputtering apparatuses (a conventional rf diode, triode and tetrode sputtering apparatus, a dc Targets Facing type sputtering (dc TF sputtering) apparatus and dual ion beam sputtering (DIB sputtering) apparatus) and investigated the differences between the sputtering methods in the deposition of the amorphous films. Secondly, in order to clarify the effect of ion bombardment to the film surface during the film deposition, Co-Ta and Co-Zr films were deposited under the codition of various amount of ion bombardment by using a DIB sputtering apparatus. As the result, it becomes clear that the ion bombardment takes an important role for the formation of the films with amorphous structures.

## EXPERIMENTAL PROCEDURE

Figure 1 shows the schematic diagrams of the sputtering apparatuses used in this work. Composite targets constructed with Fe plate and Si chips, Co and Ta plates, and Co and Zr plates as shown in Fig.2 (a), (b), (c) and (d) were used for the film preparation. The film composition was controlled by adjusting the area ratio of the each plates. Table I shows the typical sputtering conditions to deposit the amorphous films by various sputtering apparatuses. Co-Ta and Co-Zr films were deposited by using the DIB sputtering apparatus to investigate the effect of Ar ion bombardment on the film surface during sputtering on the formation of amorphous films.

Plasma potential Vp during sputtering was measured by means of a Langmuir probe method.

The composition of the film was determined by electron probe X-ray microanalysis (EPMA) and crystal structure of the films was characterized by both

 

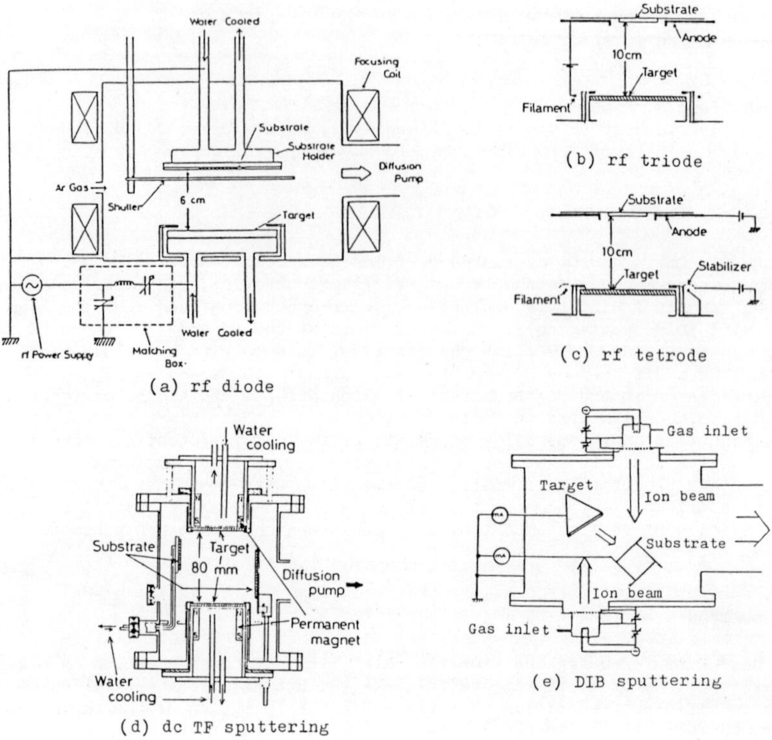


Fig.1 Schematic diagrams of sputtering apparatuses used in this work.

X-ray and electron diffractometry. Saturation magnetization 4πMs and coercive force Hc of the films at room temperature were measured with a VSM in magnetic field up to 10 kOe. Electrical resistivity of the film was measured to estimate the amorphous state of the film.

## RESULTS AND DISCUSSION

Figure 3 shows the change in 4πMs of the Fe-Si and Co-Ta films with Si and Ta content, respectively. The composition range where amorphous films are deposited by various sputtering methods are also shown in the figures. 4πMs decreases linearly as Si or Ta content increases and depends little on the film preparation methods. However, the minimum Si or Ta content to obtain amorphous films changes significantly with the sputtering meth-

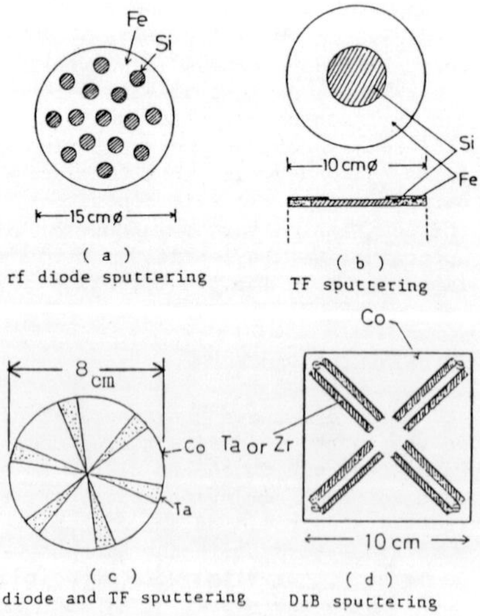


Fig.2 Composite targets used to form Fe-Si, Co-Ta and Co-Zr amorphous films in various sputtering methods.

Table I. Sputtering conditions.

| | | rf diode | rf triode | rf tetrode | dc TF | DIB |
|---|---|---|---|---|---|---|
| Applied voltage | Va (V) | ≈1000 | ≈1000 | ≈1000 | ≈600 | 400 |
| Input power | Pin (W) | 100 | 100 | 100 | 300~600 | ≳3 |
| Ar gas pressure | $P_{Ar}$(mTorr) | 5~10 | 1~10 | 1~10 | 1~7 | 0.15 |
| Substrate temperature | Ts (°C) | ≈100 | ≈100 | ≈100 | >100 | <70 |
| Deposition rate | Rd (Å/min) | 50 | 50 | 50 | 400 900 | 12 |
| Plasma potential | Vp (V) | 20 | 20 25 | 40 60 | 2 | - |

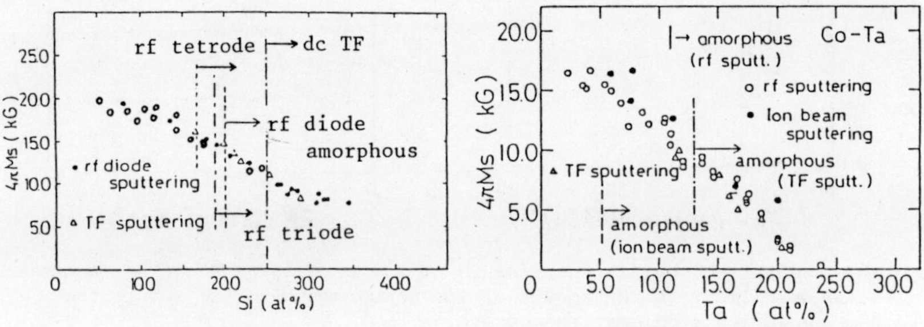


Fig.3 Dependence of 4πMs on Si and Ta content and change of the lower limit of Si and Ta content to obtain amorphous Fe-Si and Co-Ta films.

ods, that is the minimum Si content is about 25 at.% for dc TF sputtering, 20 at.% for rf diode sputtering, 19 at.% for rf triode sputtering and 17 at.% for rf tetrode sputtering. While the minimum Ta content is about 15 at.% for dc TF sputtering, about 10 at.% for rf diode sputtering and about 5 at.% for DIB sputtering without ion bombardment.

Substrate temperature, measured with a thermocouple at the substrate surface during deposition, was in the range from 100 °C to 120 °C in rf diode, rf triode and rf tetrode sputtering. While, in dc TF sputtering and DIB sputtering, the substrate temperature was maintained below 100 °C and below 70 °C, respectively. Therefore, the small value of the minimum Ta content for DIB sputtering seems to be mainly due to the low substrate temperature during film deposition. But, the large value of the minimum Si and Ta content for dc TF sputtering compared to the values of the other sputtering methods cannot be explained by the difference in the substrate temperature.

The main difference in the film deposition process between the rf diode, rf triode, rf tetrode and dc TF sputtering is in the amount of ion bombardment to the film surface during sputtering, since the plasma potential Vp during sputtering depends significantly on the sputtering methods as shown in Table I. Plasma potential Vp decreases in the order of rf tetrode sputtering, rf triode sputtering, rf diode sputtering and dc TF sputtering. The rf tetrode sputtering system used in this work had the highest plasma potential Vp, which was dominated by the anode potential maintained in the range from 40 V to 60 V in the system. Comparing the change of the minimum Si and Ta content shown in Fig.3 with the change of plasma potential Vp, it can be found that there exists a relationship between the minimum Si and Ta content and Vp, that is, the minimum Si or Ta content increases as Vp increases. These results indicate that the ion bombardment to the film surface during the growth of the film suppresses the growth of crystallites and promotes the formation of amorphous films. In order to confirm this effect of ion bombardment on the formation of amorphous films, Co-Ta and Co-Zr films were deposited with and without ion bombardment by means of DIB sputtering. Figure 4 show the change of the composition dependence of coercive force Hc of Co-Ta and Co-Zr films with the amount of ion bombardment. The composition ranges where amorphous films are

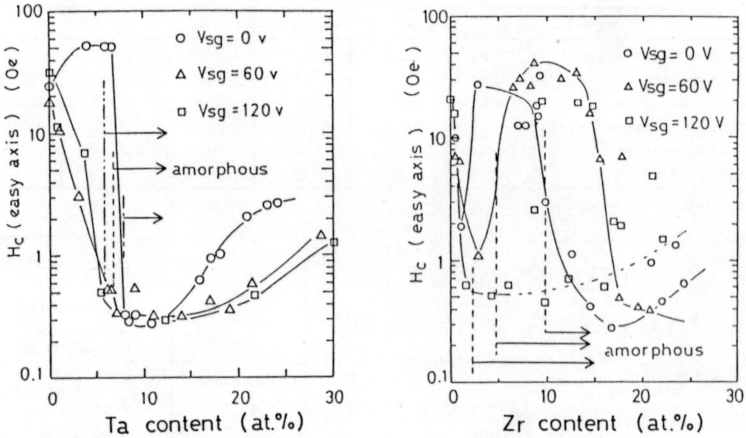


Fig.4 Change of composition dependence of coercive force Hc of Co-Ta and Co-Zr films with the amount of ion bombardment to the substrate. (Vsg is ion acceleration voltage.)

obtained, are also shown in these figures. It is evident from the figures that the minimum Ta and Zr content to form amorphous films decrease as the amount of ion bombardment increases. These results agree well with the results obtained from the deposition of Fe-Si and Co-Ta films by using various sputtering methods (see Fig.3). Therefore, it can be concluded that the ion bombardment to the film surface takes an important role to form the films with amorphous structures.

## CONCLUSION

Fe-Si and Co-Ta amorphous films have been deposited by using various sputtering methods. The lower limit of Si or Ta content to obtain amorphous films changes significantly with the sputtering methods. These differences are mainly caused by the differences in the plasma potential during sputtering which determines the amount of ion bombardment to the film surface during film growth. Namely, the larger value takes the plasma potential, the smaller value takes the lower limit of Si or Ta content. These results indicate that the ion bombardment suppresses the growth of crystallites and promotes the formation of the films with an amorphous structure. This is also confirmed by the deposition of Co-Ta and Co-Zr amorphous films under the conditions of varying amounts of ion bombardment by DIB sputtering.

## REFERENCES

(1) Y. Shimada and H. Kojima: J. Appl. Phys., 47 (1976) 4156
(2) M. Naoe, H. Yamamota and S. Yamanaka: J. Appl. Phys., 50 (1979) 606
(3) N. Heiman and N. Kazama: Phys. Rev., B17 (1978) 2215
(4) R. Krishnan, M. Tarhouni, M. Tessier and A. Gangulee: J. Appl. Phys., 53 (1982) 2243
(5) M. Naoe, H. Kazama, Y. Hoshi and S. Yamanaka: J. Appl. Phys., 53 (1982) 7846
(6) J. A. Aboaf and E. Klokholm: J. Appl. Phys., 52 (1981) 1884

PART II B

# Amorphous Alloys— Relaxation and Phase Transformation

# STRUCTURAL RELAXATION IN Fe-B GLASSES

Z. Altounian*, J.O. Strom-Olsen* and M. Olivier**
*Department of Physics, McGill University, 3600 University Street, Montreal, Quebec, Canada. H3A 2T8.
**Division of Chemistry, National Research Council, Montreal Road, Ottawa, Ontario, Canada. K1A 0R6.

## ABSTRACT

We present results on structural relaxation in Fe-B glasses. Relaxation effects were studied through measurements of the Curie temperature ($T_c$) of the glasses annealed at different temperatures ($T_a$). We observe a monotonic increase in $T_c$ with $T_a$ in all alloys due to irreversible structural relaxation. This increase continues till the start of surface/bulk crystallization. The observed decrease in $T_c$ upon further annealing is attributed to the strain introduced in the partially glassy system. We see no evidence for reversible structural relaxation in Fe-B glasses.

## INTRODUCTION

It is well known that metallic glasses on being annealed relax to a structure closer to a local minimum of free energy. Structural relaxation causes changes to many physical properties which are themselves used to monitor the relaxation process. Relaxation has been classified as of two types. The first type gives irreversible, often monotonic, changes to a property and is generally attributed to a decrease in free volume by the disappearance of voids, resulting in an increase in topological short range order (TSRO). The second type is less widely seen and is characterized by reversible changes of property - i.e. the system can be cycled between different glassy structures by annealing at different temperatures. There have been suggestions [1] that reversible relaxation is a result of changes in chemical or compositional short range order (CSRO). Both irreversible and reversible structural relaxations have been demonstrated in a number of magnetic metallic glasses, mainly through measurements of the Curie temperature, $T_c$, which is a simple and sensitive measure of the structural state. The alloys investigated were usually of the metal-metalloid type, containing more than two metallic elements (e.g. $Fe_{40}Ni_{40}B_{20}$). Quite recently structural relaxation results on binary Fe-B glasses were presented by So et al. [2] and interpreted as a reversible structural relaxation, the driving force being the deviation of the CSRO in the glass from that in crystalline $Fe_3B$.

The extensive work of Nold et al. [3] on the structure of amorphous $Fe_{80}B_{20}$ by X-ray and neutron diffraction provided for the first time reliable experimental partial structure factors. The results showed that in Fe-B glasses, the B atoms are at the centre of a coordination sphere containing nine Fe atoms with a very well defined Fe-B distance and with no B-B atoms in contact. The B coordination is independent of concentration [4] and implies that the B atoms force a special arrangement of the Fe atoms by chemical interaction rather than just fill up the Bernal holes in the dense random packing of the Fe atoms, and indeed Nold et al. [3] find that the normalized CSRO parameter is 1.0 corresponding to perfect chemical order. If these results are correct it is hard to understand how reversible relaxation due to changes in CSRO could take place in a binary metal-metalloid glass. We therefore decided to reinvestigate structural relaxation in Fe-B glasses in some detail.

EXPERIMENTAL

Fe-B buttons of 1.2-1.5 g were prepared by arc melting appropriate amounts of 99.999% pure Fe and 99.9% pure B under titanium gettered argon atmosphere. In order to ensure homogeneity, each button was melted up to five times. Sample mass losses were typically < 0.01 g after five melts. Glassy ribbons were obtained by melt spinning, under 50kPa He atmosphere, onto the surface of a polished Cu wheel at a tangential speed of 50 $ms^{-1}$. Extreme care was taken to maintain the same melt spinning parameters for all ribbons. This is a crucial step as we shall see later. The glassy ribbons were typically 1.5mm wide and 25μm thick. X-ray diffractometer scans did not show any sign of crystallinity for the compositions used in this study. Assuming that the mass change after the first arc melting is due to B loss the compositions of the glassy alloys are at worst 0.2 at. % B deficient from the nominal compositions. The crystallization temperatures, $T_x$, enthalpy changes, $\Delta H_c$, and Curie temperature, $T_c$, were determined by means of a calibrated differential scanning calorimeter (DSC) under a flow of oxygen-free argon gas. Pure Ni was used as a calibrant for $T_c$. Structural data were obtained by an automated Stöe-Nicolet diffractometer and Mössbauer spectra were obtained at room temperature in the transmission configuration. The glassy samples (20-25 mg) were annealed in the DSC at several annealing temperatures, $T_a$, for a period of 30 minutes after which $T_c$ was determined at a heating rate of 40K $min^{-1}$. The samples were then immediately cooled at a rate of 320K $min^{-1}$ from typically 5K above $T_c$ down to room temperature. Part of each sample was crystallized and the remainder then used for X-ray diffraction and/or Mössbauer spectroscopy to detect any structural changes due to the heat treatment.

RESULTS AND DISCUSSION

The glassy Fe-B compositions studied in this report are from $Fe_{87}B_{13}$ to $Fe_{78}B_{22}$. $T_x$ and $T_c$ corresponding to a heating rate of 40K $min^{-1}$ are shown in figure 1. The mode of crystallization below the eutectic composition (17.3 at. % B) involves the primary crystallization of α-Fe and the formation of tetragonal $Fe_3B$ at higher temperatures [5]. For compositions above the eutectic, tetragonal $Fe_3B$ forms simultaneously with some α-Fe. As shown in figure 1, $T_c$ increases in a smooth fashion with boron content. These results for $T_x$ and $T_c$ are similar to the data of Hasegawa and Ray [6], except that they observed a break in the $T_c$ curve with composition at about 19 at. % B. We believe this may be an artifact due to two different extrapolation techniques for obtaining $T_c$. Considering that the increase in $T_c$ with boron content is of the order 20-35K per at. % B, the $T_c$ dependence is a very useful check on the composition and homogeneity of the Fe-B glasses.

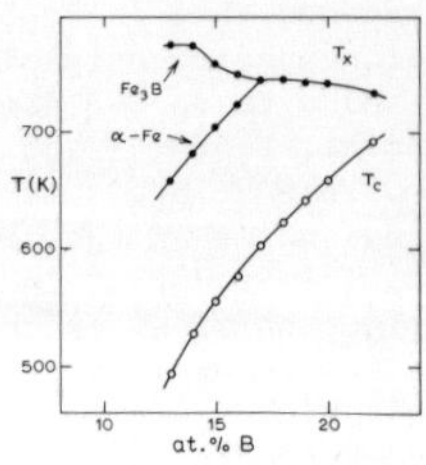


Figure 1.

Crystallization ($T_x$) and Curie ($T_c$) temperatures measured at 40K/min. for Fe-B glasses.

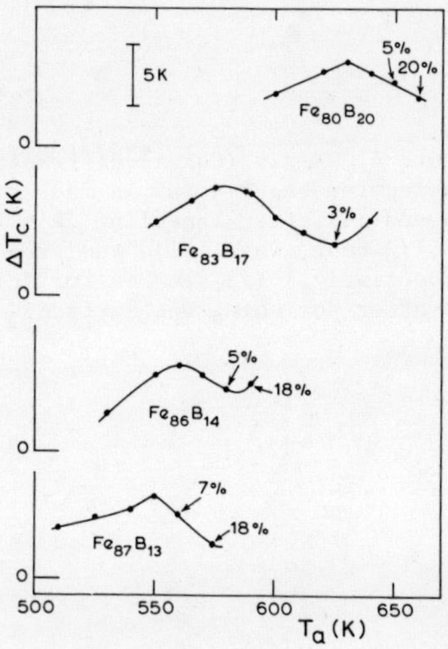


Figure 2. Change in Curie temperature $\Delta T_c$ on 1/2 hour annealing at temperature $T_a$ for some Fe-B glasses. Some of the crystallization amounts are indicated.

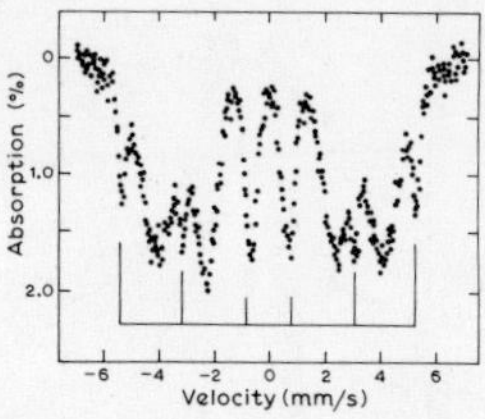


Figure 3. Room temperature Mössbauer spectrum for $Fe_{86}B_{14}$ annealed at 580K for 1/2 hr. The stick diagram marks the positions of the α-Fe absorption lines.

Figure 2 shows the change in $T_c$ as a function of $T_a$ for some representative Fe-B compositions. The variation of $T_c$ with $T_a$ is similar to the results of So et al. [2]; our interpretation of it, however, is different. The initial increase in $T_c$ with $T_a$ is generally agreed to be an irreversible relaxation effect due to the decrease of free volume. This irreversible change in TSRO continues till $\Delta T_c$ reaches a maximum value, $\Delta T_c^{max}$. Samples annealed at higher $T_a$ show a decrease in $T_c$ and at the same time signs of incipient crystallization. For example, the $Fe_{86}B_{14}$ sample, after annealing at $T_a$ = 580K, has an enthalpy of crystallization $\Delta H_c$ lower by about 5% and this decrease is entirely in the first exothermic peak - i.e. is caused by the crystallization of α-Fe. Furthermore the Mössbauer spectrum for this sample clearly shows the presence of α-Fe as shown in figure 3. From the relative intensities of the two outer absorption lines of α-Fe to that of the amorphous phase, we estimate the amount of α-Fe to be (7 ± 2) at. %, in reasonable agreement with the enthalpy change. We have indicated in figure 2 some of the crystalline fractions obtained in this way. We therefore believe that the decrease in $T_c$ after $\Delta T_c^{max}$ is related to crystallization rather than structural relaxation. Specifically we do not find any evidence of reversibility of $T_c$ with successive anneals made at temperatures corresponding to the maximum in $T_c$ and half-way to the minimum. All that the anneals show is a steady decrease towards the minimum. To investigate further the nature and type of crystallization we have performed X-ray diffraction experiments on samples annealed at temperatures where $T_c$ shows a decrease. If one uses $CuK_\alpha$ radiation, for iron-rich alloys, the surface contribution becomes a dominant part of the total scattering process. Figure 4 shows the X-ray scattering results for $Fe_{83}B_{17}$. Figure 4a is a diffraction scan for the as-made ribbon. Figure 4b and 4c show respectively, diffraction scans from the wheel-side and free-side of the same ribbon after annealing at 612.5K for 1/2 hr. Both surfaces show the effects of crystallization (α-Fe) with the free surface having considerably more amounts of crystallization.

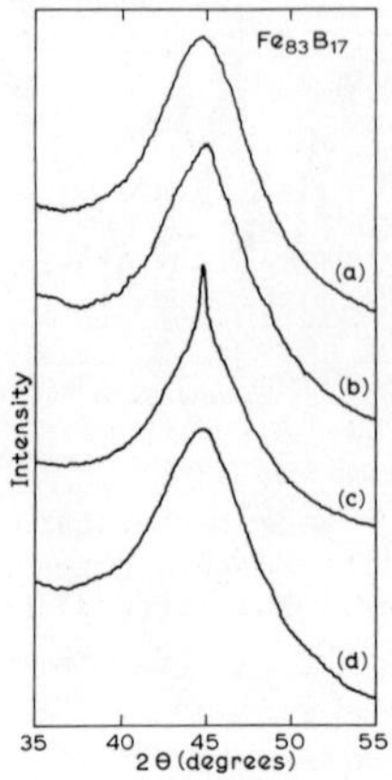


Figure 4. X-ray ($CuK_{\alpha}1$) diffractometer tracing for $Fe_{83}B_{17}$ (a) as made ribbon (b) and (c) after annealing at 612.5K for 1/2 hour, wheel side and free side respectively. (d) same as for (c) but after polishing the surface.

After polishing the surfaces, the crystalline peak disappears, as shown in figure 4d for the free-side.

If crystallization is responsible in some way for the decrease in $T_c$ then there must be a different mechanism responsible for alloys in the two sides of the eutectic composition as the mode of crystallization is different. For compositions with less than 17.3 at. % B, the first crystallization product is α-Fe. From crystallization studies [7] it is clear that the growth of α-Fe crystals is very limited. For $Fe_{84}B_{16}$, the maximum size of α-Fe is about 80nm, even after complete crystallization. This is reasonable as further growth requires long range diffusion. Therefore at each $T_a$ there is a fixed amount of α-Fe crystals present in the glass. Because Fe-B glasses are invar materials the volume expansion of the α-Fe crystals on the surface with temperature will be inhibited by the amorphous matrix. The resultant strain will result in the system experiencing an effective pressure which will lower the $T_c$ of the amorphous matrix [8]. However if $T_a$ is further increased the amount of α-Fe increases, so that the amorphous matrix becomes slightly richer in B and $T_c$ starts to increase. Note that an increase of 1 at. % B causes a 30K increase in $T_c$ at this composition range (figure 1).

For compositions above 17.3 at. % B, there is eutectic crystallization of tetragonal $Fe_3B$ and α-Fe. There is no limit to the growth size of $Fe_3B$ and thus as $Fe_3B$ crystals form and grow, the rest of the amorphous phase becomes poorer in boron content and $T_c$ decreases (figure 1).

The initial irreversible increase in $T_c$ with $T_a$ due to the decrease in free volume continues till the start of crystallization of either α-Fe or $Fe_3B$. Therefore $\Delta T_c$ will be large if $T_x$-$T_a$ is large, where $T_a$ is the annealing temperature corresponding to $T_c^{max}$. In figure 5a we have plotted $\Delta T_c^{max}$ versus $T_x$-$T_a$ in all compositions; the correlation is very good. The representative error bar includes only measurement errors. However, for $Fe_{84}B_{16}$ we have melt spun this composition an additional four times, symbolized by the crosses in the figure. The only parameter that was changed in the melt spinning process was the melt temperature from which the alloy was quenched. It is clear that $\Delta T_c^{max}$ depends critically on the melt temperature. In general, the higher the melt temperature the higher the initial $T_c$ suggesting that some relaxation has already taken place. This strongly emphasizes our earlier statement that for proper evaluation of a property change and its correlation to composition, the initial state of the

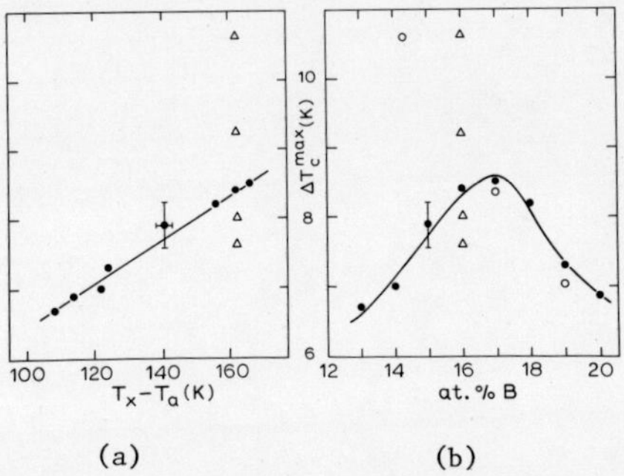


Figure 5. (a) correlation between $T_c^{max}$ and $T_x$-$T_a$ for annealed Fe-B glasses. (b) $T_c^{max}$ versus composition for annealed Fe-B glasses. The open triangles correspond to $Fe_{84}B_{16}$ glasses quenched from different melt temperatures. The open circles correspond to the data of ref.[2].

glass is of paramount importance. In fact if we plot $\Delta T_c^{max}$ versus boron content, see figure 5b, we get a completely different result from that of So et al. [2]. Their data for 30 minute anneals is reproduced in the figure. Our data simply reflects the stability of the glassy phase, showing a maximum at the eutectic composition in accordance with figure 5a.

In conclusion we believe that $T_c$ measurements in Fe-B glassy alloys do not show any evidence for reversible structural relaxation as the CSRO is already weel defined in the glassy state. The irreversible relaxation causes a monotonic increase in $T_c$ till the start of crystallization.

References


1. See for example T. Egami in Amorphous Metallic Alloys, edited by F.E. Luborsky (Butterworths, London 1983) p. 100.
2. F.F. so, R. Kaplow and R.C. O'Handley, J. Non-Crystalline Solids, 58, 285 (1983). Reversible effects in α- $Fe_{80}B_{20}$ were also reported by A.L. Greer and J.A. Leake in Proc. 3rd Int. Conf. on Rapidly Quenched Metals, edited by B. Cantor (Metals Society, London 1978) p. 299.
3. E. Nold, P. Lamparter, H. Olbrich, G. Rainer-Harbach and S. Steeb, Z. Naturforsch, 36a, 1032 (1981).
4. D.S. Boudreaux and J.M. Gregor, J. Appl. Phys., 48, 5057 (1977).
5. J.L. Walter, S.F. Bartram and R.R. Russell, Met. Trans., 9A, 803 (1978).
6. R. Hasegawa and R. Ray, J. Appl. Phys., 49, 4174 (1978).
7. J.L. Walter, S.F. Bartram and I. Mella, Mat. Sci. Eng., 36, 193 (1978).
8. K. Fukamichi, M. Kikuchi, S. Arakawa, T. Masumoto, T. Jagielinski, K.I. Arai and N. Tsuya, Solid State Commun., 27, 405 (1978).

# THERMAL STABILITY AND PHASE TRANSFORMATION IN $Hf_{1-x}Cu_x$ METALLIC GLASSES

R. SCHULZ**+, N. L. LEE+ and B. M. CLEMENS**
Physical Chemistry Department+ and Physics Department**
General Motors Research Laboratories, Warren, Michigan 48090-9055, USA

## ABSTRACT

In this paper we report the difference in the structures, physical properties and thermal behaviors of two compositions of amorphous Hf-Cu. One ($Hf_{50}Cu_{50}$) is near the eutectic composition and the other ($Hf_{67}Cu_{33}$) corresponds to an intermetallic compound. Transmission and scanning electron microscopy (TEM, SEM), differential scanning calorimetry (DSC), electrical resistivity and high-angle X-ray diffraction have been used to study these glasses in detail. Amorphous $Hf_{67}Cu_{33}$ is structurally stable up to the crystallization temperature, while $Hf_{50}Cu_{50}$ undergoes a ductile to brittle transition associated with a chemical phase segregation at temperatures far below the glass transition. An anomalous behavior in the resistivity is associated with this phase transformation.

## INTRODUCTION

Several amorphous metallic alloys with compositions near a eutectic exhibit chemical phase segregation upon annealing at temperatures below the glass transition (1). For $Zr_{50}Cu_{50}$ we recently showed (2) that, even though the phase diagram (3) shows an intermetallic compound exists at the 50-50 composition, the glass phase separates into Cu rich and Cu poor regions at temperatures far below Tg. This glass crystallizes into a mixture of $Zr_2Cu$ and $Zr_7Cu_{10}$. Since the Hf-Cu system is similar to the Zr-Cu system and no intermetallic compound exists at the 50-50 composition (4), we expect phase segregation in amorphous $Hf_{50}Cu_{50}$. In fact, since the driving force for chemical phase segregation in metallic glasses, if there is any, should be smaller than the one for crystallization, a search for phase separation represents a good test of the thermodynamic similarity between the hafnium and the zirconium based glasses. In order to examine the influence of a stable compound on crystallization behavior, we also examine the annealing and crystallization behavior of $Hf_{67}Cu_{33}$, where there is an intermetallic compound and we do not expect phase separation.

## EXPERIMENTAL PROCEDURE

The ingots were prepared by melting the appropriate amounts of high purity Cu and Hf in an arc furnace under titanium-gettered argon. They were quenched from the melt by the melt-spinning technique in the form of ribbons having a typical thickness of 25 $\mu$m. The samples were checked for amorphous structure by X-ray diffraction.

The resistivity was measured under flowing argon using a four point DC method. Thermally-induced voltages were eliminated by using computer controls to average the voltage drop across the sample for both current directions. The DSC measurments were performed on a Perkin Elmer DSC-4 under an atmosphere of high-purity-argon. Samples for transmission electron microscopy were thinned using double-sided jet polishing in a 20% nitric acid, 20% glycerin, 60% methanol solution at -30°C. The micrographs were taken on a JEOL JEM 200 C TEM/STEM operated at 100 kV. Scanning electron images (SEM) of freshly fractured samples were obtained on an ISI DS-130. The samples were annealed in evacuated sealed pyrex tubes.

## RESULTS AND DISCUSSION

Figure 1(a) shows the resistance trace of $Hf_{67}Cu_{33}$ normalized to the room temperature value for a heating rate of 15°C/min. This curve is typical of amorphous metals and shows no peculiar behavior. As shown by the DSC trace in Figure 1(a), $Hf_{67}Cu_{33}$ crystallizes in a single step at $T_x$=499°C. The heat of transformation is 10.6 cal/gram. At the crystallization temperature the resistance drops by about 18%. In comparison, the

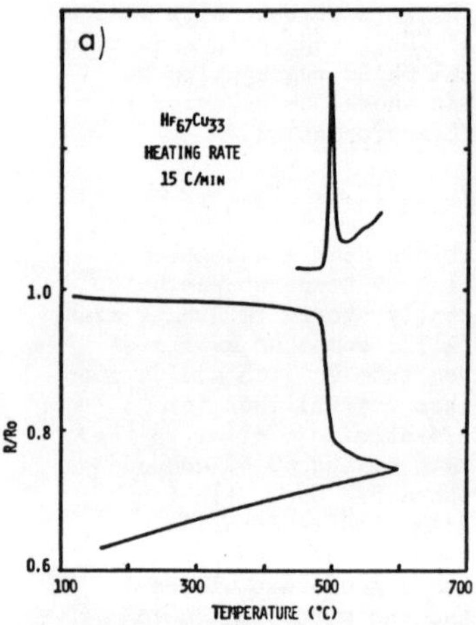


Figure 1(a). Resistance ($R/R_0$) and DSC trace (exothermic heat flow) for $Hf_{67}Cu_{33}$ as a function of temperature.

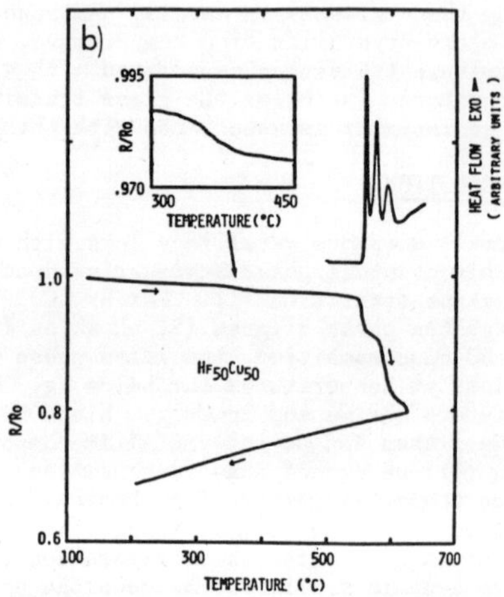


Figure 1(b). Resistance ($R/R_0$) and DSC trace (exothermic heat flow) for $Hf_{50}Cu_{50}$ as a function of temperature.

$Hf_{50}Cu_{50}$ crystallizes in a three step process (Figure 1(b)). The temperatures corresponding to each peak in the DSC trace are tabulated in Table I for a heating rate of 15°C/min.

Table I

| | $T_{x1}$ | $T_{x2}$ | $T_{x3}$ |
|---|---|---|---|
| $Hf_{67}Cu_{33}$ | 499 | -- | -- |
| $Hf_{50}Cu_{50}$ | 561 | 577 | 592 |

If we repeat the experiments at different heating rates and use the Kissinger method (5) to get the activation energy for crystallization we find, in the case of $Hf_{67}Cu_{33}$, a value of $E_a$=3.12 eV. For $Hf_{50}Cu_{50}$ the plot of $-Ln(\phi/T^2)$ versus $1/T$ (where $\phi$ is the heating rate and T the temperature of the first peak in the DSC trace) does not give a straight line as in the previous case but a broken line with $E_a$=3.71 eV for high heating rates (25, 50 and 100°C/min) and $E_a$=5.48 eV for low heating rates (2, 5, 15°C/min). This is a fairly large change (ΔE=1.77 eV) for such a small temperature interval $T_{peak}$(5°C/min) = 550°C while $T_{peak}$ (50°C/min) = 578°C. The activation energy, therefore, seems to increase with decreasing heating rate or temperature. It is difficult to draw any conclusions based only on this result. Kelton et al. (6) demonstrated recently that failure to account properly for the morphology of the growth might lead to nonlinear artifacts in the analysis of the kinetics of crystallization. Contrary to our result, they found for $Pd_{82}Si_{28}$ that, over a much larger temperature range, the activation energy for the growth velocity decreases with decreasing temperature from 3.4 eV around 395°C to 1.6 eV around 285°C. It is worth mentioning that they found the Kissinger analysis valid even in the case where surface crystallization is important. Their Kissinger plot, however, gives rise to a straight line. Experiments to determine whether partial oxidation at slow heating rates could be responsible for this behavior have not yet been performed.

Using the Piloyan method (7), the Avrami exponent characteristic of the transformation can be obtained. A plot of the logarithm of the deflection of the DSC trace versus 1/T gives a slope equal to nE where n is the Avrami exponent and E the activation energy for crystallization. In the case of $Hf_{67}Cu_{33}$ we find nE=20.3 eV and therefore n=6.5. This is a fairly large value but even more surprising are the values for $Hf_{50}Cu_{50}$. At 5°C/min the Piloyan plot gives a slope nE=38.2 eV and since $E_a$=5.48 eV, n=7. For a high heating rate (50°C/min) the slope remains the same (nE=38.1 eV) but now, since $E_a$=3.71 we find for n the remarkable value of 10. This method of determining the Avrami exponent is not very precise and requires great precision in the measurement of the experimental DSC. However, even if we allow an error of ±2, these values still remain much larger than what is usually found for amorphous metals.

A value for n of 10 was also recently reported by Altounian et al. (8) for the crystallization of amorphous $Fe_{27}Zr_{73}$. This was assumed to be representative of explosive crystallization or massive nucleation caused by self heating. This same assumption in our case would also explain the sharpness of the first DSC exotherm in Figure 1(b) compared with the second and third peaks. The increase in the observed enthalpy change during crystallization in going from low to high heating rates, as shown in Table II, can also result from this interpretation. Explosive crystallization at high heating rates will result in a more complete reaction over a smaller temperature interval. This interpretation would also imply that higher values of n indicate faster nucleation rates so that crystallization of $Hf_{50}Cu_{50}$ at high heating rates produces a much larger number of tiny crystals than low heating rate crystallization.

**Table II**

| $Hf_{50}Cu_{50}$ | 2°C/min | 5°C/min | 15°C/min | 25°C/min | 50°C/min |
|---|---|---|---|---|---|
| $\Delta H_c$ | 1.9 | 2.3 | 6.6 | 6.9 | 6.4 |

Referring again to Figure 1(b) we observe more unusual behavior in the resistance trace around 350°C. The resistance drops by about 1% between 300 and 400°C. The inset in Figure 1(b) shows an enlargement of the resistance trace in this temperature range. This change in the resistivity is also associated with a ductile to brittle phase transformation in the material. We recall that it was also in this temperature range that we observed the anomalous behavior in the resistivity of $Zr_{50}Cu_{50}$ glasses (2). In $Zr_{50}Cu_{50}$, the resistivity bump which was similar to the one observed during the early stage of Guinier-Preston zone formation was interpreted as resulting from phase separation caused by the long range diffusion of Cu in the amorphous matrix. To check for phase separation in the Hf-Cu alloys in this temperature range, we annealed a sample of $Hf_{50}Cu_{50}$ at 375°C for 12 hours and looked at the endothermic peak in the DSC trace near the glass transition temperature. Figure 2 shows the result. There is a clear indication for the existence of two amorphous phases in the annealed sample as shown by the two glass transition temperatures in Figure 2(b).

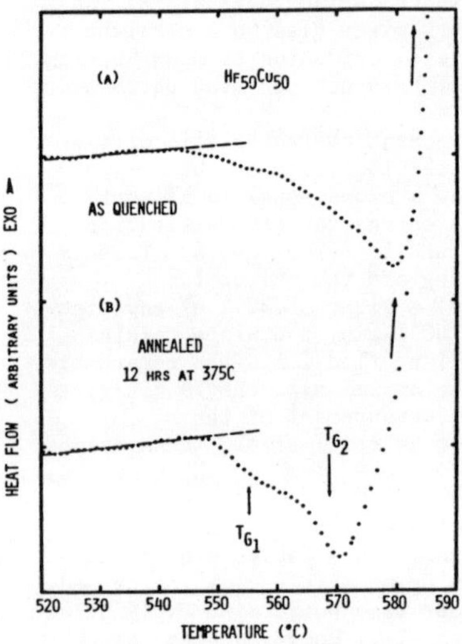


Figure 2. Enlargement of the DSC traces just below the crystallization temperature for (a) $Hf_{50}Cu_{50}$ initially as quenched and (b) previously annealed 12 hours at 375°C.

To investigate in greater detail the microstructural transformation which has taken place, high resolution scanning electron micrographs of fresh fractured samples of $Hf_{50}Cu_{50}$ were taken before and after the phase transition. In this technique, contrast can result from mechanical deformation of the sample during fracture. Figure 3(a) shows the fractured surface of a sample annealed at 325°C for 16 hours. The surface looks completely featureless, typical of homogeneous amorphous materials. The samples annealed at 425°C for 16 hours, i.e., right after the resistivity drop in Figure 1(b), exhibit a very regular line pattern with a typical interline spacing of about 750 A (Figure 3(b)). These features are similar in appearance to Walner lines which are characteristic of fracture surfaces of brittle solids such as inorganic glasses, ceramics, hard plastics, and intermetallic compounds (9). This is somewhat similar to one-dimensional spinodal decomposition. It is possible that Cu segregation may be responsible for the striations. High resolution Auger line scans will be performed in order to check this hypothesis. Finally, we annealed the sample at 500°C for 16 hours, which is 60°C below the crystallization temperature $T_x$ measured at 15°C/min (Table I). We observe very clearly several spherical microcrystals with a typical size of about 250 A. There is no trace of the previous regular pattern at this annealing temperature (Figure 3c).

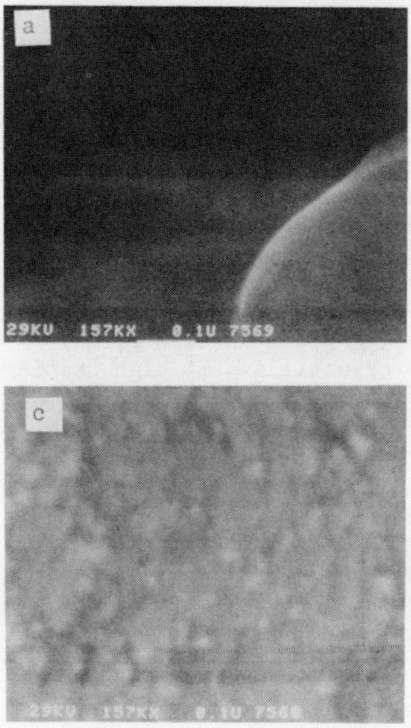


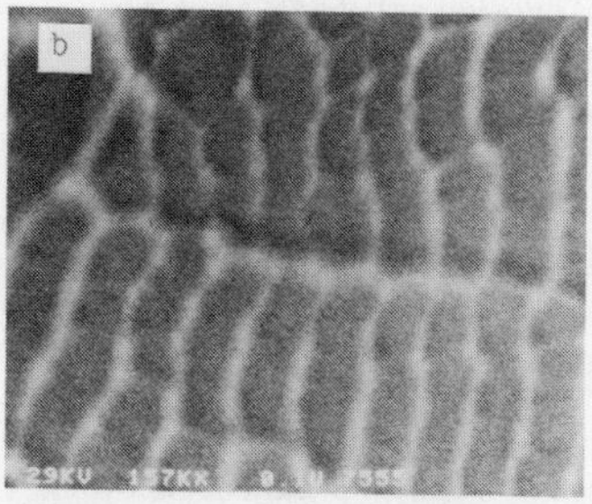


Figure 3. Scanning electron micrographs of fresh fractured samples of $Hf_{50}Cu_{50}$ annealed 16 hours at (a) 325°C, (b) 425°C, and (c) 500°C.

This structural transformation which seems to occur between 350°C and 400°C was studied using transmission electron microscopy. Here no mechanical deformation is applied to the sample. Figure 4(a) shows a TEM micrograph of $Hf_{50}Cu_{50}$ annealed 12 hours at 375°C. A regular pattern is still observed, but with a wavelength of about 1800 A. The diffraction pattern in Figure 4(b) shows two broad rings. This in conjunction with the two glass transition temperatures observed in Figure 2 strongly suggests the presence of two amorphous phases in this annealed amorphous alloy It should be pointed out that two broad diffraction rings can result from partial oxidation or microcrystalline precipitates about 20 Å in size. Further work is necessary to undeniably prove that phase separation is occurring as a precursor to crystallization in $Hf_{50}Cu_{50}$ metallic glasses. This work is currently in progress and will be presented in a separate report.

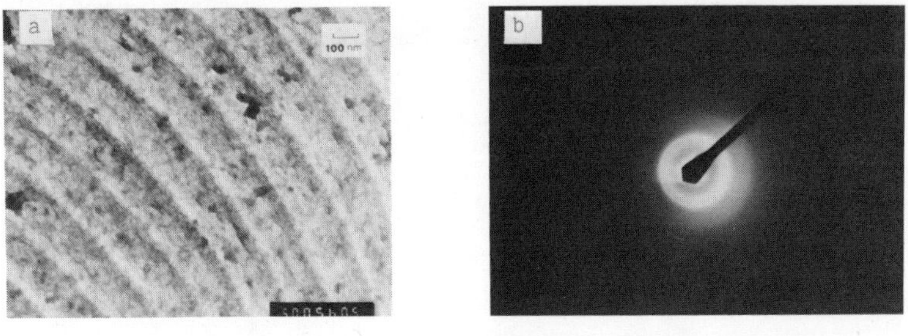


Figure 4. (a) TEM micrograph of $Hf_{50}Cu_{50}$ annealed 12 hours at 375°C and (b) selected area diffraction pattern after the same heat treatment.

## CONCLUSIONS

$Hf_{50}Cu_{50}$, like its homologue $Zr_{50}Cu_{50}$, undergoes a structural phase transformation around 350°C which gives rise to a nonlinear behavior in the variation of resistivity with increasing temperature. Differential scanning calorimetry, electron microscopy and selected area electron diffraction suggest that this transformation corresponds to an amorphous phase separation of the initially single phase glassy structure. $Hf_{67}Cu_{33}$, on the other hand, is stable all the way up to the crystallization temperature. It crystallizes in a single step into the intermetallic compound, tetragonal $Hf_2Cu$. The DSC results indicate that massive (or explosive) crystallization is taking place in this glassy alloy system and especially in $Hf_{50}Cu_{50}$ when crystallized at high heating rates.

## ACKNOWLEDGMENT

The authors would like to acknowledge the help of Harry Sturner with the transmission electron microscope, Walt Lange with the scanning electron miroscope, Melissa Wright with the DSC, and Jack Johnson with X-ray diffraction. We would also like to thank John Strom-Olsen and his group at McGill University for helping one of us (R.S.) with the sample preparation.

*Present address: Hydro Quebec Research Institute, Varennes, Quebec, Canada J0L 1P0.

## REFERENCES

1. L. E. Tanner and R. Ray, Scripta Metallurgica 14, 657 (1980).

2. R. Schulz, K. Samwer, and W. L. Johnson, J. of Non Cryst. Solids 61-62, 997 (1984).

3. M. Hansen, Constitution of Binary Alloys, 2nd Ed. (McGraw-Hill, 1958), 656.

4. W. G. Moffatt, The Handbook of Binary Phase Diagrams (General Electric 1976) 6/80; Russian Metallurgy (Metally) 2 (1974) 123.

5. H. E. Kissinger, Anal. Chem. 29, 1702 (1957).

6. K. F. Kelton and F. Spaepen, Acta Metall. 33, 455 (1985).

7. J. Colmenero, J. Ilarraz, and J. M. Barandiaran, Thermochimica Acta 35, 381 (1980).

8. Z. Altounian, C. A. Volkert, and J. O. Strom-Olsen, J. Appl. Phys. 57, 1777 (1985).

9. A. Kawashima, K. Hashimoto, and T, Masumoto, Scripta Met 14 41 (1980).

# EVIDENCE FOR THE FORMATION OF A METASTABLE CRYSTALLINE PHASE DURING CRYSTALLIZATION OF Zr-Ni METALLIC GLASSES OF APPROXIMATELY 57–63.2 at. % Zr*

C. G. MCKAMEY, D. S. EASTON, AND D. M. KROEGER
Metals and Ceramics Division, Oak Ridge National Laboratory, Oak Ridge, TN 37831

## ABSTRACT

The crystallization of rapidly solidified Zr-Ni metallic glasses has been studied by differential scanning calorimetry (DSC), x-ray diffraction (XRD), and transmission electron microscopy (TEM). Evidence has been found for a metastable phase which is the first crystalline phase to form upon heating in compositions of 57 to 63.2 at. % Zr. This phase seems to be more easily formed and is most stable in the region of 58 to 59 at. % Zr. Upon subsequent heating to higher temperatures, the metastable phase transforms to the equilibrium phases ZrNi and $Zr_2Ni$. The relationship of this new phase to phase separation at the eutectic composition of 63.5 at. % Zr is discussed in support of the conclusions drawn from low temperature specific heat studies reported previously [1,2].

## INTRODUCTION

In the Zr-Ni metallic glass system, the absence of XRD evidence for crystallization in compositions around the eutectic at 63.5 at. % Zr after heating through the first exothermic DSC peak has caused an increased interest in the nature of the transformations at these compositions [3,4]. Researchers have suggested that either a transition is occurring within the amorphous state (phase separation) [3–5] or very fine grained microcrystals are precipitating from the amorphous matrix [6].

Our previous studies on Zr-Ni glasses of compositions between 55 and 74 at. % Zr support the theory of phase separation [1,2]. From low-temperature specific heat data, evidence for two superconducting phases was observed in a splat quenched sample of composition $Zr_{62.9}Ni_{37.1}$. From measurements of the density of states at the Fermi level, superconducting transition temperatures, and ΔH on crystallization, we concluded that the transitions were representative of two phases having compositions near $Zr_{60}Ni_{40}(Zr_3Ni_2)$ and $Zr_{66.7}Ni_{33.3}(Zr_2Ni)$.

Recently a study of the crystallization processes of these alloys using differential scanning calorimetry (DSC), x-ray diffraction (XRD), and transmission electron microscopy (TEM) was conducted [7,8]. During these studies, evidence was found for the formation of a metastable phase which is the first crystalline phase to form upon heating in compositions of 57 to 63.2 at. % Zr [9]. Upon further heating to higher temperatures, the metastable phase transforms to the equilibrium phases ZrNi and $Zr_2Ni$.

This report summarizes the evidence accumulated to date for the presence of this metastable phase.

---

*Research sponsored by the Division of Materials Sciences, U.S. Department of Energy under contract DE-AC05-84OR21400 with Martin Marietta Energy Systems, Inc.

 

## EXPERIMENTAL PROCEDURES

Alloy buttons of Zr-Ni compositions between 55 and 70 at. % Zr were prepared by arc-melting high purity zirconium and nickel in an argon atmosphere on a water-cooled copper hearth. Specimens used in this study were prepared by either splat quenching, melt spinning, or electron beam evaporation. XRD studies were performed using a Philips spectrometer together with a fully automated Philips powder diffractometer (APD 3600), also using CuKα radiation. Thermal analysis was performed using a Perkin-Elmer DSC-2. TEM studies were conducted on a JEOL-100CX operating at 120 KV. Details of the sample preparation, XRD studies, and DSC procedures are presented elsewhere [8,9].

## RESULTS

The alloys were given annealing treatments of various times, temperatures, and furnace conditions (described in ref. 9). A subsequent XRD study of these alloys indicated that alloys of 58-60 at. % Zr did not crystallize directly to the equilibrium phases. A distinctive XRD pattern, which did not match either equilibrium phase, could be produced by annealing for short times (<4 hrs) at 703 K. This pattern, which we believe denotes the existence of a metastable phase, was present in samples heated for up to 840 hrs at 598 K.

XRD results of the progression of the crystallization process for melt-spun $Zr_{60}Ni_{40}$ are summarized in Fig. 1. Samples were isothermally annealed in a dynamic vacuum for twenty minutes at each temperature. The metastable phase, which is most distinct after annealing at 703 K, is the first crystalline phase to form from the amorphous structure. Annealing at higher temperatures causes the metastable pattern to weaken as the peaks for the equilibrium phases (ZrNi AND $Zr_2Ni$) develop and increase in intensity.

A detailed XRD analysis of the sample annealed at 703 K to produce the metastable phase resulted in four medium-to-strong lines and many other weak ones. The description and results of this analysis can be found in ref. 9. Although some of the lines are close to lines of the ZrNi and $Zr_2Ni$ phases, the patterns do not match.

Figure 2 shows the DSC traces produced by the crystallization of Zr-Ni metallic glasses at a heating rate of 20 K/min. As seen in this figure, the alloys of compositions 57.1 through 63.2 at. % Zr exhibited small exotherms more than 100° above the main crystallization exotherms. This phenomenon was observed for both melt-spun and splat-quenched samples made from the same ingot and from different ingots of the same composition. This high-temperature peak, as well as all other DSC peaks, was absent on subsequent cooling down and reheating. The maximum temperature of the high-temperature peak is approximately 862 K, which occurred at $Zr_{59}Ni_{41}$ (Fig. 3). The enthalpy change associated with this peak shows a maximum at 58–59 at. % Zr (Fig. 4).

XRD studies were performed on the DSC annealed samples to determine what transformations the DSC peaks represented. Samples were annealed in the DSC at 80 K/min to particular points on the DSC curve, then cooled to room temperature at 320 K/min. The high-temperature peak was found to be associated with the transformation of the metastable phase to the equilibrium phases (Fig. 5). The XRD pattern for $Zr_{55}Ni_{45}$ (Fig. 5a), heated to a temperature corresponding to the top of the single asymmetric DSC peak, exhibited only equilibrium phase formation. Annealing to higher temperatures only sharpened the features. The composition $Zr_{57.7}Ni_{42.3}$ (Fig. 5b), annealed to the top of the sharp low temperature DSC peak, produced a strong metastable XRD pattern. This pattern persisted after annealing samples to higher temperatures, until the high-temperature peak was reached.

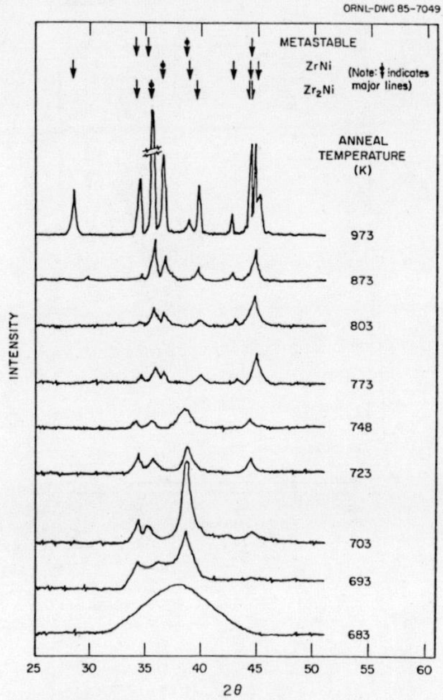


Fig. 1. X-ray diffraction spectra of $Zr_{60}Ni_{40}$ samples isothermally annealed at varying temperatures showing the progression of the crystallization process: amorphous → metastable → $ZrNi+Zr_2Ni$.

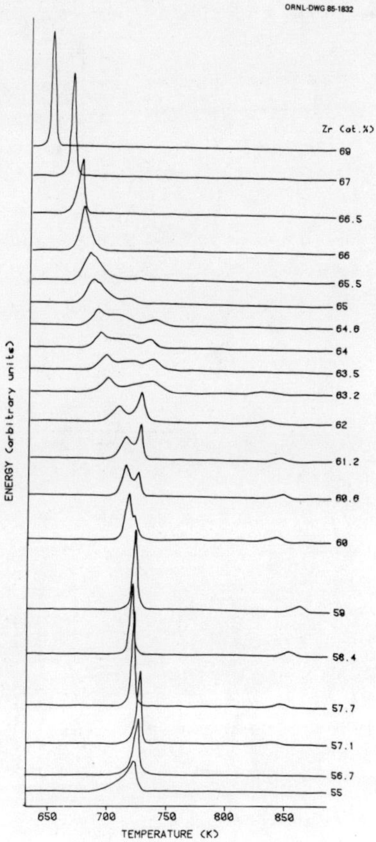


Fig. 2. DSC traces produced by the crystallization of Zr-Ni metallic glasses at a heating rate of 20 K/min.

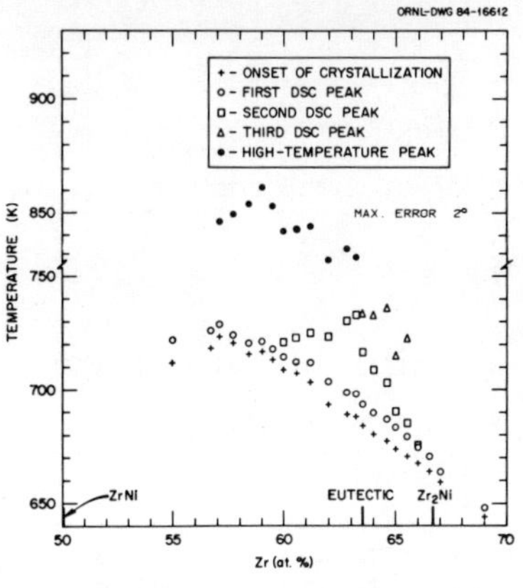


Fig. 3. Transformation temperatures as a function of composition at a heating rate of 20 K/min.

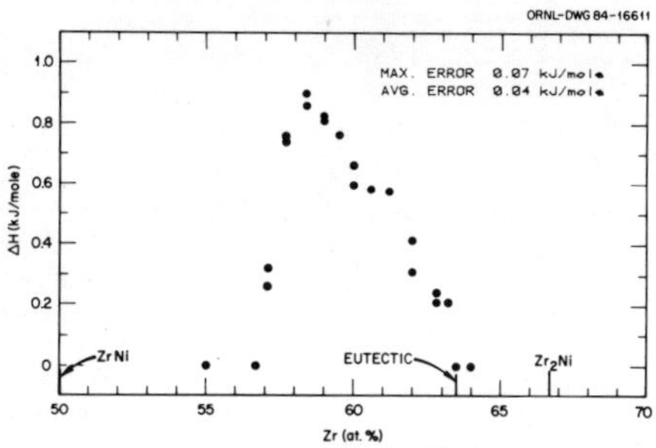


Fig. 4. The enthalpy change of the high-temperature peak as a function of composition.

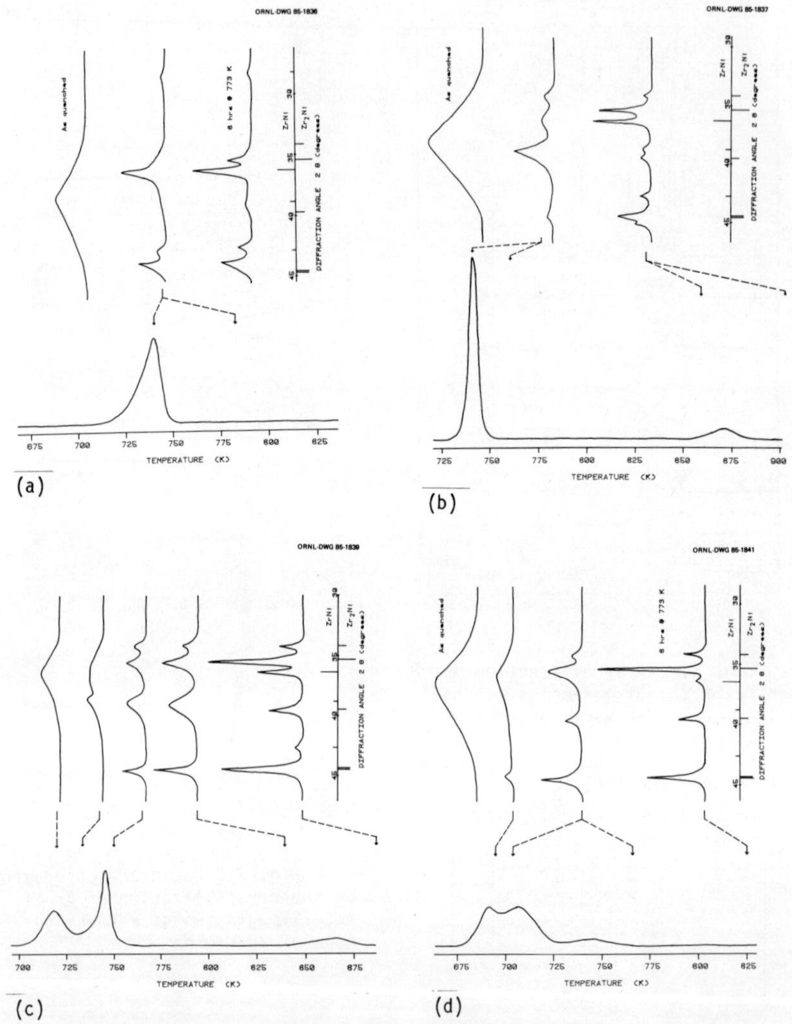


Fig. 5. XRD patterns obtained for Zr-Ni metallic glasses as a function of the progression of the crystallization process. The isochronal DSC trace is presented at a heating rate of 80 K/min. The XRD lines for the ZrNi and $Zr_2Ni$ phases are included for comparison. (a) $Zr_{55}Ni_{45}$ (b) $Zr_{57.7}Ni_{42.3}$ (c) $Zr_{62}Ni_{38}$ (d) $Zr_{65.5}Ni_{34.5}$.

Only XRD lines belonging to the two equilibrium phases were seen after annealing to or above temperature corresponding to the high-temperature peak. Two low-temperature and one high-temperature peak characterized the DSC trace of $Zr_{62}Ni_{38}$ (Fig. 5c). At the lower temperatures, the metastable phase formed first in this composition, followed by $Zr_2Ni$. After heating above the high-temperature peak, only ZrNi and $Zr_2Ni$ were seen. No high-temperature DSC peak and correspondingly no metastable phase were seen on compositions >63.5 at. % Zr, as exhibited by the XRD study of $Zr_{65.5}Ni_{34.5}$ in Fig. 5d.

As observed by TEM, the metastable phase exhibited a distinct acicular morphology compared to the equilibrium phases. The size of the needle-like metastable crystals did not exceed approximately 0.25 by 3.0 µm in size. Below ~57 at. % Zr, the first crystals to form in the amorphous matrix were diamond shaped with little internal structure (Fig. 6a). These crystals, which XRD identified as equilibrium phase, continued to nucleate and grow in size until complete crystallization was achieved. At compositions which exhibited a high-temperature peak on heating in the DSC, the first crystals formed as long needles (Fig. 6b), identified as the metastable phase. In some of these crystals, internal striations could be detected in the long direction. After heating through the high-temperature peak, the needle-like crystals were no longer seen and XRD revealed the presence of only equilibrium phases. From the eutectic composition ($Zr_{63.5}Ni_{36.5}$) through $Zr_{67}Ni_{33}$, where the XRD results indicated that $Zr_2Ni$ crystals are the first to form, the crystals exhibited a globular morphology, often with internal striations (Fig. 6c).

## DISCUSSION AND CONCLUSIONS

The evidence from XRD, DSC and TEM studies of the crystallization of Zr-Ni metallic glasses indicates that, between the composition of approximately 57 and 63.5 at. % Zr, the first crystalline phase which appears on heating is not one of the equilibrium phases, but a metastable phase first reported by Easton et al. [9]. This phase transforms to the equilibrium phases on further heating.

The results from our study of the crystallization processes of Zr-Ni metallic glasses between 55 and 70 at. % Zr, [8] combined with the results of low-temperature specific heat measurements, [1,2] indicate that phase separation may be occurring at compositions around the eutectic at 63.5 at. % Zr. The presence of the metastable phase at 58-60 at. % Zr lends further support to this hypothesis. It is approximately this composition which marks the low-zirconium boundary for the occurrence of the DSC and XRD evidence for phase separation.

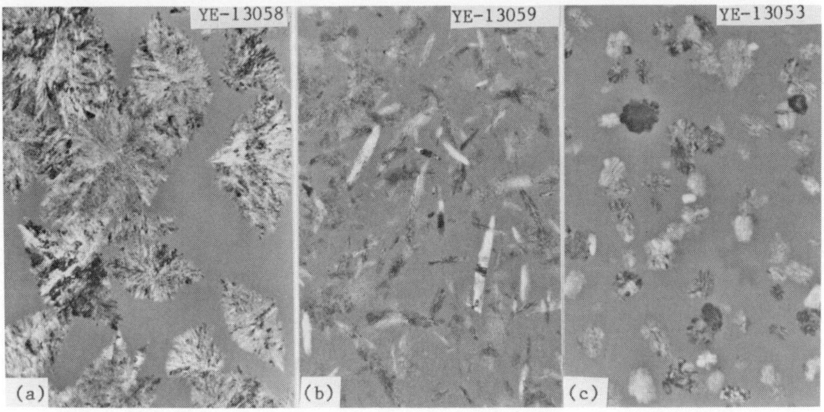


Fig. 6. TEM micrographs of the initial crystals which form upon heating in the DSC at 80 K/min from (a) $Zr_{55}Ni_{45}$, (b) $Zr_{60}Ni_{40}$, and $Zr_{65.5}Ni_{34.5}$. 30,000X

Furthermore, it seems possible that the study of the metastable phase and its role in the crystallization process may shed some light on the nature of the amorphous structure in these alloys. The results suggest that the glass structures of Zr-Ni metallic glasses in this composition range are characterized by a chemical short range order (CSRO) which is similar to, or facilitates the formation of, the crystalline phases to which they transform. Glasses of 57–59 and 66.7 at. % Zr appear to transform rapidly to the metastable and $Zr_2Ni$ phases, respectively, as evidenced by a single sharp DSC exotherm, indicating that little diffusion is necessary at these compositions. As-quenched glasses of compositions between these two phases (around the eutectic) phase separate before crystallization can proceed. Further discussion of these hypotheses are presented elsewhere [8].

REFERENCES


1. D. M. Kroeger, C. C. Koch, J. O. Scarbrough, and C. G. McKamey, Phys. Rev. B 29(3), 1199 (1984).
2. D. M. Kroeger, C. C. Koch, C. G. McKamey, and J. O. Scarbrough, J. Non-Cryst. Solids 61(2), 937 (1984).
3. K. H. J. Buschow, B. H. Verbeck, and A. G. Dirks, J. Phys. D: Appl. Phys. 14, 1087 (1981).
4. Z. Altounian, The Guo-hua, and J. O. Strom-Olsen, J. Appl. Phys. 54(6), 3111 (1983).
5. H. van Swijgenhoven, L. M. Stals, and K. H. J. Buschow, Phys. Stat. Solidi A72, 153 (1982).
6. R. Schulz, V. Matijasevic, and W. L. Johnson, Phys. Rev. B30, 6856 (1984).
7. C. G. McKamey, Masters Thesis, University of Tennessee, Knoxville, June 1985.
8. C. G. McKamey, D. M. Kroeger, D. S. Easton, and J. O. Scarbrough, submitted to J. Mater. Sci.
9. D. S. Easton, C. G. McKamey, D. M. Kroeger, and O. B. Cavin, accepted by J. Mater. Sci.

# HIGH RESOLUTION TRANSMISSION ELECTRON MICROSCOPE STUDY OF AMORPHOUS AND PARTIALLY CRYSTALLINE $Fe_{78}B_{13}Si_9$ ALLOY

A.R. BHATTI, J.C. BARRY* and B CANTOR
Department of Metallurgy & Science of Materials,
University of Oxford, Parks Road, Oxford OX1 3PH, UK
*Present Address: Centre for Solid State Science,
Arizona State University, Tempe, Arizona

## ABSTRACT

Amorphous and partially crystalline $Fe_{78}B_{13}Si_9$ alloys have been examined by high resolution transmission electron microscopy at 200kV. In lattice imaging mode, the as-quenched alloy shows small regions, with a mean size of ~ 20Å, consisting of well defined patterns of fringes. Three different types of crystals can be identified at the very earliest stages of crystallisation of the alloy.

## INTRODUCTION

Numerous models [1-6] have been developed by different workers to define the structure of amorphous alloys. Until recently, structural models of amorphous alloys which assume purely non-crystalline arrangements of atoms have been favoured [2-5]. Recent transmission electron microscopy, Mössbauer spectroscopy and Neutron diffraction experiments [7-13], however, suggest that amorphous alloys might consist of small domains of regular arrangements of atoms. In the present work, high resolution transmission electron microscopy (HRTEM) has been used to investigate (a) the initial as-quenched structure of an $Fe_{78}B_{13}Si_9$ amorphous alloy and (b) the earliest stages of crystallisation of the same alloy during heat treatment.

## EXPERIMENTAL TECHNIQUE

The as-quenched $Fe_{78}B_{13}Si_9$ alloy was in the form of ribbon, with a 30μm x 2.5cm cross-section. Ribbon samples were heat-treated at 510°C under an Ar atmosphere, at a series of annealing times chosen from previous calorimetric work [14], in order to observe isothermal crystallisation of the amorphous alloy. Thins foils of as-quenched and heat treated samples were prepared by Ar ion milling in a Gatan ion thinner. Electron microscopy was performed in a JEOL 200CX with a structural resolution of 2.5Å and an information limit of 1.7Å. Bright field phase contrast images were obtained by using a large objective aperture such that all spacings greater than 1.9Å could contribute to the image. Dark field diffraction contrast images were obtained using an objective aperture which allowed spacings greater than 8Å to contribute to the image, and by tilting the beam so as to align the segment of the diffused ring of interest with the microscope optic axis. The JEOL 200CX objective lens has a $c_s$ = 1.2mm and the images were taken at near to the optimum defocus value of -660Å, and with a beam divergence angle of 0.9mRad.

## RESULTS AND DISCUSSION

Figs.1-3 show typical high resoution bright field and dark field transmission electron micrographs of as-quenched amorphous $Fe_{78}B_{13}Si_9$ and a corresponding diffraction pattern. In the small encircled areas of the

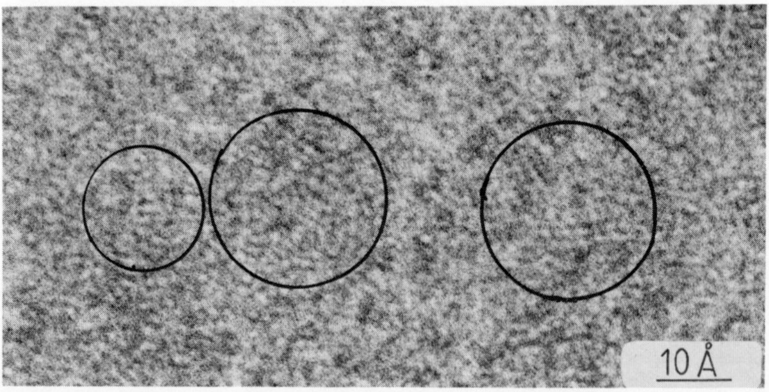


Fig.1. High resolution bright field electron micrograph of the as-quenched amorphous alloy showing small domains consisting of fringes (encircled regions)

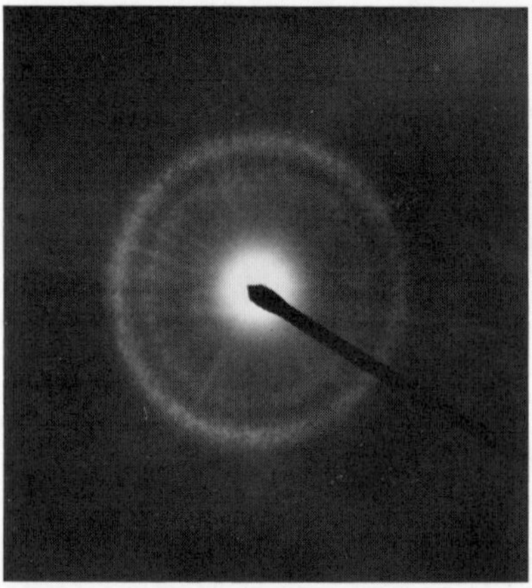

Fig.2. Diffraction pattern corresponding to Fig.(1).

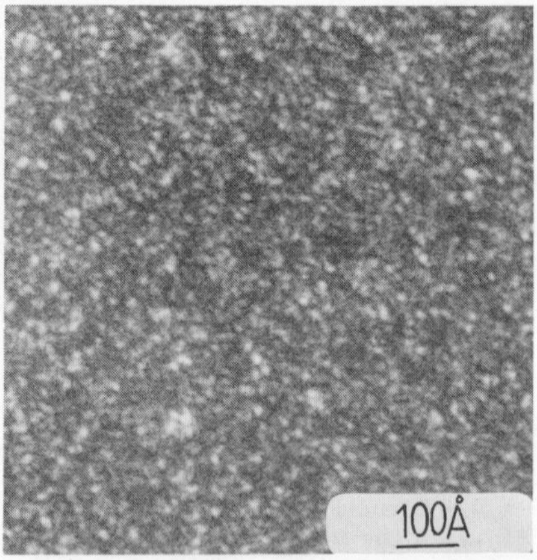


Fig.3. High resolution dark field image corresponding to Figl.(1) showing speckle contrast from small domains.

bright field micrograph fringes are visible, indicating ~ 20Å domains consisting of ordered configurations of atoms. The same domains lead to speckle contrast on dark field micrographs and spottiness on diffraction patterns, as shown in Figs.2 and 3. Similar domains were observed on a number of occasions. For observation of lattice images by transmission electron micrscopy, only rows of atoms which are very closely parallel to the incident beam give rise to fringes in the image. It is possible, therefore, that additional ordered regions exist in other areas where fringes are not observed.

It is unlikely that the fringes in Fig.1 are caused by either irradiation damage or a temperature rise in the specimen foil druing observations in the microscope, since diffraction patterns recorded before and after each through-focal series did not show any differences. It is also unlikely that the fringes in Fig.1 are a specimen preparation artefact, since low energy 3keV Ar ions were used during ion milling, and in any case similar fringes can be seen in amorphous alloys specimens prepared by ultramicrotomy and electropolishing [9,8]. Graczyk and Chaudhari [15] have shown that fringe-like images can be obtained from models of an amorphous alloy based on purely non-crystalline arrangements of atoms, when the aggregatre is viewed at particular orientations. However, the fringe-like contrast extends at most over two spacings and only rarely to a greater extent. The arrays observed in the present work occur too frequently to be likely to be explained by this effect. The images observed in Figs.1-3, therefore, seem to result from the real existence of ordered regions of atoms.

The above results suggest that the as-quenched $Fe_{78}B_{13}Si_9$ alloy contains small ~ 20Å regions with ordered configurations of atoms dispersed in the amorphous matrix. It is possible that these ordered regions can act as embryos for subsequent crystallisation of the alloy during heat-treatment. In fact, amorphous $Fe_{78}B_{13}Si_9$crystallises in a complex manner. HRTEM showed that the alloy was essentially unchanged by heat treatment at 510°C until crystallisation began after 6 minutes. Fig.4 shows a typical bright field transmission electron mirograph obtained after heat treating the alloy for 13 mins at 510°C. Three different types of crystal were found to form simultaneously from the very earliest stages of crystallisation. These three different types of crystal were identified as (a) dedritic bcc α-Fe, marked 1 in Fig.4, (b) approximately elliptical bct $Fe_3B$, marked 2 in Fig.4 and (c) spherulitic lamellar aggregates of bcc α-Fe and bct $Fe_3B$, marked 3 in Fig.-4.

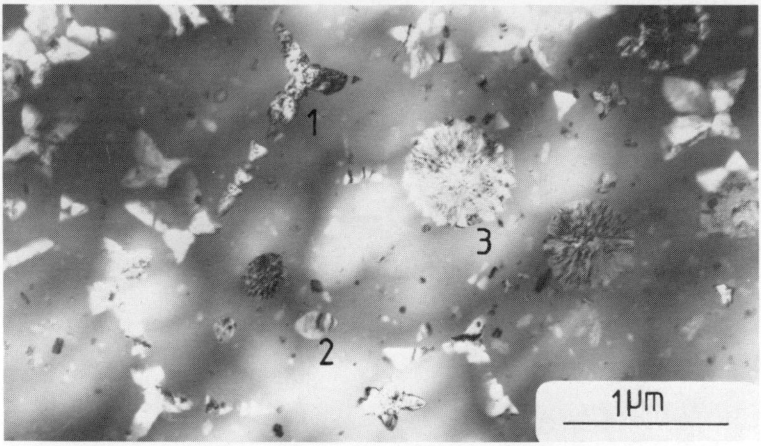


Fig.4. Transmission electron micrograph showing different crystal morphologies

Fig.5 shows a high resolution transmission electron micrograph of the interface between a dendritic α-Fe crystal and the amorphous matrix. The spacing of the lattice fringes in the dendritic α-Fe crystal is 2.02Å, corresponding to the (110) planes of bcc α-Fe. The interface between the crystal and the matrix is rather diffuse, so that the crystal is not bounded by facets of well-defined crystallographic planes.

Fig.6 shows a high resolution transmission electron micrograph of an approximately elliptical $Fe_3B$ crystal. The lattice fringes have a spacing of 4.40Å which corresponds to (200) planes in the $Fe_3B$ crystal. The interface between the $Fe_3B$ crystal and the amorphous matrix shown in Fig.6, is sharper than the interface between the bcc α-Fe crystal and the amorphous matrix shown in Fig.5. There is clear evidence of steps along the $Fe_3B$ interface. Lateral migration of these steps is a possible mechanism for the growth of this type of crystal during heat-treatment.

Fig.7 shows a lattice image from a two-phase spherulitic crystal made up of alternating lamellae of bcc α-Fe and bct $Fe_3B$. The two phase, spherulitic structure results from a eutectoid type of crystallisation reaction. Further details of the crystallisation morphologies and kinetics in amorphous $Fe_{78}B_{13}Si_9$ alloy are presented elsewhere [16].

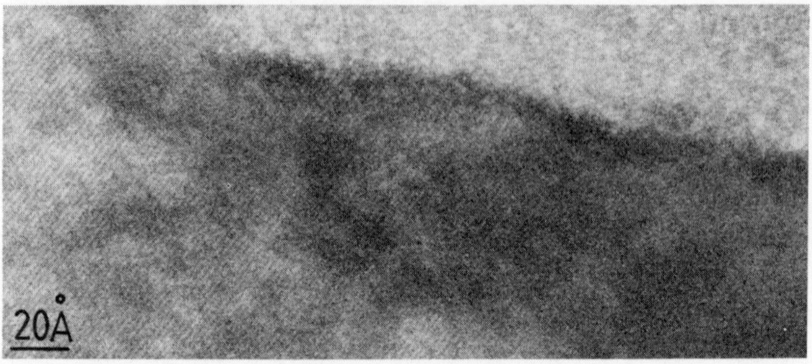


Fig.5. High resolution lattice image showing the interface beteween a dendritic bcc α-Fe crystal and amorphous matrix

Fig.6. High resolution lattice image showing the interface between an approximately elliptical bct $Fe_3B$ crystal and amorphous matrix

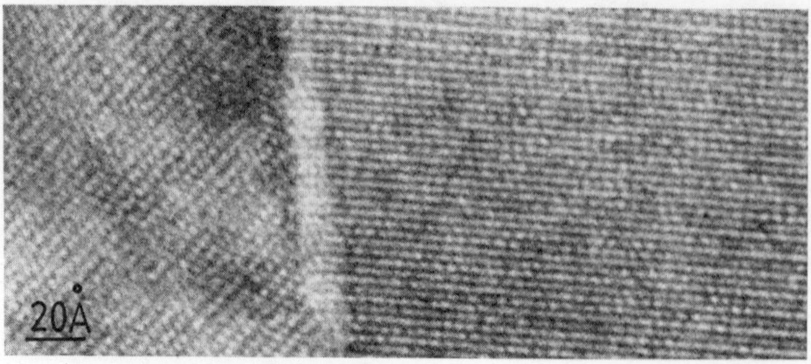


Fig.7. High resolution lattice image of a 2-phase spherulitic crystal showing the lamellar structure of bcc α-Fe and bct $Fe_3B$ crystals

## CONCLUSIONS

Small regions of regular arrangements of atoms are present in as-quenched amorphous $Fe_{78}B_{13}Si_9$ alloy. The mean size of these regions is ~ 20Å. The amorphous alloy exhibits a complex crystallisation behavour and at least three different types of crystal are formed at the earliest stages of crystallisation.

## ACKNOWLEDGEMENTS

The authors would like to thank Professor Sir Peter Hirsch and Professor J W Christian for the provision of laboratory facilities. The financial support of the Science and Engineering Research Council is gratefully acknowledged. Dr A Taube and the General Electric Corporate R and D Lab, Schenectady, USA, are thanked for supplying the as-quenched amorphous $Fe_{78}B_{13}Si_9$ alloy ribbon. We would also like to thank Professor F E Fujita and Dr P H Gaskell for helpful discussions.

## REFERENCES

1. P.H. Gaskell, J. Non-Cryst. Solids 32, 207 (1979).
2. R. Yamamoto, H. Matsuoka and Doyama, Phys. Letters. 64A, 457 (1978).
3. J.D. Bernal, Proc. Roy. Soc. London, A280, 299 (1964).
4. D.E. Polk, Acta Metall. 20 485 (1972).
5. J.L. Finney, Proc. Roy. Soc. London, A319, 479 (1970).
6. F.E. Fujita in "Rapidly Quenched Metals IV", edited by T Masumoto and K Suzuki (Japan Inst. of Metals, Sendai, 1982) p.301.
7. Y. Hirotsu and R. Akada, Jpn.J. Appl. Phys. 23, L479 (1984).
8. Y. Ishida, H. Ichinose, H. Shimada and H. Kojima, in " Rapidly Quenched Metals IV", edited by T. Masumoto and K. Suzuki (Japan Inst. of Metals, Sendai, 1982) p421.
9. J. Zweck and H. Hoffmann, in "Rapidly Quenched Metals V" edited by S. Steeb and H. Warlimont (Elsevier North Holland, Amsterdam, 1985) p.509.
10. H. Ino, H. Ichinose and K. Nagata in "Rapidly Quenched Metals V" edited by S. Steeb and H. Warlimont (Elsevier North Holland, Amsterdam, 1985) p.263.
11. R. Oshima and F.E. Fujita, Jpn. J. Appl. Phys. 20, 1 (1981).
12. P. Lamparter, W. Sperl, S. Steeb and J. Bletry, Z. Naturforsch. 37a, 1223 (1982).
13. P.H. Gaskell and D.J. Smith, J. Microscopy, 119, Pt.1, p.63 (1980).
14. M.A. Hughes and B. Cantor to be published.
15. J.F. Graczyk and P. Chaudhari, Phys. Stat. Sol. (b), 75, 593 (1976).
16. A.R. Bhatti, J.C. Barry and B. Cantor to be published.

# DEVITRIFICATION OF AN $FE_{43}CR_{25}NI_{20}B_{12}$ GLASS

TEIICHI ANDO*, YUTAKA HARA** AND N.J. GRANT**
*Toyo Kohan Co. Ltd., Visiting Scientist, Materials Processing Center, Massachusetts Institute of Technology.
**Dept. of Materials Science and Engineering, Massachusetts Institute of Technology, 77 Massachusetts Ave., Cambridge, MA 02139.

## ABSTRACT

The kinetics of bcc crystallization and subsequent transformations in an $Fe_{43}Cr_{25}Ni_{20}B_{12}$ glass were monitored by magnetic measurements and X-ray diffraction. The metastable bcc phase (ferrite) which forms first transforms to austenite and a boride during isothermal annealing. At low temperatures the ferrite grows without significant partitioning of Fe, Cr and Ni. Increasing the annealing temperature results in alloy partitioning which is directly linked with the ferrite decomposition.

## INTRODUCTION

Crystallization of a metallic glass resembles in many ways the crystallization of a highly undercooled liquid and hence gives a unique opportunity to study systematically the behavior of the "liquid" to solid transformation. The extent of crystallization is frequently measured by differential scanning calorimetry (DSC), magnetic or electrical measurements or by direct observation in TEM [1]. Among these methods, DSC gives the most accurate quantitative determination of the crystallization kinetics. A sophisticated technique has been developed and used successfully to determine the overall devitrification kinetics of $Fe_{80}B_{20}$ [2]. However, when devitrification of a metallic glass is initiated by the formation of a metastable crystalline phase followed shortly thereafter by transformation of the metastable phase to a stable phase, DSC techniques are not likely to determine the volume fraction crystallized.

Magnetic measurements may be used to separate the metastable and stable phases provided that one of the two phases is ferromagnetic. Such a condition may be satisfied in an iron alloy in which austenite and ferrite are "comparably" stable at the crystallization temperature. Low metalloid additions are preferred to guarantee glass formation and to provide sufficient subsequent ductility to permit handling and testing of the transformed products. An alloy which satisfies these conditions, an $Fe_{43}Cr_{25}Ni_{20}B_{12}$ melt spun glass, was chosen for the present work.

Thompson and Spaepan [3] showed by theoretical calculations of the nucleation in a monoatomic system that the free energy of the liquid/bcc interface is smaller than that of the liquid/fcc interface, favoring bcc nucleation over fcc nucleation when the driving forces for nucleation are comparable. Kelly et al. [4] recently demonstrated by calculations based on a subregular solution model that heterogeneous mucleation of the bcc phase may predominate in type 303 stainless steels at high undercoolings.

Nuclei forming in a highly undercooled matrix may grow massively without long range diffusion of alloying elements [5]. In the $Fe_{43}Cr_{25}Ni_{12}$ system, the fast diffusing element B may partition even at low crystallization temperatures, but substitutional elements may maintain their proportions in the composition of the bulk alloy.

The purpose of this investigation was to examine the kinetic hierarchy of crystallization and alloy partitioning during the initial and subsequent phase transformations in the $Fe_{43}Cr_{25}Ni_{20}B_{12}$ alloy.

EXPERIMENTAL PROCEDURE

Melt spun ribbons of the $Fe_{43}Cr_{25}Ni_{20}B_{12}$ glass (1.5 mm wide by 30 μm thick) were annealed for crystallization in argon filled quartz tubes at temperatures between 633 and 873K. Both isochronal and isothermal annealing was carried out. The isochronal treatment was done for 60 minutes to locate the temperature range in which bcc formation occurs readily. The isothermal annealing was performed at 666, 741 and 773K to study the kinetics of the bcc crystallization and the subsequent transformations.

The glass decomposition was monitored by measuring the magnetization of the heat treated specimens in a vibrating sample magnetometer with a maximum field strength of 7 kOe.

X-ray diffraction, TEM and STEM were also used to supplement the magnetic measurements. A JEOL 200CX was used for the TEM observation, and fine beam (2.5-5.0 nm) STEM analysis to characterize alloy partitioning was performed on an HB5 STEM.

RESULTS AND DISCUSSION

X-ray diffraction showed virtually no crystallinity in the as melt spun ribbons. Close examination by TEM confirmed that the as cast material was amorphous except in limited areas which showed some crystallinity. The as-cast material was not attracted noticeably by a small permanent magnet. A more accurate measurement showed that the magnetization was only 2.4 emu/g at 6 kOe.

Fig. 1 shows the variation of magnetization as a function of annealing temperature. The sharp increase in magnetization above 650K is caused by

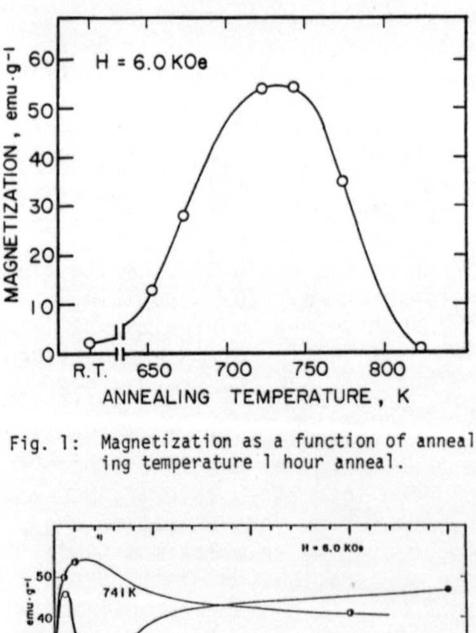


Fig. 1: Magnetization as a function of annealing temperature 1 hour anneal.

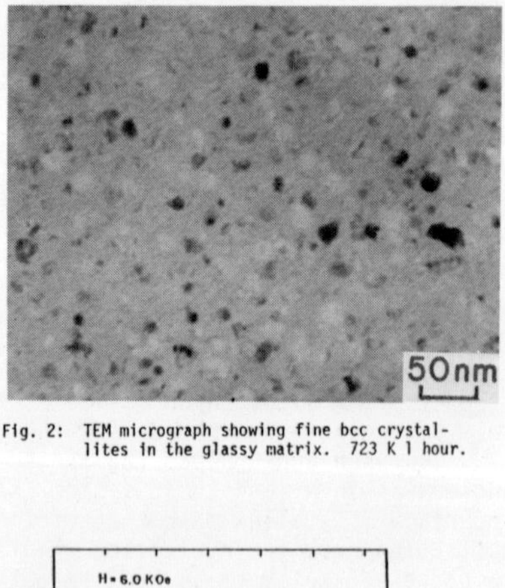


Fig. 2: TEM micrograph showing fine bcc crystallites in the glassy matrix. 723 K 1 hour.

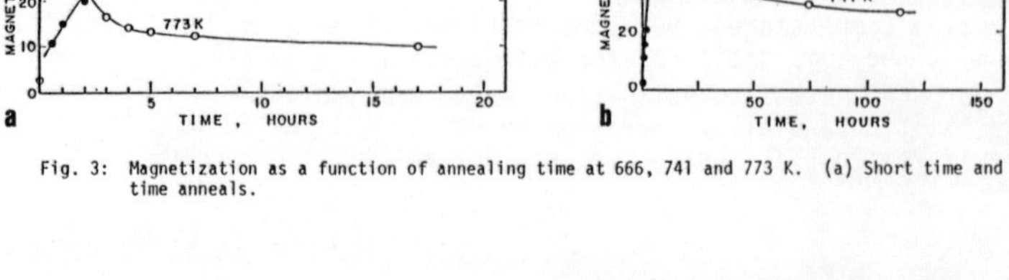


Fig. 3: Magnetization as a function of annealing time at 666, 741 and 773 K. (a) Short time and (b) long time anneals.

rapid formation of a ferromagnetic phase which was determined to be ferrite. Fig. 2 shows a TEM micrograph of the ferrite crystallites in a sample annealed at 723K. At 743K, the alloy shows a maximum magnetization of about 54.5 emu/g and begins to lose ferromagnetism at higher temperatures. The ferromagnetic behavior is completely lost again above 823K.

Fig. 3(a) and (b) shows the results of magnetic measurements on the isothermally annealed samples. At 741 and 773K, magnetization goes through a maximum as annealing time is increased. Although magnetization depends not only on the volume fraction but also on the composition of the bcc phase, the sharp decrease in magnetization at 773K after 30 minutes strongly indicates that the ferrite which forms initially transforms subsequently to austenite. The same trend is obvious at 741K except that the rate of the bcc to fcc transformation appears to exceed that of bcc formation from the glassy state at a much later stage. At 666K, however, magnetization continues to increase monotonically even after 114 hours. The increase in magnetization is very rapid up to approximately 4 hours. After 50 hours at 666K, magnetization increases only very slowly.

Fig. 4(a)-(c) shows the change in X-ray peak intensity during isothermal annealing. As expected, the transition from the bcc to the fcc structure is clearly observed for 741 and 773K by the replacement of the (110)bcc peak with the (111)fcc peak and the occurrence of the (200)fcc peak. No austenite is recognized in samples annealed at 666K. No boride peaks are detectable in the scanned angle range at any temperature shown, indicating that relatively small amounts of the boride are present even in long time annealing over this temperature range. TEM reveals, in fact, some boride crystallites in samples annealed for long times at 741 and 773K.

The results of energy dispersive X-ray flourescent analysis are summarized in Table I. Boron was not detected by the analysis. During the initial stage (2h) of crystallization at 666K, partitioning of Fe, Cr and Ni occurs to only a small extent, with slight enrichment of Fe in the crystalline phase, which is presumed from the results of X-ray diffraction to be ferrite. After 145h at 666K, alloy partitioning becomes more apparent, with depletion of Cr and enrichment of Fe in the bcc crystallites. The Ni/Fe atomic ratio in the ferrite, however, decreases only slightly indicating a low mobility of Ni at 666K.

The initially rapid and later decreased rate of increase in magnetization at 666K, Fig. 3, may be attributed to a possible change of the rate limiting element. As proposed first by Hillert [6], the growth of a precipitate in a multicomponent matrix may be controlled by the fastest diffusing element, leaving the other elements relatively unpartitioned, provided that there is sufficient driving force for the precipitation. Local equilibrium may or may not exist at the interface. Hillert [7] recently suggested that the massive transformation may be regarded as a

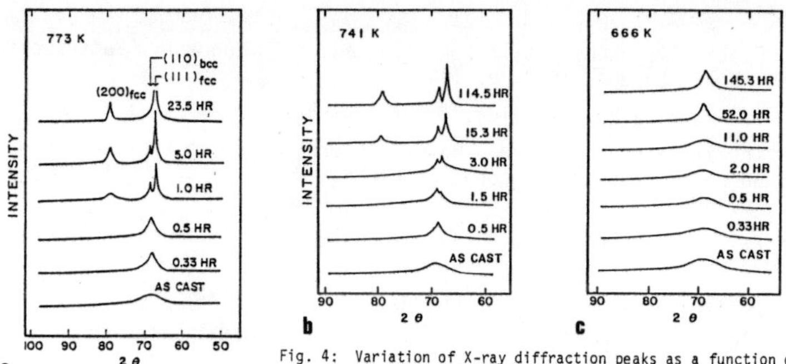


Fig. 4: Variation of X-ray diffraction peaks as a function of annealing time at (a) 773K, (b) 741K and (c) 666K.

limiting case of the precipitation described above. Elimination of B from the metallic glass, therefore, is considered to lead to a genuinely massive or polymorphic crystallization.

At higher temperatures, Ni begins to diffuse more readily and the Ni partitioning between the crystallites and the glassy matrix becomes more apparent. As a result, the decomposition of the glass, first initiated by bcc crystallization, proceeds further towards the equilibrium, where either austenite and a boride or ferrite, austenite and a boride coexist, depending on the annealing temperature.

TABLE I Alloy partitioning during the decomposition of $Fe_{43}Cr_{25}Ni_{20}B_{12}$ glass

| | | composition (at. pct.) | | |
|---|---|---|---|---|
| | | Fe | Cr | Ni |
| bulk alloy | | 48.9 | 28.4 | 22.7 |
| 666 K | crystallites | 51.1 | 27.9 | 21.0 |
| 2 h. | amorphous matrix | 46.8 | 29.0 | 24.2 |
| 666 K | crystallites | 61.8 | 13.8 | 24.4 |
| 145.3 h. | amorphous matrix | 38.4 | 29.7 | 31.9 |
| 741 K | matrix grains | 41.3 | 12.2 | 46.5 |
| 114.5h. | boride | 35.6 | 61.5 | 2.9 |

The boride phase observed in the sample annealed for 114 hours at 741K was found to be rich in Cr, Table I. The precipitation of the Cr boride appears to be balanced by the formation of the Ni rich austenite. STEM analysis indicates that both Cr and Ni are rejected from ferrite, which forms in the amorphous matrix. This seems to indicate heterogenous nucleation of the austenite and the boride at the ferrite/glass interface. Although the sequence of such transformations is still unclear, X-ray diffraction results after long annealing times at 741K confirm that the predominant phase is the austenite.

## CONCLUSIONS

1. Crystallization in an $Fe_{43}Cr_{25}Ni_{20}B_{12}$ glass is initiated by formation of a metastable bcc phase which subsequently decomposes to an fcc phase and a boride.

2. Fe, Cr and Ni do not partition significantly during the early stages of bcc precipitation at 666K. At 741K, alloy partitioning occurs readily, resulting in the decomposition of the bcc phase.

## ACKNOWLEDGEMENTS

The authors are pleased to acknowledge support by the Army Research Office and the availability of the excellent NSF-MRL analytical facilities. We express thanks to Dr. C. Ashdown for the availability of his carefully prepared amorphous melt spun ribbon.

## REFERENCES

1. M.G. Scott, Amorphous Metallic Alloys, ed. F.E. Luborsky, Butterworths & Co., Ltd., London, 1983, p. 144.

2. A.L. Greer, Acta Metallurgica, 30, 171 (1982).

3. C.V. Thompson and F. Spaepen, Materials Processing in the Reduced Gravity Environments of Space, Proceedings of MRS, Vol. 9, North Holland, N.Y., 1982, p. 603.

4. T.F. Kelly, M. Cohen and J.B. Vander Sande, Metall. Trans., 15A, 819, (1984).

5. D.A. Karlyn, J.W. Cahn and M. Cohen, Trans. TMS-AIME, 245, 197 (1969).

6. M. Hillert, Internal Report, Swedish Inst. for Metal Research, 1953.

7. M. Hillert, Metall. Trans., 15A, 411 (1984).

# MULTI-STEP MICRO-CRYSTALLIZATION STUDIES ON AMORPHOUS $Fe_{90-x}Si_xB_{10}$ ALLOYS

W.MINOR[§], R.MALMHALL*, ANNA ROIG, A. INOUE[†] AND K. V. RAO
*Dept. of Solid State Physics, Royal Institute of Technology, Stockholm, Sweden.*
§ *Guest Scientist from Warsaw University, Poland*
* *ASEA Research and Innovation, S-721 78 Västerås, Sweden*
† *Guest Scientist from Tohoku University, Sendai, 980 Japan.*

## ABSTRACT

In order to elucidate the sensitivity and information that can be obtained from Magneto-Thermo-Gravimetric, MTG, technique, we present a comparative study of our results on structure relaxation as well as crystallization sequences as a function of time and temperature of amorphous $Fe_{90-x}Si_xB_{10}$ alloys in the Fe-rich regime, with those obtained from synchrotron Energy-Dispersive-X-ray-Diffraction, EDXD, and Differential Scanning Calorimetry, DSC methods.

## INTRODUCTION

The possibility to obtain randomly distributed combination of elements in a 'glassy' metallic state by means of rapid quenching techniques has created new opportunities to investigate phase diagrams of multicomponent metallic systems in a rather fundamental way. We can study, for example, nucleation and crystal growth kinetics almost at an atomic level. The evolution of complex structures and phases can now be followed and studied dynamically in a time resolution of few minutes using high intensity X-rays from synchrotron radiation sources combined with Energy-Dispersive-X-ray Diffraction, EDXD, technique [1]. Thus, considerable research efforts in recent years have been devoted to different methods of investigating the crystallization properties of metallic glasses. Among the direct and indirect methods EDXD [2], Transmission Electron Microscopy, Mössbauer spectroscopy, Differential Scanning Calorimetry, DSC, [3], saturation magnetization, coercivity, Curie temperature, and electrical resistance [4,5] techniques are commonly used. However, it is still of interest to look for techniques that are simple and yet quite sensitive to changes in both local short range order as well as topological order which characterize the stability and crystallization phenomena in 'glassy' metals. For example, the ferromagnetic transition temperature $T_C$ is extremely sensitive to changes in the local short range order and also microstructural variations in a material at various stages of heat treatment can be readily detected because of the sensitivity of the first derivative of the magnetization around $T_C$ to such changes. We have exploited this MTG technique to study the magnetic behaviour of low-boron containing amorphous $Fe_{90-x}Si_xB_{10}$ alloys. In this paper in order to elucidate the sensitivity and the information one can obtain by the MTG method, we compare our results on structure relaxation as well as crystallization studies as a function of time and temperature with those obtained from EDXD and DSC techniques for the same samples.

## EXPERIMENTAL DETAILS

Amorphous ribbons of $Fe_{90-x}Si_xB_{10}$, with x ranging from 7 to 21, about 20 microns thick and several mms in width were produced from the constituent elements of at least 99% purity by the usual melt-spinning technique using a single roller and was made available to us by Prof. H.Matyja. A Perkin-Elmer type DSC-4 Calorimeter connected with a Thermal Analyses Data Station 3600 has been used to investigate the thermal properties. The thermo-gravimetric data were obtained on a Perkin-Elmer TGS-2 balance system with a horse-shoe magnet providing the low fields (about 5-50 Oe). The TGS-2 was also coupled with the Data Station 3600. During MTG measurements, typically two to five small pieces of the ribbons weighing about 1.5 mgms were heated in an atmosphere of continuous flow of argon at a low pressure suitable for a measuring

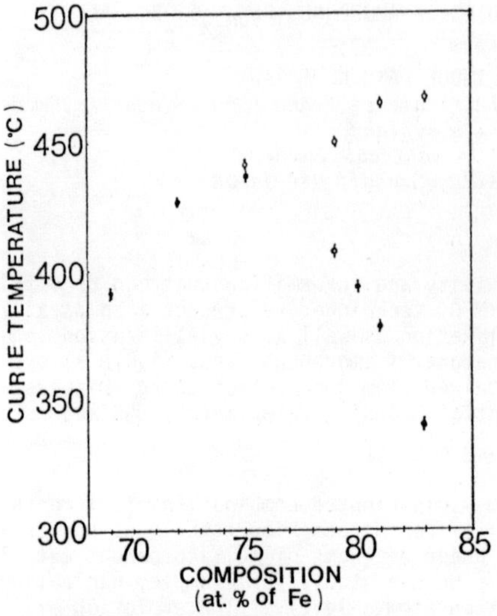


*Fig 1. Magnetic phase diagram at the Fe-rich regime of amorphous $(Fe\text{-}Si)_{90}B_{10}$ alloys (●). The open circles are the Curie temperatures achieved after 1 hr annealing at 450°C.*

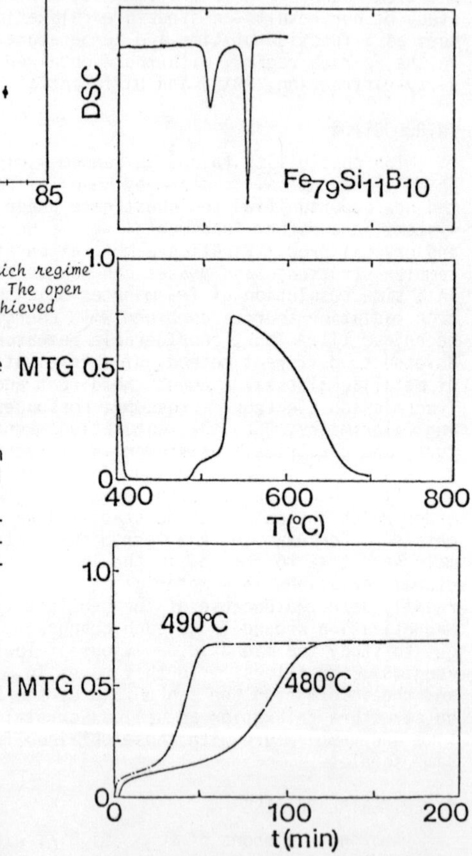


*Fig 2. Two-stage crystallization as a function of temperature as reflected in Differential Scanning calorimetric (DSC), and Magneto-Thermo-Gravimetric (MTG) techniques on heating an amorphous $Fe_{79}Si_{11}B_{10}$ alloy at 10 K/min heating rate. The dynamic evolution of the two crystallization stages are from Isothermal-MTG data taken at 480 & 490°C, just below the spontaneous crystallization temperature for the same alloy.*

sensitivity of 1 part in $10^7$. The equipment covers the range from room temperature to about 1000°C. Low field (usually 5 Oe) magnetic measurements were carried out both at a constant heating rate as well as an isothermal mode maintaining a very high stability aof the annealing temperatures. All the magnetic and DSC measurements reported here were taken at a constant heating rate of 10K/min.

RESULTS AND DISCUSSIONS

The effect of silicon substitution on the magnetic properties of the $Fe_{90-x}Si_xB_{10}$ is shown in Fig 1 where the magnetic phase diagram, $T_C$ vs Fe-concentration has been plotted. In these data $T_C$ is determined from the minimum in the first derivative of the magnetization with respect to temperature, around $T_C$. As seen in the figure, $T_C$ (closed circles for the 'as quenched' ribbons) vs Fe-concentration has a broad maximum similar to the well known unusual behaviour for amorphous Fe-B system, namely $T_C$ decreases rapidly with increasing Fe concentration.

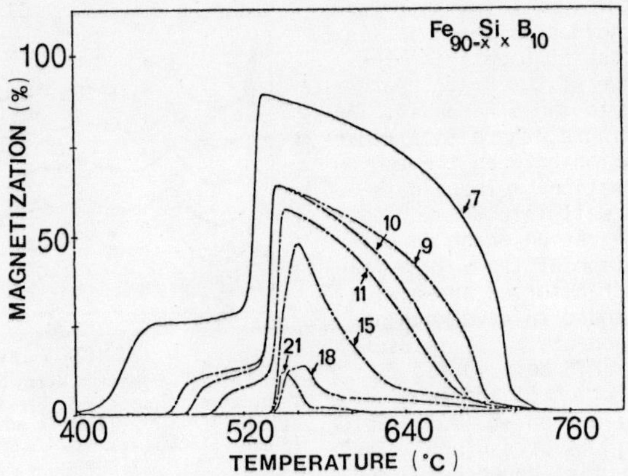


*Fig 3. Low-field, (~5 Oe), MTG data reflecting the effect of Si substitution on the crystallization processes in amorphous Fe-rich Fe-Si-B alloys.*

In the present study $T_C$ decreases by more than 100°C for change of 75 to 83 at.% Fe. Rest of the discussions in this paper will be on the properties of alloys with concentration greater than 75 at.% Fe.

In fig 2 is shown data obtained on thermal scanning a $Fe_{79}Si_{11}B_{10}$ sample at 10 K/min in both DSC and TGS measurements. Much above $T_C$ (=410°C) around which the magnetization drops to a negligible value in the MTG data, we observe two stages of spontaneous crystallization which are seen as successive abrupt increases in the magnetization and which correlate very well with the two exothermic peaks seen in the DSC data for the same heating rate. The existence of two crystallization stages are also clearly manifested in the isothermal low-field magnetic measurements, IMTG, taken at two temperatures (fig 2) just below that for the first spontaneous crystallization. These IMTG data taken at 490°C and 480°C as a function of time sensitively show the differences in the time scales for nucleation, and crystallization of each stage which evolves into identical final magnetic state. No such detailed information is possible from a DSC technique. Also note that $T_C$ is barely discernible in the DSC data in contrast to the large effect in the MTG data.

In fig 3 we summarize the effect of Si on the crystallization stages observed for these Fe-Si-B amorphous alloys. Two regimes of concentrations for the effect of Si are found regarding crystallization: 1) One with two crystallization stages wherein increasing Si from 7 to 11 at% suppresses the first stage of crystallization; 2) and the other with alloys having greater than 11 at.% Si where only the second crystallization stage presists which is also again strongly suppressed with increasing Si. These results which indicate that in these Fe-B-Si alloys the hypereutectics crystallize in two steps while the hypoeutectics crystallize in one step are consistent with the observation from DSC studies by Ramanan [3]. Comparison of quantitative estimates for the activation energies obtainsed from DSC and IMTG data will be presented elsewhere.

It is now interesting to compare earlier [2] EDXD data for the evolution of crystallization with time, and our isothermal-MTG data for the same alloy. In fig 4 the EDXD pattern for an amorphous $Fe_{83}Si_7B_{10}$ sample at room temperature after annealing it at 350°C for 18, 44, 85 and 1375 minutes is shown. Pronounced crystallization is found to evolve with time in an essentially amorphous matrix. From these studies Minor et al [2] identify that the evolving structure is mostly $\alpha$-Fe with Si dissolved in it. The onset and evolution of this first crystallization stage in this alloy for over 650 minutes at the same temperature 350°C investigated using MTG technique is shown in fig 5. The dots are the EDXD data for the time dependence of the Bragg peak corresponding to the α-Fe phase. As seen in the figure, there is a remarkable correlation between the two sets of information. These data can be fitted well into the Johnson-Mehl-Avarami equation which describes the evolution of the volume fraction of the transformed phase with time, and is found to give meaningful Avrami exponents. It is useful to recall our IMTG data in fig 2 where the evolution of the two-stage crystallization at 480 & 490°C for $Fe_{79}Si_{11}B_{10}$ is presented. The details of the time dependence of the two stages during crystallization can also be quantitatively interpreted and correlated with the EDXD data as shown above.

Finally, for alloys with Fe concentration greater than 75 at% we have monitored the variation of $T_C$ with annealing times at a given temperature. The open circles in fig 1 represent equilibrium values for $T_C$ attained after 1 hr annealing. Details of this study will be published elsewhere.

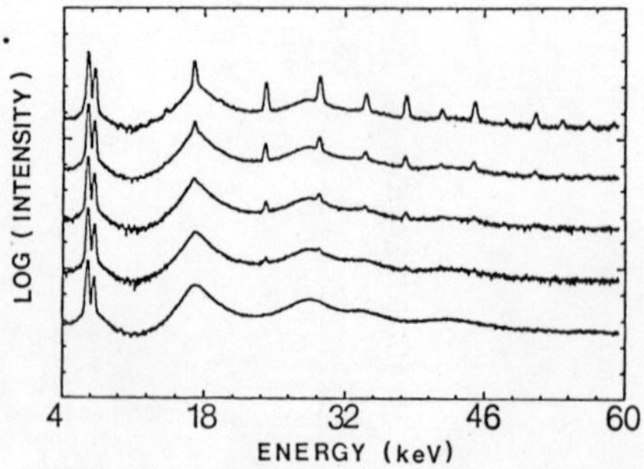


*Fig 4. Energy-Dispersive-X-ray Diffraction pattern from an amorphous $Fe_{83}Si_7B_{10}$ sample at room temperature after annealing it at 350°C for 18, 44, 85 and 1375 minutes.*

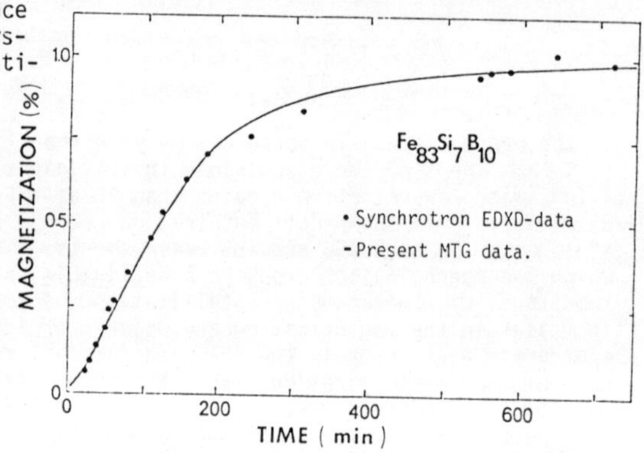


*Fig 5. Dynamic evolution of the first crystallization stage at 350°C as obtained from low-field magneto-thermo-gravimetric technique. The dots are data for the same process from Energy-Dispersive X-ray-Diffraction, EDXD.*

ACKNOWLEDGEMENTS

This work is supported by the Swedish Technical Board for Research and Development , STU.

REFERENCES

1 B. Buras, N.Nimura, and J.Staun Olsen, J. Appl. Cryst 11, 137 (1978)
2 W.Minor, B.Schoenfeld, B.Lebech, B.Buras and W.Dmowski (to be published)
3 V.R.V.Ramanan and G.E.Fish, J. Appl Phys 53, 2273 (1982)
4 See for example review by M.G.Scott in Amorphous Metallic Alloys Ed F.E. Luborsky (Butterworths, U.K. 1983) p 144.
5 M.Burrous and F.Varret, Solid State Commun. 57, 713 (1986)
T. Tarnóczi, Phys Stat Solidi (a) 87, 283 (1985)
K.A.Bertness, K.V.Rao and H.H.Liebermann, J Non Cryst Solids 61&62, 883 (1984)

PART II C

# Amorphous Alloys—Consolidation, Mechanical, and Chemical Properties

# AN INVESTIGATION OF THE BOND BETWEEN LAYERS OF CONSOLIDATED AMORPHOUS RIBBON (POWERCORE™ STRIP)

D. RAYBOULD AND A. DIEBOLD
Allied-Signal, Corporate Technology, P.O. Box 1021R, Morristown, NJ 07960

## ABSTRACT

Thin amorphous metal ribbons can be bonded together to form a thick strip which remains amorphous. This, Powercore™ strip, has similar magnetic properties to the annealed thin ribbon and so is a direct substitute for silicon steel in power transformers. Its good properties are partially a result of the nature of the bond between the layers of ribbon. This paper reports on a study of this bond by Auger microscopy and laser acoustic microscopy.

## INTRODUCTION

Thin ribbon of amorphous metals has been produced for some time in a variety of widths and lengths by Metglas Products for use in distribution transformers. Using this rapidly solidified ribbon distribution transformers ranging from 10 to 100 KVA have been constructed by different manufacturers using their own transformer designs. These have losses of about one quarter of those of a conventional transformer, confirming laboratory measurements. Thus Metglas® amorphous alloys offer substantial energy savings for pole type distribution transformers, of which there are nearly 40 million throughout the U.S.A.

Theoretically, the much larger power transformers could also benefit from the lower core losses of amorphous metals, and they are also a large market. Indeed, in most of Europe, one stacked core transformer substation is used instead of several pole mounted distribution transformers. However, for stacked core transformers, other factors become important and influence the total operating costs. In the early 80's it became clear that, especially for the larger stacked transformers, new techniques would have to be developed to allow their economical manufacture. In particular, the thin ribbon necessitates laying around 10 times more laminations than conventional silicon steel; also, the decreased amount of solid material that can be packed into a given space increases the physical size of the transformer. These problems made the cost of manufacturing a large stacked transformer so large that it affected total owning cost and reduced the potential savings of the lower core losses [1]. To overcome this obstacle EPRI funded research to lead to the economical manufacture of amorphous stacked power transformers. Three approaches were funded [1]: -

(1) Direct casting 5 to 10 mil of an inch thick 6-8" wide amorphous ribbon.

(2) Bonding 1 mil of an inch thick approximately 8" wide ribbon together to provide an optimum thickness laminate of 5 to 25 mil of an inch, while retaining the amorphous state.

---

Work on Powercore™ Strip is jointly funded by E.P.R.I. and Allied Corporation.
Metglas® is an Allied Corp. registered trademark for amorphous alloys.
Powercore™ is Allied Corp. trademark for consolidated amorphous ribbon.
EPRI - Electric Power Research Institute

(3) Stacking conventional 1 mil of an inch thick ribbon several ribbons at a time to improve productivity.

The bonding approach was carried out at Allied Corp. using heat and pressure to bond the ribbon together [2]. This program was successful and produced laminated ribbon with magnetic properties similar to those of a single ribbon. The technique produced over 100 mil of an inch thick material, but for logistic reasons 5 to 25 mil of an inch thick material is easier to produce, fig. (1). The simultaneous application of heat and pressure results in an improvement in the flatness of the ribbon so that packing factors of over 90% are consistently obtained, thus decreasing the physical size of the transformers. This material can be supplied in the annealed state so that the customer has only to cut it to the required length and stack it into a transformer. In the annealed state, the laminate is relatively ductile and readily undergoes repeat bending, fig. (2). The magnetic properties of the laminate have been studied by both Metglas Products and Westinghouse Corp. under the EPRI programs. Typical core loss values for the laminate and silicon steel are shown in Table I.

Table I

| | Amorphous Metal | | | Silicon Steel |
|---|---|---|---|---|
| | Standard single ribbon field annealed | Laminate as shipped | | M-4 |
| Thickness [mil.] | 1 | 5 | 10 | 11 |
| Typical Core Loss [w/Kg] 1.4T, 60 Hz | 0.18 | 0.21 | 0.29 | 0.90 |

Loss separation measurements confirm that the eddy current loss of the laminated strip is significantly lower than would be predicted from increasing the thickness of a single ribbon, indicating that lamination does not produce good electric contact between the layers of ribbon.

A final advantage of the laminate is the high reproducibility that is obtained in both physical and magnetic properties. The material is currently commercially available in trial quantities under the name of Powercore™ strip.

This paper is concerned with investigating the bond formed between the layers of ribbon in the laminate. This bond determines the strength of the strip and the electric contact between the ribbon and hence the magnitude of the eddy current, thus partially determining the core loss of the strip.

## EXPERIMENTAL

The laminated ribbon was produced in sizes from 2 to 20 mil (0.05 to 0.5 mm) thick and 1 to 2 mil (25 to 5- mm) wide using temperature and pressure to produce bonding [2]. The investigation showed that the actual temperature or pressure did not affect the type of bond obtained, once bonding

Fig (1) "Production" POWERCORE™ Strip 10 mil. thick.

Fig (2) Annealed POWERCORE ™ Strip, unbroken after repeat bending was attained. Metglas alloy 2605-S2, nominal composition $Fe_{78}B_{13}Si_9$ (at%) was the amorphous alloy investigated. It is an alloy intended for transformer applications.

The bond was observed by carefully delaminating the ribbons with the aid of a razor.

Surface analysis was carried out by scanning electron microscopy, SEM, and scanning Auger microscopy, SAM. For the latter, a SAM-SIMS instrument from Physics Electronics was used. The variation of composition with depth was determined by Auger depth profiling. The etch rate of the ion gun used for this was calibrated by $SiO_2$ layers of known depth. Both xenon and argon were used as the gas source for the ion gun.

The occurrence of a true bond was determined by scanning laser acoustic microscopy, SLAM, by Sonoscan Inc. This technique transmits acoustic waves through the specimen and these are observed in real time by laser interferometry. Bonded areas appear as light regions.

Crystallinity was checked for by X-ray and by DSC analysis. The continuous monitoring of magnetic properties was a more sensitive control of the retention of the amorphous state.

## RESULTS

### a. Scanning Electron Microscopy

Debonded strip was examined early in the program using a Hitachi scanning electron microscope (fig. 3). No areas of direct metal to metal bonding were found. Both the surface next to the tooling and the surfaces of the debonded ribbon showed evidence of local plastic deformation. Machining marks on the tooling were transferred to the surface of the strip and the debonded ribbon could be seen to have flowed into the irregularities of the adjoining ribbon. Light contrast areas were also found. The reason for these light contrast areas could not be determined by SEM.

### b. Auger Microscopy

Auger analysis showed that the white areas were consistently richer in oxidized silicon than their surroundings, fig (4). This silicon concentration was found to be a surface phenomenon, and fig (5) shows the disappearance of a white contrast area as the surface is etched away. Graphical depth profiles confirm the surface nature of the high silicon and oxygen concentration [3]. A detailed investigation was undertaken, which showed that the white contrast areas correspond to similar white areas on the ribbon from which they were debonded [3].

### c. Scanning Laser Acoustic Microscopy

The nature of the bond between the ribbon was investigated by laser acoustic microscopy. The bond was found not to be uniform, as can be seen from the white spots and lines representing bonding in fig (6). Lines can also be formed for reasons other than bonding in certain modes of operation of the microscope; therefore, care is necessary in interpreting data. This is especially true as the number of ribbons in the strip increases and hence the superimposition of the bonds at each layer. Fig (7) is for a typical well bonded strip. The ripple or wave pattern seen is believed to originate from the pattern seen on as cast ribbon and is a result of the casting process. The crests of these waves are high points on the ribbon and would therefore be expected to be areas of bonding or intimate contact. The light

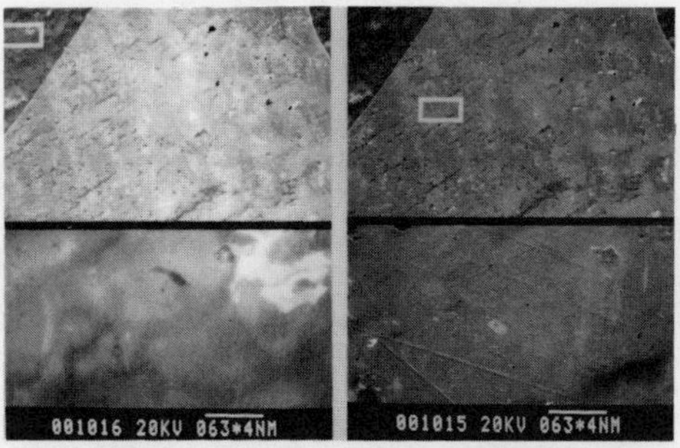


Fig (3) A scanning electron micrograph of partially debonded strip. Note, the evidence of plastic deformation, indicated by tooling marks and the white contrast areas.

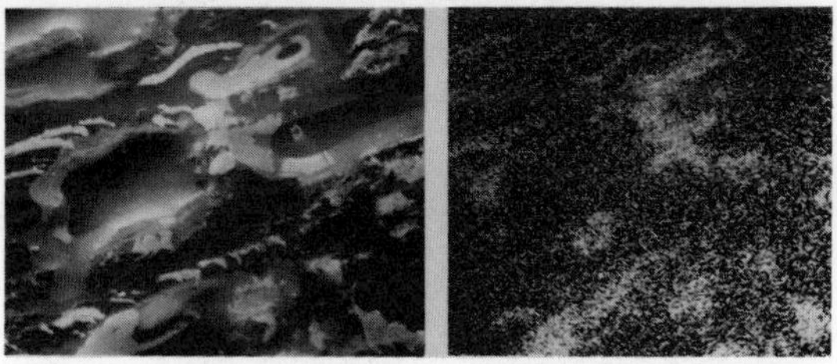

1000X 1000X

Fig (4) Auger map showing the silicon distribution, note their correspondence with white areas of S.E.M.

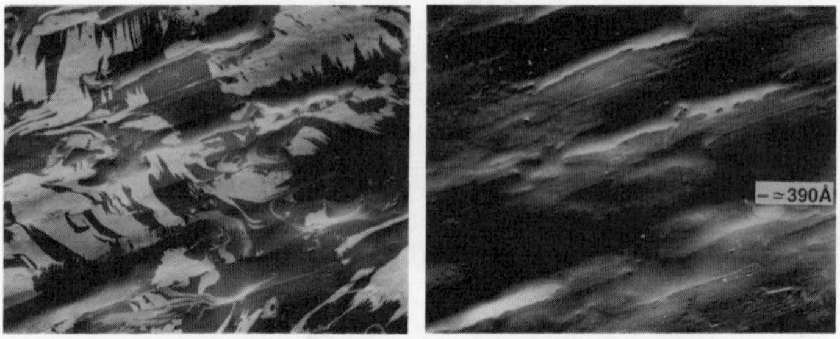


204X 204X

Fig (5) Illustrating the surface nature of the white contrast areas. Micrograph of the delaminated strip and after the oxide layer has been etched away.

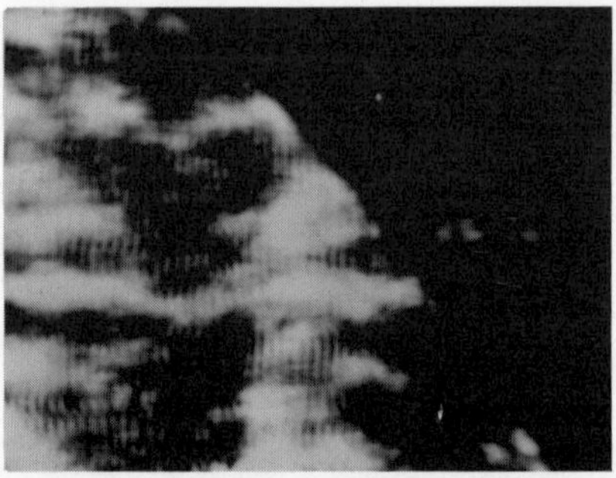

Fig (6) A scanning laser acoustic micrograph (SLAM) of as rolled 5 layer strip. White areas indicate areas of bonding. The large dark triangle is a purposely debonded layer.

Fig (7) A SLAM for a typical well bonded POWERCORE™ Strip

spots correspond in size to the areas of high silicon concentration observed by Auger microscopy, confirming that the areas of high silicon are the regions of bonding.

## DISCUSSION

The observations of the debonded ribbon by Auger microscopy and of the bonded strip by laser acoustic microscopy has allowed a model to be developed for the bond between the layers of consolidated ribbon. The bond apparently originates at high points on the ribbon and spreads as local plastic deformation brings more of the surfaces of the ribbon into contact. This plastic deformation fills the surface irregularities of the as cast ribbon producing a mechanical bond, similar to that used in Lego® building blocks. A chemical bond is produced by diffusion of silicon to the surface and the formation of a silicon oxide bond. This, however, is a localized bond. It is observed that after the strip is bent repeatedly so as to break this rigid oxide bond that the strip is still held together by the mechanical interlocking. Several more repeat bends are required before visible delamination

The local plastic deformation of the ribbon into surface irregularities of adjoining ribbon is consistent with the improved pack factor of the ribbon. Ribbon with a pack factor of less than 70% has been formed into strip with a pack factor of over 90% on many occasions. In addition, the pack factor of the initial ribbon does not appear to change the pack factor of the consolidated strip by more than 1 or 2%. The localized oxide bond is also consistent with the low eddy current losses found for the strip. The high electrical resistance of the oxide and its localization account for the poor electrical contact between the layers of ribbon.

Further work is underway on this potentially useful material, both under an EPRI program and as part of Metglas Products commercialization of amorphous metals.


## ACKNOWLEDGEMENTS

The Powercore strip was produced by T. Esposito. C. Agramonte of Sonoscan Inc. provided valuable assistance in interpreting the SLAM micrographs. The support of the management of Corp. Tech. and of Metglas Products is gratefully acknowledged, as is the EPRI funding of the Powercore program.


## REFERENCES


(1) H.R. Sheppard and E.T. Norton, "Amorphous Metal Core Power Transformers (Economic Considerations)", "Rapidly Quenched Metals, R.Q.5", editors S. Steeb and H. Warlimont, publishers Elsevier Science, 1985.

(2) J. Kushnick, D.L. Sawhney and R. Hathaway, U.S. Patent 4,529,458 granted 1985, assigned to Allied Corp.

(3) D. Raybould and A.C. Diebold, "An Investigation by Auger and Laser Acoustic Microscopy of the Bond Between Layers of Consolidated Amorphous Ribbon", J. Material Sci., in press.

# WEAR PROPERTIES OF A SHOCK CONSOLIDATED METALLIC GLASS AND GLASS-CRYSTALLINE MIXTURES

THAD VREELAND, JR., NARESH N. THADHANI, AND ANDREW H. MUTZ, Division of Engineering and Applied Science, California Institute of Technology, Pasadena, CA 91125
SUSAN P. THOMAS AND ROGER K. NIBERT, Roy C. Ingersoll Research Center, Borg-Warner Corp., Des Plaines, IL 60018

## ABSTRACT

Powder flakes prepared from 50 μm thick melt spun ribbons of Markomet 1064 ($Ni_{52.5}$ $Mo_{38}$ $Cr_8$ $B_{1.5}$ wt%) were shock consolidated in the unannealed and annealed condition. The unannealed flakes (microhardness 933 kg/mm$^2$) are amorphous while flakes annealed at 900°C for 2 hours have an fcc structure with a grain size of 0.3 μm and microhardness of 800 kg/mm$^2$. The shock consolidated amorphous powder compact (250 kJ/kg shock energy) shows no crystal peaks in an X-ray diffractometer scan. Compacts of annealed powder (400 to 600 kJ/kg shock energies) contain amorphous material (18-21%) which was rapidly quenched from the melt formed at interparticle regions during the consolidation process. The microhardness of the amorphous interparticle material is 1100 kg/mm$^2$. Wear properties of the compacts measured in low velocity pin on disk tests show low average dynamic friction values (∿0.03). The 60 hour cumulative wear appears to correlate with the energy of shock compaction and surface porosity of the compacts rather than the metallic glass content.

## INTRODUCTION

Metallic glasses, with high metalloid concentration and associated hardness, have been observed to show superior wear resistance when used as coatings on steel substrates [1]. Shock wave consolidation has been employed to produce bulk solids with metastable structures [2-5]. During consolidation, the shock energy is preferentially utilized in heating and melting interparticle surfaces [6,7]. The melted regions rapidly solidify and may form new metastable structures or retain the metastable structure of the starting powder. Shock consolidation is a unique form of thermal and mechanical processing and can be used to process metallic glass powders to form bulk amorphous solids [8,9], and crystalline powders of glass forming alloys to produce a solid mixture of metallic glass and microcrystalline material [6,10]. In this paper, we report some wear properties of a shock consolidated glass forming alloy powder containing varying amounts of glass and microcrystalline phases.

## SHOCK CONSOLIDATION

### Powder Characterization

The Markomet 1064 alloy ($Ni_{52.5}$ $Mo_{38}$ $Cr_8$ $B_{1.5}$ wt%) powder was prepared by ball milling melt spun amorphous ribbons (∿50 μm thick). Annealing of the powder at 900°C for 2 h produces a fully microcrystalline powder (confirmed by x-ray and electron diffraction). Optical microscope observations on polished and etched powder particles show that the amorphous phase resists attack by the Marbles reagent, while microcrystalline grains in the annealed powder are attacked and show a dark etching contrast [10].

## Powder Consolidation

Green compacts were prepared by static loading of the powder in steel containers and the compacts were consolidated by impact of propellent driven stainless steel flyer plates at velocities from 0.8 to 1.4 km/s. The consolidated powder compacts were recovered as ∿20 mm diameter discs ∿5 mm thick.

Optical photomicrographs of polished and etched sections of recovered compacts (plane of the shock) are shown in Fig. 1a,b. The dark etching regions in the compact of annealed powder in Fig. 1a are microcrystalline while the white nonetching regions are metallic glass (as confirmed by TEM [10]). The compact of amorphous powder, Fig. 1b shows etching at amorphous particle boundaries which did not melt sufficiently to dissolve surface contaminants. The amorphous powder compact shows no crystalline peaks in an X-ray diffractometer scan. The shock energy, glass content, and microhardness of the particles and interparticle amorphous regions are given in Table I for the four compacts used in the wear tests.

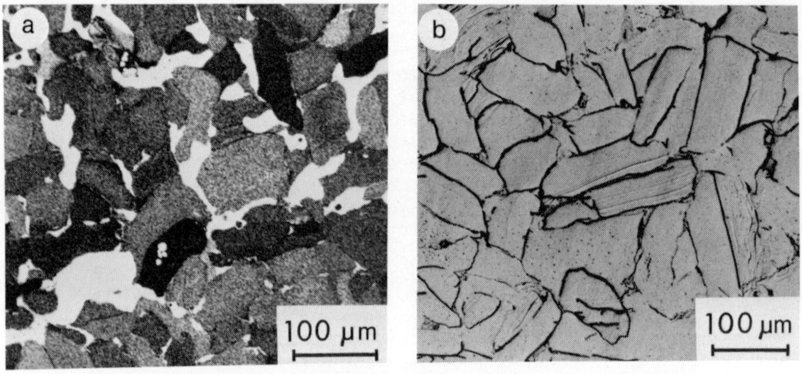


Fig. 1. Optical photomicrographs of sections (shock plane) of compacts of (a) microcrystalline powder (shot #816), and (b) amorphous powder (shot #817).

The volume fraction of metallic glass formed upon compaction was determined by measurement of the areal fraction of the nonetching glass phase in optical micrographs, and ranged from 0.18 to 0.21 with increasing shock energy. The microhardness of the glass phase formed at microcrystalline particle boundaries is greater than the hardness of the melt spun glass and the hardness of the microcrystalline particle interiors. Interparticle regions of the amorphous powder compact also exhibit a higher hardness than that in the interior of the amorphous particles.

TABLE I - CONSOLIDATION EXPERIMENTS AND RESULTS

| Shot# | Shock Energy(kJ/kg) | Glass Fraction | Microhardness($DPH,kg/mm^2$) Particle | Interparticle |
|---|---|---|---|---|
| 813(μx+a) | 592 | 0.21 | 792 ± 134 | 1095 ± 87 |
| 815(μx+a) | 467 | 0.19 | 782 ± 99 | 1060 ± 153 |
| 816(μx+a) | 419 | 0.18 | 801 ± 205 | 1146 ± 98 |
| 817(a) | ∿250 | 1.0 | 933 ± 131 | 1144 ± 75 |

μx = microcrystalline, a = amorphous

## WEAR TESTS

A low velocity friction apparatus (LVFA) [11,12], employing a three pin on disc configuration was used to obtain boundary friction test conditions which produce adhesive wear. Test samples were made from ∿3 mm discs (0.5 mm thick) of the compacts bonded to 3 mm pins. The discs were ground to match the 3 mm diameter pins with a 45° chamfer and the flat ends of the discs were mechanically polished, finishing with 5 μm diamond grit. The pins were run at 0.3 m/s on a circular track (17.5 mm diameter) on the mechanically polished surface of a disc (∿40 mm thick 52100 steel, microhardness 900 kg/mm$^2$) for a 17 h break-in period at ambient temperature prior to making measurements of friction and wear. The disc was lubricated with an automotive transmission fluid. Friction and wear values at 0.4 m/s and the highest temperature and thrust load used in the tests (149°C and 19.3 MPa pressure) are given in Table II. Wear volume was determined from optical microscope measurements of the chamfer diameter of the pins after 60 h of running. Values determined using an AISI 1095 steel pin base-line sample (microhardness 700 kg/mm$^2$) are also given in Table II.

TABLE II - LOW VELOCITY FRICTION APPARATUS RESULTS (149°C, 19.3 MPa)

| Sample # | Surface Porosity Ranking | Coefficient of Friction $\mu_{static}$ | $\mu_{dynamic}$ | $\mu_{ave\ dyn.}$ | Cumulative Pin Wear after 60 hrs. (mm$^3$) |
|---|---|---|---|---|---|
| 813 | L-M | 0.105 | 0.064 | 0.02 | 0.10 |
| 815 | M-H | 0.111 | 0.074 | 0.03 | 1.03 |
| 816 | M | 0.092 | 0.078 | 0.02 | 0.13 |
| 817 | H | 0.085 | 0.082 | 0.05 | 2.31 |
| 1095 Steel | L | 0.113 | 0.087 | 0.10 | 0.08 |

Microscopic examination of the polished surfaces of the compacts revealed some porosity due to particle pull-out. This surface porosity was observed to decrease for compacts consolidated with increasing shock energy.

Initial static and dynamic coefficient of friction values are given in Table II. There is insignificant variation in the measurements for the static friction data. The dynamic sliding friction increased with increasing load and temperature, and decreased as the sliding speed was increased. This behavior is typical for boundary friction conditions. Under the most severe test conditions (149°C, 19.3 MPa) the dynamic coefficient of friction (after the final 17 h break-in) values tend to increase with increasing porosity. Highest dynamic friction coefficient resulted with the base-line 1095 steel sample because of its greater surface roughness. The average dynamic coefficient of friction values, on the other hand are the average of the values obtained after break-in at 5 min, 1 h, and 17 h, and follow the trend for the dynamic values.

The pin volume loss in the wear tests correlates with the energy of shock compaction and surface porosity, and not with the volume fraction of metallic glass in the samples. Wear of samples #813 and #816, consolidated at high energies, was less than that of #815 and #817. Samples of shot #817 (all glass) and #815(19% glass) showed maximum wear which is attributed to particle pull-out. The increase in dynamic friction during the break-in cycle also indicates particle pull-out. In samples #813 and #816, the pin wear is equal to that of the base-line 1095 steel sample, within the limits of experimental uncertainty.

In contrast to the pin on disc test, preliminary wear test results on an all glass Markomet 1064 alloy compact in the block on disc test (at high contact pressures) indicates significantly lower wear rates than on hardened steel base-line samples. The results of the block on disc tests will be reported elsewhere.

## CONCLUSIONS

1. Shock consolidation of amorphous powder flakes of Markomet 1064 alloy with a 250 kJ/kg shock energy produces a fully amorphous compact with microhardness ranging from 933 to 1100 kg/mm$^2$. The interparticle regions of the compact exhibit the highest hardness.

2. The shock consolidation of annealed Markomet 1064 powder produces compacts with metallic glass at microcrystalline particle interfaces. The volume fraction of metallic glass increases with increasing shock energy up to 600 kJ/kg, and the microhardness of the metallic glass and microcrystalline particles is $\sim$1100 and $\sim$800 kg/mm$^2$, respectively.

3. A porosity is observed on mechanically polished surfaces of the shock consolidated compacts which is attributed to particle pull-out. The porosity decreases with increasing shock energy.

4. Low velocity wear tests (0.4 m/s) show a dynamic coefficient of friction and a 60 h cumulative wear which correlates with the energy of shock compaction rather than the metallic glass content of the compact.

5. The compact with 21% metallic glass between microcrystalline particles (consolidated with the maximum shock energy) exhibited the lowest dynamic friction, and wear comparable to that of the hardened AISI 1095 steel base-line sample.

## ACKNOWLEDGEMENTS

The shock consolidations were carried out using the facilities of Prof. Thomas Ahrens in the Caltech Seismological Laboratories and were supported in part by the National Science Foundation under Grant No. DMR-8315214 and the Caltech Program in Advanced Technologies sponsored by Aerojet General, General Motors, GTE, and TRW Laboratories. Wear Tests were performed by Mr. Marc Yesnik at the Roy C. Ingersoll Research Center.

## REFERENCES

1. M. Mehra, PhD Dissertation, California Institute of Technology, 1984.
2. D.G. Morris, Metal Sci. 16, 457 (1982).
3. D. Raybould, J. Matl. Sci. 16, 589 (1981).
4. D.G. Morris, Mat. Sci. and Eng. 57, 187 (1983).
5. P. Kasiraj, PhD Dissertation, California Institute of Technology, 1984.
6. T. Vreeland, Jr., P. Kasiraj, A.H. Mutz, and N.N. Thadhani, in Proc. of Int. Conf. on Metallurgical Applications of Shock-Wave and High-Strain-Rate Phenomena, edited by L.E. Murr (to be published by Marcel Dekker Publishing Company, New York, March 1986).
7. R. Schwarz, P. Kasiraj and T. Vreeland, Jr., source cited in Ref. 6 (to be published).
8. P. Kasiraj, D. Kostka, T. Vreeland, Jr., and T.J. Ahrens, J. of Non-Cryst. Sol. 61 and 62, 967 (1984).
9. T. Vreeland, Jr., P. Kasiraj, and T.J. Ahrens, in Proc. of MRS Symposium on Rapidly Solidified Metastable Materials, edited by B.H. Kear and B.C. Giessen (Elsevier Science Publishers, New York, 1984), p. 139.
10. N.N. Thadhani, A.H. Mutz, P. Kasiraj, T. Vreeland, Jr., source cited in Ref. 6 (to be published).
11. M.L. Haviland and J.T. Rodgers, Lubrication Engr. 17, 110 (1961).
12. C. Albertson and G. Wolfram, ASLE Trans. 12, 77 (1969).

# CORRELATION BETWEEN FRACTURE MECHANICAL AND ATOM PROBE INVESTIGATIONS ON METALLIC GLASS RIBBON

CALVO M., MENAND A.*, OSTERSTOCK F. AND CHERMANT J.L.
Equipe Matériaux-Microstructure de l'U.A. 251,
ISMRa-Université, 14032 Caen Cedex, France
* Equipe Microscopie Ionique, U.A. 808, Faculté des Sciences et Techniques de Rouen, 76130 Mont Saint-Aignan, France

## ABSTRACT

Metallic glasses have good mechanical and magnetical properties. However they embrittle after annealing and/or low temperature testing. In this paper, we describe the rupture of atom probe tips using fracture mechanical measurements. We found boron rich areas which seem to be interconnected and more numerous after annealing and act as crack nuclei. The good correlation between the calculated and the measured size of the critical defect seem to confirm that the boron segregations are on the origin of embrittlement and gives good ground for further straightforward investigations.

## EMBRITTLEMENT IN TERMS OF FRACTURE MECHANICS

In a previous study [1], we have shown that metallic glass ribbons follow perfectly the concepts of fracture mechanics in plane stress.

SEM observations have allowed us to link the decrease of the toughness, and so the embrittlement during heat treatment, with the change of the fracture surface and the rupture conditions from plane stress to plane strain [2] [3]. The continuous evolution from plane stress to plane strain means an embrittlement and thus a decrease of the plastic zone size at the crack tip. Its size and shape can be calculated from the $K_c$ values [1][4]. The embrittlement is probably to be attributed to short or long range rearrangements inside the material during the heat treatment.

## EMBRITTLEMENT IN TERMS OF FIELD ION MICROSCOPY AND ATOM PROBE

Atom probe (AP) provides a method for quantitative microanalysis with very high spatial resolution. A metallic specimen, prepared in the shape of a very thin tip, can be then analysed atom per atom. Such analysis have been achieved in the last few years on metallic glasses [5][6].

Piller and Haasen [5] have studied a FeNiB alloy in field ion microscopy (FIM), performing also some mechanical experiments. They have detected boron rich areas and an increase in size and number of these areas with increasing embrittlement. The boron content on the fracture surface of the tip is higher (Fig. 13 of [5]). The authors have concluded to a phase separation, with a $(FeNi)_3B$ amorphous phase, which volume fraction increases with annealing time and temperature. These areas should be responsible for the embrittlement.

We have also detected such areas of high boron content (often more than 25 at. %). These areas seem to be more numerous after heat treatment, and also interconnected, as it can be observed in comparing Fig. 1a and 1b. Fig. 2 reveals a regroupment of such areas, detected for an as-annealed specimen. The AP available in Rouen has made it also possible to detect variations in iron and nickel contents. The high boron content observed after the tip rupture in the AP indicates the important role of such boron rich areas towards the mechanical behaviour.

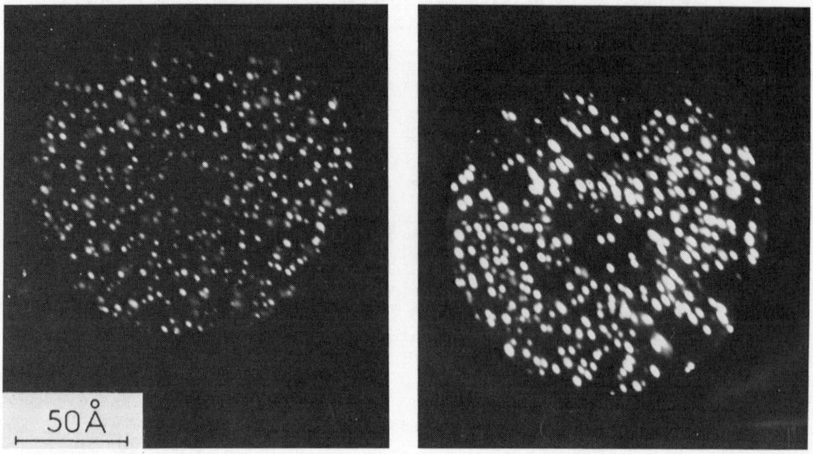


a) V = 8.84 kV. b) V = 8.82 kV.

Fig. 1. Neon field ion images of $Fe_{40}Ni_{40}B_{20}$ amorphous alloy as-quenched (a) and after annealing 2 h45 at 300°C (b). The dark and bright areas are more pronounced after annealing. The bright areas (boron rich) seem to be interconnected.

However the very high apparent boron levels (40 % to 100 %) found after tip rupture must be taken with care. As Menand et al [7] have shown the value of the evaporation field is higher for boron atoms than for nickel and iron.

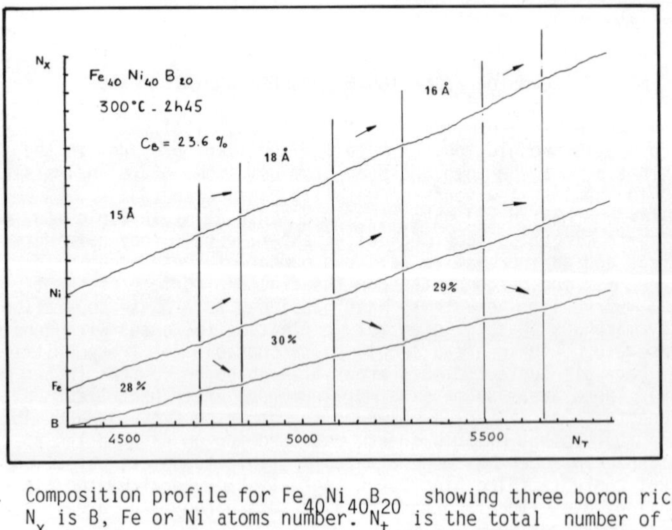


Fig. 2. Composition profile for $Fe_{40}Ni_{40}B_{20}$ showing three boron rich areas. $N_x$ is B, Fe or Ni atoms number. $N_t$ is the total number of detected atoms.

Experiments performed on a $FeB_{20}$ specimen with a pulse to DC field ratio variing from 7 % to 20 %, have shown that nominal boron concentration is only obtained for pulse fractions higher than 15 % ; the apparent boron concentration reachs 66 % for a pulse fraction of 7 %. In this latter case most of the iron atoms are preferantially field evaporated at DC field and therefore not detected.

The discrepancy between the observed boron concentrations : (i) for the boron enriched regions into the specimen (25 % to 30 %) and (ii) after tip rupture (40 % to 100 %), can be explained in a simple way as follows : as the field strength required for the evaporation of boron rich areas is higher, preferential evaporation of iron and nickel atoms at the steady state field occurs. This phenomenon leads, therefore, to the measurement of a more pronounced boron enrichment than the actual value.

These pronounced preferential evaporation effects are an additional strong evidence of the existence of boron enriched region where tip rupture occurs.

## THE TIP REPRESENTED AS A CENTER CRACK TENSILE (CCT) PANEL

The AP measurements have revealed in the as-quenched specimens the presence of boron rich areas about 20 A large. After heat treatment they are more numerous and seem to be interconnected in some case. These interconnected "chains" are formed over a distance of a few 100 A. They have been observed in the analysis cylinder but can be supposed to exist everywhere inside the tip. Tip rupture could then occur if such brittle areas can act as crack nuclei. Mechanical calculations allow such an interpretation [8].

The tip can indeed be represented as a CCT panel, similar to the ones we used for $K_c$ measurements [1][2] (see Fig. 3). The applied stress is due to the electrical field. It has has already been calculated by P. Birdseye [9]. Using a value of $K_c$ 1 to 3 MPa m we obtain for the critical crack length a value of 300 to 500 A, which is of the order of magnitude of the largest observed interconnected boron rich areas. The value of $K_c$ has been evaluated from mea-

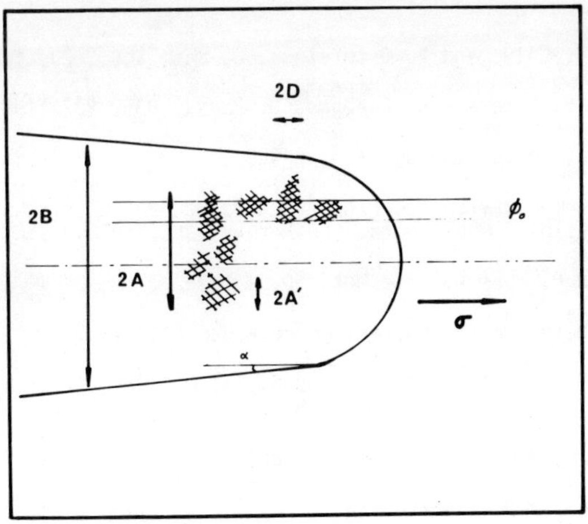


Fig. 3. Schematic representation of the tip and the interconnected areas as a CCT panel.
2B is the tip diameter, 2A and 2A' the lengths of the regroupment and of the areas, 2D the analysis depth, Øo the analysis diameter.

surements made at 77K. The temperature in the AP is 20K. The $K_c$ may thus be slightly overestimated and the critical crack size smaller than the calculated value.

## DISCUSSION AND CONCLUSION

The very good agreement we have obtained in the above calculations indicate that the boron segregation can be supposed to be directly linked to the embrittlement of $Fe_{40}Ni_{40}B_{20}$ and may be metallic glasses where such segregations can occur.

Furthermore the material should become more brittle and more homogeneous, in the mechanical sense, the more these areas become numerous and interlinked. In a previously published paper [2] we had tried to link the increase of brittleness, the plastic zone size decrease, a supposed increase of density of the slip planes and the change of specimen rupture surface. We had interpreted these observations by a higher density of equivalent slip planes. It should correspond to an homogeneous critical shear stress and strain in the bulk of the material.

In the case of metallic glasses where boron segregation can be observed (like in FeNiB), the equivalent slip planes could be determined by the boron rich regroupments or "network".

This could be an interpretation of the embrittlement of such metallic glasses and has to be linked and compared with the structure modells which exist in the litterature.

## ACKNOWLEDGMENTS

Part of this work has been done while M. Calvo was in the Institut für Werkstoffkunde und Werkstofftechnik in Clausthal Zellerfeld (FRG) with the support of CNRS (A.I.E. 85 375).

## REFERENCES

1. W. Henning, M. Calvo and F. Osterstock, J. Mat. Sci., 20, 1889 (1985).

2. M. Calvo and F. Osterstock, Czechos. J. Phys., B35, 337 (1985).

3. M. Calvo, DEA, Université de Caen, (1983).

4. D. Broek, Elementary Engineering Fracture Mechanics (Sijthoff and Noordhoff International Publishers, the Netherlands, 1978), p. 11.

5. J. Piller and P. Haasen, Acta Met. 30, 1 (1982).

6. A. Menand, Thèse d'Etat, Université de Rouen (1984).

7. A. Menand, M. Bouet, C. Martin and J. Gallot, Proc. 29th FIM Symposium Almquist and Wiksell, Stockholm, 1982.

8. M. Calvo and A. Menand, to be published.

9. P.J. Birdseye and D.A. Smith, Surface Sci., 23, 198 (1970).

# CATALYTIC PROPERTIES OF GLASSY AND CRYSTALLINE TRANSITION METAL ALLOYS PRODUCED BY RSP*

S.S. Mahmoud+, D.A. Forsyth and B.C. Giessen
Department of Chemistry and Materials Science Division, Barnett Institute, Northeastern University, Boston, MA 02115

## ABSTRACT

The catalytic activity and selectivity of $Pd_{80}Si_{20}$, $Pd_{35}Zr_{65}$, $Ni_{30}Zr_{70}$, $Fe_{40}Ni_{40}B_{20}$, $Fe_{85}B_{15}$ and $Cu_{50}Zr_{50}$ alloy ribbons in their amorphous and crystalline states were studied in the hydrogenation of 1-hexene and phenylethyne, dehydrogenation of methylcyclohexane and dehydrocyclization of n-heptane. $Pd_{80}Si_{20}$ was found to be the most active of the alloy catalysts studied. Most alloys were more active amorphous than in the crystalline state; however, the catalytic selectivities of the amorphous and crystalline states of each alloy were identical.

## INTRODUCTION

Recently, metallic glasses have been the subject of studies concerning the potential for their use as hydrogenation catalysts in the hope that their properties might differ from those of the currently used crystalline metal catalysts in their activities and selectivities. The expectation that differences might be found between amorphous and crystalline catalysts of the same composition is based on the results of studies by Somorjai and others who have found the catalysis of some reactions to be "structure sensitive" (1). Thus, Yokoyama et al. have studied the catalytic activity of fifteen amorphous alloy ribbons based on Fe-Ni containing P and/or B as glass forming additives with respect to their catalytic effect in the hydrogenation of CO; they found that C1 and C3 hydrocarbons were the principal products, and noted that the catalytic activity of the amorphous alloys was several to several hundred times higher than that of the crystalline alloys of the same composition (2,3). Previously, Smith et al. studied the catalytic activity and selectivity of splat cooled amorphous $Pd_{80}Si_{20}$ with respect to the deuteration of cyclododecene; they found that the amorphous splats were more selective than crystalline palladium in catalyzing addition rather than exchange (4).

The work reported in the present paper is part of an extended study of the catalytic properties of melt spun alloy ribbons with respect to hydrogenation and related reactions; an early account of this work was given in Ref. 5. The use of NMR as an analytical tool in some of these studies is reported in Ref. 6; a note on additional hydrogenation catalysts is in preparation.

* Communication No. 278 from the Barnett Institute
+ Present Address: Department of Chemistry, Brandeis University, Waltham, MA 02235

The present paper deals with the catalytic activity and selectivity of six transition metal alloys with respect to three different reactions (hydrogenation, dehydrogenation and dehydrocyclization) and reports on the difference (or absence thereof) in the catalytic behavior of the amorphous and crystalline states of each alloy.

## EXPERIMENTAL

Catalyst preparation: Transition metal alloys with the six compositions given in Table I were prepared as small (~1g) buttons in an arc furnace, then melt spun using a modified Pond-Maddin inert gas chamber melt spinner [7] to produce amorphous ribbons 20-50 μm thick, 2-5 mm wide and several meters long. To obtain the alloys in their crystalline form,

Table I. Catalytic activities and selectivities in the hydrogenation of 1-hexene at T = 50°C.

| Catalyst | % Conversion[a] after 2 hrs. | % Conversion[a] after 8 hrs. | Selectivity[b] coefficient |
|---|---|---|---|
| Am. $Pd_{80}Si_{20}$ | 100.0 | 100.0[c] | 2.5 |
| Cryst. $Pd_{80}Si_{20}$ | 42.4 | 100.0[c] | 2.5 |
| Am. $Pd_{35}Zr_{65}$ | 17.0 | 100.0[c] | 1.3 |
| Cryst. $Pd_{35}Zr_{65}$ | 0.0 | 25.3[d] | 1.3 |
| Am. $Ni_{30}Zr_{70}$ | 7.7 | 95.6 | 1.6 |
| Cryst. $Ni_{30}Zr_{70}$ | 2.2 | 80.9 | 1.6 |
| Am. $Ni_{40}Fe_{40}B_{20}$ | 0.0 | 3.8 | — |
| Cryst. $Ni_{40}Fe_{40}B_{20}$[e] | 1.2 | 10.3 | — |
| Am. $Fe_{85}B_{15}$ | 0.0 | 1.1 | — |
| Cryst. $Fe_{85}B_{15}$ | 0.0 | 0.0 | — |
| Am. $Cu_{50}Zr_{50}$ | 0.0 | 0.0 | — |
| Cryst. $Cu_{50}Zr_{50}$ | 0.0 | 0.0 | — |

a Catalyst activities are given as percentage of conversion after two and eight hours.

b Selectivity coefficients are given as the ratio of isomerization to saturation at 10% saturation; a high selectivity coefficient indicates a more isomerization-selective catalyst.

c Reaction was completed earlier.

d After 10 hours.

e Extensive fragmentation observed.

the amorphous alloys were heated under vacuum in sealed quartz tubes for 12 hours at or above their crystallization temperatures. The amorphous structure or crystallinity of the ribbons was confirmed by X-ray diffractometry as described in detail elsewhere [8].

Hydrogenation reactions: The reactions were carried out in a glass hydrogenator equipped with a pressure transducer to measure hydrogen absorption by pressure change; the hydrogenator was kept in a constant temperature oil bath. 5.0 ml of 1-hexene dissolved in 20-30 ml of 1,4-dioxane and 1.0 g of catalyst were placed in the reaction vessel and held at the reaction temperature (generally 50$^oC$) under ~1.3 atm hydrogen or deuterium for 72 hrs. The reactions were monitored analytically by taking 1 ml samples of the reaction mixtures and examining them by $^{13}C$ FT-NMR and/or gas chromatography after different time periods.

Dehydrogenation and dehydrocyclization reactions: These reactions were carried out in a stainless steel fixed-bed flow reactor equipped with a preheating chamber which had been designed and built for these reactions. To operate the reactor, ~1g of catalyst was placed on a quartz wool catalyst bed, the reactor was heated to the desired temperature (200-550$^oC$) under helium atmosphere, and, upon reaching that temperature, the substrate (methylcyclohexane or n-heptane) was passed through the reactor in a helium flow via a helium bubbling injector. The reaction products were examined by gas chromatography.

Analytical Methods: We summarize in the following the techniques used for product analysis and give some details: 1) For the hydrogenation reactions, $H_2$ pressure drop measurements were used to determine the degree of saturation during reaction; 2) $^{13}C$ FT NMR was used to determine the amounts of D-H exchange, isomerization and saturation in the hydrogenation of 1-hexene; 3) A Varian 920 Aerograph GC equipped with 30 ft x 1/8 inch Supelco column (23% SP-1700 on chromosorb PAW 80/100) was used for analyzing 1-hexene hydrogenation products; 4) A Hewlett Packard F-and-M GC with 8 ft x 1/8 inch Supelco column (GP 10% TCEP on chromosorb PAW 100/120) and FID detector were used for the analysis of the dehydrogenation of methylcyclohexane and dehydrocyclization of n-heptane.

## RESULTS AND DISCUSSION

1. Hydrogenation of 1-hexene: In the hydrogenation of 1-hexene over all alloys showing activity in the time periods reported, the products obtained were n-hexane, cis- and trans-2-hexene, and trans-3-hexene. Figure 1 presents the results of the hydrogenation of 1-hexene over $Pd_{35}Zr_{65}$ alloy ribbons; the percentages of the unsaturated components present (1-hexene, cis- and trans-2-hexene and trans-3-hexene) are plotted against percent saturation (% n-hexane).

Several observations can be made. First, trans-2-hexene is the major unsaturated product at all levels of saturation. Second, the ratios of the unsaturated intermediate products to each other remain approximately constant as the reaction proceeds. Third, the ratio of saturated product (n-hexane) to intermediate unsaturated products remains constant until the starting material is consumed; during this time period, the total amount of intermediate products formed is larger than the amount of hexane formed, i.e., isomerization is preferred over saturation in this stage of the process. Fourth, when all starting material is consumed, saturation proceeds at the expense of the intermediate products, with the ratio of intermediate products again remaining constant.

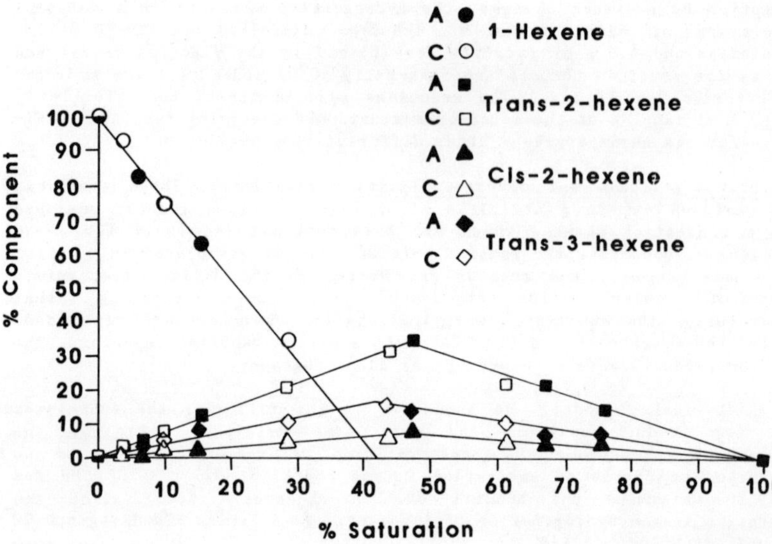


Figure 1. Deuteration of 1-hexene over amorphous (A) and crystalline (C) $Pd_{35}Zr_{65}$.

Turning next to the relative effectiveness of different catalysts with respect to this reaction, Table I shows the catalytic selectivities (ratio of percent isomerization to saturation at 10% saturation) and the catalytic activities (percent conversion after 2 and 8 hours of reaction) of the six alloys chosen.

We first review the selectivities. Where they could be obtained (i.e., where the activity was sufficiently high), the catalytic selectivities of these alloys all show a preference of isomerization over saturation. Each alloy has the same selectivity in its amorphous and crystalline state, i.e., at any saturation level both structural modifications of each catalyst produced the same amount of each component (see Figure 1), with the preference of isomerization over saturation that had been noted above. This preference may be rationalized by assuming that these alloys are hydrogen-poor catalysts, i.e., that there is little hydrogen available on their surfaces for reaction; this may be due to either a high or low (but not a moderate) hydrogen - to - catalyst surface bonding energy [9-12]. Turning to the activities of the six alloys, it is seen that $Pd_{80}Si_{20}$ is most active, $Pd_{35}Zr_{65}$ and $Ni_{30}Zr_{70}$ are of moderate activity and $Ni_{40}Fe_{40}B_{20}$, $Fe_{85}B_{15}$ and $Cu_{50}Zr_{50}$ have low catalytic activities. Also, it seems that the amorphous state of each alloy is more active than its crystalline counterpart, except where fragmentation of the crystalline alloy is extensive, as in the case of the brittle crystalline $Ni_{40}Fe_{40}B_{20}$.

2. <u>Hydrogenation of phenylethyne:</u> In the hydrogenation of phenylethyne 1, phenylethene 2 was the initial product with ~95% yield, followed by phenylethane 3 as the final product.

$$\underset{\underline{1}}{Ph - C \equiv C - H} \underset{}{\overset{H_2}{\rightleftharpoons}} \underset{\underline{2}}{Ph - CH = CH_2} \overset{H_2}{\rightleftharpoons} \underset{\underline{3}}{Ph - CH_2 - CH_3}$$

Figure 2 shows the hydrogenation of phenylethyne over $Pd_{35}Zr_{65}$; at the point of complete consumption of phenylethyne, ~52.5% of the total deuterium to be consumed to saturation has been consumed to produce ~95% phenylethene and ~5% phenylethane. The behavior of all six catalysts with respect to this reaction is shown in Table II; the selectivity is given as the ratio of phenylethane to phenylethene formed at the point of disappearance of phenylethyne.

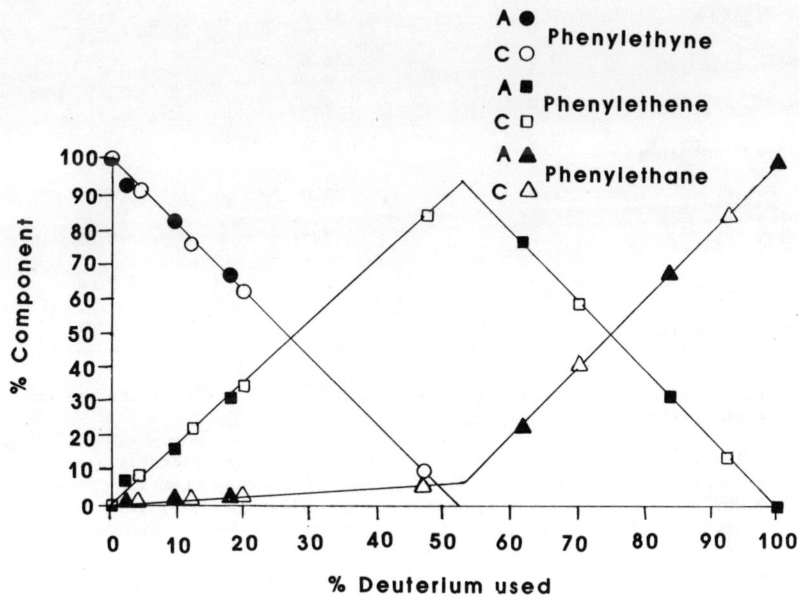


Figure 2. Deuteration of phenylethyne over amorphous (A) and crystalline (C) $Pd_{35}Zr_{65}$.

All alloys showed similar selectivities, strongly favoring partial saturation to phenylethene over complete saturation to phenylethane. Again, the amorphous and crystalline forms of each alloy showed identical selectivities, i.e., at any percentage of deuterium consumed, both catalyst modifications produced the same percentages of products.

Table 2 also shows the activities of the catalysts as percent conversion of phenylethyne to other products after 2 and 8 hours; again, $Pd_{80}Si_{20}$ was the most active alloy, while all other alloys were of moderate to low activity.

Table II. Catalytic activities and selectivities in the hydrogenation of phenylethyne at 50°C.

| Catalyst | % Conversion[a] after 2 hrs. | % Conversion[a] after 8 hrs. | Selectivity[b] coefficient |
|---|---|---|---|
| Am. $Pd_{80}Si_{20}$ | 53.0 | 100.0 | 0.03 |
| Cryst. $Pd_{80}Si_{20}$ | 12.8 | 67.4[c] | 0.03 |
| Am. $Pd_{35}Zr_{65}$ | 8.5 | 38.0 | 0.08 |
| Cryst. $Pd_{35}Zr_{65}$ | 7.4 | 32.9 | 0.08 |
| Am. $Ni_{30}Zr_{70}$ | 0.0 | 12.5 | 0.12 |
| Cryst. $Ni_{30}Zr_{70}$[d] | 0.0 | 43.0 | 0.12 |
| Am. $Ni_{40}Fe_{40}B_{20}$ | 2.9 | 9.6 | 0.12 |
| Cryst. $Ni_{40}Fe_{40}B_{20}$[d] | 0.8 | 11.4 | 0.12 |
| Am. $Fe_{85}B_{15}$ | 2.6 | 18.0 | 0.03 |
| Cryst. $Fe_{85}B_{15}$ | 0.0 | 3.7 | 0.03 |
| Am. $Cu_{50}Zr_{50}$ | 0.0 | 0.6 | 0.04 |
| Cryst. $Cu_{50}Zr_{50}$ | 0.0 | 7.6 | 0.04 |

a Catalytic activities are given as amounts of conversion after two and eight hours.

b Selectivity coefficients are given as ratios of phenylethane to phenylethene at the time when phenylethyne concentration reaches 0.0%; a higher selectivity coefficient indicates relatively higher selectivity towards complete saturation (production of phenylethane) rather than towards formation of phenylethene.

c After seven hours.

d Extensive fragmentation observed.

3. Dehydrogenation and dehydrocyclization reactions: Methylcyclohexane was chosen as a dehydrogenation substrate in the expectation that toluene and benzene would be the major products in this reaction. For the same reason, n-heptane was chosen as a dehydrocyclization substrate. To carry out these reactions, high temperatures (up to ~550°C) were used; temperature changes have a significant effect.

We report here only general trends observed for these reactions, with emphasis on the effects of changes of temperature and hydrogen partial pressure.

The effect of temperature is characterized by a maximum in the activity at about 350°C, at lower temperature, product yields are reduced because of slower reaction kinetics, while at higher temperatures, cracking and coke formation increase, which in turn lower the catalyst activity. The effect of hydrogen pressure on selectivity is seen, e.g., in the dehydrogenation of methylcyclohexane over $Pd_{80}Si_{20}$: when a hydrogen flow is provided to maintain a partial $H_2$ pressure of 250 mm Hg, the ratio of aromatics to other products is 1:1; when no hydrogen is supplied to the reaction vessel, this ratio increases to 2:1, showing that aromatization is more favored at lower hydrogen concentrations. A similar behavior was observed in the dehydrocyclization of n-heptane, where the ratio of aromatics to other products was 0.5:1 when hydrogen was supplied and 1:1 when hydrogen was cut off.

## SUMMARY

Some of the alloys ribbons studied were moderately (but not strongly) selective for the reactions studied. Further, the selectivities of the alloys studied were found to be almost identical in the glassy and crystalline states.

As to their activities, amorphous ribbons were not very active, due to the small surface area (~0.1 $m^2/g$) of these materials; however, amorphous alloys were somewhat more active than the crystalline alloys of the same composition. This effect is probably due to differences in the surface structures of amorphous and crystalline alloys, specifically, a decrease in the <u>density</u> of active sites on crystallization. However, it appears that the <u>type</u> of such defects is unchanged in this process, resulting in the structure independence of the selectivity.

Among further studies required, surface analysis of the ribbon surfaces before and during catalysis are especially important.

## ACKNOWLEDGMENT

We are pleased to thank the Dow Corp. for their support of these studies and acknowledge secretarial assistance by Ms. Jana Volf.

## REFERENCES

1. R.K. Herz, W.D. Gilespie, E.E. Petersen, and G.H. Somorjai, J. Catal., 67, 371-386 (1981), and references therein.

2. A. Yokoyama, H. Komiyama, H. Inoue, T. Masumoto and H.M. Kimura, J. Catal., 68, 355 (1981).

3. H. Komiyoma, A. Yokoyama, H. Inoue, T. Masumoto and H.M. Kimura, Chem. Lett., Chem. Soc. Jap. 195 (1983).

4. W.E. Brower, M.S. Matyjaszczyk, T.L. Pettit, and G.B. Smith, Nature, 301, 497 (1983).

5. B.C. Giessen, S.S. Mahmoud, D.A. Forsyth, and M. Hediger, in Rapidly Solidified Amorphous and Crystalline Alloys, Mater. Res. Soc. Symp. Proc. Vol. 8, B.H. Kear, B.C. Giessen, and M. Cohen, Eds., North Holland, New York, N.Y., p. 255 (1982).

6. D.A. Forsyth, M. Hediger, S.S. Mahmoud, and B.C. Giessen, Anal. Chem. 54, 1896 (1982).

7. R. Pond, Jr. and R. Maddin, Trans. Met. Soc., AIME, 245, 2475 (1969).

8. S.S. Mahmoud, Ph.D. Thesis, Northeastern University, Boston, MA (1985).

9. R.L. Burwell, Jr., in Catalysis Progress in Research, F. Basolo and R.L. Burwell, Jr. Eds., Plenum Press, New York, NY, 1973.

10. G.A. Somorjai, "Chemistry in Two Dimensions: Surfaces", Cornell University Press, Ithaca, NY, p. 389, 1981.

11. W.J.M. Rootsaert and W.H.M. Sachtler, Z. Phys. Chem., 26, 16 (2960).

12. J.H. Sinfelt, Adv. Catal., 23, 91 (1973).

PART II D

# Amorphous Alloys—Magnetic Properties

# MAGNETIC RELAXATION AND STRUCTURAL TRANSFORMATION IN METALLIC GLASSES

R.C. O'HANDLEY
Massachusetts Institute of Technology, Department of Materials Science and Engineering, Cambridge, MA 02139

## ABSTRACT

Current understanding of the local atomic structure of amorphous materials is reviewed. Some results of probing short-range order by selected techniques are cited to illustrate the degree of uniformity that exists on a local scale. Observations of phase separation and phase changes are described with particular emphasis on a temperature driven, reversible transformation of the local structure observed magnetically in several cobalt-base glasses. The manifestations and implications of such transformations within the glassy state are examined. Several examples pointing to quasi-crystalline and to non-crystalline (non-space filling) local structures are given.

## INTRODUCTION

The most direct characterization of atomic arrangements in glasses have come from scattering studies (X-ray, neutron, electron). Such studies point to the absence of long-range order and to the existence of a certain statistical regularity in local coordination of metallic glasses. However, local atomic coordinates or even the angular arrangements of atoms (local bond orientational order) in amorphous solids cannot be uniquely determined from these experiments because of the random orientation of the local structural units. Scattering measurements provide statistical descriptions of the atomic arrangements, and thus must be supplemented by three-dimensional structural models and by experiments sensitive to local symmetry. A variety of magnetic measurements are effective symmetry-sensitive probes. Here we focus on two of these, magnetic relaxation and magnetic studies of structural transformations, and consider what they tell us about local atomic order in metallic glasses.

## LOCAL ATOMIC ORDER

Egami et al [1] have developed a model of local structural defects in metallic glasses based upon two unique ways in which atomic configurations can fluctuate or deviate from the average disorder characteristic of the glassy state. One fluctuation defines a local shear defect and the other a local density fluctuation. The former defects have a directional property and can be associated with localized strain dipoles.

Fig. 1 shows shematic 2-D representations of possible atomic configurations involved in a) the reorientation of a local shear stress by metalloid hopping between two local minima (double well potential) and b) short-range translation of a positive density fluctuation about a metalloid atom as it hops between two local minima. These processes involve changes in local atom configurations that a) have a directional shear character and can therefore couple directly to an applied uniaxial stress or couple indirectly through magnetostriction to changes in magnetization orientation or b) have no directional character and move only to lower their total energy.

Presumably, local shear reorientation can occur without the creation or annihilation of free volume. These stress dipoles can be oriented by the application of a directed mechanical or magnetic force and revert to their original orientation upon its removal or with a suitable change in its direction. Thus such changes in local structural anisotropy are essentially reversible, low-activation barrier processes (Fig. 1a). On the other hand, translation or mutual annihilation of compressive and dilational local

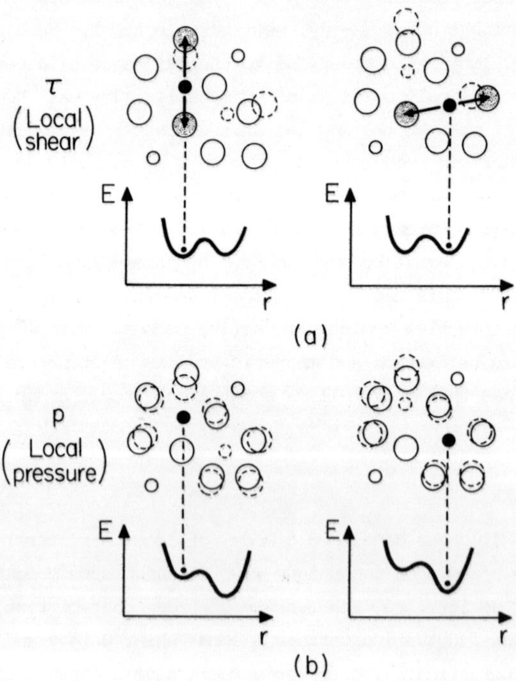


Fig. 1 Schematic 2-representations of atomic configurations about defects in amorphous alloys. a) strain dipole and its reorientation, b) density fluctuation and its diffusion.

defects occurs not under the stimulus of an applied force but in response to an internal stress or composition gradient. Thus such changes alter the free volume, do work, and are irreversible (Fig. 1b).

Kronmuller [2] has shown that reorientational processes (a) exhibit a different characteristic time dependence $\Delta\mu/\mu \propto \exp[-t/\tau]$ compared to diffusional or hopping processes (b) which go as $\Delta\mu/\mu \propto t^{-3/2}$ $(t > \tau_D)$. Here $\mu(t)$ is the time-dependent property change (e.g. magnetic permeability) accompanying the relaxation process and $\tau_D$ is the Debye relaxation time.

Allia et al [3] have confirmed the predominance of local shear stress reorientation (rather than diffusion of density fluctuations) in magnetic relaxation effects. They derived and verified the coupling of these defects to the direction of magnetization via the magnetostriction:

$$\Delta\mu/\mu = A(T)\ \lambda_s^{\ 2}\ \langle\tau^2\rangle/M_s$$

It is very important to move beyond modelling the local structure in glasses with hypothetical atomic configurations and work toward obtaining experimental determinations of specific local orderings.

## Well-defined defects

Collins et al [4] have studied magnetic after effects in magnetostrictive and in non-magnetostrictive glasses. They have resolved three separate processes at 150°, 230° and 290°C in the isochronal relaxation spectrum of amorphous $Fe_{76}B_{16}Si_8$ (Fig. 2a). Moreover, the time dependence of the peaks allows identification of those at 150° and 230°C with reorientation of shear stress dipoles while that at 290°C is largely diffusional in nature. In light of Allia's work, we then assume the reorientational processes to depend on the direction of magnetization through the magnetostriction (which is $\lambda_s = 32 \times 10^{-6}$ for this Fe-base glass).

This identification of the different origins for the various peaks is supported by the fact that the reorientational peaks are <u>not</u> observed in cobalt-base glasses for which the magnetostriction is essentially zero ($\lambda_s = -2 \times 10^{-6}$), (see Fig. 2b). That is, the directional strain-defects active in these two low-temperature processes can only couple to the magnetic field excitation, or, conversely, can only have their reorientation detected magnetically, when the magneto-elastic coupling is non-zero as it is in the Fe-base alloy.

It is important to note that these data identify relaxation processes due to different, well-defined <u>defects</u> in amorphous alloys. Earlier results [5] have also indicated a finite number of distinguishable processes involved in relaxation of metallic glasses.

These data are cited as an example of how indirect probes of local atom configurations can reveal the existence of various types of structural defects and perhaps various types of local atomic orderings in metallic glasses. In this sense they add to what we know from scattering studies. However, relaxation measurements have not yet been able to identify exact atomic configurations and are able only to classify the defect processes into fairly broad categories. Magnetic relaxation studies point to the existence of fairly well-defined, (as opposed to random) defect configurations in glasses. We now address the issue of whether such discrete defects reflect well-defined local atomic orderings as well as just what these orderings might be.

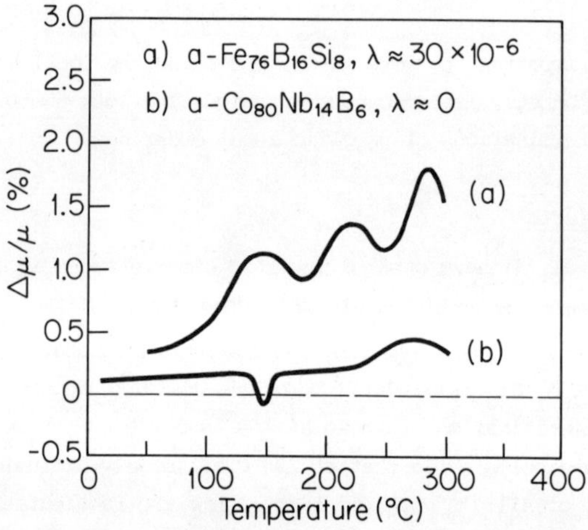


Fig. 2 Magnetic permeability after effects, isochronal spectra for amorphous alloys. a) magnetostriction $Fe_{76}B_{16}Si_8$ amorphous alloy relaxation spectrum showing three discrete processes. b) non-magnetostrictive $Co_{80}Nb_{14}B_6$ amorphous alloy spectrum showing suppression of all processes, particularly the two low-temperature processes.

## Discrete local configurations

Certain arguments and considerable data can be cited in support of local atomic order in metallic glasses that resembles either nearly-crystalline short-range order (SRO) (The so-called quasi-crystalline model) [6] on the one hand or non-crystalline SRO (e.g. icosahedral) on the other. For example, the now-classic NMR experiments of Panissod et al [7] indicate that in metallic glasses of composition close to that of a stable crystalline compound, the metalloid environment shows the symmetry of the crystalline state (cubic, orthorhomic and tetragonal cases are demonstrated). On the other hand, the earliest dense-random-packing-of-hard-spheres (DRPHS) models point to the stability of non-crystalline tetrahedral and icosahedral structures based on local economy of packing, [9] while more recent molecular dynamics studies argue convincingly for the stability of local five-fold bond orientational order. [10]

It has long been appreciated that the short-range atomic order in liquids can reflect that of the compounds stable at lower temperatures [11], and, moreover, that above a eutectic point a tense coexistence - rather than a compromise - exists between the local orders favored in the hypo- and hyper-eutectic regions. [12] Johnson et al have demonstrated the coexistence of competing SRO's in metallic glasses with direct structural observations. [12] In other cases he has shown the possibility of crossing boundaries between two distinct amorphous phases (call them $\alpha_1$ and $\alpha_2$) by varying composition in a series of glassy alloys. [13] Lashmore et al [14] have given evidence of polymorphism (coexistence of two different structures at a given composition) in amorphous NiP alloys made by different processes.

It is against this background that we have studied the possibility [15] of crossing such $\alpha_1$ - $\alpha_2$ phase boundaries in a single glassy alloy by varying temperature (Fig. 3). Indeed, some indirect experimental evidence exists to support this possibility. [16-18] The question that is more difficult to address is the nature of the local orders in the two amorphous phases $\alpha_1$ and $\alpha_2$. Two possibilities are considered for the amorphous short-range orders: quasi-crystalline (e.g. nearly hcp or fcc local order) and non-crystalline (e.g. icosahedral SRO). [19]

Below we review some of the evidence for reversible structural transformations between two well-defined local configurations in glasses. We speculate about possible local glassy 'phases'. Given recent discoveries related to icosahedral phases, [20] and in particular their accessibility from some amorphous compositions, [21,22] our results take on new significance and, I believe, offer new insight.

## $\alpha_1$ - $\alpha_2$ TRANSFORMATIONS

We first cite some data that imply the possibility of transforming reversibly from one fairly-well-defined local atomic order to another in certain metallic glasses.

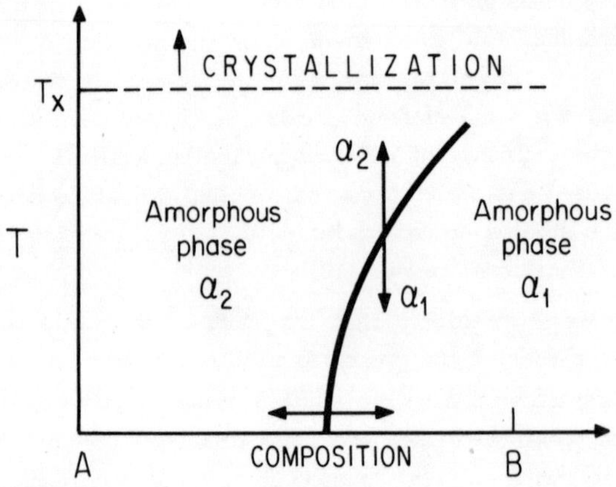


Fig. 3 Hypothetical phase diagram for amorphous alloys of composition $A_{1-x}B_x$. If two different local amorphous orders exist in different composition ranges, it is possible that one of these e.g. $\alpha_1$, may transform reversibly to amorphous phase $\alpha_2$ upon heating below crystallization.

Second-order evidence for first-order transformation

This section derives its heading from the fact that magnetic measurements give an indirect view of local order. The connection between local order and observables in the case of our measurements of approach to magnetic saturation [16] and magnetostriction [17] is not as well defined [23] as in other local magnetic probes such as NMR [8] or Mossbauer spectroscopy although its basis is the same. We believe the simpler methods we have employed are justified because of their greater sensitivity to the subtle changes believed to be involved in $\alpha_1$ - $\alpha_2$ transformations.

The approach to magnetic satururation (Fig. 4a) in a polycrystalline or amorphous material is a sensitive, albeit indirect, measure of the distribution in orientation and in strength of the local magnetic anisotropy. For an applied field $H_a$ near the 'knee' of the magnetization curve a decrease in $M(H_a)$ implies an increase in strength of the local anisotropy. [17] We have observed, in several Co-Nb-B glasses, a thermo-magnetic hysteresis in $M(H_a,T)$ indicating reversible (to an extent) changes in local anisotropy strength (Fig. 4b). [16,17] The low $M(H_a,T)$ phase at low temperature is assumed therefore to have a higher local magnetic anisotropy and hence a lower symmetry than the high $M(H_a)$ phase. The saturation magnetization is observed to be invariant through the transformation which is centered at $T_o$.

Thermodynamically, $T_o$ = 100$^o$C in amorphous $Co_{80}Nb_{14}B_6$ while the glass

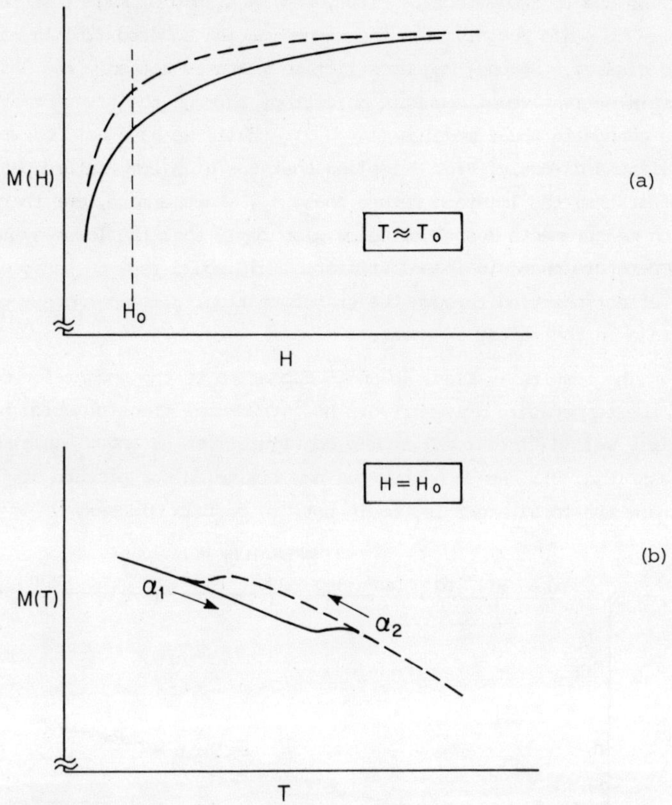


Fig. 4 a) Schematic approach to magnetic saturation M(H) for two slightly different local symmetries. b) Magnetization, below saturation, vs. temperature schematically depicting observation of low temperature amorphous transformation observed in several Co-base glasses.

crystallizes only above 400°C. The kinetics of the transformation are such that the full width $\Delta T$ of the thermomagnetic hysteresis in this composition is of order 100°C when the measurements are taken at 1°/min. Faster (slower) heating rates increase (decrease) $\Delta T$.

Similar thermomagnetic hysteresis has been observed by careful measurement of magnetostriction in amorphous $Co_{82}Nb_{10}B_8$ (Fig. 5). [24,25] It is important to compare the data in this figure with that in Fig. 4. The higher $T_o$ temperature observed for amorphous $Co_{82}Nb_{10}B_8$ is due to the higher ratio of cobalt/niobium. The narrower hysteresis associated with the transformation in Fig. 5 is due to the slower rate of heating necessitated by the magnetostriction measurement. The larger fractional change in $\lambda$ through the transformation ($\Delta\lambda/\lambda$ = 8%) compared to that for magnetization ($\Delta M/M$ = 2-3%)

may be due to two effects. First, the magnitude of magnetostriction is known to scale roughly with the magnetization squared [26] in many metallic glasses. Second, magnetostriction measures not only the change in the strain derivative of magnetic anisotropy through the transformation but also the change in shear modulus ($\lambda \propto G^{-1}$). While we have not measured shear stiffness directly, Fig. 5 implies that the high temperature phase $\alpha_2$ is softer than the low-temperature phase $\alpha_1$. Furthermore, our thermal expansion measurements described below also imply that the lower-symmetry low-temperature phase is indeed stiffer. This would tend to enhance the magnetic effect observed through the transformation, allowing more magnetoelastic strain in the softer $\alpha_2$ phase.

The results in Figs. 4b and 5 appear to be the sought-for manifestation of the temperature driven reversible structural transformation between two fairly well defined local atomic configurations $\alpha_1$ and $\alpha_2$ in a metallic glass (Fig. 3). As such they put new limits on our picture of just how random the local order is, or is not, in certain glasses.

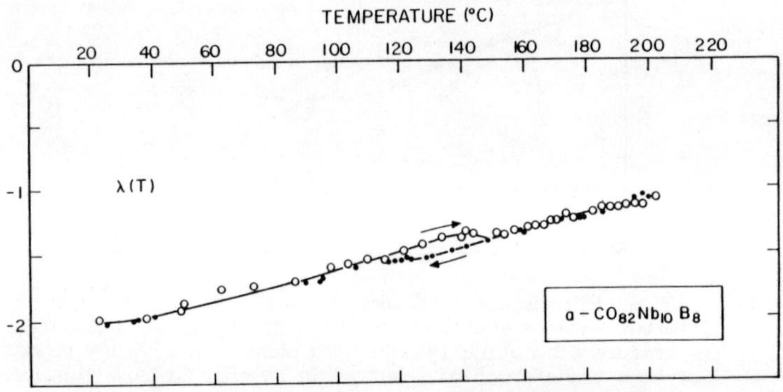


Fig. 5 Transformation Effect measured in magnetostriction well below crystallization analagous to schematic shown in Fig. 3b.

Another perspective

The magnetic measurements reviewed above suggest a change in the symmetry of the environment seen by the local magnetic moment. The macroscopic manifestation of this symmetry change has been averaged over the random orientation of the local order. There also appears to be evidence of a change in structure through this transformation that is independent of the anisotropy imposed by an applied magnetic field.

We have measured thermal expansion, creep and creep recovery [18] in certain Co-Nb-B glasses known to manifest local structural transformations.

If we call the initial, low-temperature, amorphous phase $\alpha_1$ and the final, high-temperature ($T > 200^oC$), phase $\alpha_2$, we can make several statements about the differences in their local order based on magnetic and mechanical observations of the transformation. As stated previously, the magnetic observation of the transformation implies that amorphous phase $\alpha_1$ is characterized by a local order of lower symmetry than phase $\alpha_2$. Thermal expansion measurements show $\alpha_1$ to have a higher temperature coefficient of expansion and hence probably less free volume than $\alpha_2$. Again by inference, we take this to indicate that $\alpha_1$ is stiffer than $\alpha_2$ (shear modulus $G_1 > G_2$). As stated earlier, this is consistent with the magnetostrictive observation of the transformation.

Creep, and creep recovery after an initial period of stress pretreatment, can be used to infer further refinements in our picture of amorphous phases $\alpha_1$ and $\alpha_2$. It is noted here that the observations of creep recovery in metallic glasses subjected to an initial stress treatment represent the mechanical analog of the magnetic after effects observed in structural relaxation of strain dipoles which have been initially oriented by an applied magnetic field.

These observations are summarized in Fig. 6 which shows the temperature dependence of the free energy for the two local configurations $\alpha_1$ and $\alpha_2$. This figure also includes in tabular form a list of characteristics distinguishing amorphous phases $\alpha_1$ and $\alpha_2$.

## Electronic basis for $\alpha_1$ - $\alpha_2$ transformation

In an attempt to consider the role of electronic structure on local atomic order, we have calculated the molecular orbital eigenvalues and wave functions for a variety of clusters. First, to model the putative $\alpha_1$ - $\alpha_2$ transformation, we have focussed on the highest occupied molecular orbital (HOMO) for a cubo-octahedron (basis for fcc structure) of cobalt atoms. [19] This result provides a firm basis for comparison with our more recent calculations on icosahedral cobalt clusters. [24,27] The main results of such a comparison are illustrated in Fig. 7. The calculations show that the slight shear of atomic coordinates in going from close packed crystalline (CN = 12), high symmetry, fcc local order to more densely packed, (CN = 12) non-crystalline higher symmetry icosahedral, local order has little effect on the density of states. Furthermore, the calculated magnetic moments, for the two clusters are essentially the same: $\mu_{CO} = 1.56$ (fcc) and $\mu_{CO} = 1.61$ (icos). This is consistent with the fact that both phases $\alpha_1$ and $\alpha_2$ appear from our measurements to saturate at the same value of magnetization. However, the slight increase in symmetry on going to an icosahedral phase may be the key to the easier approach to saturation inferred for the higher temperature phase $\alpha_2$ (see Fig. 4). No overall stabilization of either phase is evident from these calculations. Thus an easy $\alpha_1 - \alpha_2$ transformation having no effect on saturation moment but showing easier magnetization in the higher symmetry $\alpha_2$ phase appears possible.

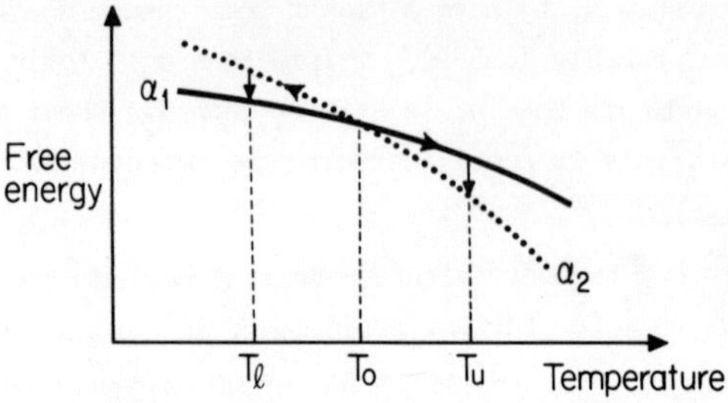


| PROPERTY | PHASE $\alpha_1$ | PHASE $\alpha_2$ | BASIS |
|---|---|---|---|
| Magnetic Anisotropy | Higher | Lower | M(H) |
| Local Symmetry | Lower | Higher | Magnetic Anisotropy |
| Free Volume | Lower | Higher | Thermal Expansion |
| Configurational Entropy | Lower | Higher | Thermal Expansion |
| Shear Stiffness | Higher | Lower | Thermal Expansion and Magnetostriction |

Fig. 6 Free energy G vs. temperature for two phases postulated for amorphous $\alpha_1 - \alpha_2$ transformation. $T_o = 100^oC$ is the central temperature of the transformation and its width, kinetically determined, $\Delta T$ is of order $100^oC$ for heating rates of order $1^o$ $min^{-1}$. Also tabulated are some characteristics of the two glassy phases inferred from various measurements.

## QUASI-CRYSTALS

We have reviewed some of the evidence for the existence of a plurality of amorphous phases with well-defined, distinct short-range orders as well as considered the evidence for temperature-driven, reversible transformations among them. We turn again to the question of the detailed nature of the local order.

### Materials classification

Metastable, non-crystalline materials may be classified in perspective with other materials according to their short-range and long-range order as in Table I.

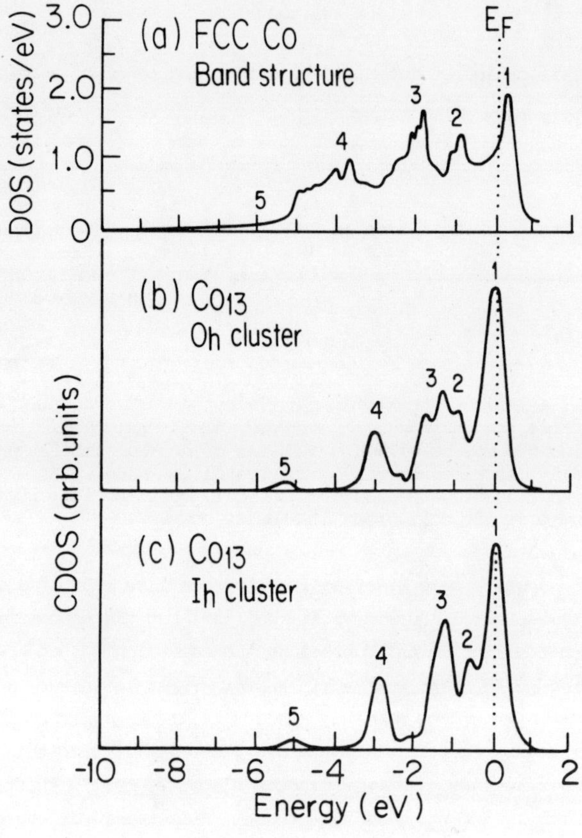


Fig. 7 Comparison of valence band electronic structure for a) fcc cobalt with calculations for 13-atom clusters of b) $O_h$ and c) $I_h$ symmetry.

If the SRO is crystalline and persists over distances greater than about one micrometer, we have a macrocrystal, either polycrystalline or single crystal. If the SRO is crystalline and persists only below this range before another crystallite (having a different structure and/or orientation) is reached, the crystalline material is often called microcrystalline. (Some carry this categorization one step finer and define nanocrystalline materials as those whose short-range crystalline order persists up to only 50 Å or so). But basically these materials are all crystalline. If the short-range order only approximates that of some crystalline phase but is frustrated beyond about ten Angstroms, i.e. there is no long-range order, then we have an amorphous alloy (type I, Table I). In all of the cases described so far, the local order may resemble that of an equilibrium phase, or given non-equilibrium processing, it may approach that of a metastable phase.

Table I

Classification of materials by nature and range of atomic order

| Short-range order | Range of SRO | Long-range order | Range of LRO | Material Classification |
|---|---|---|---|---|
| | | | | Crystalline materials |
| crystalline[a] | > 1.0μm | Same as SRO | > 1.0μm | Macrocrystal |
| crystalline[a] | < 1.0μ | Same as SRO | < 1.0μm | Microcrystal |
| | | | | Non-crystalline materials |
| nearly cryst[a,b] | = 10 Å | No long-range order | | Amorphous I |
| non-cryst. | = 10 Å | No long-range order | | Amorphous II |
| non-cryst. | ? | Quasi periodic[c] | = 0.1 μm | Quasi-crystal |

a) Crystalline SRO may be either equilbrium or non-equilibrium phase.
b) Historically described by the quasicrystalline model.[28]
c) There is no long-range periodicity but only quasi-periodicity. Long-range bond-orientational order does exist however.

When the local atomic ordering is very much like that of a crystalline phase (sometimes called 'quasicrystalline')[28] but the processing is sufficiently non-equilibrium as to preclude the establishment of crystalline order much beyond 10 Å (Amorphous I), the electronic structure and fundamental physical properties are very similar to those of the corresponding crystalline state. Many metallic glasses probably fit into this category.[8] However, the properties of some amorphous alloys suggest a short-range order unlike that of any known crystalline phase. Such materials are categorized in Table I as being of type II. Because of the difficulty of direct structural determinations on a scale of 10 Å, it is often unclear whether the short-range order of a metallic glass is nearly-crystalline (Amorphous I) or non-crystalline (Amorphous II).

Finally, it has been observed recently that a class of materials exists which have non-crystalline SRO (perhaps like glass II), however this order - at least its orientation - is coherent over several hundred Angstroms. The translational ordering is only quasi-periodic (the result of a superposition of two incommensurate lattices).[29] The atomic order in these 'quasi-crystals' has been reported to be icosahedral (5-fold symmetry)[20,30] or dodecahedral (10-fold symmetry[31] Quasicrystals appear to be an intermediate phase between glasses and microcrystalline materials both in terms of the range of their local order as well as in their inferred free energy. Some amorphous phases transform on heating to the icosahedral phase before crystallizing.[21,22]

We may, therefore, further distinguish amorphous alloys I and II by the hypothetical free energy curves represented in Fig. 8. Type I devitrifies directly to the crystalline state. This presumably reflects its predisposition for crystalline long-range order due to its nearly crystalline initial local configuration. Type II, on the other hand, devitrifies to the quasi-crystal (e.g. icosahedral) state initially. This reflects a growth of its non-crystalline orientational correlations out to several hundred Angstroms. Upon further heating this metastable - and, in 3D, inherently frustrated - structure, transforms to the crystalline state. Comparing Figs. 6 and 8 we then identify amorphous Type I with amorphous phase $\alpha_1$ and amorphous Type II with $\alpha_2$.

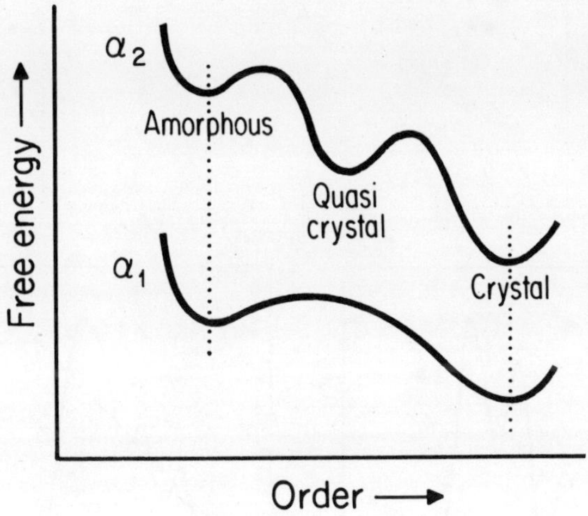


Fig. 8 Hypothetical free energy curves versus order parameter for two types of amorphous alloys $\alpha_1$ (Glass I) and $\alpha_2$ (Glass II). Allowed transformation paths are shown by arrows.

## Electronic structure of quasi-crystals

Above we described electronic structure calculations on cobalt clusters designed to model the hypothetical $\alpha_1$ - $\alpha_2$ amorphous phase transformation. Similar comparative calculations for aluminum clusters in fcc and icosahedral symmetry serve as an electronic basis for models of quasicrystalline phases. They reveal several interesting features, two in particular are relevant here. Crystalline fcc aluminum is a free electron metal with a density of states that increases from the bottom of the valence band to $E_F$ as $D(E) \propto E^{1/2}$ (this is the case both experimentally[32] and in band structure[33] or cluster calculations[34] as shown in Fig. 9a). However, in the icosahedral clusters[35] the overall energy of the manifold of states decreases significantly (compared to fcc) and the density of states increases faster than expected for a 3-D free-electron metal, exhibiting a peak near $E_F$: $D(E) \propto E^n$ with n > 1 and $D(E_F)_{I_h} = C \times D(E_F)_{O_h}$ (where $C \cong 1.5$ if n = 1 and $C \cong 2$ if n = 3/2) (see Fig. 9). Thus the icosahedral cluster (local order) is a decidedly more stable ground state than the crystalline case and icosahedral local symmetry is even more strongly favored at high temperatures by an anomalously strong electronic entropy due to the large value of $D(E_F)$.

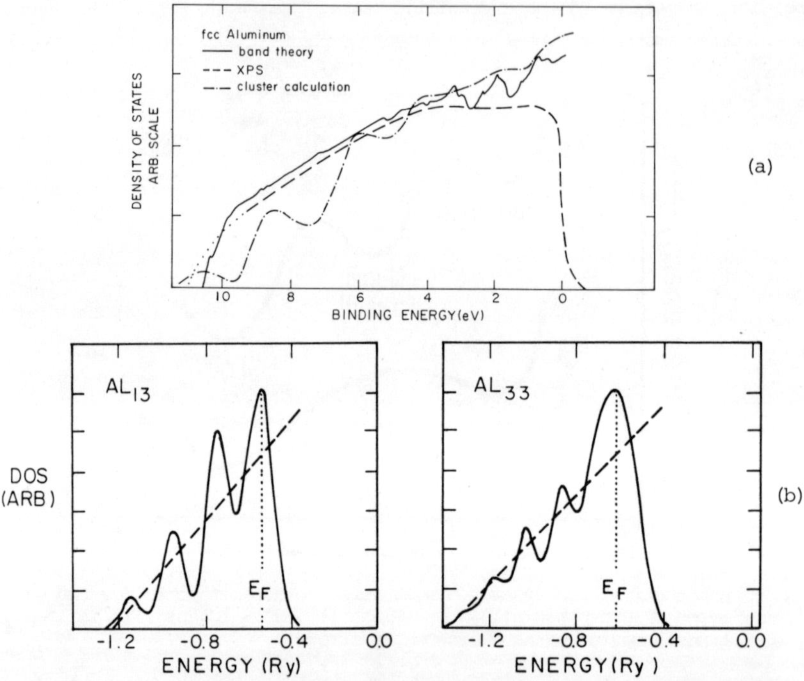


Fig. 9 Electronic structure of close packed aluminum. a) Experimental (dashed), band structure (solid) and cluster result (dot-dashed) for fcc Al. b) theoretical results for 13- and 33-atom icosahedral Al clusters showing strong stabilization of valence band manifold and sharply increased density of states at $E_F$ for icosahedral Al compared to crystalline Al.

Substitution of Mn for the central aluminum atom in the icosahedral clusters increases the total energy and introduces strong, destabilizing antibonding states at $E_F$. This result argues for a non-central, Mn site in icosahedral Al-Mn and Al-Mn-Si quasicrystals. This important result indicates an electronic motivation for preferring atomic models of quasicrystals based on Mackay icosahedra[36] or a Al-Mn-Si structure[37] with Mn atoms outside the Al-icosahedron rather than for models based on Mn-centered Al icosahedra.[20,31]

Summary

We have reviewed some of the evidence for discrete types of defects and defect relaxation processes in amorphous alloys. We have also reviewed evidence that the local structure, rather than being random, can be characterized by well-defined local atom configurations, either nearly-crystalline, $\alpha_1$, or patently non-crystalline, $\alpha_2$ (e.g. icosahedral) in nature. Furthermore, considerable evidence exists suggesting that some metallic glasses can be driven thermally between two different, amorphous local orders.

The recent discoveries related to icosahedral phases has helped define the nature and limits of SRO in amorphous alloys. It seems likely that those metallic glasses which exhibit local non-crystalline atomic arrangements ($\alpha_2$) can be transformed to the quasi-crystalline state prior to crystallization. However those metallic glasses showing evidence of nearly-crystalline local order ($\alpha_1$), probably transform directly to the crystalline state.

ACKNOWLEDEMENT

Our relaxation studies are supported by a grant from 3M Company and different aspects of the work on amorphous transformations are supported by ONR (Contract No. N00014-83-K-0364) and NSF (Grant No. DMR 8318829).

References

1. T. Egami and D. Srolovitz, J. Phys. F: Metal Phys. 12, 2141; T. Egami, Rep. Prog. Phys. 47, 1601 (1984).

2. H. Kronmuller, "Nachwirkung in Ferromagnetika", Springer Tracts in Natural Phil., Vol. 12 (1968); N. Moser and H. Kronmuller, J. Magn. and Magn. Mater. 19, 275 (1980).

3. P. Allia and F. Vinai, Phys. Rev. B 26, 6141 (1982).

4. A. Collins, R.C. O'Handley, N.J. Grant, J. Mag. Magn. Mater. (1986).

5. J. Cost and J.T. Stanley, J. Non-Cryst. Sol. 61-62, 799 (1984); A. Hernando and Riveiro, Phys. Rev. B 32, 5102 (1985).

6. P.H. Gaskell, J. Phys. C: Sol. St. Phys. 12, 4337 (1979).

7. C. Bryant and J.J. Burton, phys. stat. sol. B 85, 393 (1978).

8. P. Panissod, D. Alliaga Guerra, A. Ammamou, J. Durand, W.L. Johnson, Phys. Rev. Lett. 44, 1465 (1980).

9. J.D. Bernal, Proc. Roy. Inst. 37, 355 (1959); G.S. Cargill in Solid State Phys.

10. P.J. Steinhardt, D.R. Nelson, M. Ronchetti, Phys. Rev. B 28, 784 (1983).

11. R. Kaplow, S.L. Strong and B.L. Averbach, in Local Atomic Arrangements Studied by X-ray Diffraction, edited by J.B. Cohen and J.E. Hilliard, Metallurgical Society Conference B6, (Gordon and Breach, New York, 1965) p. 159; J.J. Gilman, Phil. Mag. B 37, 577 (1978).

12. C.O. Kim and W.L. Johnson, Phys. Rev. B 23, 143 (1981); A. Mak, K. Samwer, W.L. Johnson, Phys. Lett. 98A, 353 (1983).

13. W.L. Johnson and S.T. Hopkins, Solid State Comm. 43, 537 (1982).

14. D.S. Lashmore, L.H. Bennett, H.E. Schone, P. Gustafson, R.E. Watson, Phys. Rev. Lett. 48, 1760 (1982).

15. R.C. O'Handley and N.J. Grant, Physica 119B, 173 (1983).

16. B.W. Corb, R.C. O'Handley, J. Megusar, N.J. Grant, Phys. Rev. Lett. 51, 1386 (1983); R.C. O'Handley, B.W. Corb, J. Megusar, N.J. Grant, J. Non-Cryst. Sol. 61-61, 773 (1984).

17. R.C. O'Handley, B.W. Corb, N.J. Grant, J. Appl. Phys. 55, 1808 (1984).

18. E. Girt, A. Mitra Ghemawat, R.C. O'Handley N.J. Grant, Proceedings of Third International Conference on Structure of Non-Crystalline Material, Grenoble, July 1985.

19. M.E. Eberhart, R.C. O'Handley, K.H. Johnson, Phys. Rev. B 29, 1097 (1984).

20. D. Shechtman, I. Blecht, D. Gratias and J.W. Cahn, Phys. Rev. Lett. 53, 1951 (1984).

21. D.A. Lilienfeld, M. Nastassi, H.H. Johnson, D.G. Ast, J.W. Mayer, Phys. Rev. Lett. 55, 1587 (1985).

22. S.J. Poon, A.J. Drehman and K.R. Lawless, Phys. Rev. Lett. 55, 2324 (1985).

23. R.C. O'Handley and N.J. Grant in Rapidly Quenched Metals, edited by H.H. Steeb and H. Warlimont, (Elsevier Science Publishers, Amsterdam, 1985) p. 1125.

24. R.C. O'Handley, M.E. McHenry, M.E. Eberhart, K.H. Johnson, N.J. Grant, in Amorphous Materials, edited by R. Jaffee ACTA MET, 1985.

25. R.C. O'Handley and K. Dickson, (unpublished data).

26. R.C. O'Handley, Solid State Comm. 21, 1119 (1977).

27. M.E. McHenry, R.C. O'Handley, K.H. Johnson, N.J. Grant, J. Mag. and Magn. Mater. (1986).

28. It is unfortunate that the recently discovered phases which have non-crystalline local orders (five-fold or ten-fold rotational symmetry which persists over hundreds of Angstroms), have been dubbed 'quasi-crystals'. The related term 'quasicrystalline' has long been used to describe the nearly crystalline local order inferred to exist in many amorphous alloys. Its use goes back to Gaskell.[6] In order to avoid confusion, we will refer to the first phase as quasicrystalline glasses or as described by the quasicrystalline model. The latter, metastable, non-crystalline, non-glassy phases, will be called 'quasi-crystals' or icosahedral phases.

29. D. Levine and P. Steinhardt, Phys. Rev. Lett. 53, 2477 (1984).

30. P.A. Bancel, P.A. Heiney, P.W. Stephens, A.I. Goldman, P.M. Horn, Phys. Rev. Lett. 54, 2422 (1985).

31. L. Bendersky, Phys. Rev. Lett. 55, 1461 (1985).

32. P. Steiner, H. Holscht, W. Steffen, S. Hufner, Z. Phys. B 38, 191 (1980).

33. V.L. Moruzzi, J.F. Janak, A.R. Williams, in Calculated Electronic Properties of Metals, (Pergamon Press, New York, 1978) p. 52.

34. D.R. Salahub and R.P. Messmer, Phys. Rev. B 16, 2526 (1977).

35. M.E. McHenry, M.E. Eberhart, R.C. O'Handley, K.H. Johnson, Phys. Rev. Lett. 56, 81, 1986.

36. A.L. Mackay, Physica 114 A, 609 (1982).

37. P. Guyot and M. Audier, Phil. Mag. B 52, L15 (1985).

# MAGNETIC DOMAIN STRUCTURES IN CERIUM-DOPED $Fe_{80}B_{20}$, $Fe_{80}B_{16}Si_2C_2$, AND $Fe_{78}B_{13}Si_9$ GLASSY RIBBONS*

G. S. CANRIGHT** AND D. M. KROEGER***
**University of Tennessee, Department of Physics, Knoxville, TN 37996
***Metals and Ceramics Division, Oak Ridge National Laboratory, Oak Ridge, TN 37831

## ABSTRACT

The addition of cerium in parts-per-million quantities has been shown to be effective in inhibiting the annealing embrittlement of melt-spun amorphous ribbons [1]. The effects of cerium doping on magnetic domain structures are reported in this paper. Domain structures have been observed using the Bitter colloid technique on the free surfaces of as-quenched ribbons, as well as on both surfaces of ribbons which have been thinned by electropolishing. In the binary and quaternary alloys, the optimum (in terms of mechanical properties) cerium content gives ribbons which are apparently free of quenched-in stresses, since they lack the commonly observed "maze domains" which are found for both undoped and excessively doped ribbons. In contrast, the ternary alloy, which shows little decrease in annealing embrittlement versus doping, is free of maze domains in the as-quenched condition, with or without doping. The effect on the former two alloys is interpreted in terms of enhanced stress-relief during the quench.

## INTRODUCTION

Iron-based metallic glasses exhibit a combination of high magnetization and low ac losses which makes them attractive candidate materials for power-handling magnetic cores. In previous work [1,2] we have shown that the tendency of these materials to embrittle upon annealing can, in some cases, be eliminated by doping with an appropriate amount (~75 atomic ppm) of cerium. Some effects of doping with cerium on the magnetic properties and magnetic domain structure were also seen [2,3]. In this paper we report on further studies of the magnetic domain structures versus doping for melt-spun ribbons of three alloys: $Fe_{80}B_{20}$ (binary), $Fe_{78}B_{13}Si_9$ (ternary), and $Fe_{80}B_{16}Si_2C_2$ (quaternary).

## EXPERIMENTAL

Ribbons were made by melt-spinning in helium gas; details of alloy preparation and melt-spinning are described elsewhere [2]. Cerium concentrations in the ribbons were found to be ~40% of those in the ingots, as determined by energy dispersive X-ray fluorescence analysis [2]. Concentrations reported in this paper are nominal values, i.e., ingot concentrations. Magnetic domain structures were observed using the Bitter technique [4] on samples ~2 cm long. Electropolishing was done using 1.5 cm samples. The electrolyte was one-third nitric acid and two-thirds methanol; current density was ~70 mA/cm$^2$. Domain structures were observed for at least four thinned samples, and numerous unthinned samples, of each alloy.

---

*Research sponsored by the Office of Energy Storage and Distribution, Electric Energy Systems Program, U.S. Department of Energy, under contract DE-AC05-84OR21400 with Martin Marietta Energy Systems, Inc.

As-quenched domain patterns for various cerium dopings are shown for the quaternary alloy in Fig. 1; the ribbon axis is horizontal in this and all other figures. We will use the term "optimally doped" to mean that doping which was most effective (~75 atomic ppm) at preventing annealing embrittlement (i.e., until crystallization); "overdoped" then implies a significantly higher level of doping (~300 atomic ppm) [2]. The patterns shown are typical. The optimally doped quaternary samples were generally free of maze domains [5], while both undoped and overdoped samples showed a relatively high density of maze domains. The longitudinal structure of Fig. 1(b) was commonly seen, but transverse domains were also seen, as were more irregular planar domains.

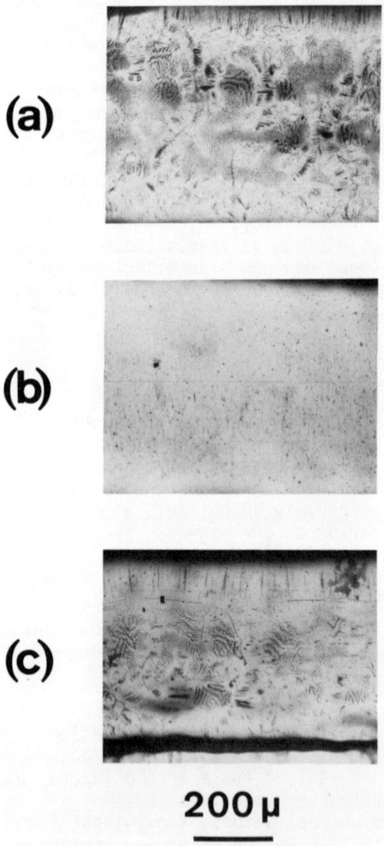


Fig. 1. Domain structures of as-quenched $Fe_{80}B_{16}Si_2C_2$ alloy. (a) undoped; (b) 75 atomic ppm Ce; (c) 260 atomic ppm Ce.

A similar variation of maze domains with doping may be seen in Fig. 2. Relative to the quaternary samples, the binary samples showed fewer maze domains in both undoped and overdoped samples, with the maze domains tending to appear near the ribbon edges. In some optimally doped $Fe_{80}B_{20}$ samples, as shown in Fig. 2(b), the domain walls were broad and not easily seen by the Bitter technique, indicating that the underlying anisotropy is small. For optimally doped binary samples, as with the quaternary, domains were almost invariably planar, with no apparent preference for direction in the plane.

In the ternary samples, domain structures showed no clear variation with doping, up to 100 atomic ppm cerium. Figure 3 shows typical undoped and 75 ppm cerium samples, containing only planar domains. Maze domains were infrequently seen for all doping levels, including zero cerium. It is of interest that the $Fe_{78}B_{13}Si_9$ alloy also showed no change in tendency to embrittle versus doping [2].

Thinned samples of quaternary alloy were examined to determine if the observed effect was confined to the surface. Figure 4 shows samples thinned from the free surface of the ribbon. [The lower, very straight line in Fig. 4(b) is due to an impression left by a profilometer.] Clearly the patterns seen in Fig. 1 characterize the bulk of the material, as found previously for undoped ribbons of $Fe_{80}B_{20}$ [6], and $Fe_{40}Ni_{40}P_{14}B_6$ [7].

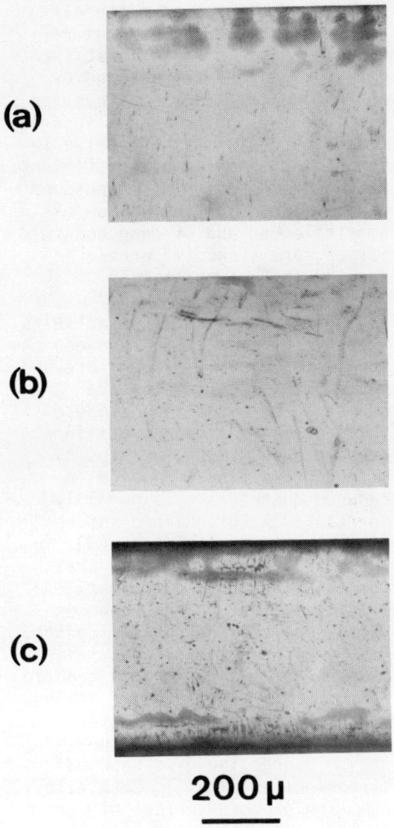


Fig. 2. Domain structures of as-quenched $Fe_{80}B_{20}$ plus (a) Zero Ce; (b) 50 atomic ppm Ce; (c) 300 atomic ppm Ce.

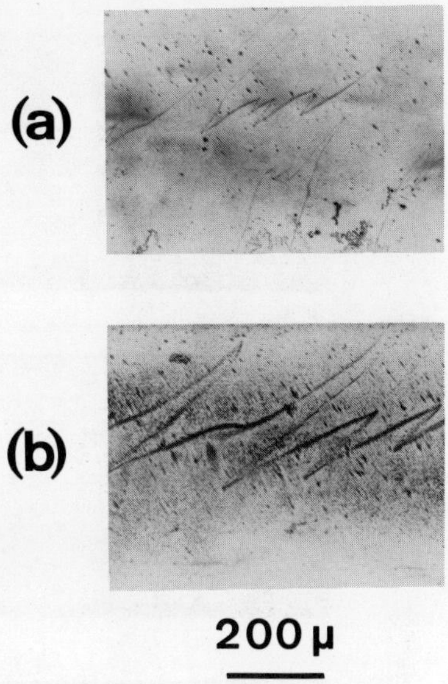


Fig. 3. Domain structures of $Fe_{78}B_{13}Si_9$ plus (a) Zero Ce; (b) 75 atomic ppm Ce.

Maze domains were seen in overdoped material to the maximum depth examined (6 µm).

Domain observations were also obtained on samples of quaternary alloy, thinned from the wheel side; this side is normally too rough to permit observations on unthinned samples. Undoped material showed maze domains to a depth of at least 4 µm, while optimally doped samples showed longitudinal or transverse planar domains to 5 µm. Observations of overdoped samples were not conclusive: some showed evidence of maze domains, but in most cases it proved impossible to resolve any domain structure.

DISCUSSION

The presence of maze domains is normally due to residual stresses in the as-quenched ribbon [5,6]. Maze domains in overdoped ribbons are not confined to the surface; and observations of annealed, overdoped samples show that the maze domains are eliminated by annealing. Hence, in these samples, there is no reason to suppose a different origin for the observed maze domains.

(a)

(b)

(c)

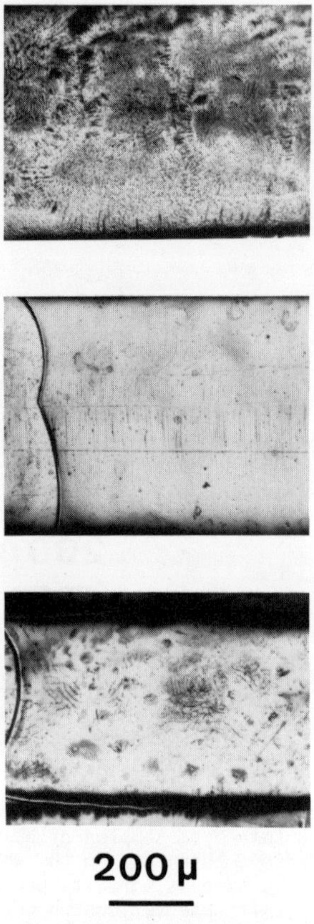


Fig. 4. Domain structures of $Fe_{80}B_{16}Si_2C_2$, thinned from the free surface. (a) undoped, –5 μm; (b) 75 ppm Ce, –4 μm; (c) 260 ppm Ce, –3 μm.

Similarly, optimally doped samples of binary and quaternary alloy appear to be devoid of residual stresses, or at least such stresses are greatly reduced by optimal doping. These alloys are also the two, of the three studied, for which doping was effective in inhibiting annealing embrittlement. Hence the two observed effects of doping, i.e., the inhibition of embrittlement and of quenched-in stress, are strongly correlated for the three compositions investigated.

Quenched-in stresses arise as a result of competition between the stress-relief (diffusion) rate and the cooling rate, during the quench. One might then suppose that optimal doping results in a reduced quench rate, perhaps by affecting wetting. However, a reduced quench rate is normally correlated with enhanced susceptibility to embrittlement [8]. Hence it seems more likely that, for the binary and quaternary alloys, optimal doping enhances diffusion rates during the quench. Our measurements of stress-relief rates after the quench [2] tend to support this idea; optimal doping enhanced stress-relief in the binary alloy, but not in the ternary. Overdoping gave reduced stress-relief rates in both alloys.

The expected effect of cerium is to react with impurities (O,S) in the melt, resulting in insoluble precipitates. The optimal doping level is consistent with complete removal of the oxygen and sulfur from our alloys [2]. Thus it seems that diffusion rates in these alloys may be sensitive to small amounts of impurities.

REFERENCES


1. D. M. Kroeger, C. C. Koch, C. G. McKamey, J. O. Scarbrough, and R. A. Padgett, RQM 5, Wurzburg, FRG, Aug. 1984, 1369.
2. D. M. Kroeger, G. S. Canright, C. G. McKamey, D. S. Easton, and J. O. Scarbrough, Acta Met., to be published.
3. G. S Canright and D. M. Kroeger, IEEE Trans. Magnetics, to be published.
4. C. Kittel, Phys. Rev. 76, 1527 (1949).
5. J. D. Livingston, Phys. Status Solidi A 56, 637 (1979).
6. H. Kronmüller and W. Fernengel, Phys. Status Solidi A 64, 593 (1981).
7. M. Tejedor and B. Hernando, J. Phys. D 13, 1709 (1980).
8. R. Gerling and R. Wagner, Scripta Met. 17, 1129 (1983).

# MAGNETIC AFTER EFFECTS AND MAGNETOSTRICTION OF FE-BASE METALLIC GLASSES

A.K. COLLINS, R.C. O'HANDLEY AND N.J. GRANT
Department of Materials Science and Engineering, Massachusetts Institute of Technology, Cambridge, MA 02139

ABSTRACT

Magnetic after effects (MAE's) have been studied in Fe-base and Co-base metallic glasses. Fe-base amorphous alloys show large MAE's due in part to the high magnetoelastic energy (large $\lambda_s$) whereas Co-base alloys exhibit much smaller MAE's due to their low magnetostriction. A detailed study of the contribution of the magnetoelastic energy to the MAE's was done on the $Fe_{83-x}Cr_xB_{17}$ and $Fe_{83-x}Nb_xB_{17}$ glass systems.

## INTRODUCTION

The magnetic after effect of the permeability (disaccomodation) is observed in both crystalline and amorphous ferromagnets. In both cases the nature of the effect is structural. In crystalline materials the magnetic relaxation is due to alignment of well defined defects or impurities with the magnetization direction[1,2]. This process has a discrete activation energy and it takes place within a narrow temperature range. In amorphous alloys however, the nature of the defects is less well defined and the migration of the defect structures is characterized by a broad spectrum of activation energies. Consequently, the temperature range over which the relaxation takes place is broad as well[3].

In both crystalline and amorphous materials the free energy of the system (ferromagnet and defect) can be expressed as a sum of the elastic, magnetoelastic, and anisotropy energy[4]. The defect alignment process during magnetization changes all three of these contributions to differing degrees.

In this paper we report the contribution of the magnetoelastic energy to the MAE's in Fe-base alloys. A detailed study was performed on the $Fe_{83-x}Cr_xB_{17}$ and $Fe_{83-x}Nb_xB_{17}$ glass systems. These glass systems were chosen because they are based mostly on one transition metal, Fe, and they show a variation in the saturation magnetostriction constant $\lambda_s$ from $30 \times 10^{-6}$ to $2 - 3 \times 10^{-6}$ as a function of Cr or Nb content.

## EXPERIMENTAL

Two series of amorphous alloys $Fe_{83-x}Cr_xB_{17}$ and $Fe_{83-x}Nb_xB_{17}$ were prepared by melt spinning. The amorphous state was identified by X-ray scattering.

Magnetic relaxation was measured using a Gen Rad Digibridge after the sample's initial state was set by application of a dc saturating pulse[5]. The isothermal relaxation was measured for T varying between 25$^{o}$C and 300$^{o}$C. The isochronals were constructed using

$$\frac{\delta\mu}{\mu} = \frac{\mu[t_1] - \mu[t_2]}{\mu[t_1]} \tag{1}$$

where $\mu[t_1]$ and $\mu[t_2]$ are the permeability values at $t_1$ = 20 sec and $t_2$ = 60 sec.

Magnetostriction measurements were performed by means of the strain gauge method and the magnetization was measured by using a vibrating sample magnetometer. From the magnetization versus temperature behavior we obtained the Curie temperature values.

## RESULTS AND DISCUSSION

The isochronal relaxation spectra of amorphous $Fe_{76}B_{16}Si_8$ and

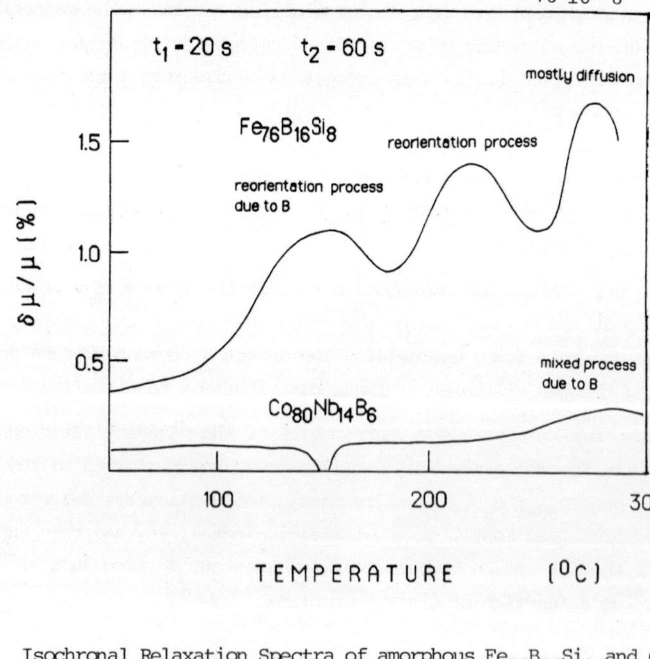


Fig. 1 Isochronal Relaxation Spectra of amorphous $Fe_{76}B_{16}Si_8$ and $Co_{80}Nb_{14}B_6$ alloys.

$Co_{80}Nb_{14}B_6$ are shown in figure 1. The $Fe_{76}B_{16}Si_8$ relaxation spectrum displays three discrete processes. The nature of these processes was identified earlier[5]. The relaxation spectrum of $Co_{80}Nb_{14}B_6$ displays only one maximum, that centered at 300°C. A comparison between the behavior of the two materials reveals an interesting result. The amplitude of the relaxation is stronger in the Fe-base glass and there are more processes taking place even at low temperatures in the Fe-base glass than is the case for the Co-base glass.

Since the magnetoelastic effects in Fe-base glasses ($\lambda_s = 30 \times 10^{-6}$) are stronger than in Co-base glasses ($\lambda_s = 0$), we conclude that the observed phenomenon is due to the fact that the contribution of the magnetoelastic energy to the MAE's is crucial. In order to investigate this point we proceeded with the study of two series of Fe-base glasses. The interesting feature of these glasses is that we can vary the saturation magnetostriction value from $30 \times 10^{-6}$ to $2 - 3 \times 10^{-6}$ by adding a small amount of either Cr or Nb. In this way we have a glass based mostly on Fe but with variable $\lambda_s$.

The MAE's of these metallic glasses are shown in figures 2 and 3. We observe in both spectra again three discrete relaxation processes that peak at 150, 230, and 280°C. The two processes which occur at the higher temperatures 230 and 280°C disappear quickly as we add more Cr or Nb. This

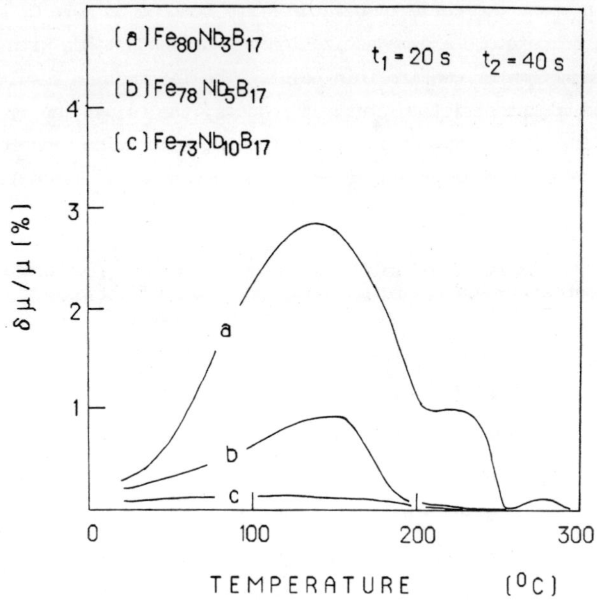


Fig. 2 Isochronal Relaxation Spectra of amorphous $Fe_{83-x}Nb_xB_{17}$.

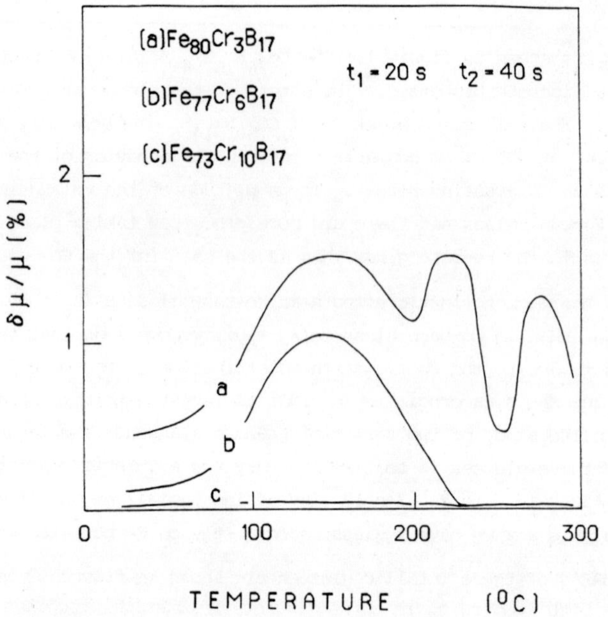


Fig. 3 Isochronal Relaxation Spectra of amorphous $Fe_{83-x}Cr_xB_{17}$.

could happen because the Curie temperatures are reduced as more Cr or Nb is added. Curie temperature and magnetization measurements taken with a vibrating sample magnetometer confirm this point. The interesting result though is that as the magnetostriction constant reduces, the relaxation amplitude becomes smaller. Room temperature values for the saturation magnetostriction, magnetization and after effect amplitude are shown in Table I.

Table I

Room-temperature values of saturation magnetization $I_s$, saturation magnetostriction $\lambda_s$ and permeability after effect $\delta\mu/\mu$ of $Fe_{83-x}Nb_xB_{12}$ and $Fe_{83-x}Cr_xB_{12}$

| SAMPLE | $\delta\mu/\mu$ (%) | $I_s$ (emu/gr) | $\lambda_s \times 10^{-6}$ |
|---|---|---|---|
| $Fe_{80}Cr_3B_{17}$ | 0.39 | 163 | 19 |
| $Fe_{78}Cr_5B_{17}$ | 0.24 | 135 | 13 |
| $Fe_{77}Cr_6B_{17}$ | 0.11 | 118 | 7 |
| $Fe_{73}Cr_{10}B_{17}$ | 0.03 | 78 | 3 |
| $Fe_{80}Nb_3B_{17}$ | 0.35 | 137 | 16 |
| $Fe_{78}Nb_5B_{17}$ | 0.24 | 87 | 12 |
| $Fe_{73}Nb_{10}B_{17}$ | 0.09 | 50 | 4 |

P. Allia and F. Vinai[6] proposed a model describing the relation between after effects and magnetostriction. According to their view the magnetic after effects are due to the existence of a magnetostrictive interaction between regions of local shear stress and the magnetization direction. Egami et al.[7] had previously defined such defects as regions where the local shear stress $\tau$ exceeded some mean value by more than a certain amount. The main result of the proposed model can be summarized in the following equation

$$\frac{\delta\mu}{\mu} = A[T]\,\frac{\lambda_s^2 < \tau^2 >}{I_s} \tag{2}$$

where $\delta\mu/\mu$ is the relative change in the permeability, $\lambda_s$ is the magnetostriction constant, $I_s$ is the saturation magnetization value, $<\tau^2>$ is the second moment of the shear stress fluctuations and A[T] is a temperature dependent term describing the total number of defects and the remaining constants.

According to this result, plotting $(\delta\mu/\mu) \times I_s$ versus $\lambda_s^2$ should give a straight line. The result was proven for a series of different alloys by P. Allia and F. Vinai[6]. It is interesting to demonstrate the more rigorous applicability of the above relation for a series of alloys that have mainly the same composition and where $\lambda_s$ varies in a controlled way. Figure 4 shows the plot of $(\delta\mu/\mu) \times I_s$ versus $\lambda_s^2$ for the data shown in Table I. Indeed we observe a good agreement with the proposed model.

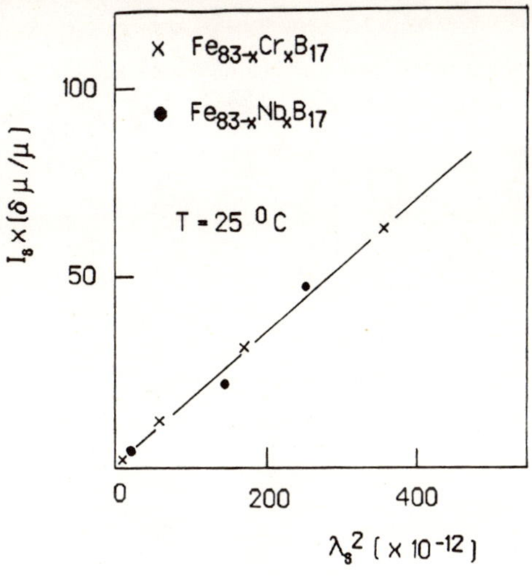


Fig. 4 Plot of $(\delta\mu/\mu) \times I_s$ as a function of $\lambda_s^2$.

## CONCLUSIONS

The MAE's are of a structural nature and not limited to a specific composition. They are observed when defects couple to the magnetization direction through magnetoelastic or magnetic anisotropic energies. The magnetoelastic coupling dominates the relaxation process centered at 150°.

## ACKNOWLEDGEMENTS

We would like to acknowledge that this work is supported by the 3M-company.

## REFERENCES

1. G. Richter, Ann. Physik 29, 605 (1938); 32, 683 (1938).
2. H. Kronmuller, "Nachwirkung in Ferromagnetika", Springer Verlag, Berlin, Heidelberg, New York, (1968).
3. N. Moser and H. Kronmuller, J. Magn. Magn. Mat. 19, 275 (1980).
4. G. DeVries, Physica 25, 1211 (1959).
5. A.K. Collins, R.C. O'Handley, N.J. Grant, J. Magn. Magn. Mat. 54-57, (1986).
6. P. Allia and F. Vinai, Phys. Rev. B 26, 11 (1982).
7. T. Egami, K. Maeda, V. Vitek, Philos. Mag. 41, 883 (1980).

# COMPOSITION DEPENDENCE OF MAGNETIC SHIELDING PROPERTIES FOR AMORPHOUS RIBBONS $(FeNi)_{78}Mo_4B_{17}Si$

Der-Ray Huang, Wen-Huu Dow, Pei-Chih Yao and Shu-En Hsu
Materials R&D Center, Chung Shan Institute of Science and Technology
P. O. Box 1-26, Lungtan, Taiwan 32500, R.O.C.

## ABSTRACT

Amorphous ribbons $(FeNi)_{78}Mo_4B_{17}Si$, made by melt-spinning technique were prepared as magnetic shielding materials. Cylindrical shields (~0.05mm thick) were fabricated by wrapping ribbons on 20cm long pyrex tubes with 45mm diameter. DC and AC (60Hz, 400Hz, 800Hz, 1 kHz) magnetic shielding data were obtained by following the modified ASTM (A698-74) procedure. For DC magnetic fields, Ni-rich shields showed maximum shielding values at lower applied fields (0.1 Oe~1 Oe), while Fe-rich shields showed maximum shielding values at higher applied fields (1~5 Oe). For AC magnetic fields, the magnetic fields, the magnetic shielding effectiveness was also influenced by skin effect at shields' surfaces. By comparing the data to calculations from shielding theory, there is about 8% magnetic shielding effectiveness induced by skin effect at 1 kHz.

## INTRODUCTION

For DC and low frequency magnetic shielding, a high permeability material is very effective for magnetic shielding. Because of their excellent soft magnetic and mechanical properties, iron-nickel base amorphous ribbons have been used as good magnetic shields in some electric devices. Amorphous ribbons $(Fe_{78-x}Ni_x)\ Mo_4B_{17}Si$ with different concentration of Fe and Ni were investigated by measuring their magnetic, electric and magnetic shielding properties. The DC magnetic shielding effectiveness was primarily a function of shield's permeability, thickness and shape, while AC magnetic shielding effectiveness was affected not only by shunting effect, as for DC fields, but also by skin effect which induced an induction opposite the applied field.

## EXPERIMENT

Amorphous ribbons of the composition $(Fe_{78-x}Ni_x)\ Mo_4B_{17}Si$ $(0 \leq x \leq 78)$ were made by melt-spinning technique. The ribbons used in this experiment were about 25μm in thickness and 20mm in width. The magnetization was measured as a function of the applied field $(0 \leq H \leq 10$ kOe) in a vibrating sample magnetometer (VSM).[1] This method was applicable to a very small sample and gave satisfactory results. The lower field soft magnetic properties such as coercive force (Hc), remanence (Br) and permeability (μ) were measured by winding multi-layer toroids with long ribbons (~200cm). A toroid sample with average diameter about 5cm was wound with 200 turn primary coils and 150~200 turn secondary coils. DC hysteresis curves were plotted with a recorder and AC (60Hz, 400Hz, 800Hz, 1 kHz) B-H loops were obtained with an oscilloscope. The values of soft magnetic properties were the averages over 20 different samples.

The shielding samples were fabricated by wrapping amorphous ribbons on 20cm long pyrex tubes with diameter 45mm. The thickness of shields were about 0.05mm. DC and AC (60Hz, 400Hz, 800Hz, 1 kHz) magnetic shielding data were obtained by following the procedures described in ASTM standard A698-74 with slight modification. A pair of Helmhotz coils with 30cm in diameter was used for testing cylindrical shields. [2][3] The magnetic shielding effectiveness was defined as $S_H = 20 \log_{10} Ho/Hi$ (dB) [4][5], where Ho is the magnetic field strength measured in the center of the coils, and Hi is the magnetic field strength measured in the shield.

## RESULTS AND DISCUSSION

In order to select a good magnetic shielding material for DC and very low frequency magnetic fields, we must prelimarily estimate the following significant factors--permeability and magnetization. The initial magnetization curve of amorphous ribbons $(FeNi)_{78}Mo_4B_{17}Si$ ($0 \leq x \leq 78$) measured by VSM were plotted as Fig. 1. We can estimate the induced magnetization (or

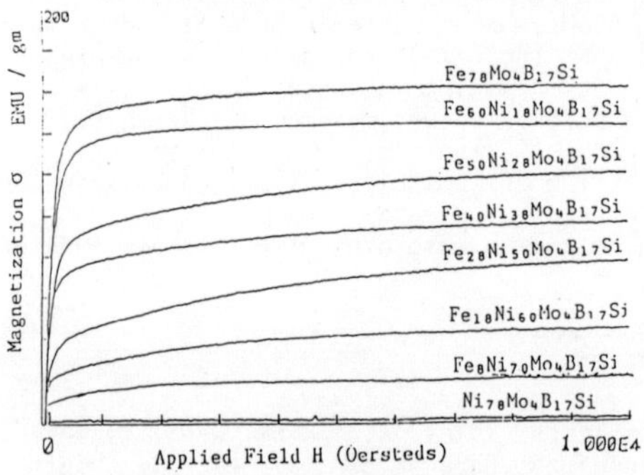


Fig. 1 Composition dependence of initial magnetization curve for amorphous ribbons $(FeNi)_{78}Mo_4B_{17}Si$.

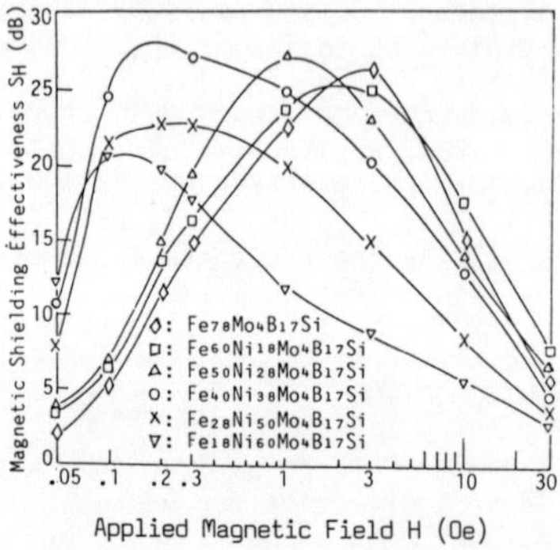


Fig. 2 DC magnetic shielding effectiveness as a function of applied fields for amorphous shields.

induction) of shielding materials under various applied fields. The Ni-rich amorphous ribbons can magnetized at lower fields with higher initial permeabilities, while Fe-rich amorphous ribbons showed lower initial permeabilities. The saturation magnetization $\sigma_s$ (at 10 kOe) obviously varied as function of Fe-content (Table I). For $Fe_{78}Mo_4B_{17}Si$, the saturation magnetization (at 10 kOe) equals 163 emu/gm, but for $Ni_{78}Mo_4B_{17}Si$, the value is nearly zero.

Table I

| $(Fe_{78-x}Ni_x)Mo_4B_{17}Si$ x = | 0 | 18 | 28 | 38 | 50 | 60 | 70 | 78 |
|---|---|---|---|---|---|---|---|---|
| Saturation Magnetization (emu/gm) | 163 | 130.2 | 121.1 | 97.2 | 77.9 | 44.7 | 16.2 | 1.22 |
| Resistivity (μΩ-cm) | 140 | 147 | 152 | 158 | 163 | 166 | 170 | |

Fig. 2 indicates the DC magnetic shielding effectiveness as a function of applied field (0.05 Oe~30 Oe) for amorphous shields. Ni-rich shields show the maximum values at lower applied fields (0.1 Oe~30 Oe), while Fe-rich shields show the maximum vlaues at higher applied field (1 Oe~5 Oe).

For AC fields, the magnetic shielding phenomenon is a combination of two effects: 1) Shunting effect--at lower frequency, the applied magnetic field is shunted through the low reluctance shield (as for a DC field); 2) Skin effect--at high frequency, eddy currents are induced in the shield which oppose the applied field. Therefore, for lower frequencies, the initial permeability of shielding material was the most important factor of shield. For higher frequencies, since eddy currents are induced as an exponential function of frequency, the resistivity ρ was also an important factor of magnetic shield.

The initial permeability of amorphous ribbons $(FeNi)_{78}Mo_4B_{17}Si$ are shown in Fig. 3. Ni-rich ribbons show higher initial permeabilities and with maximum permeability at lower applied fields (0.1 Oe~1 Oe), while Fe-rich ribbons show maximum permeability at higher applied fields (0.5 Oe~5 Oe). Fig. 4 shows AC 60Hz magnetic shielding effectiveness under applied fields of 0.1 Oe~30 Oe. The characteristics of the curves are similar to those of Fig. 3.

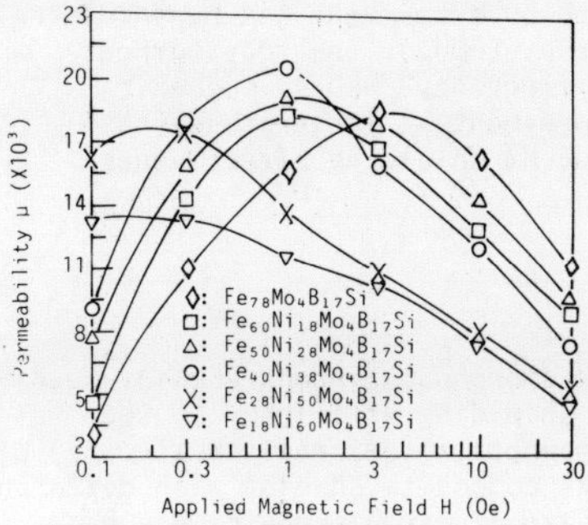


Fig. 3 Permeabilities as a function of AC (60Hz) peak magnetic field for amorphous $(FeNi)_{78}Mo_4B_{17}Si$.

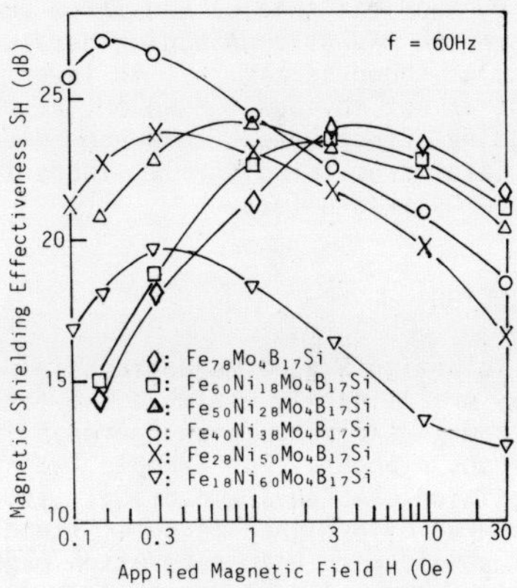


Fig. 4 AC 60Hz Magnetic shielding effectiveness as a function of applied field for amorphous ribbon $(FeNi)_{78}Mo_4B_{17}Si$.

Generally, AC permeability decreased as the frequency increased. From Fig. 5, we can find this phenomenon for amorphous $(Fe_{78-x}Ni_x)\ Mo_4B_{17}Si$ $(0\le x\le 78)$.

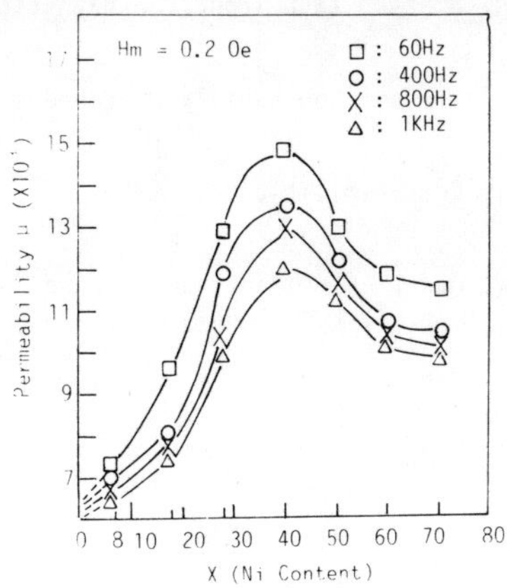


Fig. 5 Composition dependence of permeability for amorphous ribbons $(FeNi)_{78}Mo_4B_{17}Si$.

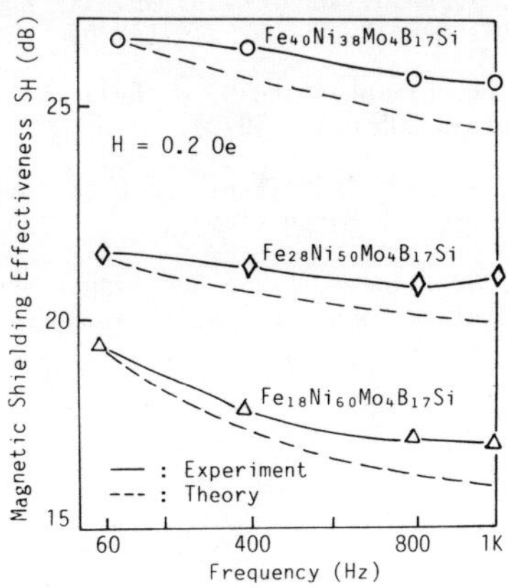


Fig. 6 Comparison of AC magnetic shielding effectiveness by experiment and by theory.

For a thin-shell cylindrical shield with transverse applied field, the magnetic effectiveness derived from shunting effect shielding theory is approximately given as:

$$S_H = 1 + \frac{1}{2}\frac{\mu t}{R} \quad (1)$$

μ: permeability of shielding material
t: thickness of shield
R: radius of shield

With cylindrical shields made from amorphous ribbons $(FeNi)_{78}Mo_4B_{17}Si$, we can compare the magnetic shielding data obtained by experiment and by shielding theory as shown in Fig. 6. At lower frequency (60Hz), the eddy current effect is not obvious. However, at higher frequency (1 kHz), magnetic shielding effectiveness measured in our experiment is larger than those calculated from formula (1). There is about 8% shielding effectiveness induced by skin effect.

## CONCLUSION

To design a good magnetic shield using amorphous ribbons $(FeNi)_{78}Mo_4B_{17}Si$, we may preliminarily estimate the magnetic shielding efficiency by the electromagnetic properties (permeability, magnetization, resistivity,...) of amorphous ribbons. For example, we can make composite shields with different composition of $(FeNi)_{78}Mo_4B_{17}Si$. If a precise electronic device has been interfered by a higher magnetic field (>10 Oe), we may fabricate a multi-layer shield with high-saturation magnetization amorphous ribbons $Fe_{78}Mo_4B_{17}Si$ in the outer shells, and with high-initial permeability amorphous ribbons $Fe_{40}Ni_{38}Mo_4B_{17}Si$ in the inner shells.

## REFERENCES

1. S. N. Kaul, IEEE Trans. on Magnetics, Vol. MAG-17, No. 2, 1208 March (1981).

2. L. I. Mendelsohn, E. A. Nesbitt and G. R. Bretts, IEEE Trans. on Magnetics, Vol. MAG-12, No. 6, 924 (1976).

3. J. P. Dismukes and G. J. Sellers, Third Int. Conf. on Rapidly Quenched Metal, 205 July (1978).

4. A. Thomas, IEEE Trans. on Electromagnetic Compatibility, Vol. EMC-10, No. 1, 142 March (1968).

5. G. Greifinger, P. G. Greifinger and L. W. Hart, IEEE Trans. on Electromagnetic Compatibility, Vol. EMC-23, No. 1, p. 1, Feb. (1981).

# STUDY OF THE ORIGIN OF THE MAGNETIC ANISOTROPY IN RF SEQUENTIAL CO-SPUTTERED $Co_{100-X}B_X$ AND $Co_{74}Fe_6B_{20}$ THIN FILMS

D. Y. KIM* AND R. M. WALSER*
*University of Texas, Electrical and Computer Engineering Dept., Austin, Texas 78712

## ABSTRACT

We prepared thin, amorphous, compositionally modulated films (CMF) by RF sequential co-sputtering with Co, B, and Fe targets. In principle this technique can produce controlled, small (<15 Å) scale modulations in composition. The in-plane anisotropies and coercivities of the CMF were more than one magnitude larger than those of typical thin, homogeneous, amorphous, alloy films (AF). The increase in anisotropy was growth induced and could be reduced by annealing to values comparable to those of amorphous AF. The coercivity, however, could not be significantly decreased by annealing, except in CMF with the highest boron concentration. The stability of the coercivity might, therefore, be an observable consequence of the small scale composition heterogeneity.

## INTRODUCTION

In RF sequential co-sputtering, CMF are made by rotating substrates under two or more continuously excited sputtering targets. In the absence of interdiffusion, an ideal modulation wavelength can be computed from the rotational substrate velocity and the target deposition rates. Sputtered, amorphous, CMF with coherent modulation wavelength 15-100 Å have been reported [1] to have properties that are uniquely different from thin films of either of the constituent elements, or of the alloy. On the other hand, sputtered, amorphous CMF with modulation wavelength <15 Å have not been as extensively studied and there are important questions as to whether this small scale composition modulation exists in the CMF, and whether it can have observable effects on their magnetic properties. In this work we prepared Co-B binary component CMF and Co-Fe-B ternary component CMF (referred to hereafter as BCMF and TCMF respectively) with small modulation wavelength ≈10-15 Å. Their magnetic properties were compared with those reported for other amorphous AF produced by sputtering alloy targets at normal incidence onto a stationary substrate. BCMF were prepared with average compositions Co(100-x)B(x) with 15<x<39. The only TCMF prepared had an average composition Co(74)Fe(6)B(20) which, in AF, has small magnetostriction. Its magnetic properties were contrasted with that of the magnetostrictive BCMF to qualitatively determine the influence of magnetostriction on their behavior.

## EXPERIMENTAL

The CMF were deposited by RF sequential co-sputtering onto substrates rotating at 2 RPM on a water cooled substrate under two or three continuously excited 5 inch diameter targets of Co, Fe and B. Deposition rates of ≈8 Å, ≈1 Å, and 1.5- 4.0 Å per rotation for Co, Fe, and B respectively, were used to produce BC and TC CMF 100 nm-200 nm thick. The substrates were 25 mm in diameter glass slides, or silicon wafers which were used for AES analysis. Some of the silicon wafers had thin (500 Å) carbon films which were used in the TEM/TED studies. The target to substrate distance was 3 1/8 inches and a uniformity aperture plate was located 0.9 inch above the substrate. An adjustable RF bias voltage could be applied to the anode plate. The background and argon sputter gas pressures were $4 \times 10^{-7}$ Torr and 10 mTorr respectively. The average CMF composition was analyzed by ICP-AES and depth profiled by AES.

The magnetic properties of the CMF were measured with a thin film hysteresis loop tracer. The coercivity was obtained from the hysteresis loop along the easy axis which was taken as the direction with minimum coercivity. The anisotropy field was taken as the intersection of the hard axis loop, at a small drive field, with the saturation magnetization.

There are two important differences in the deposition of AF and CMF; (1) the AF has no composition modulation , and (2) the AF is usually deposited at normal incidence, while the CMF is deposited at oblique incidence. From the thickness profiles of a film deposited on a stationary substrate, the maximum angle of incidence Ω (measured from the anode-cathode normal) in the standard CMF deposition mode was approximately 60 degrees. With this large angle of incidence, anisotropic growth morphologies are expected [2]. To investigate we surrounded some substrates with small glass or stainless steel cylinders (1 3/8 inch diameter and ≈ 0.9 inch high) and reduced to $\Omega \approx 20^o$. CMF with $\Omega \approx 20^o$ and $\Omega \approx 60^o$ were simultaneously deposited for comparison.

RESULTS

The coercivity and in-plane anisotropy of the as-deposited BCMF (with $\Omega \approx 60^o$) are shown in Figs. 1a and 1b for depositions made with, and without substrate bias, respectively. Both parameters are more than 10 times than those reported for other typical AF [3]. From TED it was determined that all of the BCMF with compositions in this range were amorphous. Their saturation magnetiztions decreased linearly with % B over this range and agreed with values reported for AF to within 15% [4]. Both the coercivity and anisotropy were minimized in BCMF deposited at an RF substrate bias of 50-75 volts. The anisotropy monotonically decreased with % B except for (1) large variations in the BCMF deposited without substrate bias, and (2) occasional large increases for BCMF compositions close to the eutectic composition at ≈18 at % B) and, also near ≈25 at % B. The latter were usually accompanied by a decrease in the measured saturation magnetization, possibly indicating the presence of an out-of-plane component of the magnetization.

Typical easy and hard axis hysteresis loops are shown in Fig. 2a and 2b for a BCMF with average composition Co(82)B(18) deposited without RF bias, and at $\Omega \approx 60^o$ and $\Omega \approx 20^o$ respectively. Restricting the angle of incidence essentially eliminated the in-plane anisotropy and the easy and hard axis loops were virtually indistinguishable. In the case of the magnetostrictive BCMF, the easy axis coercivity was decreased in the films deposited at $\Omega \approx 20^o$. This observation was independent of the RF substrate bias. The hysteresis of the TCMF also became isotropic with restricted Ω, but their coercivity increased substantially. Similar changes were observed for the TCMF deposited with optimum substrate bias. The BCMF and TCMF deposited at $\Omega \approx 20^o$ were isotropic in the plane and, following saturation, their remanence was ≈90% of the saturation magnetization for loops in any direction. This behavior suggests the possibility of a rotatable anisotropy in which the easy axis assumes the direction of the last,large, applied field [5].

TEM micrographs of BCMF and TCMF deposited at $\Omega \approx 60^o$ show [Fig.3] morphological features arising from angle-of-incidence, self-shadowing effects that have been observed in many

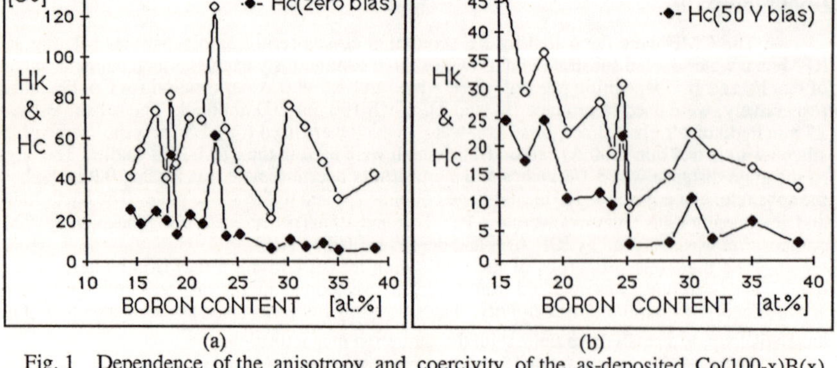


(a) (b)

Fig. 1 Dependence of the anisotropy and coercivity of the as-deposited Co(100-x)B(x) BCMF on the boron content with (a) and without bias (b).

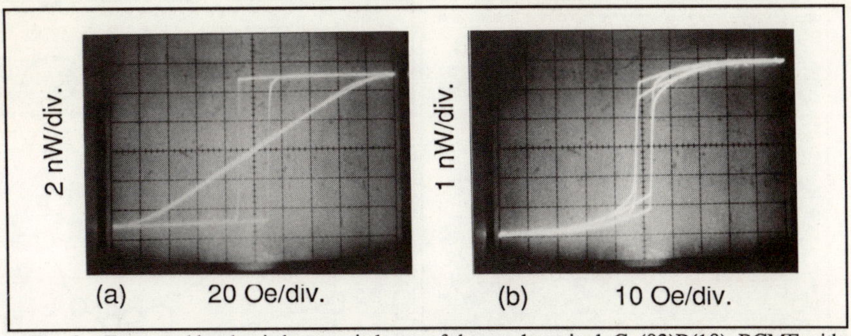


Fig. 2 Easy and hard axis hysteresis loops of the as-deposited Co(82)B(18) BCMF with zero bias, and with (a) $\Omega \approx 60^o$ and (b) $\Omega \approx 20^o$.

thin film studies [2]. Growth anisotropies of this type can, in principal give rise to shape demagnetizing fields comparable to the anisotropies measured for CMF deposited at $\Omega \approx 60^o$ [6].

Growth induced anisotropies were found in both BCMF and TCMF. However, at zero bias their easy axes were perpendicular and parallel, respectively, to the direction of the beam. The magnitude of the anisotropy of all CMF decreased with increasing substrate bias. With increasing bias the easy axis of TCMF rotated to become perpendicular to the direction of the beam at the RF bias that minimized the anisotropy. The easy axis of the BCMF did not change orientation with bias.

TCMF deposited at different RF bias voltages were AES depth profiled to determine their O, N and Ar content. In Fig. 4 the results show that the rotation of the easy axis is correlated with changes in the % O and % Ar in the CMF. The minimum in the % O and the anisotopy occur at the same RF bias voltage, as observed in previous studies [7]. Both quantities increased with still further increases in the RF bias and the easy axis again rotated toward the beam direction.

The thermal stability of the CMF was studied by isochronal annealing in flowing $N_2$ for 30 minutes at temperatures up to 450 °C. TED analysis confirmed that all the CMF had crystallized at T<450 °C. At temperatures below the recrystallization temperature , anisotropy of the BCMF tended to decrease, but their coercivities either increased, or remained constant. The only exception noted was for the BCMF with the largest %B which also had the smallest coercivity of the as-deposited BCMF. The coercivity of this film was significantly decreased by anneals below the crystallization temperature. The TCMF were more stable with annealing and their anisotropy and coercivity changed little below the crystallization temperature. A large increase in both the coercivity and the anisotropy was observed for all CMF at 450 °C where they had crystallized.

## DISCUSSION

The as-deposited and annealed CMF with modulation wavelength ≈10-15Å were found to be magnetically much harder than those typical of AF. Their large growth-induced anisotropy could be eliminated by cutting off the most oblique part of the angular distribution of the incident beam. They were also substantially reduced by annealing below their crystallization temperatures. Their coercivity, however, was thermally stable in the amorphous state, except for BCMF with the highest % B. These conclusions were nearly identical for both BCMF and TCMF.

The easy axis of the the low magnetostrictive TCMF, deposited without RF bias, was parallel to the beam direction. With increasing RF bias the orientation of the easy axis changed and was aligned perpendicular to the beam incidence for the RF bias that minimized the oxygen incorporated in the film. This may indicate the development of an orthogonal morphological feature that, in the absence of significant stress anisotropy in the low magnetostictive TCMF, could cause the easy anisotropy axis to reorient. It is not difficult to envision how the relative growth rates of competing parallel and perpendicular morphological features could be controlled by the amount of oxygen present.

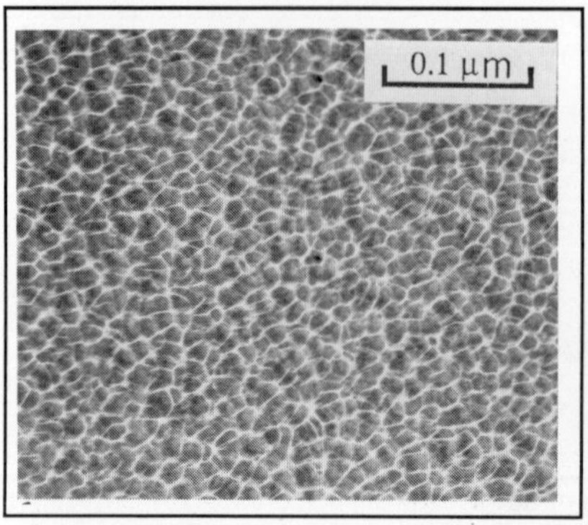


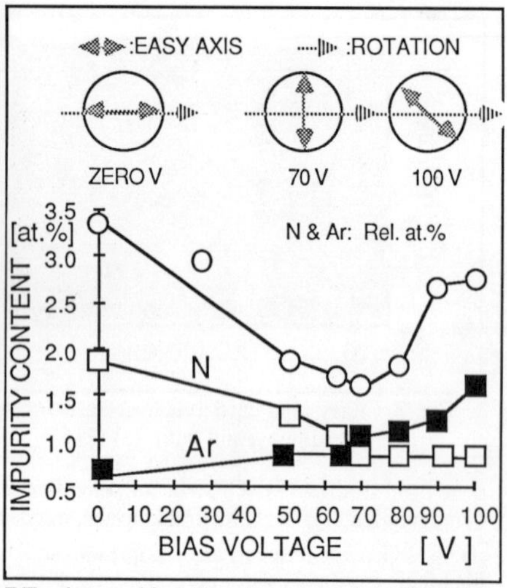


Fig. 3 (Above) TEM micrograph of a 500 Å thick Co(74)Fe(6)B(20) TCMF deposited without RF bias and $\Omega \approx 60^{\circ}$.

Fig. 4 (Right) Variation in the O, Ar, and N impurity concentrations in TCMF with RF substrate bias. Inserts show orientation of easy axis at three different RF bias voltages.

The easy axis of the magnetostrictive BCMF was, however, perpendicular to the beam incidence for all RF substrate bias. This suggests that, in this case, there may be a stress anisotropy field aligned with the perpendicular, growth-induced, shape anisotropy field. These fields could lock in a perpendicular easy axis if their sum exceeded that of competing, parallel, shape anisotropy fields.

We also observed that the coercivities of small modulation wavelength CMF are much larger than those of typical AF. The stability of their enhanced coercivity in the amorphous state could be associated with the small scale composition heterogeneity. Recently, a similar suggestion has been made in connection with the stability of a enhanced vertical anisotropy in Co-Pd CMF with comparable modulation wavelenghs [8].

## ACKNOWLEDGEMENTS

We would like to thank Dr. Philip Trouilloud for useful conversations and his careful review of this manuscript. We also gratefully acknowledge ONR Grant # N00014-84-K-0453 and (DARPA) USAF Contract # F33657-84-C-2058 for support of this research.

## REFERENCES

1. N. S. Kazama , et. al., IEEE Trans. Magn. MAG-20, 1296 (1984); J. Magn. Magn. Mat. 35, 86 (1983).
2. N. G. Nakhodkin and A. I. Shalderan, Thin Solid Films 10, 109 (1972); A. G. Dirks and H. J. Leamy, J. Appl. Phys. 49, 3430 (1978).
3. R. J. Kobliska, et. al., Appl. Phys. Lett. 33, 473 (1978); Y. Shimada, et. al., IEEE Trans. Magn. MAG-17, 1199 (1981).
4. T. R. McGuire, J. A. Aboaf, and E. Klokholm, IEEE Trans.Magn. MAG-16, 905 (1980); J. P. Malozemoff, et. al., J. Appl. Phys. 50, 5885 (1979).
5. R. F. Soohoo, Magnetic Thin Films , (Harper and Row, New York, 1965), p.125.
6. W. Metzdorf and H. E. Wiehl, Phys. Stat. Sol. 17, 285 (1966).
7. W. N. Hammer and K. Y. Ahn, J. Vac. Sci. Technol. 17, 804 (1980); H. J. Leamy and A. G. Dirks, J. Appl. Phys. 50, 2871 (1979).
8. P. J. Carcia, A. D. Meinhaldt, and A. Suna, Appl. Phys. Lett. 47, 178 (1985).

# CHARACTERIZATION OF INHOMOGENEITIES IN AMORPHOUS SUPERCONDUCTORS

Esen E.Alp, S.K.Malik(1), Y.Lepetre, P.A.,Montano(2), and G.K.Shenoy
Materials Science and Technology Division, Argonne National Laboratory, Argonne, Ill. 60439

ABSTRACT

The structure and superconducting properties of rapidly quenched Zr-Ni and Zr-V alloys have been studied by XRD,EXAFS spectroscopy,TEM,electron diffraction,ED, electrical resistivity and ac susceptibility measurements. Amorphous $Zr_{66}Ni_{34}$ ribbons quenched with different cooling rates were also examined by flux pinning measurements. Characteristic differences are observed in the flux pinning mechanisms of microcrystalline and amorphous samples. The temperature dependence of upper critical fields have been measured down to 0.4 K and the results are analyzed in terms of WHH theory.

## INTRODUCTION

The superconducting properties of amorphous transtion metal alloys are recieving increased attention [1,2]. Superconductivity is becoming an effective tool to study the effect of atomistic scale disorder on the electronic behaviour of amorphous transition metal alloys [1]. However, it should be mentioned that measurements made at different laboratories on similar alloys do not necessarily agree with each other, and there is a reproducibility problem. [3]. This is mainly because of variations in the cooling rate and the subsequent thermal relaxation following the production of these alloys. Such variations will result in microstructures that are microcrystalline with a grain size of few hundred angstroms to amorphous phase with crystalline inhomogeneities embedded in it. These inhomogeneities may vary in size and might affect the superconducting behaviour of the amorphous matrix. Since the amorphous phase itself may have compositional as well as density variations at scales comparable to the coherence length of the superconducting electrons [4], it is necessary to investigate the effect of these different type of inhomogeneities on the superconducting behavior of amorphous metals.

## EXPERIMENTAL

The amorphous ribbons were produced by melt-spinning technique using alloy buttons prepared in an arc-furnace. Different cooling rates could be obtained by changing the wheel speed onto which the molten metal is ejected. Three different Zr-Ni and two different Zr-V alloys were prepared with wheel speeds between 40-60 m/sec, with nominal compositions of $Zr_{66}Ni_{34}$, and $Zr_{34}V_{66}$. These samples failed to give crystalline peaks as examined by X-ray diffraction. The superconducting transition temperatures are determined using ac susceptibility and dc resistivity. Transition widths are found to be higher when detected by ac susceptibility. This became even more pronounced when magnetic field was applied. Moreover, for Zr-V samples more than one transi- tion were observed, indicating the inhomogenous character of the samples. X-ray diffraction measurements were carried out using Cu $K_\alpha$ radiation on a Rigaku scanning diffractometer, equipped with a diffracted beam monochromator. Crystallization experiments were done by Perkin-Elmer DTA 1700 unit. Transmission Electron Microscopy studies were carried out on samples thinned using ion-milling techniques. The sample holder was held at $LN_2$ temperature during the ion-milling procedure. EXAFS measurements were made at Stanford Synchrotron Radiation Laboratory.

(1) Permanent address : Tata Institute of Fundamental Research, Bombay,India
(2) Dept. of Physics, West Virginia University,Morgantown, WV

RESULTS AND DISCUSSION

The thicknesses, the transition temperatures, and the resistivities of the Zr-Ni samples studied are given in Table 1. Of the three samples studied, sample No 1 gave a broad electron diffraction band corresponding to an average interatomic distance of 2.56 Å. However, for sample No 2, an electron diffraction pattern consisting of sharp dots, with irregular interplanar distances was obtained. This was also the case for Zr-V samples. The microstructure and the electron diffraction patterns of Zr-Ni samples are given in Fig.1. X-ray diffraction experiments resulted in a broad peak at $37^{o}$ for $Zr_{66}Ni_{34}$. Exothermic crystallization peaks were observed for all the Zr-Ni samples, as shown in Fig. 2. They were all ductile and could be bent over 180 degrees. Non-existence of sharp diffraction lines in XRD, crystallization peak in DSC, and the extent of ductility can be taken as an indication of amorphicity. However, the differences in superconducting transition temperature $T_c$, upper critical fields $H_{c2}(T)$ and critical current densities $J_c$ indicate that the structural parameters at the scale of the coherence length are different. The $H_{c2}$ for Zr-V samples are shown in Fig.3. Two different slopes observed in Fig 3 (a) is an indication of an inhomogenous sample, and those can be detected with ac susceptibility measurements.In Fig.4 the upper critical fields as a function of temperature are given. The solid lines are fits obtained by using standard WHH theory. In the limit of $\lambda_{so}>\lambda_{tr}$ and $\lambda_{tr}>>1$, $H_{c2}(T)$ is given by the following implicit equation:

$$\ln(1/t) = (1/2+i\lambda_{so}/4\gamma)\Psi\{1/2+(h+\lambda_{so}/2+i\gamma)/2t\} + (1/2-i\lambda_{so}/4\gamma)\Psi\{1/2+(h+\lambda_{so}/2-i\gamma)/2t\} - \Psi(1/2), \quad (1)$$

where $t=T/T_c$, and $\gamma=[(\alpha h)^2-\lambda_{so}^2/4]^{1/2}$, $\alpha$ is the Maki parameter (symbols have the same meaning as in Ref.6). The $\lambda_{so}$ values obtained from these fits are given in Table 1. We have also measured the magnetoresistance in the normal state up to 7 T and did not observe any change. We interpret the fact that $H_{c2}(T)$ can be fit with Eq.(1) without invoking unreasonable slopes as an indication of no enhancenment in $H_{c2}(T)$. This is consistent with the reports that the enhancements in $H_{c2}(T)$ as a result of disorder induced localization effects are accompanied by negative magnetoresistance[7].

No 1 No 2 No 3

Fig. 1. Electron micrographs of $Zr_{66}Ni_{34}$ samples. The elctron diffraction patterns are shown in the insert.

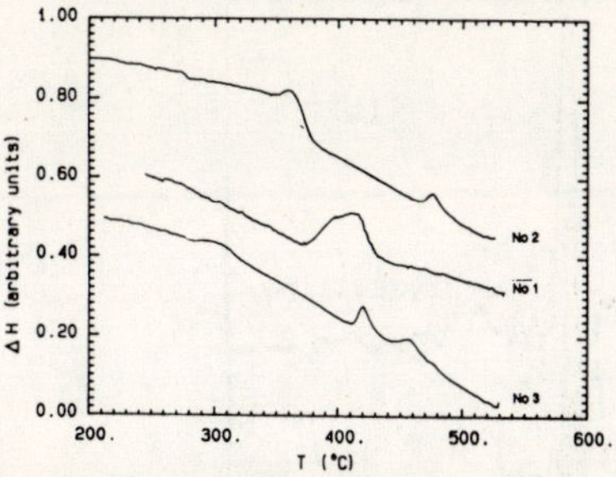


Fig.2 Crystallization behavior of different $Zr_{66}Ni_{34}$ alloys

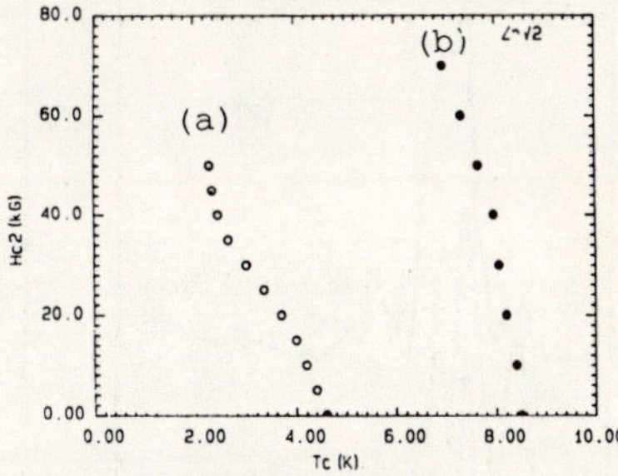


Fig.3 Upper critical field values versus temperature for $Zr_{34}V_{66}$ a) rapidly quenched, b) crystalline

Flux pinning measurements were carried out at T=1.7 K. Since $T_c$'s for three Zr-Ni samples were different, temperature scaling is done by dividing $F_p$ by the maximum value it reaches for each sample [8]. The $H_{c2}$'s are determined by extrapolating the linear portion of $J_c{*}H^{1/4}$ vs. H curve to the zero current density. The results are then plotted , as shown in Fig.5. The peak positions (h=0.3) and the general apparence for the amorphous samples (No:1 and 3) are similar, whereas sample No 2 peaks at h=0.1 and drops off very sharply.

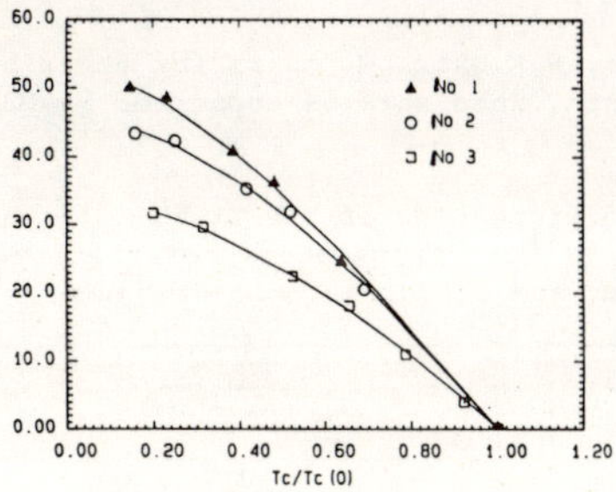


Fig.4. Upper critical field values versus reduced temperature for $Zr_{66}Ni_{34}$, solid lines are the fits described.

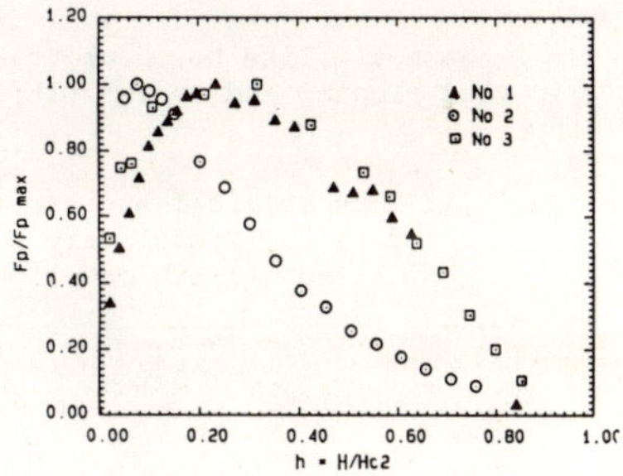


Fig.5 The ratio of the pinning force versus reduced field.

We have used the EXAFS technique to look at short range order in these glasses at the scale 2-4 Å. The Fourier transforms of $\chi(k).k^3$ for Ni- and Zr-edge spectra are given in Fig.6. The interatomic distances for crystalline $Zr_2Ni$ and Sample No 1 and 2 are given in Table 1. The data analysis is underway to determine the coordination numbers. Although, EXAFS measurements are important in determining the short-range order in amorphous metals, due to the limited wave number k-range (3-15 $Å^{-1}$), information crresponding to medium range order (k<3 $Å^{-1}$) is not available. In this respect, small angle x-ray and neutron scattering measurements should be more helpful.

## CONCLUSIONS

We have shown that the superconducting properties of rapidly quenched Zr- alloys are strongly dependent on the inhomogeneities in the material. Flux pinning measurements can be effectively used to distinguish microcrystallinity from an amorphous matrix. The temperature dependence of upper critical fields measurements are useful in characterizing the samples.

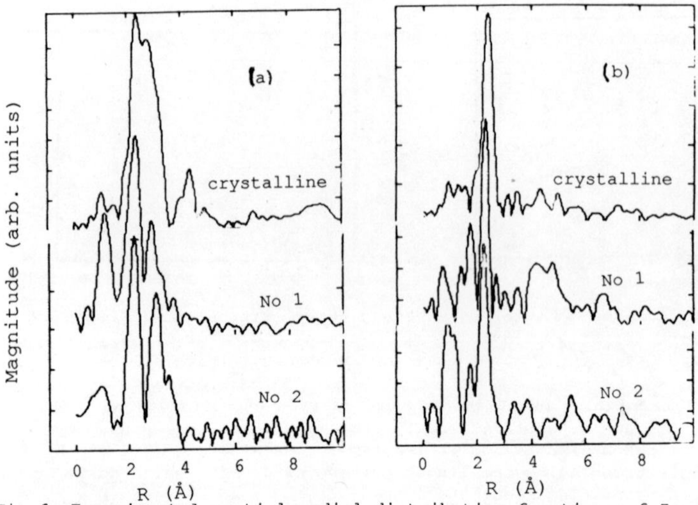


Fig 6. Experimental partial radial distribution functions of $Zr_{66}Ni_{34}$ obtained from a) Zr-K-edge, b) Ni-K-edge absorbtion measurements.


ACKNOWLEDGMENTS

We would like to thank to Dr.D Capone, W.K.Kwok, H.Sowers for their help in flux pinning and resistivity measurements. This work is supported by US DOE.


Table 1. Some physical and superconducting properties of the Zr-Ni alloys studied. Symbols are d:thickness, ρ: room temperature resistivity, λ: spin-orbit coupling parameter, and r: interatomic distance

| Sample | | d (μm) | ρ(3) (μΩ-cm) | $T_c$(5) (K) | $-dH_{c2}/dT)_{T=Tc}$ (kG/K) | λ | $r_{Ni-Ni}$ (Å) | $r_{Ni-Zr}$ (Å) |
|---|---|---|---|---|---|---|---|---|
| $Zr_{34}Ni_{66}$ | No 1 | 19.3 | 156. | 2.62 | 29.2 | 149. | 2.55 | 2.62 |
| | No 2 | 21.1 | 158. | 2.41 | 32.8 | 2.88 | 2.54 | 2.74 |
| | No 3 | 16.8 | 246. | 1.91 | 30.7 | 2.42 | -- | -- |
| $Zr_2Ni$ crystalline | | | | 1.18 | 3.3 | | 2.66 | 2.74 |
| $Zr_{34}V_{66}$ | No 1 | | | 4.6(*) | 35.7 | | | |
| | No 2 | | | 4.1(*) | | | | |
| $ZrV_2$ crystalline | | | | 8.5 | 57.0 | | | |

(*) Two different $T_c$'s were observed with ac susceptibility technique, second one being at lower temperature (≅3.1 K).


REFERENCES

1. W.L.Johnson, Glassy Metals 1, Eds. H.J.Guntherodt, and H.Beck, Springer-Verlag, Berlin (1981), and S.J.Poon, in Amorphous Metallic Alloys, Ed. F.E. Luborsky, Butterworths, London (1983).
2. Z.Altounian, J.O.Strom-Olsen, Phsy.Rev.B, 27, 4149 (1983).
3. M.G.Karkut,and R.R.Hake, Phys.Rev.B, 28, 1396 (1983).
4. W.L.Carter, S.J.Poon,G.W.Hull,Jr.,and T.H.Geballe,Sol.St.Comm., 39, 41(1981).
5. N.R.Werthamer, E.Helfand, and P.C.Hohenberg, Phys.Rev. 147, 295 (1966) .
6. M.Decroux and O.Fischer, in Superconductivity in Ternary Compounds II, Eds. M.B.Maple and O.Fischer,Springer-Verlag (1982).
7. L.Coffey, K.Levin,and K.A.Muttalib, Phys.Rev.B, 32, 4382 (1985)
8. E.J.Kramer, J.Appl.Phys., 44, 1360 (1972).

# MAGNETO-OPTICAL PROPERTIES OF AMORPHOUS TbFe ALLOYS

M. Mansuripur, M. Ruane, P. Wolniansky, S. Chase, R. Rosenvold
Boston University, College of Engineering, 110 Cummington Street, Boston, MA 02215.

## ABSTRACT

Hysteresis loops and anisotropy energy constants are measured in a magneto-optical system that combines Kerr rotation and ellipticity to enhance signal strength. Temperature dependence of the polar Kerr effect is compared with the magnetization of the iron subnetwork in the mean-field approximation and good agreement is obtained. Perpendicular magnetic anisotropy is studied by magneto-optical methods, yielding the first two coefficients of the series expansion of anisotropy energy in terms of the angle of deviation from the easy axis.

## 1. INTRODUCTION

Amorphous rare earth-transition metal (RE-TM) alloys are now considered the most promising media for erasable optical data storage applications [1-3]. In this paper magneto-optical measurements on TbFe alloys, prepared by sputtering onto quartz substrates, are compared with theoretical models. Section 2 describes hysteresis loop measurements and compares temperature dependence of the magneto-optical Kerr effect and the iron subnetwork magnetization derived from a mean-field theory. Results of anistropy measurements are given in Section 3 where we show that second order terms must be included in the expression for uniaxial anisotropy energy.

## 2. HYSTERESIS LOOP MEASUREMENTS

Magneto-optical measurements of hysteresis loops versus temperature were performed in a differential detection system using HeNe light at $\lambda = 633$ nm. All samples were sputter-deposited on quartz to a thickness of about 1000 Å and in-situ overcoated with 300 Å of $SiO_2$. The laser was incident on each sample from the overcoated side while an axial electromagnet created a 0.1 Hz sinusoidal magnetic field of amplitude $H_{max}$. The three samples described here (samples I, II and III) were analyzed by X-ray fluorescence and found to be $Tb_{28.3}Fe_{71.7}$, $Tb_{20.3}Fe_{79.7}$ and $(Tb_{17}Fe_{83})_{94}Cu_6$, respectively.

The magneto-optic effect at red or near infrared wavelengths is believed to be mainly the result of interaction between light and the transition metal subnetwork in RE-TM alloys. To a first order approximation, the Kerr rotation angle $\theta_k$ is proportional to the magnetization of the TM subnetwork:

$$M_{Fe}(T) = \beta\theta_k(T) \tag{1}$$

where $\beta$ is the proportionality constant. The iron subnetwork magnetization can be calculated from the mean-field theory [4].

Fig. 1 shows the mean-field results for $(Tb_{29}Fe_{71})_{85}Ar_{15}$ (solid lines). The points (x's) represent measured $\theta_k(T)$ for sample I with $\beta$=2500. The hysteresis loops are square at all temperatures up to the Curie point and the behavior of $H_c$ vs. temperature is characteristic of Tb-rich samples with no compensation point temperature.

Although the mean-field theory and measured values of $\theta_k$ agree extremely well, a second independent test of the model considers data obtained from measurements of saturation magnetization $M_s$ vs. temperature. Data for a $Tb_{29}Fe_{71}$ sputter-deposited sample[2] is shown by o's in Fig.1. That sample has a slightly higher Curie temperature and its magnetization lies above our theoretical $M_s(T)$ curve. The differences could be attributed to the different amounts of film impurities; if the assumed amount of Ar in the sample is reduced to 5%, the mean-field model would give an exact match.

The sample II hysteresis loops are also square at all temperatures; $H_c$ vs. T is characteristic of an Fe-rich sample with a compensation point below room temperature. Fig. 2 shows theoretical net and subnetwork magnetizations for $(Tb_{20.5}Fe_{79.5})_{88}Ar_{12}$ (lines). The x's are measured $\theta_k$ data with $\beta$=2500; the o's are the $M_s$ data of a $Tb_{21}Fe_{79}$ sample[3]. Reducing the assumed amount of Ar to 5% will again give a better match to the $M_s$ data.

The hysteresis loops for sample III are characteristic of Fe-rich samples, but lack squareness. This is most likely due to the breakdown of magnetization into striped domains or the existence of in-plane magnetized regions. Coercivity cannot be meaningfully defined for this sample and the Kerr angle $\theta_k$ must be measured at saturation ($H = H_{max}$). In order to compare the data with the mean-field model predictions it would be best to have separate measurements of $M_s$ vs. T. Data on a composition close to that of our sample could not be found; data for $Tb_{14}Fe_{86}$ [3] was used. Fig. 3 shows model calculations on $(Tb_{13}Fe_{87})_{93}Ar_7$ along with the $M_s(T)$ data (o's) of [3]. The x's are measured $\theta_k$ for sample III with $\beta$=2500. They fall slightly below the $M_{Fe}(T)$ curve; this is expected since our sample has less iron than the one for which the curve was calculated. Generally, however, theoretical and experimental results agree well.

These comparisons tend to support the assumptions made in the mean-field model[4], particularly the postulate of antiferromagnetic coupling of a fraction of iron ions. More data is required, however, for a conclusive verification.

## 3. MEASUREMENTS OF MAGNETIC ANISOTROPY

Perhaps the most important property of amorphous RE-TM alloys for magneto-optical storage applications is the existence of uniaxial magnetic anisotropy in the direction perpendicular to the plane of the films. The energy associated with this magnetic anisotropy can be expressed as

$$E_K(\phi) = K_1\sin^2\phi + K_2\sin^4\phi + \dots \tag{2}$$

where $\phi$ is the angle of deviation of the magnetization vector from the normal to the plane of the film and $K_1$, $K_2$, ... are anisotropy energy constants. The uniaxial nature of anisotropy is responsible for the absence of terms with odd powers of $\phi$ in Equation 2. Now consider a thin film of amorphous RE-TM alloy with its magnetization saturated to $M_s$ in the perpendicular direction. If an external magnetic field H is applied in the plane of the sample, the magnetization will assume a new orientation in which the angle of $M_s$ with the normal to the plane is $\phi_o$. To determine $\phi_o$ we must minimize the total magnetic energy as a function of $\phi$. The energy is usually the sum of external field energy, demagnetizing energy, and the anisotropy energy:

$$E(\phi) = -HM_s\sin\phi + 2\pi M_s^2\cos^2\phi + K_1\sin^2\phi + K_2\sin^4\phi + \dots \tag{3}$$

Define x, $\alpha$, and $\beta$:

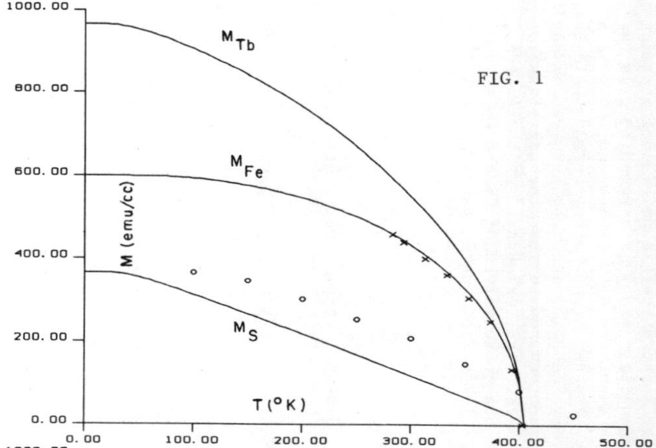


FIG. 1

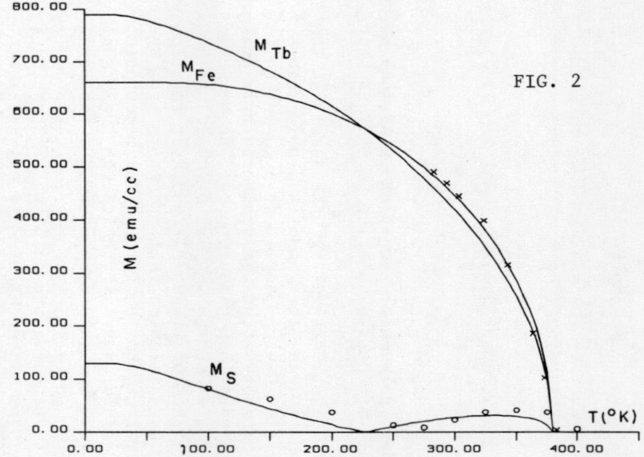


FIG. 2

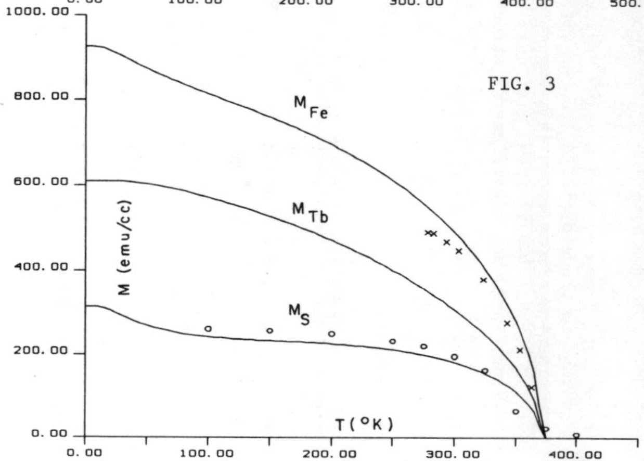


FIG. 3

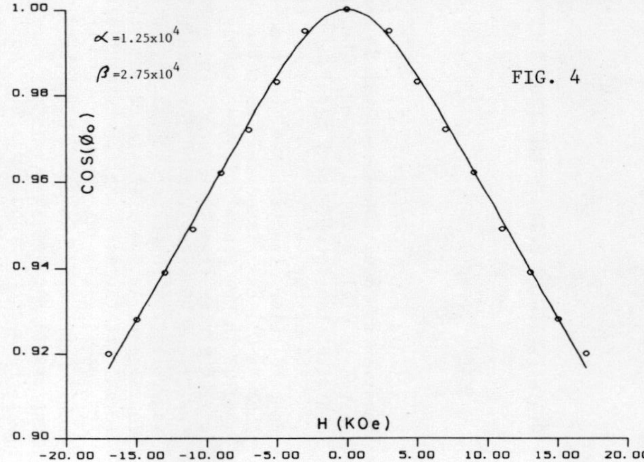


FIG. 4

$$x = \cos\phi_o \qquad \alpha = (K_1/M_s) - 2\pi M_s \qquad \beta = K_2/M_s$$

Minimization of $E(\phi)$ in Equation 3 then yields:

$$4\beta^2x^6 - 4\beta(\alpha+3\beta)x^4 + (\alpha+2\beta)(\alpha+6\beta)x^2 - (\alpha+2\beta)^2 + (H/2)^2 = 0 \qquad (4)$$

Equation 4 has retained only the first two terms from the expansion of Equation 2. Equation 4, a third order polynomial equation in $x^2$, can be solved to yield x vs. H curves. The only acceptable solution is the one that minimizes the total energy $E(\phi_o)$.

Anisotropy energy is evaluated by first saturating the sample in the perpendicular direction and then placing it in the gap of a 2-pole magnet. There a 0.1 Hz sinusoidal magnetic field having amplitude $H_{max}$=17 KOe is applied in the plane of the sample. Magnetization is monitored by a Kerr effect system like that used in hysteresis loop measurements. The measured normalized MO signal vs. H for sample I is shown in Fig. 4 (o's). At H=0 magnetization is perpendicular to the plane of the sample and Kerr rotation is largest. As the field increases the magnetization vector departs from the normal and the Kerr angle, being proportional to the perpendicular component of magnetization, decreases. If the field were strong enough the signal would reach zero; at our $H=H_{max}$ the signal has only dropped by 8%. The curve represents $x=\cos\phi_o$ vs. H.

Parameters $\alpha$ and $\beta$ can be found by matching the solution of Equation 4 with the experimental data. The first term in Equation 2 is not sufficient for the explanation of the observed behavior. Fig. 4 shows a good match between the experimental data and the theoretical model (line) assuming $\alpha=1.25 \times 10^4$, $\beta=2.75 \times 10^4$. If the mean-field model is assumed to give the correct value for $M_s$ then at T = 23°C we must have $M_s$ = 115 emu/cc, $K_1 = +1.5 \times 10^6$ erg/cc and $K_2 = +3.2 \times 10^6$ erg/cc. Whether higher order terms should be included in the expression for anisotropy energy cannot be answered by our experiment; a more powerful magnet is required to move the magnetization vector further away from its easy axis. The existence of these terms, however, would not significantly affect the above values of $K_1$ and $K_2$.

Thanks are due to Dr. Alan Bell of IBM's San Jose Research Laboratories for providing the TbFe samples.

REFERENCES

1. P. Chaudhari, J. Cuomo, and R. Gambino, Appl. Phys. Lett., 22, 337 (1973).

2. Y. Mimura, N. Imamura, and T. Kobayashi, IEEE Trans. Magnet., 12, 779 (1976).

3. Y. Mimura, N. Imamura, T. Kobayashi, A. Okada, and Y. Kushiro, J. Appl. Phys., 49, 1208 (1978).

4. M. Mansuripur, M.F. Ruane, and M.N. Horenstein, SPIE, 529, 25 (1985).

5. K. Ohashi, H. Tsuji, S. Tsunashima, and S. Uchiyama, Jap. J. Appl. Phys., 19, 1333 (1980).

PART III

# Crystalline Magnetic Materials

# RAPIDLY SOLIDIFIED ALLOYS FOR PERMANENT MAGNETS

G.C. HADJIPANAYIS
Department of Physics, Kansas State University, Manhattan, KS 66605

## ABSTRACT

The magnetic and structural properties of melt-spun Fe-R-B alloys are reviewed. The hard magnetic properties of these alloys are attributed to the presence of a highly anisotropic tetragonal phase $Fe_{14}R_2B$ which in the case of ribbons occurs in very fine grains. The interaction of domain walls with this fine microstructure leads to the observed high coercivities. The effects of partial substitution of iron, rare-earth and boron on the formation of $Fe_{14}R_2B$ and on its magnetic properties are also discussed.

## INTRODUCTION

Research on iron-rare-earth magnets begun in early 70's when Rhyne[1] and his coworkers produced amorphous $TbFe_2$ alloys by rapid dc sputtering. Later Clark[2] observed large coercivities on the same materials at cryogenic temperatures. Subsequently he succeeded in producing large coercivities at room temperature in crystallized $TbFe_2$ samples demonstrating thus the potential of these alloys for permanent magnet development. Large coercive fields were also obtained in other rare-earth based systems (amorphous[3] or crystallized[4-6]) but with energy products not exceeding 8 MGOe.

In 1982 a major breakthrough took place when Hadjipanayis et al.[7-8] were able for the first time to obtain coercivities in excess of 10 kOe and an energy product of around 12 MGOe in a cobalt-free melt-spun Fe-Pr-BSi sample. The hard magnetic properties of the samples were attributed to the formation of a highly anisotropic tetragonal $Fe_{20}Pr_3B$ phase which is now properly indexed[9] as $Fe_{14}R_2B$. Later it was shown[20] that the $Fe_{14}R_2B$ phase is formed with most of the rare-earth elements.[10-12] The same phase was found to exist in as-cast alloys and Sagawa et al.[13] produced the first Fe-Nd-B magnets with energy products up to 35 MGOe. Narasimhan et al.[14] were able to increase the energy product up to 45 MGOe and at present these magnets are made commercially by several companies.

The objectives of this report are:

(i) to discuss briefly our early studies of rare-earth-iron based alloys that had led to the discovery of Fe-R-B magnets,

(ii) to summarize the magnetic and structural properties of melt-spun Fe-R-B alloys,

(iii) to discuss the effects of iron substitution with Co,Mn,Ni and neodymium substitutions with other rare-earths on the magnetic properties,

(iv) to present some preliminary data on the effects of metalloid substitutions, and

(v) to discuss the origin of magnetic hardening in the melt-spun alloys.

## DISCOVERY OF Fe-R-B ALLOYS

Our early studies on iron-rare-earth alloys were focused on the ternary systems Fe-R-Ga[6], $Fe(R_1R_2)_2$ and $Fe-(R_1R_2)$ alloys[15] with $R_1$≡Tb and $R_2$≡Pr,Nd,Sm. Rapid solidification by means of melt-spinning was used together with a subsequent heat treatment of the as-quenched samples to produce metastable phases and special microstructures that could not be produced otherwise. The coercivity, $H_c$, of those samples was relatively high (up to 4 kOe) but the magnetization was relatively small (because of

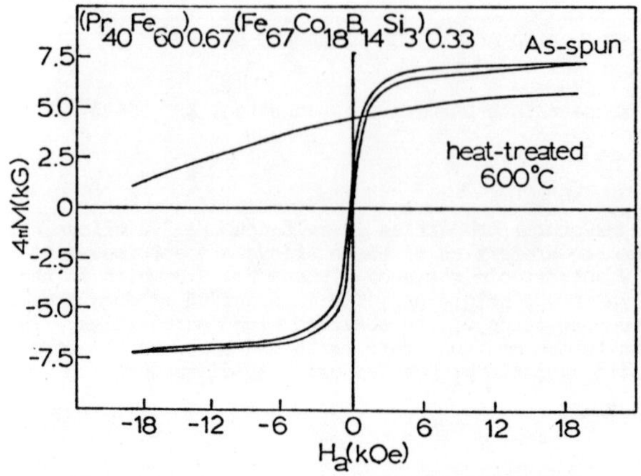


Fig. 1. Hysteresis loops of a Fe(Co)-Pr-BSi sample.

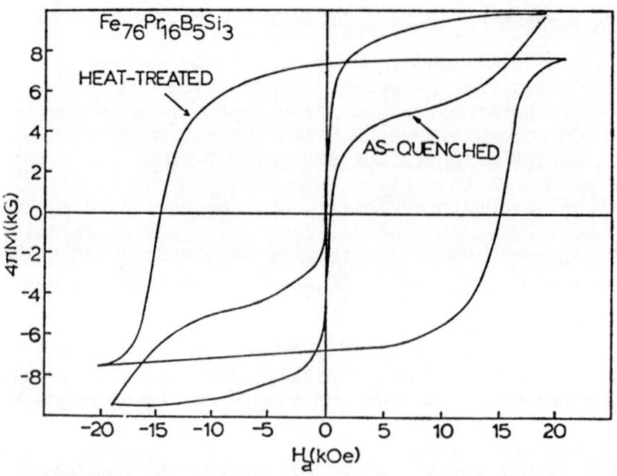


Fig. 2. Hysteresis loops in Fe-Pr-BSi.

the low Fe-content, 30-40 at%) limiting thus the energy product, $(BH)_m$, to very low values. Attempts to enhance the magnetic moment by increasing the Fe-content failed because $H_c$ was significantly reduced in the Fe-rich alloys. This behavior changed completely, however, when a certain amount of metalloid was added into the system.[15] After an <u>in-situ</u> heat treatment in the vibrating sample magnetometer (M vs T up to <u>700°C in</u> a field of ~1 kOe)) a $Fe_{62}Co_6Pr_{26.8}B_{4.2}Si_1$ sample developed a coercivity which could not be measured with an ordinary electromagnet (Fig. 1). In that heat-treatment the amorphous sample crystallized into the $Fe_{14}Pr_2(BSi)$ phase with a Curie temperature around 300°C. That was the beginning of the new generation cobalt-free permanent magnets.[8] After some systematic studies, the composition was simplified to Fe-Pr-(BSi) and the loop shown in Fig. 2 was obtained. Studies over a wide range of composition showed that the best magnetic properties can be obtained with a composition $Fe_{77}Pr_{15}(BSi)_8$. This composition leads mostly to the formation of tetragonal $Fe_{14}Pr_2(BSi)$ phase together with some minor high Pr content and boron-rich phases $(Fe_4Pr_{1.11}(BSi)_4$. Likewise the tetragonal $Fe_{14}R_2B$ phase was found to exist in alloys containing neodymium (Fig. 3). This phase is not metastable but it is rather stabilized by the addition of boron.

Table I. Magnetic Properties of Fe-R-B Alloys

| Compound | $T_C^{am}$ (°C) | $T_C^{cr}$ (°C) | $I_s^*$ (T) | $H_A^*$ (kOe) | $H_C^{**}$ (kOe) |
|---|---|---|---|---|---|
| $Fe_{77}Y_{15}B_8$ | 86 | 287 | 1.03 | 30–40 | 2.8 |
| $Fe_{77}Ce_{15}B_8$ | 90 | 150 | 1.38 | 50–60 | 1.0 |
| $Fe_{77}La_{15}B_8$ | 135 | 310 | 1.40 | 30 | 0.5 |
| $Fe_{77}Pr_{15}B_8$ | 145 | 296 | 1.47 | 290 | 6–10 |
| $Fe_{77}Nd_{15}B_8$ | 163 | 310 | 1.58 | 320 | 10–15 |
| $Fe_{77}Sm_{15}B_8$ | 206 | 338 | 1.40 | basal | 1.0 |
| $Fe_{77}Gd_{15}B_8$ | 293 | 380 | 0.79 | 40–50 | 3.5 |
| $Fe_{77}Tb_{15}B_8$ | 166 | 342 | 0.57 | >200 | >16 |
| $Fe_{77}Dy_{15}B_8$ | 201 | 315 | 0.48 | >200 | >16 |
| $Fe_{77}Ho_{15}B_8$ | 184 | 301 | 0.47 | – | 12 |
| $Fe_{77}Er_{15}B_8$ | 97 | 280 | 0.59 | basal | 0.7 |

*Values of saturation magnetization $I_s$ and anisotropy field $H_A$ at 4.2 K obtained from Sinnema et al.[19]

**The coercivity values given are obtained after the M vs T experiment. They are not optimized.

exchange interactions. Also the Curie temperatures of both the amorphous and crystalline $Fe_{14}R_2B$ phases are found to scale with the DeGennes factor indicating significant contributions from indirect exchange interactions between rare-earth and Fe moments.

The measured coercivities are found to be highest for the alloys containing Tb and Dy. This observation is consistent with anisotropy field data obtained in crystalline alloys showing values exceeding 200 kOe for the Tb and Dy samples (Table I). The coercivity is also sensitive to microstructure and as we shall see below the microstructure in melt-spun samples is ideal for magnetic hardening. Accurate values of anisotropy and saturation magnetization are difficult to obtain in ribbons because of the random distribution of fine $Fe_{14}R_2B$ crystallites. Such studies are usually made on single crystals or aligned powders. However in ribbons the fine $Fe_{14}R_2B$ crystallites have a size below $1\mu$ making it extremely difficult to prepare powders with a size of single grain. For the latter case, high-field magnetization data (up to 160 kOe) at 4.2 K were used and a fit to the law of approach to saturation ($M=M_s(1-A/H^2)$) gave the values of $M_s$ = 1.4, 0.45T and $H_A$ = 180, 190 kOe for Nd and Tb samples, respectively. The lower values of $M_s$ and $H_A$ obtained in Fe-Tb-B are probably due to the presence of other minor non-magnetic phases and to the fact that the magnetic fields used are much lower than the anisotropy fields. In Fe-Nd-B, because of the spin reorientation behavior, the $1/H^3$ term must be included in the law of approach to saturation to find both $K_1$ and $K_2$.

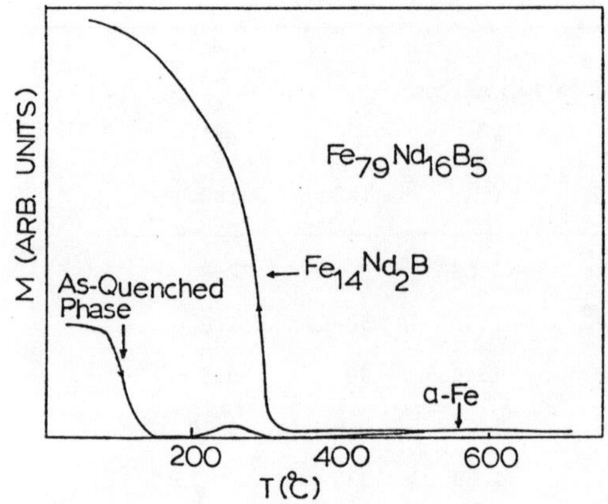


Fig. 3. M vs T in Fe-Nd-B.

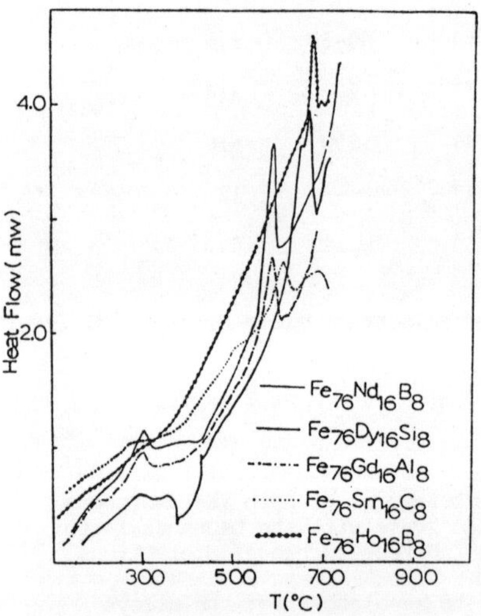


Fig. 4. Differential scanning calorimetry data in Fe-R-M alloys.

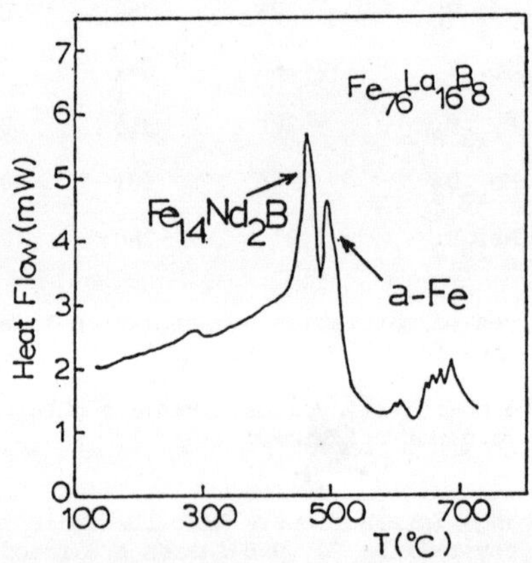


Fig. 5. Differential scanning calorimetry data for Fe-La-B.

MAGNETIC PROPERTIES OF Fe-R-B

Similar studies were made in other iron-rare-earth boron alloys[16] and the results are listed in Table I. Differential scanning calorimetry measurements showed that the crystallization temperatures $T_{cr}$, are all in the range of 560-675°C (Fig. 4). The higher $T_{cr}$ occurs in alloys containing heavy rare-earth metals indicating that they are more stable thermally. Most of the melt-spun samples crystallized into the $Fe_{14}R_2B$ phase with the exception of Fe-La-B which shows a[17-18] two step crystallization where the crystallization to $Fe_{14}La_2B$ is followed by a transformation to a-Fe (Fig. 5).

The high Curie temperatures observed in alloys containing nonmagnetic Y and La suggest that most of the contributions come from direct Fe-Fe

Spin Reorientation

The crystallized Fe-Nd-B ribbons showed a peak in the ac susceptibility data at around 120 K (Fig. 6). This peak has also been observed in aligned Fe-Nd-B powders[20] and $Fe_{14}Nd_2B$ single crystals and has been attributed to a tilt of the easy axis of magnetization away from the c-axis below 140 K. The tilt reaches the value of 30° at 4.2 K. The effect has also been confirmed with neutron diffraction[21] and Mössbauer studies.[22] It is attributed to competing anisotropies in the different Nd sites. The effect is not observed in samples that contain Pr.

Transition Metal Substitutions

The effects of partial Fe substitution with Co,Mn and Ni have been examined[23] and the results are shown in Table II. It is found that the Curie temperature increases steadily with Co substitution while it decreases

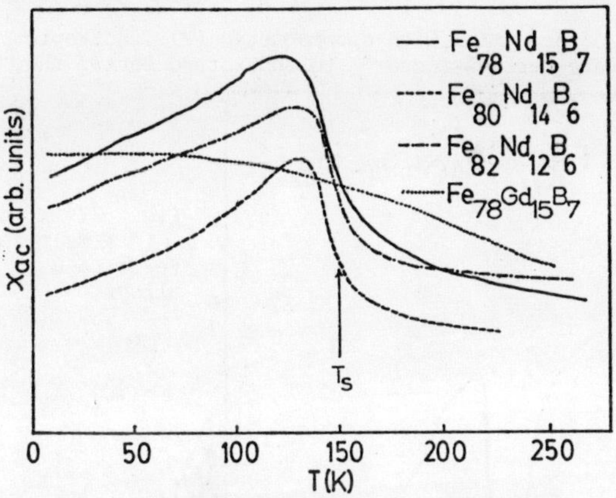


Fig. 6. Ac susceptibility data in Fe-R-B alloys.

Table II. Effects of Partial Fe Substitution on the Magnetic Properties of $Fe_{78-x}Tm_xPr_{16}B_6$ Alloys

| Sample | $T_c^{am}$ | $T_c^{cr}$ | $H_c^*$ |
|---|---|---|---|
| TM x | (°C) | (°C) | (kOe) |
| Co 1. | 120±10 | 310±10 | 7–10 |
| 3 | 140 | 320 | 7–10 |
| 5 | 224 | 360 | 7–10 |
| 40 | | 704 | 7 |
| 78 | | >720 | 13.7 |
| Ni 1 | 138 | 326 | 5–7 |
| 3 | 147 | 325 | |
| 5 | 187 | 332 | |
| 7 | 178 | 315 | |
| Mn 1 | 65 | 235 | 5–7 |
| 3 | 70 | 235 | 5–7 |
| 5 | 60 | 211 | 5–7 |

*The values of coercivity shown are obtained after the MvsT experiment.

drastically with the substitution at Mn which couples antiferromagnetically with Fe. In the case of Ni, an initial increase of $T_c$ is observed and this is followed by a large decrease for higher Ni concentrations. Similar observations were made by Narasimhan[14] and the results can be explained with the energy band theory. The increase in $T_c$ for Co substituted alloys leads to smaller temperature coefficients of remanence. It is interesting to note that for Co substitutions the Curie temperature of the 14:2:1 phase is increased above the crystallization temperature (Fig. 7) so that magnetic annealing is possible for these alloys (crystallization in the presence of a magnetic field). However, no significant results have been observed for Fe-Co-Nd-B after a magnetic annealing in a field of 10 kOe. The Co,Mn,Ni substituted alloys when crystallized showed again large $H_c$ (Fig. 8) indicating that the anisotropy of the substituted alloys is still high. In the case of Co-rich Co-Nd-B alloys large coercive fields could also be obtained consistent with the fact that the anisotropy field of the $Co_{14}Nd_2B$ phase[24] is comparable to that of $Fe_{14}R_2B$.

## Rare-Earth Metal Substitutions

Partial substitution of neodymium with non-magnetic (Y), isotropic (Gd) and anisotropic (Tb) rare-earth was made[25] to understand better the

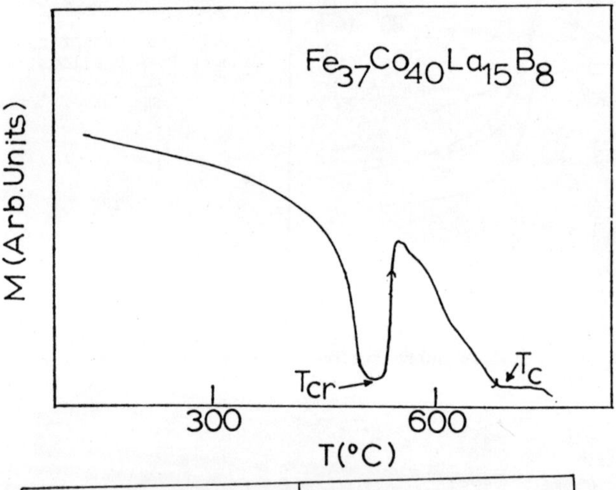


Fig. 7. M vs T in a melt-spun Fe-Co-La-B alloys.

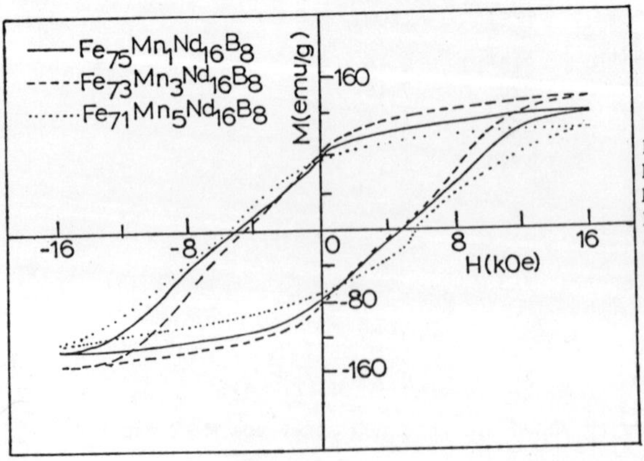


Fig. 8. Hysteresis loops of Mn and Ni substituted Fe(Mn,Ni)-Nd-B alloys.

effects of exchange and anisotropy on the magnetic properties of the mixed alloys. The Curie temperatures of the mixed rare-earth 14:2:1 phases are found to be an average of the Curie temperatures of the individual rare-earth 14:2:1 phases. The increase in anisotropy by heavy rare-earth substitutions is reflected in an increase of $H_c$ (Fig. 9). Substitutions of Y and Gd on the other hand, lead to a reduction in coercivity. Present studies include the mixture of Nd with Sm and Er which show easy plane magnetization in the 14:2:1 phase. This combination of rare-earths is expected to lead to special magnetic structures with the easy axis at an angle between the c-axis and the basal plane.

Metalloid Substitutions

The $Fe_{14}R_2M$ phase is found to be formed only with boron as the metalloid with the exception of carbon that allows formation of a similar phase when used in $Fe_{14}Gd_2C$. The Curie temperature of this phase is found to be about 260°C with a coercivity of about 2 kOe and an anisotropy field of 15 kOe. However, the 14:2:1 phase is still formed after partial substitution of boron with other metalloids. In some cases the phase is formed even with 75% substitution of boron (Fig. 10). This is not possible in the as-cast

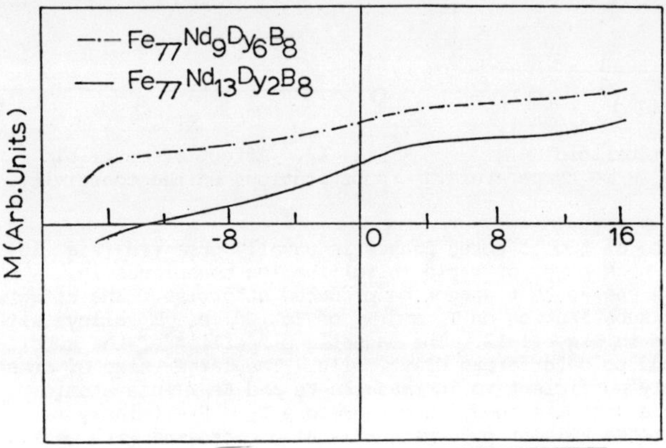


Fig. 9. Hysteresis loops of rare-earth substituted Fe-Nd(R)-B alloys.

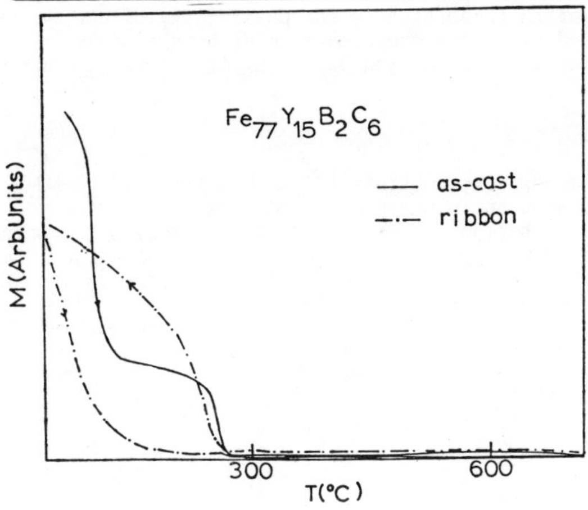


Fig. 10. M vs T in carbon substituted Fe-Y-B(C) alloys.

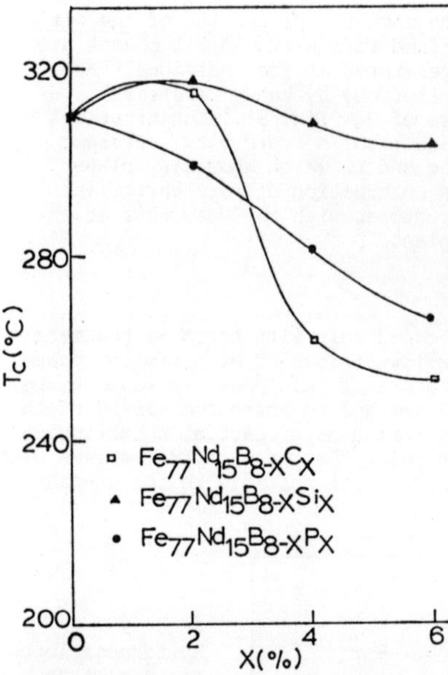


Fig. 11. Effect of metalloid substitutions on the Curie temperature.

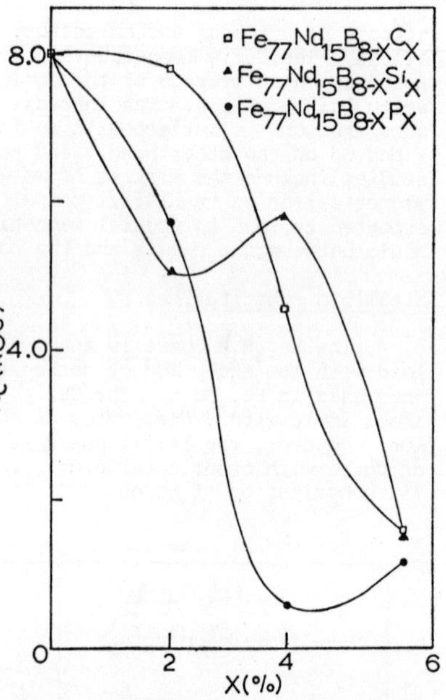


Fig. 12. Effect of metalloid substitutions on the coercivity.

alloys where a mixture of two or three phases is usually observed (Fig. 10). This again shows the high power of rapid solidification techniques in producing metalstable phases that cannot be produced otherwise. The effects of partial metalloid substitution on $T_c$ and $H_c$ of $Fe_{78}Nd_{15}B_{7-x}M_x$ alloys with M≡C,Si,P,Ge are shown in Fig. 11-12. The magnetic properties of the substituted alloys are found to deteriorate drastically. The larger size of other metalloids is probably sufficient to increase Fe-Fe and Fe-R interatomic distances leading to a decrease in $J_{ex}$ and therefore $T_c$. Preliminary studies indicate that the anisotropy constant is also decreased with metalloid substitution. However in these studies the possibility of charge transfer must be considered since the other metalloids have a higher valency. Mössbauer studies are now in progress to examine this hypothesis.

## MICROSTRUCTURE

Transmission electron microscope studies[26] on heat-treated Fe-Nd-B ribbons showed the same phases as in sintered Fe-Nd-B magnets[27]; the tetragonal $Fe_{14}Nd_2B$ phase, a tetragonal $Fe_4Nd_{1+\varepsilon}B_4$ phase, a high Nd content phase and some a-Fe. The secondary phases are not distributed around the 14:2:1 grains but rather they are found at intersections of several $Fe_{14}R_2B$ grains. The only difference observed in the ribbons is the much finer size of the grains. The grain size in melt-spun ribbons ranges from 250-4000 A (Fig. 13) whereas that in sintered magnets is much greater than 1$\mu$ (10-50$\mu$). In Fe-Tb-B the $Fe_{14}Tb_2B$ grains are even finer with a size approximately 100A. This fine microstructure is again characteristic of rapid solidification processes and leads to optimum coercivities which are much higher than those observed in sintered magnets. (This is true when the ribbons consist mostly of the $Fe_{14}R_2B$ phase. If a-Fe or another soft phase is present the coercivity is decreased substantially).

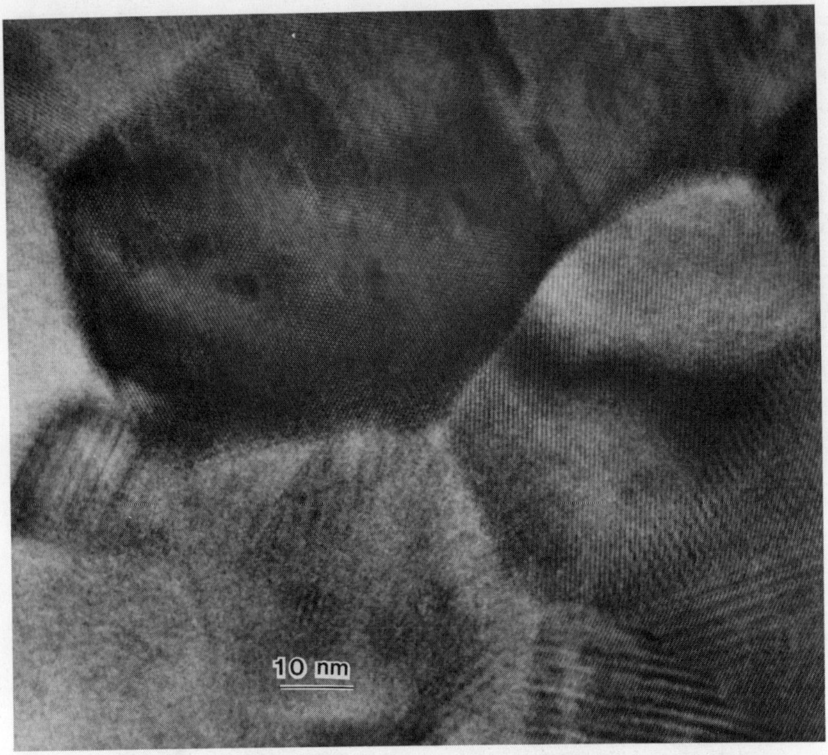


Fig. 13. Microstructure of crystallized Fe-Nd-B ribbons showing $Fe_{14}Nd_2B$ grains.

Lorentz microscope studies show the presence of domain walls (Fig. 14). In the sintered magnets each grain contains several magnetic domains.[26] In ribbons,[27] however, the smaller grains are basically single domains while the larger ones contain several domains. The size of the domains in Fe-Nd-B is around 0.3$\mu$ and is approximately equal to that reported by Livingston.[28] In both cases domain walls end at grain boundaries (Fig. 14) and it is believed that they are pinned there.

## MAGNETIC HARDENING STUDIES

In an attempt to understand the origin of the large coercive fields in these alloys, the microstructure was correlated with the magnetic properties.[29] The initial magnetization curves and the hysteresis loops for both sintered and melt-spun Fe-Nd-B magnets are shown in Fig. 15. In sintered magnets the initial magnetization increases rather drastically at low fields and the samples saturate at a field of 10 kOe. However, the initial magnetization of melt-spun samples increases rather slowly with magnetic field and is far from saturation in the available field of 17 kOe. The field dependence of coercivity (Fig. 16) is also slightly different in the two cases with a stronger dependence in ribbons. However, in both cases

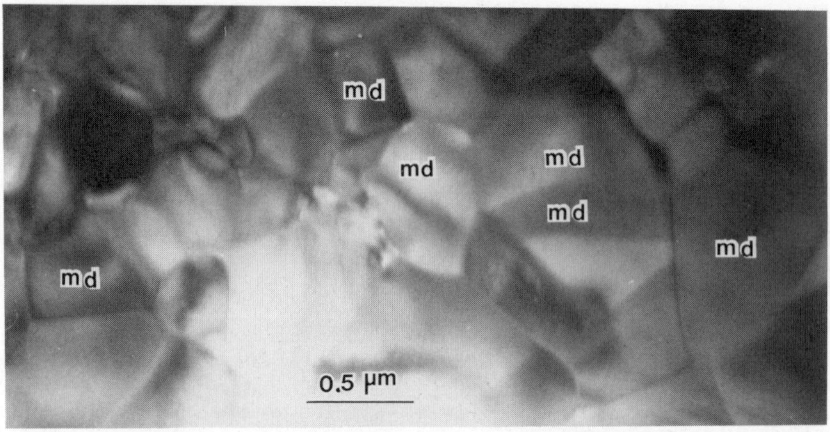


a

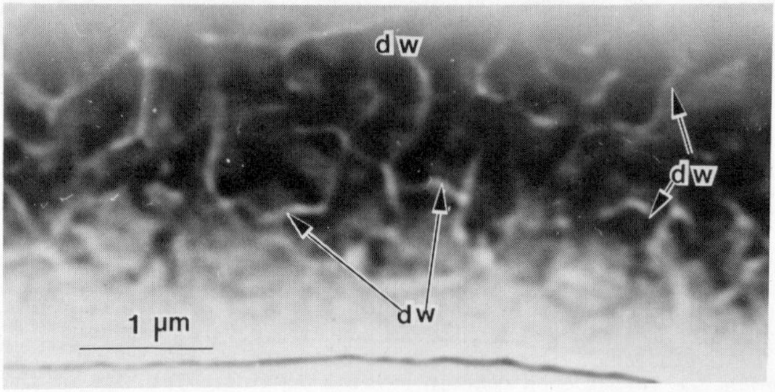


b

Fig. 14. Lorentz microscopy in Fe-Nd-B ribbons (a) displaced aperture method indicating magnetic domains (md), (b) overfocused picture showing domain walls (dw).

two different regions are observed with the lower field dependence due to possibly pinning inside the grains (lower values of $H_c$) and the higher field dependence due to domain wall pinning at grain boundaries. In ribbons the smaller grains are single domain but they do not behave as such because they are so close together that a pseudo-domain wall is formed because of exchange interactions. Therefore when a field is applied the domain walls move rather easily inside the larger grains but with more difficulty in the smaller ones because of domain wall pinning at grain boundaries leading to the behavior observed in Fig. 15.

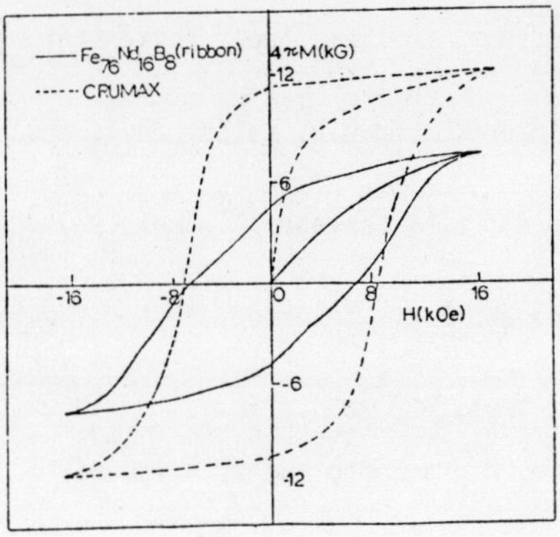


Fig. 15. Initial magnetization curves and hysteresis loops in Fe-Nd-B alloys.

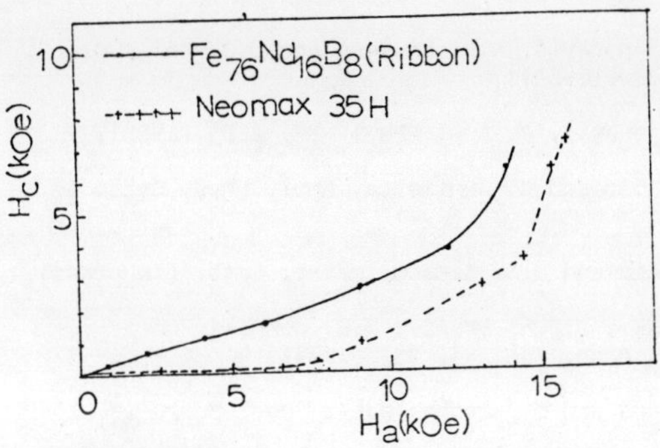


Fig. 16. Field dependence of coercivity in Fe-Nd-B alloys.

ACKNOWLEDGMENTS

This work is supported jointly by the Office of Naval Research and U.S. Army Research.

REFERENCES

1. J.J. Rhyne, in Handbook of Physics and Chemistry of Rare Earths, edited by K.A. Gschneider and L. Eyring (North Holland, Amsterdam 1979) pp. 259-294.

2. A.E. Clark, Appl. Phys. Lett. 23, 642 (1973).

3. G. Hadjipanayis, S.G. Cornelison, J.M. Gerber and D.J. Sellmyer, J. Magn. Magn. Mat. 21, 101 (1980).

4. N.C. Koon and B.N. Das, Appl. Phys. Lett. 39, 840 (1981).

5. J.J. Croat, IEEE Trans. Magn. MAG-18, 1442 (1982).

6. G. Hadjipanayis, S.H. Wollins, R.C. Hazelton, K.R. Lawless, R. Prestipino and D.J. Sellmyer, J. Appl. Phys. 53, 7780 (1982).

7. G.C. Hadjipanayis, R.C. Hazelton and R.K. Lawless, "Investigation of Crystalline Iron-Platinum-Nickel and Amorphous Rare-Earth-Iron Alloys for Permanent Magnets," Final Technical Report, Office of Naval Research (March, 1983).

8. G.C. Hadjipanayis, R.C. Hazelton, K.R. Lawless, Appl. Phys. Lett. 43, 797 (1983).

9. J.F. Herbst, J.J. Croat, F.P. Pinkerton and W.B. Yelon, Phys. Rev. B29, 4176 (1984).

10. G.C. Hadjipanayis, R.C. Hazelton and K.R. Lawless, J. Appl. Phys. 55, 2073 (1984).

11. J.J. Croat, J.F. Herbst, R.W. Lee and F.E. Pinkerton, J. Appl. Phys. 55, 2078 (1984).

12. N.C. Koon and B.N. Das, J. Appl. Phys. 55, 2078 (1984).

13. M. Sagawa, S. Fujimura, N. Togawa, H. Yamamoto and Y. Matsuura, J. Appl. Phys. 55, 2083 (1984).

14. K.S.V.L. Narasimhan, J. Appl. Phys. 57, 4081 (1985).

15. R.C. Hazelton, G.C. Hadjipanayis, K.R. Lawless and D.J. Sellmyer, J. Magn. Magn. Mater, 40, 278 (1984).

16. Y.F. Tao and G.C. Hadjipanayis, J. Appl. Phys. 57, 4103 (1985).

17. G.C. Hadjipanayis, Y.F. Tao and K. Gudimetta, Appl. Phys. Lett. 47, 757 (1985).

18. H.H. Stadelmaier, N.C. Liu and N.A. Elmasm, Mater, Lett. (in press).

19. S. Sinnema, R.J. Radwanski, J.J.M. Franse, D.B. deMooij and K.H.J. Buschow, J. Magn. Magn. Mat. 44, 333 (1984).

20. H. Oesterreicher, F. Spada and C. Abache, Mater. Res. Bull. 19, 1069 (1984).

21. D. Givord, H.S. Li, J.M. Moreau and P. Teuaud, International Conference on Magnetism, San Francisco (1985).

22. N.C. Koon, E. Callen, J. Foster, A. Martinez and S. Segnam, International Conference on Magnetism, San Francisco (1985).

23. G.C. Hadjipanayis, C.P. Wong and Y.F. Tao, in Rapidly Quenched Metal, S. Steeb, H. Warlimont (eds) Elsevier Science Publishers B.V. (1985).

24. K.H.J. Buschow, D.B. deMooij, S. Sinnema, R.J. Radwanski and J.J.M. Franse, J. Magn. Magn. Mater. 51, 211 (1985).

25. K.V. Rao, Y.F. Tao and G.C. Hadjipanayis, This Conference Proceedings.

26. G.C. Hadjipanayis, K.R. Lawless and R.C. Dickenson, J. Appl. Phys. 57, 4097 (1985).

27. G.C. Hadjipanayis, R.C. Dickenson and K.R. Lawless, International Conference on Magnetism, San Francisco (1985).

28. J.D. Livingston, J. Appl. Phys. 57, 4137 (1985).

29. G.C. Hadjipanayis, Proceedings of Fourth International Symposium on Magnetic Anisotropy and Coercivity in Rare-Earth Transition Metal Alloys, Dayton, Ohio (1985).

# AN ANALYTICAL ELECTRON MICROSCOPY CHARACTERIZATION OF MELT-SPUN IRON/RARE-EARTH/BORON MAGNETIC MATERIALS

**R.C. Dickenson & K.R. Lawless**
Dept. of Materials Science, Univ. of Virginia, Charlottesville VA 22901

**G.C. Hadjipanayis**
Dept. of Physics, Kansas St. Univ., Manhattan KN 66506

Iron/rare-earth/boron permanent magnet materials have recently been delveloped to reduce the need for the strategic element cobalt, which was previously the primary component of high-energy magnets. These materials are generally produced by annealing rapidly solidified ribbons or by conventional powder metallurgy techniques. This paper will report results from an analytical electron microscopy characterization undertaken to establish the relationship between the magnetic properties and the microstructure of two iron/rare-earth/boron (Fe/RE/B) alloys. Ribbons of $Fe_{75}Pr_{15}B_{10}$ and $Fe_{77}Tb_{15}B_8$ were produced by melt-spinning. To obtain optimum magnetic properties, both alloys were then annealed at 700°C, the FePrB ribbons for 6 minutes and the FeTbB ribbons for 90 minutes. Foils for transmission electron microscopy were prepared by ion-milling the ribbons on a cold stage and examined using a Philips 400T TEM/STEM equipped with an energy dispersive x-ray unit.

**$Fe_{75}Pr_{15}B_{10}$** In the as-spun state $Fe_{75}Pr_{15}B_{10}$ was found to have a finely crystalline structure (Fig. 1a), but the grain size varied substantially with distance from the wheel side of the ribbon. Selected area diffraction (SAD, Fig. 1b) was used to identify the primary crystalline phase as α-Fe. Two diffuse rings were observed with d-spacings of 0.33nm and 0.30nm. The 0.33nm d-spacing has been interpreted as the combination of the strongest reflection from several Pr-borates, and the 0.30nm d-spacing as an unidentified Pr-rich oxide with a glassy character which is stabilized by some boron content [1]. In addition, relatively large (25nm) crystallites of a FCC Pr-rich phase (a=1.04nm) were observed, though not shown here. The overall composition of this alloy is shown in the EDS spectrum (Fig. 1c).

The FePrB ribbons were annealed to promote growth of the hard magnetic phase $Fe_{14}Pr_2B$ (tetragonal: a=0.88nm, c=1.22nm). An overview of the resulting structure (Fig.2a) revealed that the $Fe_{14}Pr_2B$ phase was the major constituent with an average grain size of 85nm. The corresponding SAD pattern (Fig. 2b) confirms the presence of the hard magnetic phase. The $Fe_{14}Pr_2B$ phase, identified by CBED (Fig.3b) and EDS (Fig. 3c), crystallized as equiaxed grains as shown in a [110] oriented bright field image (Fig. 3a). Lattice imaging (Fig. 3a) revealed few planar defects in the $Fe_{14}Pr_2B$ phase, but some small 'inclusions' (dark spots) were observed. These inclusions were postulated to be α-Fe precipitates.

Small amounts of at least two additional phases were observed, generally as smaller grains (<50nm) at $Fe_{14}Pr_2B$ grain boundaries. The most prevalent of these was a Pr-rich phase occuring as very small crystallites at $Fe_{14}Pr_2B$ grain boundary junctions. In comparable Nd-based alloys, Nd-rich phases have been identified by various authors as FCC (a=0.52nm), BCC (a=0.29nm), BCC (a=1.14nm), DHCP (a=0.37nm, c=1.20nm) and amorphous [2]. The microstructure and microchemistry of the $Fe_{14}RE_2B$ grain boundary regions are complex and strongly dependent on the thermal history, so observations vary with alloy preparation. The Pr-rich grain boundary phase in this study was observed to be about 50nm in size and generally showed planar defects (Fig. 4a). This phase was identified by CBED (Fig. 4b) to be cubic with a=0.29nm which is consistent with α-Fe. EDS (Fig. 4c) indicated that this phase had a high concentration of Pr, but α-Fe cannot dissolve an appreciable amount of Pr. It has been suggested, though not yet confirmed, that the BCC phase is α-Fe and the high Pr content is explained by the presence of a Pr-rich oxide which was co-precipitated with α-Fe from a single glassy parent phase [3]. These grain

Mat. Res. Soc. Symp. Proc. Vol. 58. 

boundary regions are thought to aid in pinning magnetic domain walls.

The second minor constituent (Fig. 5a) shows a high density of planar defects and was slightly richer in Pr than nominal (Fig 5b). This phase was identified as $Fe_4Pr_1B_4$ (tetragonal: a=0.71nm, c=2.74nm) which would make little contribution to the coercivity because of its low Curie temperature.

**$Fe_{77}Tb_{15}B_8$**

The Tb based alloy required a much longer annealing time to generate the desired magnetic coercivity, but higher coercivity was obtained (>17kOe) than from Nd or Pr based alloys. In the as-spun state, the ribbons were observed to have a glassy structure in the regions closest to the quenching wheel becoming crystalline (Fig. 6a) with an increasing grain size (Fig. 7a) as the distance from the wheel became greater. SAD (Fig. 6b) from the region in Figure 6a showed $\alpha$-Fe and possibly Tb-borates and Tb-rich oxides, which was exactly analogous to the Pr-based alloy. The region shown in Figure 7a consisted primarily of very small grains (<100nm) of $Fe_{14}Tb_2B$ as identified by CBED (Fig. 7b).

In the annealed state, the structure consisted primarily of a polycrystalline matrix which contained large grains of a heavily faulted phase. $\alpha$-Fe was a major constituent of the matrix, but the other constituents have not been identified as yet. The primary cryatalline phase existed as large grains (100-500nm) which showed a very high density of planar faults (Fig. 8a) and contained slightly more Tb than the nominal composition. CBED from two orientations (Fig. 8b, 8c) revealed a complex crystal structure which we have identified as tetragonal (a=0.50nm, c=2.84nm). This tetragonal structure is distinctly different from that of the $Fe_{14}RE_2B$ phase (responsible for the hard magnetic properties in other Fe-RE-B alloys) which rarely exhibits planar faulting. A more thorough characterization of this promising system has been undertaken.

This work was sponsored by the Office of Naval Research.

1. N.A. El-Masry and H.H. Stadelmaier, **Mater. Lett.** 3 (9,10), 405-408 (1985).
2. K.R. Lawless and R.C. Dickenson, Proceedings of the 1st Beijing Conference and Exhibition on Instrumental Analysis, Beijing, CN, 1985.
3. H.H. Stadelmaier and N.A. El-Masry, Proceedings of the 4th International Symposium on Magnetic Anisotropy and Coercivity in Rare Earth-Transition Metal Alloys, Dayton, OH, 613-631 (1985).

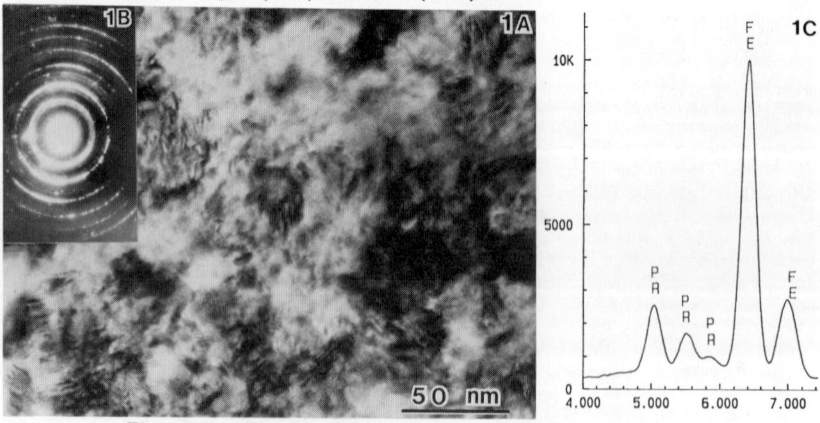


**Figure 1 - $Fe_{75}Pr_{15}B_{10}$ As-Spun**
a. Bright field micro-crystalline region.
b. SAD showing $\alpha$-Fe rings, Pr-borates and amorphous Pr-oxide.
c. EDS showing overall composition of alloy.

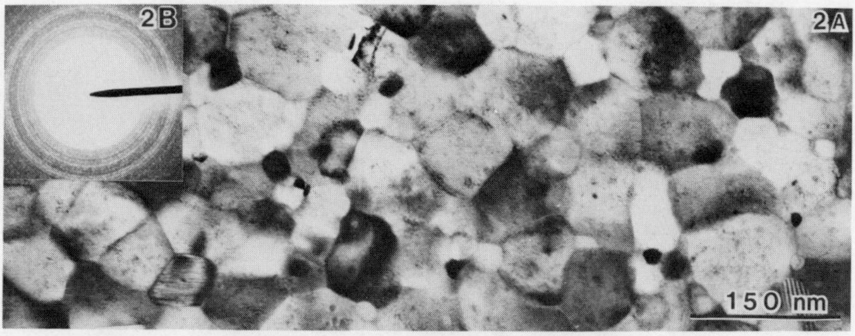


**Figure 2 - $Fe_{75}Pr_{15}B_{10}$ Annealed 700° - 5 minutes.**
a. Bright field showing overall structure, primarily $Fe_{14}Pr_2B$.
b. SAD corresponding to 1a.

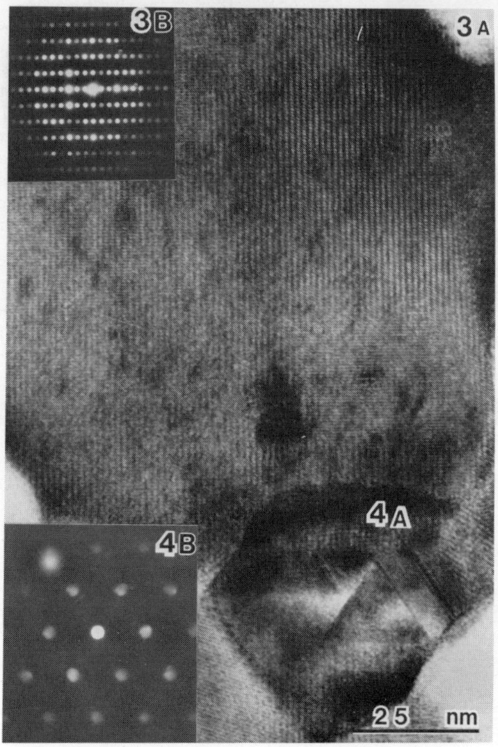


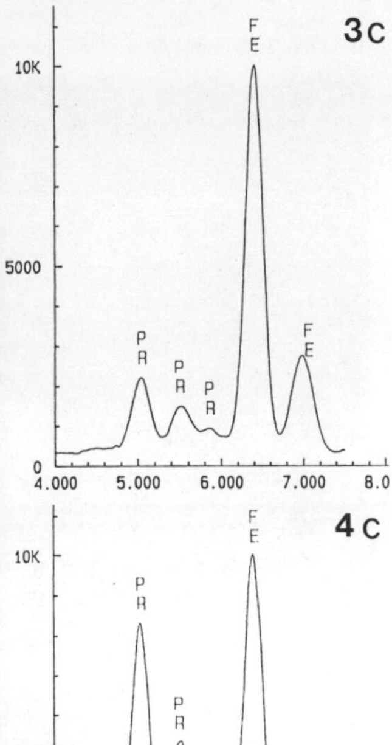


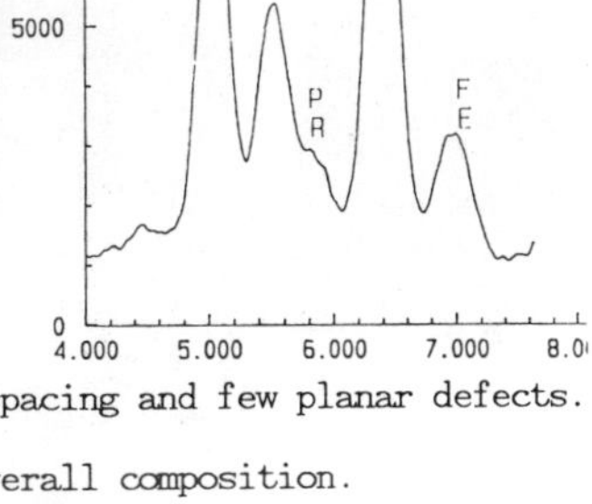


**Figure 3 - $Fe_{14}Pr_2B$ Grain**
a. Bright field lattice image showing 1.22nm spacing and few planar defects.
b. Corresponding CBED (110).
c. Corresponding EDS, slightly less Pr than overall composition.

**Figure 4 - Intergranular Pr-Rich Crystallite**
a. Bright field showing planar faulting.
b. [111] CBED corresponding to $\alpha$-Fe (cubic: a=0.29nm).
c. Corresponding EDS indicating presence of Pr-rich oxide.

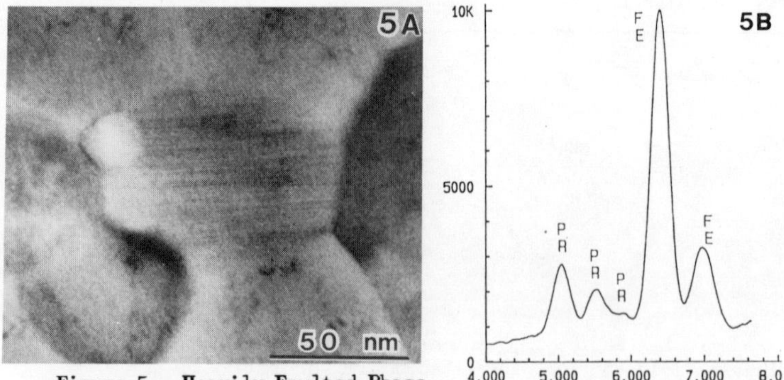


**Figure 5 - Heavily Faulted Phase**
a. Bright field showing heavily faulted crystallite, probably $Fe_4Pr_1B_4$.
b. Corresponding EDS showing slightly more Pr than overall composition.

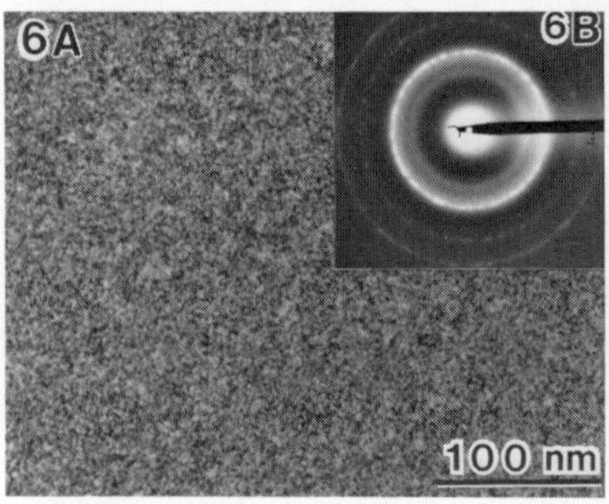


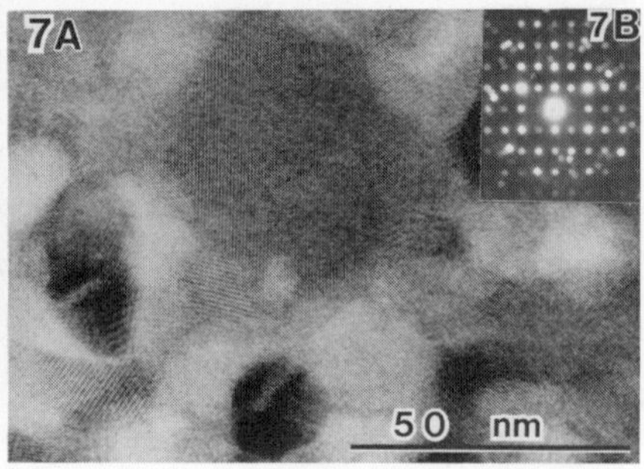


**Figure 6 - $Fe_{77}Tb_{15}B_8$ As-Spun**
a. Bright field showing the beginning of the crystalline region.
b. SAD showing α-Fe and possibly amorphous Tb-borates and Tb-rich oxide.

**Figure 7 - $Fe_{77}Tb_{15}B_8$ As-Spun**
a. Bright field showing region farthest from wheel, primarily $Fe_{14}Tb_2B$.
b. CBED identifying $Fe_{14}Tb_2B$ phase.

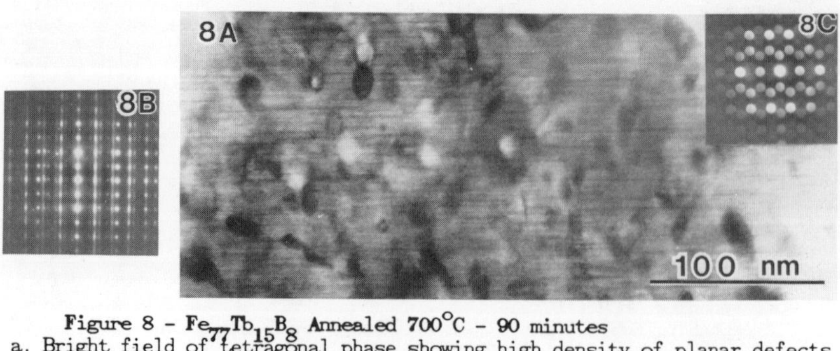


**Figure 8 - $Fe_{77}Tb_{15}B_8$ Annealed 700°C - 90 minutes**
a. Bright field of tetragonal phase showing high density of planar defects.
b. CBED showing streaking from planar defects.
c. CBED showing complexity of crystal structure.

# CRYSTALLIZATION AND MAGNETIC STUDIES IN MELT-SPUN $Fe_{77}Nd_{15-x}R_xB_8$ RIBBONS.

K.V.RAO, R.MALMHÄLL*, KRISHNA GUDIMETTA§ AND G.C.HADJIPANAYIS§
*Dept. of Solid State Physics, Royal Inst. of Technology, Stockholm, Sweden*
* *ASEA Research & Innovation, S-721 78 Västerås, Sweden*
§ *Dept. of Physics, Kansas State University, Manhattan, KS 66506, USA*

## ABSTRACT

Low field (∿5 Oe) Magneto-Thermo-Gravimetric technique has been used to study the effects of substitution of Nd with Y, Gd, and Dy in melt-spun $Fe_{77}Nd_{15}B_8$ alloys, and also to monitor the changes in the magnetic properties on heat treatment to achieve optimum magnetic hardening in these alloys. From VSM-measurements preliminary studies of the effect of substitution for Nd on the hysteretic loops are presented.

## INTRODUCTION

It is by now well recognized that cobalt-free Fe-Nd-B-based permanent magnet materials with energy products higher than ever (>40 MGOe) are expected to have a significant impact on the 'magnet world' in the years to follow. The outstanding magnetic properties, possible only in the ternary alloys, are mainly due to the highly anisotropic tetragonal $Fe_{14}R_2B$ phase, where R is a rare-earth element [1-3]. In these ternary transition metal-rare earth based alloys both 3d and 4f magnetism exist. The 3d exchange stabilizes the magnetic ordering temperature to well above room temperature, while the 3d-4f magnetic interactions provide the strong 4f anisotropy. However, it is found that the tetragonal phase by itself cannot produce the observed high coercivity (and hence the high energy product). So, magnetic hardening by suitable heat treatment or otherwise is a necessary procedure. Methods to enhance the anisotropy, raise the Curie temperature (which at present is around 300°C) of the tetragonal phase, and determination of the relevant phases developed while 'sintering' from a powder compact [3] or consolidating from rapidly solidified ribbons [4] are at present some of the areas of considerable research activity. From this point of view melt-spinning or rapid solidification technique is most useful, because one can then try a wider range of alloy combinations without the constraints of equilibrium phase diagrams, and also by choosing the appropriate parameters of spin-casting enhance the fraction of the formation of phases with high anisotropy fields. In other words, the coercivity $H_C$ is a very strong function of the quench rate which can be exploited by the melt-spinning technique. In order to monitor the changes in the magnetic properties and choose appropriate parameters for heat treatment, low field thermo-magnetic measurements are very useful. Such measurements at low fields of about 5 Oe are very sensitive to changes in the magnetic properties and reveal features often missed in studies made by conventional techniques like VSM or conventional magnetometry. Thus Magneto-Thermo-Gravimetry, (MTG), technique is simple and perhaps ideal for studies of alloy development, thermal stability and determination of the appropriate heat treatment conditions in developing new materials via rapid quenching. We have used the MTG technique to study the effect of substituting Y, Gd, and Dy for Nd in $Fe_{77}Nd_{15}B_8$ melt-spun alloys. In order to study the role of contributions from non-magnetic Y, s-state Gd, and Dy on the anisotropy, from VSM measurements, we also present our preliminary hysteresis loop data for these alloys.

## EXPERIMENTAL DETAILS

The as cast $Fe_{77}Nd_{15-x}R_xB_8$ ingots, where R = Y, Gd, and Dy, and x = 2,4,6,10, in the form of buttons of 5-6 gms were made by arc melting under high purity argon atmosphere. From these ingots the rapid quenched melt-spun ribbons were prepared under almost identical conditions in an argon atmosphere. The thermomagnetic studies were carried out in a Perkin-Elmer TGS-2 system in applied fields of about 5 Oe in the plane of the ribbons. Hysteretic loop measurements were made at room temperature with a vibration sample magnetometer, VSM, in applied fields up to 16 kOe after heat treating the 'as quenched' samples at 700°C for about 15 minutes.

## RESULTS AND DISCUSSIONS

As an example for the temperature dependence of the low field magnetization of the $Fe_{77}Nd_{15-x}R_xB_8$ system, in fig 1 data for the $Fe_{77}Nd_9Y_6B_8$ sample is shown. For an 'as quenched' sample with increasing temperatures, after

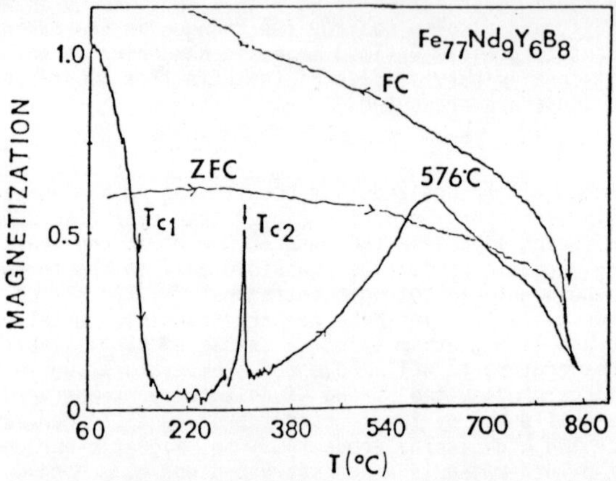


*Fig 1 Temperature dependence of low field (~5Oe) magnetization for the Fe-Nd-Y-B sample in the as quenched state and both Zero-Field-Cooled (ZFC) and Field cooled (FC) states below 850°C.*

a sharp drop in the magnetization (normalized to its value at room temperature) around a transition temperature $T_{C1}$ the observed remnant magnetization just above $T_{C1}$ is an indication of the presence of other micro-crystalline phases. In the case of this specific alloy the existence of a small amount of tetragonal phase with an unusually sharp peak in the magnetization just below the transition temperature $T_{C2}$ is easily identifiable. Thus, because of the anisotropy effects, which in some way must be related to the sharp peak around $T_{C2}$ along with the role of the microstructure of the alloy, the relative amounts of the two $T_{C1}$ and $T_{C2}$ phases cannot be readily estimated. At any rate the usual simple linear extrapolation of the magnetization data is not a safe procedure to take in estimating the relative phases present. The phase corresponding to $T_{C1}$ is probably 'glassy' if not microcrystalline which then begins to reconstitute itself into various metastable phases upon heating to above 350°C with the maximum microstructural changes occuring around 576°C. Heating the sample to above 750°C only results is the segregation of more and more of α-Fe, the soft magnetic phase which is to be avoided. Both the field cooled, FC, and zero-field cooled, ZFC, data for the magnetization in the same applied field below 850°C show the predominance of α-Fe. DC-hysteretic loop studies clearly indicate a complex but low coercivity material on such a heat treatment above 750°C Thus, low field magnetization studies as illustrated in fig 1 are very useful in quickly determining some of the useful parameters like temperature and time for achieving suitable magnetic hardening in these materials[5].

Figure 2 is an example to illustrate the differences observed in the enhancement of the tetragonal phase on substituting 6 at.% Y, Dy, and Gd when the ribbons are prepared under identical conditions. The values of $T_{C1}$ and $T_{C2}$ obtained for the respective alloys that were studied are given in table I. As seen in the table, it appears that while changes in $T_{C1}$ and $T_{C2}$ are not so significant improvements over the value for the Fe-Nd-B alloy on substitution of Nd with Y or Dy, $T_{C2}$ for the tetragonal phase is found to be increased by about 50°C with 10 at.% Gd. However, from a practical point of view alloying with Dy is found to be more beneficial as we shall discuss later with the hysteresis studies.

Now that we have some information on the role of Y, Gd and Dy substitution and the conditions for heat treatment, we describe some features of the development of coercive fields on annealing these samples. We find that the reconstitution and crystallization behaviour described above is roughly the same for all the alloys. Figure 5 illustrates the room temperature hysteretic properties in applied fields up to 16 kOe achieved after heat treating the 'as quenched' alloys for 15 minutes at 700°C. As seen in fig 5, upon substitution with Y and Gd, the following observations can be made regarding the effects on the magnetic properties of $Fe_{77}Nd_{15}B_8$: 1) With Y as the substituent, the intrinsic coercivity is reduced substantially but with a non-linear concentration dependence on Y; Also, it appears that initially on adding Y, a magnetically soft phase develops in the material which results in a sigmoidal hysteretic loop indicating a mixture of soft and a hard phase which does not saturate even at 16 kOe. The situation changes a little with 6 at.% Y in which a closed loop is obtained with a higher value for the coercivity although it never ach ves the value for the initial Fe-Nd-B alloy. Also note that for the 2 at.% Y alloy an asymmetric loop is observed at the onset of introducing the soft phase.
2) substitution of Nd with Gd appears to modify the morphology rather than effect any of the magnetic properties. Initially the coercivity increases up to 10 kOe with little change in the saturation magnetization, but the loop remains almost symmetric although some effects of mixed magnetic phases are easily noted in the shape of the loop.
3) Dysprosium on the other hand produces large coercivities, in excess of 16 kOe for the alloys with over 10 at.% Dy, So much so that due to the limitations of the available fields it was not possible to saturate or measure a closed loop. The values of the coercivities achieved in the case of different alloys are given in table I. These results

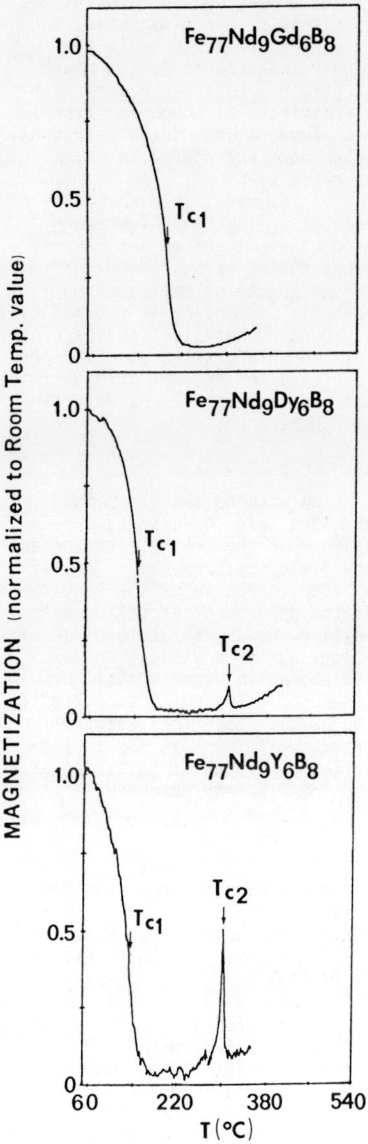


*Fig 2 Temperature dependence of the low-field magnetization of 'as quenched' Fe-Nd-B alloys with 6 at.% Y, Dy, and Gd substitution for Nd.*

can be understood in terms of the fact that Y is non-magnetic, Gd is an s-state ion, while the highly anisotropic Dy-ion makes it possible to achieve very high coercivities although the transition temperature of the tetragonal phase does not change in any dramatic way.

In summary, from the point of view of achieving a high energy product permanent magnet material, substitution of Dy for Nd in $Fe_{77}Nd_{15}B_8$ alloy modifies the magnetic properties in a beneficial way. Further detailed studies on the relationship between observed coercivities and the anisotropy fields in these systems will be reported elsewhere.

ACKNOWLEDGEMENTS

We wish to thank J.S.Muñoz, and Anna Roig for extensive help in some of the crystallization and low-field measurements. One of us (GCH) appreciates the hospitality at the department of solid state physics, Royal Institute of Technology during a visitation for collaborative experimental work. Research in Sweden is supported by the Swedish Technical Research Board, while work at KSU is supported by Office of Naval Research and the U.S.Army Research.

REFERENCES

1 J.F.Herbst, J.J.Croat, P.E.Pinkerton & W.B.Yelon, Phys Rev B29 4176 (1984).
2 G.C.Hadjipanayis, R.C.Hazelton, K.R.Lawless, Appl Phys Lett 43 797 (1983).
3 M. Sagawa et al J. Appl. Phys. 55, 2083 (1984).
4 R.W.Lee, E.G.Brewer and N.A. Schaffel, IEEE-Trans Magn. MAG-21, 1958 (1985)
5 K.V.Rao, R.Malmhäll & G.C.Hadjipanayis, J Physique C6 C6-229(1985)

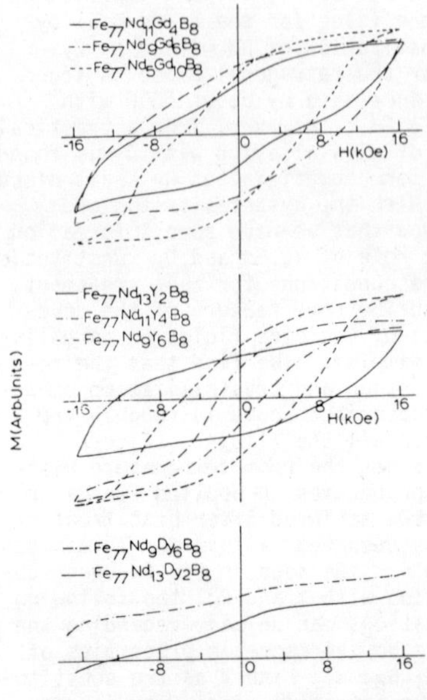


*Fig 3 Hysteresis loop data for Fe-Nd-B alloys with Dy, Y and Gd substitution for Nd after annealing the samples at 700°C for 15 minutes.*

*Table I: Transition temperatures $T_{c1}$, $T_{c2}$ (see text) and coercivity $H_c$ for $Fe_{77}Nd_{15-x}R_xB_8$ alloys.*

| | x | $T_{C2}$ | $T_{C1}$ | $H_c$ (kOe) |
|---|---|---|---|---|
| | 2 | 308°C | 147°C | 12.7 |
| | 4 | 313 | 180 | 11.0 |
| $Fe_{77}Nd_{15-x}Dy_xB_8$ | 6 | 315 | 175 | >16.4 |
| | 10 | 317 | 190 | >>16.4 |
| | 2 | 291 | 166 | 10.6§ |
| | 4 | 316 | 174 | 2.05 |
| $Fe_{77}Nd_{15-x}Y_xB_8$ | 6 | 322 | 158 | 5.7 |
| | 10 | 328 | 174 | 3.4 |
| | 2 | 333 | 194 | 4.95 |
| | 4 | 340 | 140 | 7.15§ |
| $Fe_{77}Nd_{15-x}Gd_xB_8$ | 6 | 349 | 195 | 10.3 |
| | 10 | 368 | 229 | 5.5 |

§ - displaced hystereses loops.

# MAGNETIC EFFECT ON THE PHASE EQUILIBRIA OF FCC(Fe,Ni) ALLOYS AT LOW TEMPERATURES

JEN-CHWEN LIN, YING-YU CHUANG AND Y. AUSTIN CHANG
Department of Metallurgical and Mineral Engineering, University of Wisconsin-Madison, Madison, WI 53706

## ABSTRACT

Phase stabilities of iron alloys at low temperatures are strongly infuenced by magnetic effect. The appearance of certain type of equilibria is often due entirely to magnetic contribution to the Gibbs energy of the pertinent phase. The appearance of the stable and metastable equilibria in fcc(Fe,Ni) alloys are discussed in terms of the magnetic interaction.

## INTRODUCTION

We have recently analyzed the thermodynamic and phase equilibrium data of Fe-Ni and Fe-Cr from the liquid phase down to about 500 K [1,2]. The thermodynamic properties of the liquid and various solid phases were fixed primarily by the thermochemical data measured at high temperatures. At high temperatures, the phases are paramagnetic and the measured values represent the chemical contribution to the total Gibbs energies of these phases. With decreasing temperatures, these phases become magnetic. The contributions to the thermodynamic properties of these phases due to magnetic ordering were described using the equations proposed by Chuang, Schmid and Chang [3]. The proposed equations contain two parameters, the Curie temperature $T_C$ and the mean magnetic moment per atom $\beta$. Values of $T_C$ and $\beta$ for the pertinent phases in Fe-Ni and Fe-Cr are available in the literature. From a knowledge of $T_C$ and $\beta$, the magnetic portions of the Gibbs energies were obtained and added to the chemical terms. Knowing the Gibbs energies of all pertinent phases, phase equilibria, both stable and metastable ones, were calculated for Fe-Ni and Fe-Cr. In the present communication, we wish to limit our discussion to the magnetically induced phase equilibria in Fe-Ni at low temperatures and how they can be used to rationalize some of the results obtained under electron irradiation.

## THE Fe-Ni BINARY

Fig. 1 shows the calculated Fe-Ni phase diagram from 1200 to 500 K as compared to the data reported in the literature [4-9]. With the exception of a few data points by Owen and Sully [8], there is good agreement between the calculated and experimentally measured phase boundaries. Phase separation of the fcc phase to $\gamma_1$ (paramagnetic) and $\gamma_2$ (ferromagnetic) is due to the magnetic Gibbs energy of the phase. The nonmagnetic Gibbs energy of the fcc phase is concave everywhere (holds water) within the composition range of interest. But the compositional variations of the magnetic Gibbs energy are such when it is added to the nonmagnetic term, a hump is produced in the total Gibbs energy curve. This hump causes the phase separation. Fig. 2 shows the same phase diagram without any of the experimental data. Moreover, the $T_0(\alpha+\gamma)$ line is shown as well as metastable extensions of the various phase boundaries and of $T_C^{\gamma}$. The spinodals for the $\gamma_1+\gamma_2$ gap are also shown. Since some of Gibbs energy equations used are valid down to 500 K, extension of the various phase boundaries, $T_0(\alpha+\gamma)$ and the spinodals to lower temperatures are subjected to larger uncertainties. The well-known Invar iron-nickel alloys have compositions falling within the calculated miscibility gap. Many different kinds of data on the Invar alloys have been reported in the literature. Before rationalizing some of these data in terms of the miscibility gap, let us first discuss the appearance of the spinodals in the gap.

 

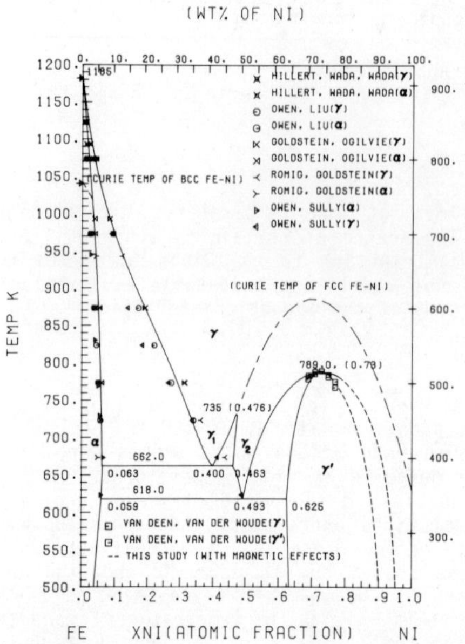


Fig. 1 The calculated Fe-Ni phase diagram from 1200 to 500 K as compared to experimental data. The symbols $\alpha$, $\gamma$ ($\gamma_1$, $\gamma_2$) and $\gamma'$ denote the bcc fcc (paramagnetic, ferromagnetic) and ordered $Cu_3Au$ ($L1_2$) phases.

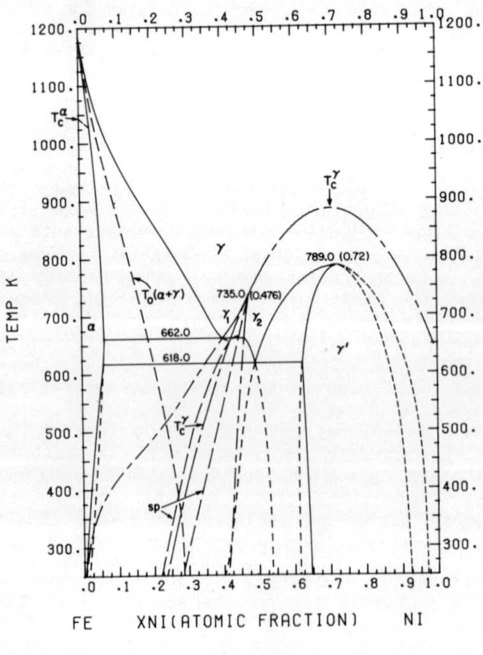


Fig. 2 The calculated Fe-Ni phase diagram from 1200 to 300 K showing also the $T_0$ ($\alpha+\gamma$) line, metastable extensions of $T_C^\gamma$, phase boundaries and the spinodals for the fcc phase.

Using a Landau expansion of the free energy, Allen and Cahn [10,11] discussed the phase equilibria in the vicinity of a tricritical point such as the one displayed in Figs. 1 and 2. The tricritical point is the temperature where the Curie temperature curve intercepts the miscibility gap of the fcc phase at 735 K and 47.6 at% Ni. Slightly below the tricritical temperature, there is one spinodal, that for the ordered phase, in agreement with Allen and Cahn's analysis. However, at lower temperatures, the other spinodal for the disordered phase also appears. We will show that this spinodal does appear using the relative stability function introduced by Schmid, Chuang and Chang [12]. This function is defined as

$$^{re}St \equiv \left(\frac{\partial^2 G}{\partial x^2}\right) / \left(\frac{\partial^2 \, {}^{id}G}{\partial x^2}\right) \qquad (1)$$

where G is the Gibbs energy of the phase, $^{id}G$ is that for ideal mixing and x is the composition in atom fraction. This function is unity for the pure component elements and is zero at the spinodals. Fig. 3 shows schematically the relative stability functions at five different temperatures for the type of equilibria exhibited by Fe-Ni. At $T_1$ (higher than the tricritical point) $^{re}St$ is positive everywhere and is discontinuous at $x(T_c)$. At $T_2$, the tricritical point, $^{re}St$ for the ordered phase reaches zero, resulting in the appearance of the spinodal. At $T_3$ (slightly lower than the tricritical point), $^{re}St$ for the disordered phase is positive from x=0 to $x(T_c)$ but $^{re}St$ for the ordered phase varies from a negative value to unity with the spinodal point at a concentration higher than $x(T_c)$. At $T_4$, $^{re}St$ for the disordered phase becomes zero at $x(T_c)$, and the spinodal for the disordered phase also appears. At $T_5$, $^{re}St$ for the disordered phase decreases from unity to a negative value at $x(T_c)$ and $^{re}St$ for the ordered phase increases from a negative value at $x(T_c)$ to unity.

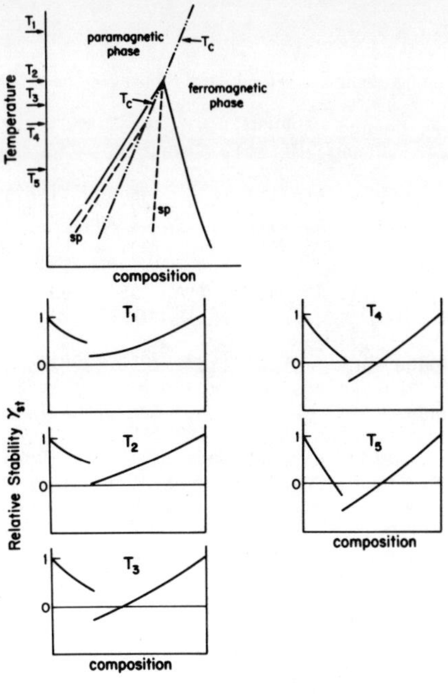


Fig. 3 The relative stability function as a function of composition at temperatures higher and lower than the tri-critical point.

Next, we will discuss the results of Chamberod, Laugier and Penisson [13] and Chamberod, Rechenberg and De Tournemine [14]. Four alloys containing 30, 32, 35 and 40 at% Ni were subjected to 3 Mev electron irradiation from a Van de Graaf accelerator at 353 K (80°C) with ~$10^{19}$ electron/$cm^2$. Mossbauer spectra of these alloys showed two phases, one iron-rich being paramagnetic and the other nickel-rich being ferromagnetic. X-ray diffraction shows an fcc structure with the lattice parameters varying linearly with composition. These alloys lost their Invar effect, i.e. their coefficients of thermal expansion increase with temperature. These results are consistent with the calculated phase boundaries of $\gamma_1+\gamma_2$. Electron irradiation produced large amounts of defects in the lattice which enhanced solid-state diffusion, resulting in phase separation of the fcc phase. According to the calculated diagram, $\gamma_1+\gamma_2$ is not a stable mixture but metastable with respect to $\alpha$ and $\gamma'$. Nevertheless, it is reasonable to expect that phase separation of the fcc phase occurs first prior to reaching the stable mixture. Chamberod et al. also carried out irradiation studies at 523 K (250°C) for alloys containing less than 40 at% Ni and more than 42 at% Ni. To the author's surprise , the alloys with 40 at% Ni and less showed more of the $\gamma_2$ phase than the alloys with nickel contents higher than 42 at%. Yet according to Fig. 2, this is what to be expected since the iron-rich alloys fall within the spinodals. The higher nickel concentration alloys must decompose via a mechanism of nucleation and growth. Chamberod et al. carried out many other types of experiments and most of their results may be interpreted readily in terms of the diagram given in Fig. 2.

ACKNOWLEDGEMENT

The authors wish to thank the National Science Foundation for financial support through Grant No. NSF-DMR-83-10529.

REFERENCES

1. Y.-Y. Chuang, Y. A. Chang, R. Schmid and J.-C. Lin, Metall. Trans. A, in print (1986).
2. Y.-Y. Chuang, J.-C. Lin and Y. A. Chang, CALPHAD, in print (1986).
3. Y.-Y. Chuang, R. Schmid and Y. A. Chang, Metall. Trans. A., 16A, 153 (1985).
4. M. Hillert, T. Wada and H. Wada, J. Iron Steel Inst., 205, 539 (1967).
5. E. A. Owen and Y. H. Liu, J. Iron Steel Inst., 163, 132 (1949).
6. J. I. Goldstein and R. E. Ogilvie, Trans. Met. Soc. AIME, 233, 2083 (1965).
7. A. D. Romig and J. I. Goldstein, Metall. Trans. A, 11A, 1151 (1980).
8. E. A. Owen and A. H. Sully, Phil. Mag., 27, 614 (1939), 31, 314 (1941).
9. J. K. Van Deen and F. Van Der Woude, Acta Metall., 29, 1255 (1981).
10. S. M. Allen and J. W. Cahn, in Proceedings of the Materials Research Society Symposium, Phase Diagram, ed. by L. H. Bennett, T. B. Massalski and B. C. Giessen, North Holland, N. Y., 195 (1983).
11. S. M. Allen and J. W. Cahn, Bull. Alloy Phase Diagrams, 3, 287 (1982).
12. R. Schmid, Y.-Y. Chuang and Y. A. Chang, Calphad, in print (1986).
13. A Chamberod, J. Laugier and S. M. Penisson, J. Magn. Magn. Mat., 10 139 (1979).
14. A. Chamberod, H. Rechenberg and R. De Tournemine, ICM-73, 3, 170 (1974).

# NEW MICROSTRUCTURAL MODEL FOR SPUTTERED Co-Cr MAGNETIC THIN FILMS

YOSHIRO NIIMURA AND MASAHIKO NAOE
Tokyo Institute of Technology, Dept. of Electrical & Electronic Engineering
2-12-1, Oh-okayama, Meguro-ku, Tokyo 152, JAPAN

## ABSTRACT

Rapidly solidified Co-Cr thin films with a perpedicular magnetic anisotropy have been extensively investigated as one of the most promising media for perpendicular magnetic recording system. Most of these films used to be prepared by conventional RF diode sputtering and vacuum evaporation methods. We have developed the Facing Targets Sputtering apparatus which is able to deposite the magnetic thin films without the bombardment by high energy particles to the substrate along with the capabilities to deposit films under wide range of argon gas pressure at high deposition rate. It has been found that the texture and crystallographic structure of the sputtered films are primarily dependent on the argon gas pressure and the films prepared by the Facing Targets Sputtering exhibit quite different properties from the films prepared by conventional methods. The films deposited in the argon gas at the low pressure on the plasma-free substrate using the Facing Targets Sputtering are composed of the fine polyhedral grains packed densely from the initial growth layer to the surface layer of the film. They show exellent c-axis orientation of hcp crystallites, smoother surface textures and the superior perpendicular magnetic anisotropy as compared with the films composed of so-called columnar grains deposited in the argon gas at the higher pressures using the conventional RF diode sputtering. These results imply that the sputtering apparatus which is capable to deposit Co-Cr films in the argon gas at relatively low pressure is indispensable for the preparation of superior perpendicular magnetic recording media. Therefore, the Facing Targets Sputtering apparatus seems to be quite suitable for this purpose. Further, we have proposed a new micro structural model with fine polyhedral grains obtained by our sputtering method, because this model will have the ideal structural characteristics for ultra high density perpendicular magnetic recording.

## INTRODUCTION

In perpendicular magnetic recording system, the demagnetizing field approaches to zero with the shorter wave length and theoretically the ultimate linear recording density is limited only by the width of the 180° magnetic wall [1]. Rapidly solidified Co-Cr thin films have been considered to be one of the most suitable media for this recording system. The perpendicular magnetic anisotropy of sputtered Co-Cr films have been supposed to be caused by the crystalline anisotropy with a perpendicular c-axis orientation of hcp crystallites in the film [2]. So-called columnar structures grown perpendicularly to the substrate plane have been observed among most of the films prepared by the RF diode sputtering and vacuum evaporation [3]. In general, it has been believed that these columnar structures are related to the perpendicular magnetic anisotropy of Co-Cr films and each columnar grain is a single crystal of hcp whose c-axis coincides with the columnar axis and behaves as a single domain particle. Further, it has been considered that the columnar particles play a very important role in sustaining the magnetization perpendicular to the film plane.

## EXPERIMENTAL

## FACING TARGETS SPUTTERING SYSTEM

Figure 1 shows a schematic diagram of the Facing Targets Sputtering (FTS) apparatus used in this study. A pair of Co-Cr alloy disks of 100 mm in the diameter are placed parallel each other as facing targets and the substrates are set vertically to the target plane. The permanent magnets are mounted along the rear edge of both targets as shown in this figure. The magnetic field in the space between the facing targets is about 100 - 160 gauss when the distance between the targets is about 100 - 125 mm. Figure 2 shows the actual distribution of magnetic flux between the targets. By this unique configuration, the high energy $\gamma$-electrons are confined in the space between the targets and the ionization of the argon gas is promoted. Consequently, various kind of films can be deposited at high deposition rate even in argon gas as low pressure as $10^{-4}$ torr. In the RF diode sputtering, films can not be deposited in such a low pressure since the stable discharge will never be maintained. As the surface of of the growing film is not exposed to the bombardment by high energy particles, the substrate temperature is kept much lower during the deposition than that in the RF diode sputtering.

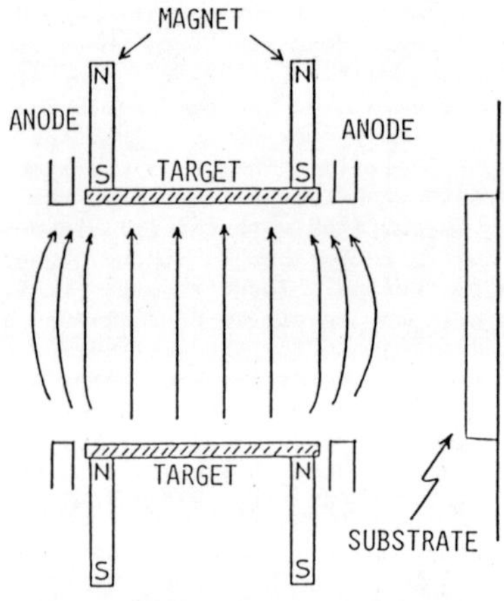


Fig.1 Schematic diagram of Facing Targets type of Sputtering apparatus.

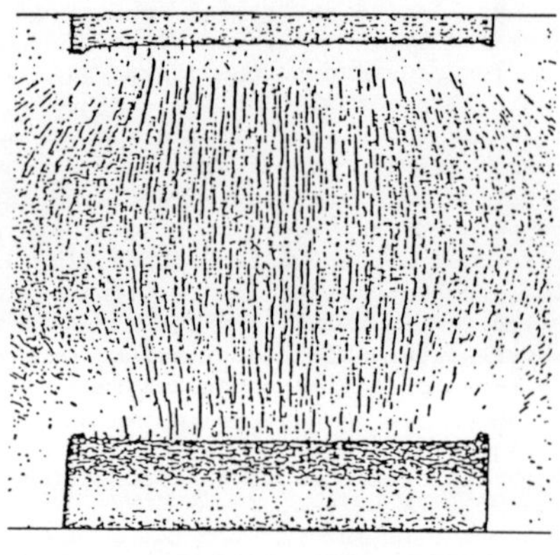

Fig.2 The distribution of magnetic flux formed between the facing Co-Cr targets.

## Experimental procedures

A pair of 21at%Cr-Co alloy disks were used as the targets suitable for depositing the films composed of stable hcp crystallites. Specimen films were deposited on various substrate of glass, Si, $SiO_2$/Si and polyimide at the deposition rate of 500 - 5,000 Å/min. The dependence of the crystal structure and other properties on substrate temperature $T_s$ and argon pressure $P_{Ar}$ have been investigated. The crystal structures were analysed by X-ray diffractometry and the dispersion of c-axis orientation $\Delta\theta_{50}$ was evaluated by measuring the half width value of the rocking curve of (002) peak. The Young's modulus $E_f$ was determined by measuring the mechanical resonance frequencies in the vibrating reed method. The cleaved cross section of the films were observed by the ultra high resolution SEM with LaB6 filament. The magnetic properties were measured by VSM.

## RESULTS AND DISCUSSION

Figure 3 shows the dependence of the Young's modulus $E_f$ on substrate temperature $T_s$ at $P_{Ar}$ of 2 mtorr. $E_f$ exhibits a slight dependence on $T_s$ but is almost same as that of 21at%Cr-Co bulk. On the contrary, $E_f$ is strongly dependent on the argon gas pressure $p_{Ar}$ in high range as shown in Fig.4.

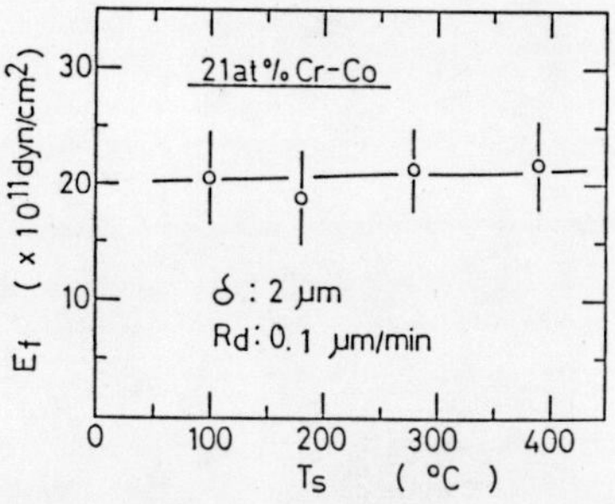


Fig.3 Dependence of Young's modulus $E_f$ on substrate temperature $T_s$.

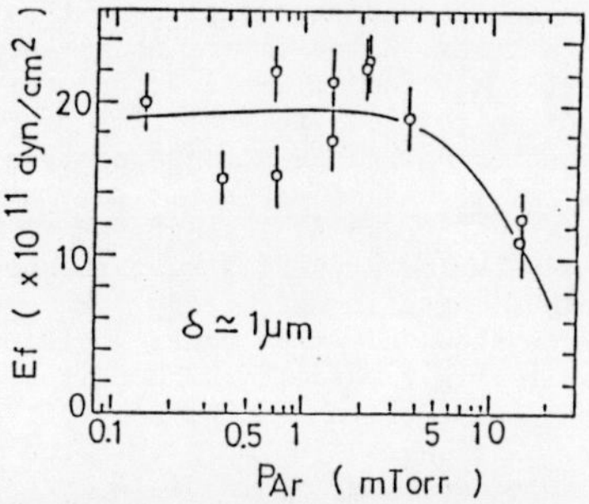


Fig.4 Dependence of Young's modulus $E_f$ on argon pressure $P_{Ar}$.

The cleaved cross section of each film was carefully observed by high resolution SEM. The drastic differences were observed on grain structures with the change of the argon pressure rather than the change of the substrate temperature. For example, clear columnar grains are observed for the films deposited at $P_{Ar}$ as high as $10^{-2}$ torr. Such textures become less distinctive and finally disappear as $P_{Ar}$ decreases down to the order of $10^{-4}$ torr. Figures 5 and 6 show the typical micrographs of cleaved cross sections of the films deposited at $P_{Ar}$ of $1.5\times10^{-4}$ torr and $P_{Ar}$ of $1.5\times10^{-2}$ torr, respectively. The film deposited at the low $P_{Ar}$ range is composed of the fine polyhedral micro grains packed densely from the initial growth layer of the film as shown in Fig 5. These films show exellent c-axis dispersion $\Delta\theta_{50}$ of less than 2.5 degrees and also the large perpendicular magnetic anisotropy field $Hk_{\perp}$ of more than 5.5 KOe. Thus, such fine polyhedral micro grains may give these films superior recording characteristics as the high density perpendicular magnetic recording media.

Fig.5 Cross sectional SEM view of Co-Cr film deposited at $P_{Ar}$ of $1.5\times10^{-4}$ torr.

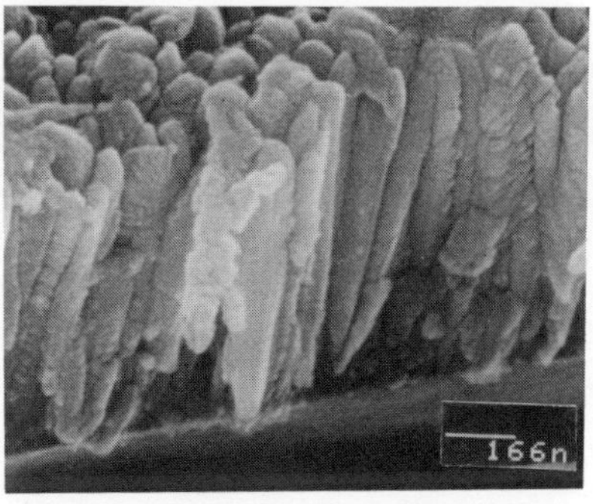


Fig.6 Cross sectional SEM view of Co-Cr film deposited at $P_{Ar}$ of $1.5\times10^{-2}$ torr.

However, the film deposited at high $P_{Ar}$ range appears to be composed of clear columnar grains which have grown perpendicularly to the film plane as shown in Fig.6. Many voids are seen at the boundaries between the columnar grains, and thus, the existence of these boundaries may be the cause of the decrease of $E_f$ shown in Fig.4. These films exhibit the low perpendicular magnetic anisotropy field $Hk_{\perp}$ attributed to a poor c-axis orientation of the hcp crystallites in the film. As already reported [4], the columnar structures are often observed in the films prepared by sputtering and vacuum evaporation when the argon gas pressure is relatively high. For example, sputtered Fe-Cu films with fcc crystal structure show in-plane magnetic anisotropy. However, it also shows clear columnar grains as shown in Fig.7 when it is deposited at argon gas pressure as high as $1\times10^{-2}$ torr. These results imply that the so-called columnar grains in sputtered Co-Cr thin films have no positive relationship to the c-axis orientation of hcp crystallites in the films and have no important role in perpendicular magnetic recording.

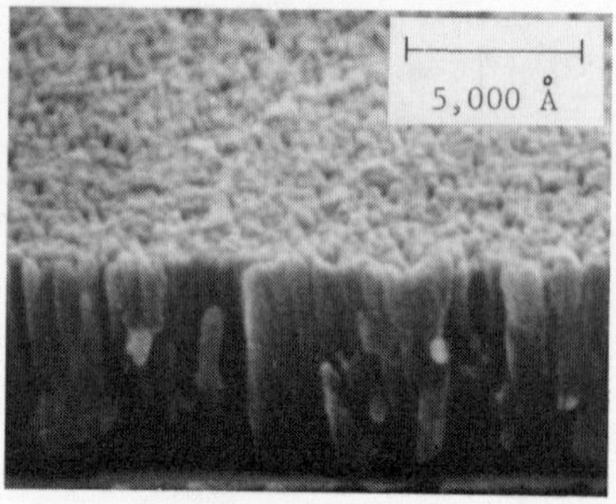


Fig.7 Cross sectional SEM view of Fe-Cu film deposited at $P_{Ar}$ of $1\times10^{-2}$ torr.

CONCLUSION

In this study, it has been found that the micro structure of Co-Cr films is primarily dependent on argon gas pressure. The so-called clear columnar grains observed in RF diode sputtered Co-Cr films may be undesirable microstructure for high density perpendicular magnetic recording. On the other hand, the micro structure of Co-Cr films prepared by Facing Targets Sputtering apparatus is apparently different from that of the films prepared by other methods. Since the films are deposited on the plasma-free substrate in FTS method, it is capable to form the denser and more homogeneous films with a superior c-axis orientation of hcp crystallites as compared with the conventional RF diode sputtering method. These Co-Cr films with fine polyhedral grains is considered to be more suitable for ultra high density perpendicular magnetic recording than the films with relatively large columnar grains.

References

[1] S. Iwasaki, " Perpendicular magnetic recording ", IEEE Trans. Magn, vol. MAG-16 (1980)71

[2] S. Iwasaki, K. Ouchi and N. Honda, " Studies of the perpendicular magnetization mode in Co-Cr sputtered films ", IEEE Trans. on Magn., vol. MAG-16 (1980)1111

[3] R. Sugita, " Co-Cr perpendicular magnetic recording tape by vacuum deposition", IEEE Trans. on Magn., vol. MAG-20 (1984)687

[4] S. Craig and G.L.Harding, " Effects of argon pressure and substrate temperature on the structure and properties of sputtered copper films ", J. Vac. Sci. Technol., vol.19, no.2, (1981)205.

PART IV

# Quasicrystals

# FORMATION OF QUASICRYSTALS IN RAPIDLY SOLIDIFIED Al ALLOYS

ROBERT J. SCHAEFER* AND LEONID A. BENDERSKY**
*Metallurgy Division, National Bureau of Standards, Gaithersburg, MD
**The Johns Hopkins University, Baltimore, MD: Guest Worker at National Bureau of Standards, Gaithersburg, MD

## ABSTRACT

Electron beam surface melting has been used to study Al-Mn and Al-Mn-Si alloys subjected to a wide range of solidification conditions. Several of the reported equilibrium intermetallic phases are not found even at moderate growth rates. Beyond a composition-dependent critical velocity the equilibrium phases are all replaced by the quasicrystalline icosahedral and decagonal (T) phases. The icosahedral phase is favored over the T phase by higher solidification velocities. The addition of Si to Al-Mn alloys eliminates the T phase, but does not significantly facilitate the formation of the icosahedral phase by electron beam melting because the ternary $\alpha$ and $\beta$ phases of Al-Mn-Si are able to grow rapidly into the electron beam melts.

## INTRODUCTION

The icosahedral phase of Al-Mn was first reported [1,2] to have been found in melt-spun ribbons containing 25 wt% Mn, which is approximately the composition of the orthorhombic phase $Al_6Mn$ [3]. Icosahedral point group symmetry implies that the material cannot have lattice translational invariance, and several different descriptions of its possible structure have been developed [1,4-8]. Although the structure cannot be periodic as in the case of a normal crystal, it can be quasiperiodic and therefore the term "quasicrystal" [4] has been used to describe this material.

More recently, another quasicrystalline phase in the Al-Mn system has been described [9]. This phase has decagonal symmetry--one ten-fold rotation axis, along which it is periodic with parameter 1.24 nm, and perpendicular to which it is quasiperiodic. This phase has been referred to as the T phase. It was seen by Sastry et al. in Al-Pd in 1978 [10] and Shechtman et al. in 1984 [11] but its crystallography was not fully identified until its close relationship to the icosahedral phase became evident.

Since the initial reports of these phases, they have been reported in melt-spun alloys of several additional systems. The icosahedral phase has been reported in several other Al-transition metal systems [1,12] as well as $Mg_{32}(Al,Zn)_{49}$ [13], $Mg_4CuAl_6$ [14], $Cd_3Cu_4$ [15], $Pd_3USi$ [16], $(Ti_{1-x}V_x)_2Ni$ [17] and ternary alloys of Al-Mn-Si [18,19]. The T phase has been found in Al-Pd and Al-Pt [12] as well as in Al-Mn-Ge [15]. In several of these systems (Mg-Al-Zn, Mg-Cu-Al, Cd-Cu) the icosahedral phase was predicted to be formed by rapid solidification because the equilibrium structure contains icosahedral groupings of atoms within its unit cell. Elser and Henley [19] have proposed that the very stable icosahedral phase in Al-Mn-Si is composed of rhombus-shaped building blocks which are present within the ternary $\alpha$ phase of Al-Mn-Si. This hypothesis might also be extended to account for the presence of the icosahedral phase in binary Al-Mn alloys, since metastable phases of binary alloy systems can frequently be identified as extensions of phase fields from ternary systems [20]. In addition, similarities have been cited between the structure of icosahedral or T-phase Al-Mn and the orthorhombic $Al_6Mn$ phase [21] or the hexagonal [22] $Al_4Mn$ phase [23]. Thus in many cases it has been implied that the existence of an icosahedral phase in rapidly solidified material is correlated to some type of structural similarity to an equilibrium phase.

In addition, Sachdev and Nelson [24] have suggested a structural relationship between icosahedral quasicrystals and icosahedral clusters in metallic glasses. This concept could be used to account for the formation of extremely small icosahedral grains from metallic glasses by annealing [16] and ion beam mixing processes [25,26].

If icosahedral phases are a modified form of equilibrium phases, it should be possible to identify the solidification conditions which lead to the icosahedral form instead of the equilibrium form, and to observe a transition from one form to the other as solidification conditions are varied. With the melt-spinning process, however, it is difficult to vary the solidification conditions in a controllable and well-characterized manner. We have therefore used electron beam surface melting to study the conditions under which the different phases form as functions of the scanning velocity of the electron beam and the composition of the alloy.

In Al-Mn alloys, we earlier found [27] that in melt-spun ribbons at the composition of $Al_6Mn$ a large amount of $\alpha$-Al is present between the icosahedral crystals. If the Mn content is increased in an effort to eliminate the free Al, almost single phase icosahedral material can be formed at about 34 wt% Mn, but only in the thinnest areas of the ribbons. In most parts of the ribbons, if the Mn concentration is high enough to eliminate the free Al, the T phase predominates the microstructure. We also found, however, that the addition of Si to the alloys eliminates the T phase, with the result that almost single phase icosahedral material could be formed in ribbons of Al-33 wt% Mn-5 wt% Si. We have therefore used both binary Al-Mn and ternary Al-Mn-Si alloys in our electron beam studies: the most intensively studied alloy was Al-25 wt% Mn.

## EXPERIMENTAL

Alloy buttons or bars were prepared by arc melting of 99.99% Al and 99.97% Mn, the total weight of each sample being approximately 20 grams. The samples were turned over and remelted several times to minimize segregation. The top and bottom surfaces of each sample were ground flat and they were then clamped to a water-cooled copper chill plate for electron beam surface melting.

Surface melting was carried out in a vacuum of $10^{-3}$ Pa using a 25 kV beam of electrons which had a diameter of approximately 1 mm at the sample surface. The electron beam was scanned across the sample surface at velocities ranging from 0.025 cm/sec to 200 cm/sec to produce a wide range of solidification conditions. The beam current varied from 8 mA for slowly scanned melts to 33mA for melts scanned at 200 cm/sec. In most cases a series of overlapping melt passes was made to give a strip of surface-melted material approximately 5 mm wide. In the case of melts scanned at velocities greater than about 3 cm/sec, the time in the liquid state is too brief for convection and diffusion to eliminate the concentration differences due to microsegregation in the arc-melted alloys. Therefore the entire sample surface was initially scanned with overlapping melts at 2.5 cm/sec, producing a greatly refined microstructure for subsequent high-speed melts.

X-ray diffraction of the melted surfaces was carried out using copper K-$\alpha$ radiation. X-ray results were used to evaluate the approximate phase composition of the melts.

Samples were then polished parallel to the original surface for metallographic examination: only enough material was ground off to remove the roughness that was generated by the melting operation, and the samples were treated with Keller's etch. Subsequently, thin slices were removed parallel to the surface and thinned for TEM study.

## SURFACE MELTING PHENOMENA

In an earlier study [28], electron beam surface melting was used to produce a wide range of solidification velocities in Ag-Cu alloys, and the resulting microstructures were determined and correlated to crystal growth theories. The microstructural evidence indicated that in this alloy the solid-liquid interface at many compositions and velocities forms a relatively smooth surface. As a result, it could be concluded that the local velocity of solidification at a point on the interface was simply the scan velocity times the cosine of the angle between the scan velocity vector and the normal to the solid-liquid interface. Thus once the shape of the melt pool was known, the local interface velocities could be determined.

For alloys such as Ag-Cu with relatively smooth solid-liquid interfaces, the shape of the solidifying part of the solid-liquid interface can frequently be determined from microstructural observation. Alternatively, numerical computations such as those of Kou et al. [29] can sometimes be used to determine the shape of the melt pool. Unfortunately, detailed calculations of melt pool shapes have not been carried out for high scan rates, where the shape of the melt pool becomes severely distorted.

In most concentrated alloy systems (with the exception of some eutectics), it is not realistic to describe the solidification front at the back of the moving melt pool as a simple surface advancing normal to itself. If the solidification front is dendritic in nature, the dendrites generally grow in specific crystallographic directions rather than propagating normal to the freezing point isotherm. The directions of dendrite growth are frequently determined by the orientations of crystals in the substrate, upon which the dendrites grow expitaxially. Some dendrites may thus be oriented favorably to follow the motion of the melt pool, while others have orientations which require them to grow much faster to keep up with the melt pool. Constitutional supercooling will occur ahead of the advancing dendrites, especially at high scan velocities where the trailing edge of the melt pool falls far behind the moving heat source and the gradients within the melt become low. If the primary solidification phase is a facetted intermetallic compound, considerable additional supercooling may be required to produce a finite rate of solidification. As a result, the entire trailing part of the melt zone may become highly supercooled and when the crystals do grow into it they may grow with random directions rather than following temperature gradients. There is then no way that heat flow analysis and metallography can combine to accurately determine the growth rate of these crystals. Figure 1 shows a melt pass at 0.5 cm/sec on the Al-25 wt% Mn alloy, and it is seen that large plate-like crystals have spread through the melt zone with relatively little correlation to the direction in which the melt zone was moving. In several places portions of the liquid have become isolated by these plate-like crystals and other phases have then nucleated within these isolated regions. For none of these crystals can the thermal conditions or growth kinetics be accurately determined.

In cases where large undercooling develops in the trailing part of the melt pool, the various phenomena associated with undercooled liquids can occur; an amorphous solid can be formed, or crystals of a stable or metastable phase can be nucleated. Figure 2 illustrates schematically how nucleation can occur in a moving melt zone when the crystals growing from the substrate are unable to grow fast enough to keep up with the nucleation temperature isotherm. In the case of sufficiently concentrated Al-Mn alloys, we have found that at high scan velocities quasicrystals can nucleate. For the reasons discussed above, however, it is not usualy possible to correlate microstructural details closely to either solidification velocities or cooling rates.

## RESULTS

An extremely wide range of microstructures was found in the electron beam melts of the Al-25 wt% Mn alloy as the melt scan velocity changed from 0.25 to 200 cm/sec. The microstructural change does not consist of a simple refinement of scale, as might be found in an alloy with a simple cellular microstructure. Instead, it is found that at different velocities, completely different phases and morphologies are found.

Several different versions of the Al-Mn phase diagram have been published [30,31], and there is considerable disagreement over which phases are present at equilibrium. Godecke and Koster [31] claim that some of the phases shown by Taylor [30] are metastable and it is clear that some of the phases form only very slowly: it is therefore to be expected that the phases produced by rapid solidification will not be predictable from the equilibrium phase diagram. Taylor's phase diagram shows a series of peritectics which suggests that slow soldification of a 25 wt% Mn alloy would take place by primary solidification of the $\phi$ phase, followed by successive coatings of $\mu$, $Al_4Mn$, and $Al_6Mn$ before finally reaching the eutectic with fcc Al. The $\phi$, $\mu$, and $Al_4Mn$ phases are all hexagonal but the structures of the last two are not known in detail. In melt-spun ribbons, all of these intermetallic phases are bypassed, thus allowing the nucleation and growth of the icosahedral phase. In our electron beam melts we found no evidence that the $\phi$ and $\mu$ phase occurred, but at low velocities we found the $Al_4Mn$ and $Al_6Mn$ phases, as well as the $\delta$ phase [32] which is sometimes called $Al_{11}Mn_4$ [33]. As the electron beam scan velocity increased, these phases either ceased to grow or their structure became greatly modified. At sufficiently high scan velocities, the T and icosahedral phases formed.

Figure 3 shows a microstructure seen in the slowest melt scans, 0.25 cm/sec. Large plates of a primary phase are widely spaced, and between them grow partially facetted dendrites with hollow squarish tips, a morphology characteristic of the orthorhombic $Al_6Mn$ phase. The primary phase has not been positively identified in this melt, but its morphology suggests that it is the same as the plate-like crystals of $\delta$ seen at 2.5 cm/sec (below).

As the melt scan velocity is increased, the single large primary plates develop into colonies of smaller plates which diverge as they grow (Figure 4). TEM observation of these colonies reveals that the primary plates consist of the $\delta$ phase, and that they are coated with a surface layer of $Al_6Mn$ in a matrix of $\alpha$-Al (Figure 5). The $\delta$ plates and their $Al_6Mn$ surface coatings have a specific crystallographic relationship, described in the Appendix. The $\phi$, $\mu$, and $Al_4Mn$ phases are thus bypassed in this solidification sequence.

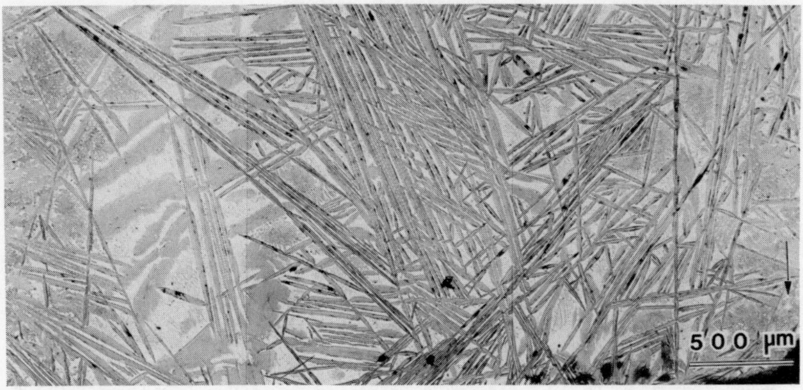


Figure 1. Melt-zone microstructure of an Al-25 wt% Mn alloy scanned at 0.25 cm/sec in the direction of the arrow.

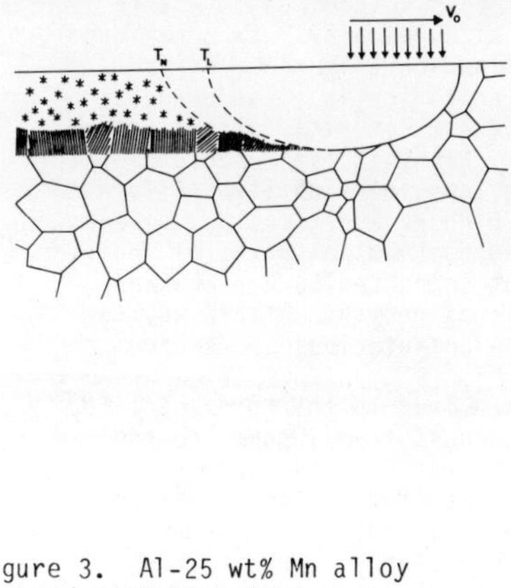


Figure 2. Surface melting by a rapidly scanned electron beam. $T_L$ = liquidus temperature, $T_N$ = nucleation temperature.

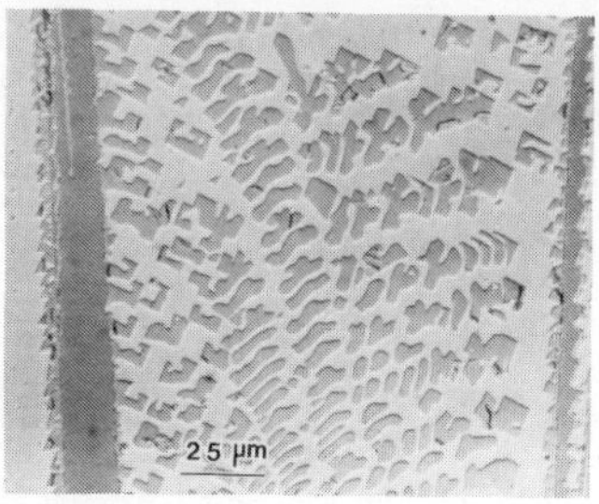


Figure 3. Al-25 wt% Mn alloy scanned at 0.25 cm/sec.

Figure 4. Al-25 wt% Mn alloy scanned at 2.5 cm/sec.

The $Al_4Mn$ phase is found only infrequently and with rather irregular distribution within the melts. It grows with increasing imperfection as the growth velocity increases. We have previously reported [23] a close crystallographic relationship between the $Al_4Mn$ phase and a modification described as the T" structure. The T" structure appears to be a highly disordered form of the $Al_4Mn$ phase and the two forms are found in different parts of the same crystal (Figure 6). High-resolution lattice imaging shows an array of dark spots which in some regions of the crystal are arranged in a regular hexagonal pattern corresponding to the basal plane of the unit cell of the $Al_4Mn$ structure: selected area diffraction from these regions shows the reciprocal lattice of $Al_4Mn$. In other parts of the crystal the array of dark spots is highly disordered, and selected area diffraction from these regions shows the T" reciprocal lattice.

Also seen in Figure 4, between the colonies of plates, are regions containing a very finely subdivided microstructure, granular in appearance. As the melt scan velocity increases from 2.5 to 10 cm/sec, the colonies of $\delta$ plus $Al_6Mn$ become confined more and more to the edges of the melt trails and the granular microstructure takes their place. X-ray diffraction of the 10 cm/sec melt trails, which consist almost 100% of the granular microstructure, shows only the T phase and $\alpha$-Al, and it is concluded that this microstructure consists of T phase grains in a matrix of $\alpha$-Al.

The T-phase grains are typically blocky cylinders: their structure is described in detail in another paper of this volume [34]. Within a given area they are remarkably uniform in size, usually 1 to 2 $\mu$m long and about 1 $\mu$m in diameter. At low magnification (Figure 7) it is apparent that there is a dendrite-like organization in these granular areas, and in melts scanned at 2.5 cm/sec these dendritic arrays extend for distances of 100 $\mu$m or more. Moreover, within these dendritic arrays one finds pentagonal features or regions in which five different sets of dendrite directions can be seen. Higher magnification, however, reveals not a dendrite with a stem and branches, but an array of individual T-phase crystals (Figure 8), apparently not connected to one another except where they have impinged as a result of growth. Within any one dendrite array, a limited number of T-phase orientations are present (Figure 9).

As the melt scan velocity was increased from 10 to 200 cm/sec, the phase composition of the 25 wt% Mn alloy changed from T phase to the icosahedral phase. This change did not take place in the same manner as the change from the $\delta$ + $Al_6Mn$ plate colonies to the T phase. Whereas that change occurred by a reduction in the volume occupied by the plate microstructure and its replacement by the granular T microstructure, it was found that between 10 and 200 cm/sec the melt zone contained an intimate mixture of icosahedral and T phase, with the latter growing epitaxially on the former. The decagonal axis of the T phase grew as an extension of one of the 5-fold axes of the icosahedral phase, and a single icosahedral crystal could therefore generate six different orientations of the T phase. The icosahedral phase grew in the form of dendrites but the T phase was found only in the form of blocky cylinders. The icosahedral dendrites in these melts were much larger than the 1-2 $\mu$m size typical of melt spun ribbons: examination of 100 cm/sec scans shows grains 10-30 $\mu$m in diameter, many of them containing conspicuous pentagonal figures (Figure 10).

Electron beam melts were also carried out on alloys containing 14, 18, 22, and 30 wt% Mn. In the 14 wt% Mn alloy, $Al_6Mn$ dendrites grew throughout the melt zone of 1 cm/sec scans, but in 2.5 cm/sec scans a cellular supersaturated solid solution of Mn in Al was formed, with no evidence of quasicrystal formation.

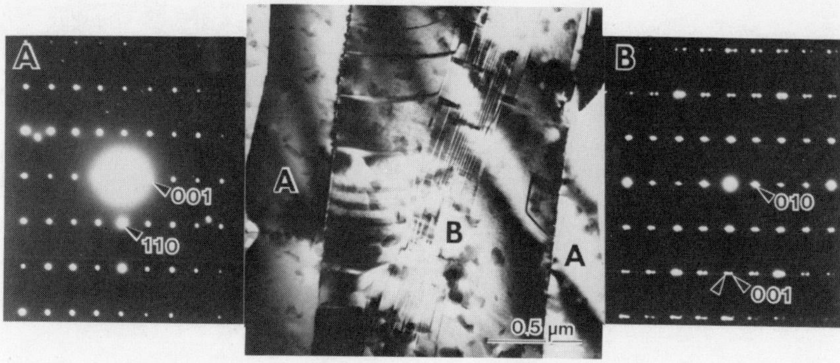


Figure 5. $\gamma$-phase plate (B) with coating of $Al_6Mn$ (A).

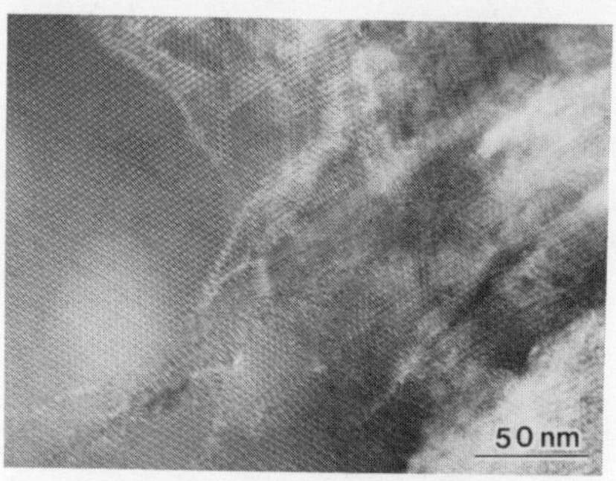


Figure 6. Lattice image showing periodic $Al_4Mn$ phase on left and irregular T" structure on right.

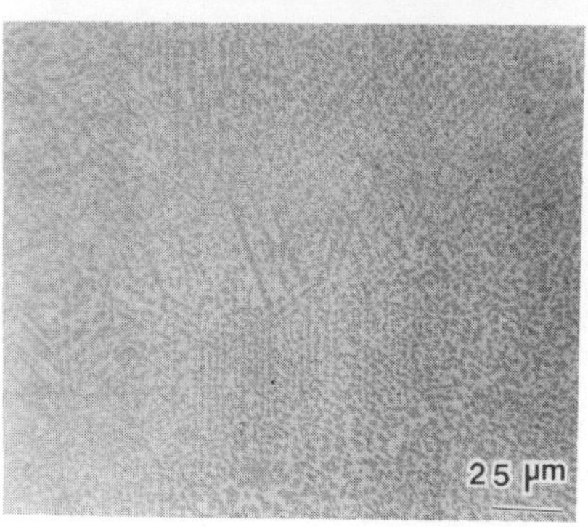


Figure 7. Dendrite-like structure of T phase. Al-25 wt% Mn scanned at 2.5 cm/sec.

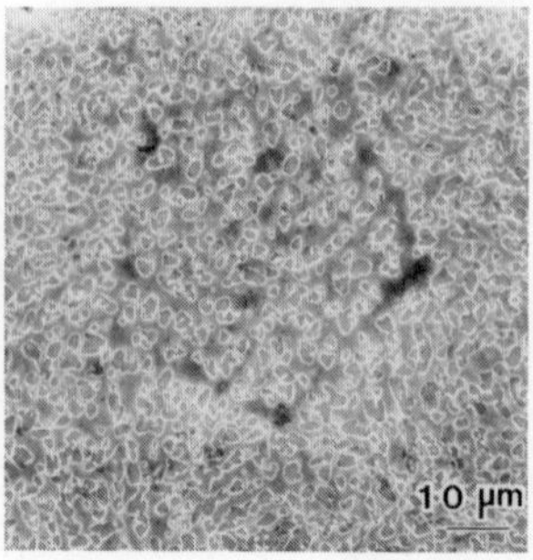


Figure 8. Pentagonal feature among T-phase crystals. From the same melt scan as Figure 7.

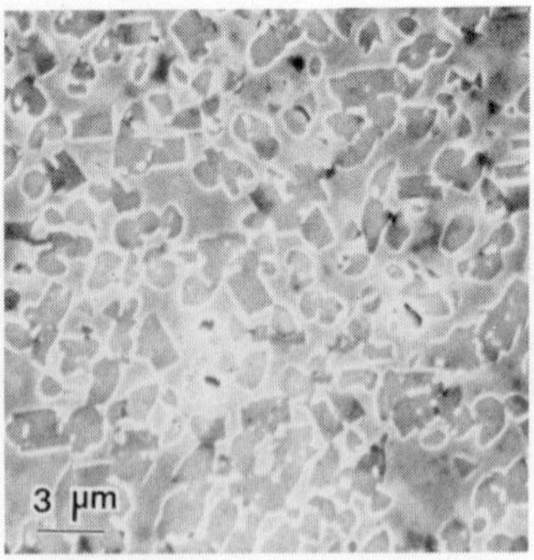


Figure 9. T phase crystals showing limited range of orientations. From the same melt scan as Figure 7.

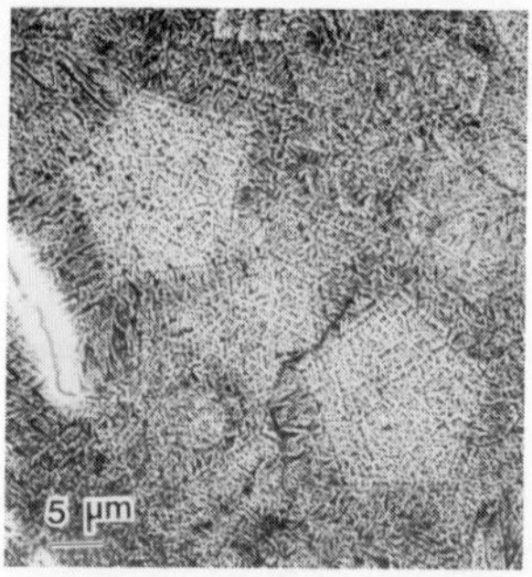


Figure 10. Icosahedral crystals in Al-25 wt% Mn scanned at 100 cm/sec.

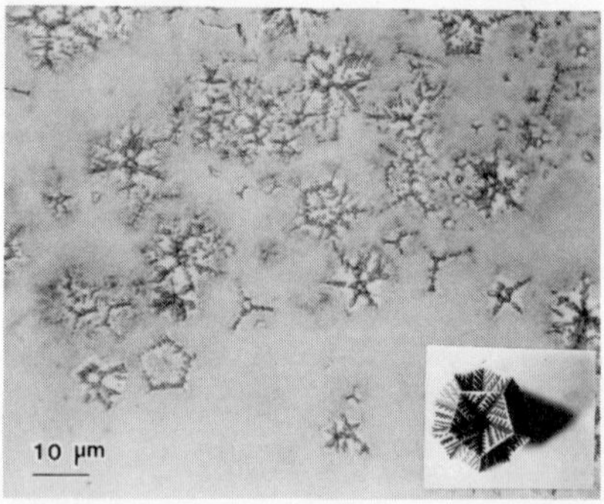


Figure 11. Icosahedral crystals in Al-18 wt% Mn scanned at 50 cm/sec, with a model illustrating the dendritic structure.

In the 18 wt% Mn alloy, $Al_6Mn$ dendrites filled most of the melt zone of 2.5 cm/sec scans, but near the center of the melts large numbers of icosahedral grains nucleated ahead of the advancing $Al_6Mn$ dendrites. At higher velocities, the $Al_6Mn$ dendrites did not penetrate significantly into the melt zone, and the icosahedral phase predominated. At this composition, very little T phase is formed and the icosahedral grains have a morphology which clearly reveals their unusual symmetry. This morphology results from a strong preference for the icosahedral dendrites to grow along their three-fold symmetry axes, and is made especially conspicuous because of the lack of T-phase overgrowth and because the number of icosahedral crystals at this composition is fairly low, so that many of them grow without interference from their neighbors. Figure 11 shows an array of icosahedral crystals in a 50 cm/sec melt zone of the 18 wt% Mn alloy, where several are oriented such that their 5-fold symmetry is shown while others are oriented to show other symmetry elements. Also shown is a model of the structure of these crystals.

The melt zone microstructures of the 22 and 30 wt% Mn alloys were qualitatively similar to those of the 25 wt% Mn alloy. The minimum scan speed for quasicrystal formation was <10 cm/sec in the 22 wt% Mn alloy and 50 cm/sec in the 30 wt% Mn alloy.

Surface melting was also carried out on alloys containing 5 wt% Si, with 18, 27, and 33 wt% Mn. The ternary β phase dominated the melt zone of the alloy containing 33 wt% Mn at velocities up to 50 cm/sec, with some icosahedral phase being formed in 100 cm/sec melts. In the alloy with 27 wt% Mn, the icosahedral phase started to replace the β phase at 50 cm/sec and almost completely replaced it at 200 cm/sec. In the alloy containing 18 wt% Mn, the slowest melts contained a mixture of the ternary α and β phases, but the β phase was no longer present at 50 cm/sec and between 50 and 500 cm/sec the α phase was replaced by the icosahedral phase. Thus considerably higher velocities were required to produce quasicrystals in the Si-containing alloys, as compared to the Al-Mn binaries.

The T phase was not seen in the alloys containing Si nor were the other phases of the Al-Mn binary system, with the possible exception that the ternary β phase is apparently the same as the binary φ phase.

## DISCUSSION

Even though the complex geometry of solidification in these alloys precludes a quantitative correlation of the observed microstructures to solidification velocities or cooling rates, one can observe the sequence of microstructural changes which leads to quasicrystal formation. The φ and μ phases are apparently too slow in their growth kinetics to ever propagate in melts moving even at only 0.25 cm/sec, and the $Al_4Mn$ phase is seen only to a very limited extent. $Al_6Mn$ is the primary solidification phase in alloys containing ⩽ 18 wt% Mn, but its growth velocity is limited to about 1 cm/sec in a 14 wt% Mn alloy and 2.5 cm/sec in an 18 wt% Mn alloy. For melt scan velocities greater than these values, the icosahedral phase nucleates ahead of the $Al_6Mn$ dendrites in the 18 wt% Mn alloy but not in the 14 wt% Mn alloy.

In alloys containing ⩾ 22 wt% Mn, the primary solidification phase is the β phase at low velocities; as long as it is able to grow, it is coated by a layer of $Al_6Mn$, but the other intermetallic phases are again bypassed. The limiting scan velocity at which these phases can fill the melt zones increases from < 10 cm/sec in the 22 wt% Mn alloy to 25 cm/sec in the 30 wt% Mn alloy. The increase in the limiting scan velocity at which the equilibrium intermetallic phases can grow, from 1 cm/sec in the 14 wt% Mn alloy to 25 cm/sec in the 30 wt% Mn alloy, is a natural result as the composition of the liquid becomes closer to the composition of the primary growing phase. When these velocities are exceeded, the quasicrystal phases appear, first the T phase but with the icosahedral phase predominating at the fastest scan rates.

Icosahedral crystals grow with a dendritic structure which is best seen in the Al-18 wt% Mn alloy. For appropriately oriented crystals, an immediately recognizable feature of this structure is the pentagon shape, sometimes with dendrite arms extending outward, as shown in Figure 11. Similar structures are seen in the 100 cm/sec scans of the Al-25 wt% Mn alloy (Figure 10).

The arrays of T-phase crystals seen in slower scans on the Al-25 wt% Mn alloy strongly suggest that there was an initial growth of icosahedral dendrites with the same morphology but a larger size than those seen in the 100 cm/sec scans. Due to the lower cooling rates in the slow scans, the T-phase crystals then nucleated and grew epitaxially on the icosahedral dendrites, which subsequently dissolved. This process would leave only six different orientations for the T-phase particles which originate from a single icosahedral dendrite. It would also account for their uniformity in size within a local area by providing a mechanism for almost simultaneous nucleation followed by slow growth. Finally, it would account for the arrangement of isolated T-phase crystallites into dendrite-like arrays similar to those of the icosahedral phase seen at higher scan velocities.

## CONCLUSIONS

In traveling melt zones, the quasicrystalline icosahedral and decagonal (T) phases nucleate when a sufficiently large supercooling is developed ahead of the crystalline phases growing up from the substrate. In binary Al-Mn alloys, the melt scan velocity required to produce this condition increases as the Mn concentration increases.

The most rapidly growing equilibrium phase in Al-25 wt% Mn alloys is the $\delta$ phase: it is coated with a layer of $Al_6Mn$, with a specific orientation relation.

The icosahedral phase grows as dendrites which show characteristic pentagonal features.

At high scan rates the icosahedral dendrites retain their icosahedral structure, but at low scan rates they are replaced by epitaxially nucleated T phase.

Although the addition of Si to Al-Mn alloys greatly facilitates the production of the icosahedral phase in melt-spun ribbons, it is not correspondingly effective in electron beam surface melts.

## APPENDIX

The primary plates coated with a second phase which form diverging colonies at low scan velocities (Figure 4), were analyzed using TEM and STEM/EDS. Figure 5 shows a bright field image and selected area diffraction patterns (SADPs) taken from the left and right side coating phase and from the central plate. The latter shows a defected structure, whereas no defects are observed in the coating. Analysis of the SADPs from the coatings shows that they are the orthorhombic $Al_6Mn$ phase, both with a [110] zone axis orientation (with slight deviations due to the thin film buckling). STEM analysis of these phases was carried out, measuring several of the central plates and coating layers. Assuming that the coating has the stoichiometric composition of the $Al_6Mn$ phase and using it as a standard, the central plate composition was estimated to be $Al_{3.1}Mn$. The phase closest in composition to that estimated is the $Al_{11}Mn_4$ $\kappa$ phase [30,33] (triclinic, $P\bar{1}$). Indeed, SADP from the central plate (Figure 5) can be described as a superposition of two patterns of the $\kappa$ phase at a [100] zone axis orientation, mirror related. The mirror plane is parallel to (010) $\kappa$ phase. The pattern probably results from the domain structure observed in the central plate.

Based on the SADPs of Figure 5, an orientation relationship between the $Al_6Mn$ coatings and the primary $\kappa$ phase can be established. The orientation relationship is:

$$(001)Al_6Mn \,||\, (010)\kappa$$

$$[\bar{1}10]Al_6Mn \,||\, [100]\kappa$$

This orientation relationship can be understood by comparing the structures of the (001) plane of $Al_6Mn$ and the (010) plane of the $\kappa$ phase, as shown in the literature [35]. For $Al_6Mn$ the structure can be regarded as (001) layers of a pentagon-triangle net ($A_1$) and a $6^3$ net. The $\kappa$-$Al_{11}Mn_4$ structure contains a stacking of the pentagon-triangle layer ($A_1$) and a rumpled $3^2434$ net ($A_2$). The stacking sequence is $XA_1A_2A_2A_1XA_1A_2A_2A_1$. The pentagon-triangle layers ($A_1$) of $Al_6Mn$ and $Al_{11}Mn_4$ are very similar in topology and dimension (Figure A1) when they are oriented according to the relation given above. Therefore, deposition of (001) layers of $Al_6Mn$ instead of $Al_{11}Mn_4$ can occur with a good coherency. This close crystallographic relationship can also account for the bypassing of the peritectic reactions with the $Al_4Mn$ and $\mu$ phases in favor of $Al_6Mn$.

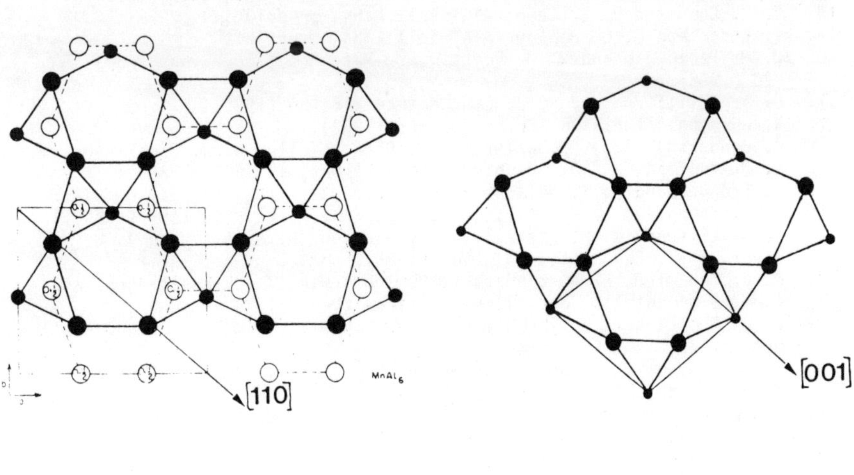


Figure A1. Structures of $Al_6Mn$ and $\kappa$-phase, oriented according to the relation given in the text.

ACKNOWLEDGEMENTS

The authors thank DARPA for financial support of this work. They also thank F. S. Biancaniello, D. Carrick, and C. Brady for sample preparation, C. Olson for SEM work, and J. W. Cahn, D. Shechtman, and W. J. Boettinger for many stimulating discussions.

REFERENCES

1. D. Shechtman and I. Blech, Met. Trans. 16A, 1005 (1985).
2. D. Shechtman, I. Blech, D. Gratias, and J. W. Cahn, Phys. Rev. Lett. 53, 1951 (1984).
3. A. D. I. Nicol, Acta Cryst. 6, 285 (1953).
4. D. Levine and P. Steinhardt, Phys. Rev. Lett. 53, 2477 (1984).
5. P. Bak, Phys. Rev. Lett. 54, 1517 (1985).
6. V. Elser, AT&T Bell Labs, preprint.
7. M. Duneau and A. Katz, Phys. Rev. Lett. 54, 2688 (1985).
8. A. L. Mackay, Nature 315, 636 (1985).
9. L. Bendersky, Phys. Rev. Lett. 55, 909 (1985).
10. G. V. S. Sastry, C. Suryanarayana, M. Van Sande and G. Van Tendeloo, Mat. Res. Bull. 13, 1065 (1978).
11. D. Shechtman, R. J. Schaefer, and F. S. Biancaniello, Met. Trans. 15A, 1987 (1984).
12. P. A. Bancel, P. A. Heiney, P. W. Stephens, A. I. Goodman, and P. M. Horn, Phys. Rev. Lett. 54, 2422 (1985).
13. P. Ramachandrarao and G. V. S. Sastry, Pramana 24, (1985).
14. G. V. S. Sastry, P. Ramachandrarao, and T. R. Anantharaman, preprint.
15. L. A. Bendersky and F. S. Biancaniello, unpublished research.
16. S. J. Poon, A. J. Drehman, and K. R. Lawless, Phys. Rev. Lett. 55, 2324 (1985).
17. Z. Zhang, H. Q. Ye, and K. H. Kuo, Phil. Mag. Lett., submitted.
18. C. H. Chen and H. S. Chen, AT&T Bell Labs, preprint.
19. V. Elser and C. L. Henley, AT&T Bell Labs, preprint.
20. J. H. Perepezko and W. J. Boettinger, Mat. Res. Soc. Symp. Proc. 19, 223 (1983).
21. L. A. Busill and and P. J. Lin, Nature 316, 50 (1985).
22. W. Hofmann, Aluminium Berlin 20, 865 (1938).
23. L. Bendersky, R. J. Schaefer, F. S. Biancaniello, W. J. Boettinger, M. J. Kaufman, and D. Shechtman, Scripta Met. 19, 909 (1985).
24. S. Sachdev and D. R. Nelson, Phys. Rev. B 32, 4592 (1985).
25. J. A. Knapp and D. M. Follstaedt, Phys. Rev. Lett. 55, 1591 (1985).
26. D. A. Lilienfeld, M. Nastasi, H. H. Johnson, D. G. Ast, and J. W. Mayer, Phys. Rev. Lett. 55, 1587 (1985).
27. R. J. Schaefer, L. A. Bendersky, D. Shechtman, W. J. Boettinger, and F. S. Biancaniello, Met. Trans. A, submitted.
28. W. J. Boettinger, D. Shechtman, R. J. Schaefer, and F. S. Biancaniello, Met. Trans. 15A, 55 (1984).
29. S. Kou, S. C. Hsu, and R. Mehrabian, Met. Trans. 12B, 33 (1981).
30. M. A. Taylor, Acta Met. 8, 256 (1960).
31. T. Godecke and W. Koster, Z. Metall. 62, 727 (1971).
32. K. Yoshida, J. Phys. Soc. Japan 32, 431 (1972).
33. J. A. Bland, Acta Cryst. 11, 236 (1958).
34. L. A. Bendersky, this volume.
35. W. B. Pearson, The Crystal Chemistry and Physics of Metals and Alloys, Wiley-Interscience, New York (1972).

# TRANSPORT PROPERTY STUDIES ON ICOSAHEDRAL SYSTEMS: RAPIDLY QUENCHED Al-V, Al-Mn, AND Al-Fe-Mn.

K. V. RAO, N. KARPE, R. MALMHÄLL*, H. U. ÅSTRÖM, AND H. S. CHEN†
*Dept. of Solid State Physics, Royal Inst. of Technology, Stockholm, Sweden.*
* *ASEA Research and Innovation, S-721 78 Västerås, Sweden.*
† *Visiting Scientist from: AT&T-Bell Laboratories, Murray Hill, NJ07974, USA.*

## ABSTRACT

We report the first Hall, and electrical resistivity study on rapidly quenched icosahedral Al-based alloys. The Hall coefficient is found to be negative and remains so for i(Al-Mn) alloys even for higher Mn concentrations. The magnitude and temperature coefficient of the electrical resistivity behaviour in these alloys is similar to those observed for amorphous metallic alloys. Thus, the conduction process in i(Al-based alloys) is essentially electronic but with a carrier mobility an order of magnitude smaller than that known for crystalline aluminium.

## INTRODUCTION

Rapid solidification technology, RST, has opened a new, exciting, and rapidly expanding field of considerable scientific and technological interest. A metallurgist has for the first time a capability to expand over a wider range the manipulation of composition and internal structure of metals. With this new 'molecular switch' totally new alloys with unique combination of properties are now feasible. At higher rates of quenching ($10^6$ K/sec) one obtains the amorphous or 'glassy' state which is considered to represent a new state of matter with remarkable mechanical and physical properties. The interpretation of the properties of these metallic glasses poses a particular challenge to the theorist, since the understanding of solids hitherto has been based on crystalline solids with translational invariance. Recently, on a rather fundamental basis there have been some theoretical evidence that the local structure of liquids and metallic glasses can be described as icosahedral [1]. Icosahedral co-ordination shells have been shown [2] to lead naturally to peaks at the appropriate distances in the radial distribution function, even obtaining the 'split-second-peak' structure commonly observed via X-ray scattering in metallic glasses. That crystals can never have five-fold axes of symmetry has been a part of doctrine of crystallography ever since Federov classified the admissible symmetry groups in 1895. Recently, this myth has lost grounds with the identification of icosahedral point group symmetry in the metastable phase of rapidly quenched Al-Mn alloys [3] which argues for long range orientational order but with no translational periodicity. It now seems quite plausible to consider the transformations from gaseous to the liquid to crystalline phases as a continuous change towards lower symmetry, i.e: from a spherical environment of the gas to the icosahedral one in liquid and then to the crystalline environment of a lower symmetry [4]. It is thus natural that most of the theoretical studies on these exciting new class of 'quasicrystals' have focussed on their symmetry properties, simulation of the diffraction patterns and comparison with experimental structure studies. Phenomenological Landau theories [5] invoking incommensurate density waves have also been proposed for explaining the origin of the 'quasi'-phase. Along with extensive work on exploring the structural properties of new alloy systems that exhibit icosahedral symmetry, we have been carrying out systematic studies of electrical, thermal, and magnetic properties. In this paper, we present what we believe is the first simultaneous measurements of Hall, and electrical resistivity of melt-spun $Al_{1-x}Mn_x$, $Al_{84}V_{16}$, and $Al_{84}Fe_7Mn_7$ alloys with icosahedral symmetry.

## EXPERIMENTAL DETAILS

The respective ingots of $Al_{1-x}Mn_x$ (with $x = 0.12; 0.14; 0.16; 0.18$), $Al_{84}V_{16}$ and $Al_{84}Fe_7Mn_7$ were prepared by induction melting the appropriate mixture of the individual component elements of high purity in boron-nitride crucibles in argon atmosphere. Thin metallic ribbons about 1 mm wide and around 15 microns thick were melt-spun on a copper wheel of about 20 cms in diameter, rotating at 2000 rpm in an argon atmosphere. The specimen were examined with a 200 kV Scanning Transmission Electron Microscope (JEOL 200 CX) equipped with a high take off angle EDS detector for STEM X-ray analyses. STEM measurements were conducted on thin areas of the 'as quenched' ribbons, thus avoiding possible artifacts of damage and modification of the structure and/or composition during sample preparation procedure. The Hall and electrical resistivity measurements were carried out simultaneously by using a double-ac technique [6]. In this method the current through the sample as well as the applied magnetic field have a low frequency dependence $f_1$ and $f_2$ respectively. The amplitudes corresponding to the sum frequency $(f_1 + f_2)$, and $f_1$ are then directly related to the Hall, and electrical resistivities. Such a simultaneous ac-measurement at the same temperature minimizes possible errors due to differences in samples, contacts, and thermal gradients etc., when comparing the resistivities. Typically the dimensions of our 'as quenched' ribbon samples were $10 \times 0.7 \times 0.01$ mm$^3$. Various types of sample contacts have been tried [7], although for the data presented here contacts to the samples were made using silver-epoxy. Due to the possible existence of pin-holes, surface roughness, and geometrical defects in these brittle melt-spun samples we estimate the absolute magnitude of our data to be subject to a maximum error of about 20%. However, the relative precision for repeated measurements on the samples are to better than 1%.

## RESULTS AND DISCUSSIONS

Selected random samples from each batch of melt-spun alloys were first investigated on the TEM to ensure that the microstructure was indeed icosahedral. The observed electron diffraction patterns showed 2-, 3-, and 5-fold patterns identical to those reported [1] for i(Al-Mn) alloys. Figure 1 is an example of a typical diffraction pattern observed along the 5-fold axis of symmetry for i($Al_{84}V_{16}$). Notice that the intensities of the diffraction spots are much more uniform compared to that usually observed for i(Al-Mn). Also, we find that the i(Al-V) alloys contains homogeneous and uniform grain sizes up to 2 microns. In the case of i(Al-Mn) alloys produced under identical conditions, a considerable distribution in grain sizes (60 nm to 2 microns) is observed as seen in fig 2 for the Al-Mn sample. STEM microanlyses studies of these two alloys for determining the chemical composition in and around the grains of the morphology shown in fig 2 reveal the following: 1) in the 'as quenched' i(Al-Mn) ribbons there is no evidence for any other crystalline phases or segregated Al. In fact, even in the intermediate region between the icosahedral grains we observe a smaller but finite concentration of Mn dissolved in Al, whereas within a grain nominal composition of Mn is found. Thus in these melt-spun systems normal X-ray data alone is not sufficient to identify the form in which Al is present. Additional evidence for the absence of pure Al-phase will be discussed again later on.
2) in i($Al_{84}V_{16}$), while the material is predominantly icosahedral, some isolated grains of crystalline Al-V-Si phase have been identified [8] between the spherulites. It is thus obvious that a detailed knowledge of the morphology of the samples being studied is very important for any quantitative understanding of the transport properties of these systems.

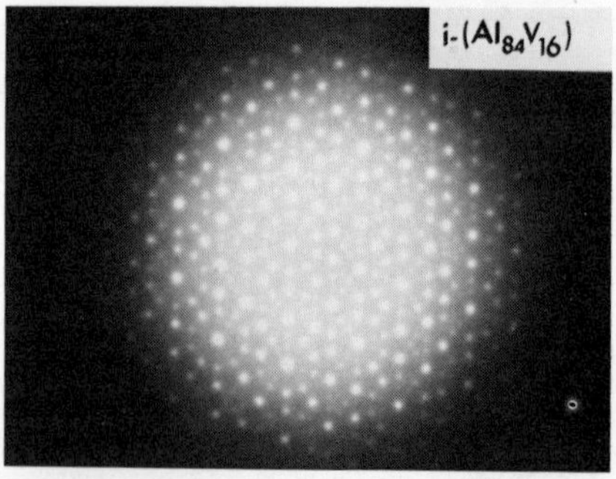


*Fig1: Selected area five-fold symmetry electron diffraction pattern for i($Al_{84}V_{16}$).*

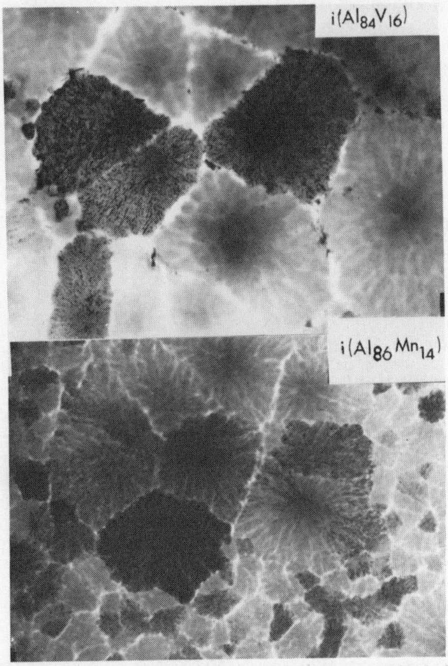


*Fig 2: Electron micrograph of rapidly solidified $i(Al_{84}V_{16})$ and $i(Al_{86}Mn_{14})$, showing rosette-shaped 'quasicrystal' spherulites.*

The room temperature Hall coefficient and the electrical resistivity data for $i(Al_{1-x}Mn_x)$, $i(Al_{84}V_{16})$, and $i(Al_{84}Fe_7Mn_7)$ alloys are presented in figs 3 & 4, and also tabulated in Table I. With increasing Mn concentration (12 to 18 at.%) the electrical resistivity is found to increase monotonically from 55 μΩ cm to 270 μΩcm. The observed magnitude of the resistivities, as well as preliminary studies of temperature coefficients over a fairly wide temperature range are found to be very similar to the behaviour known for amorphous materials [9]. For the i(Al-V) and i(Al-Fe-Mn) alloys the respective resistivities are 77 and 59 μΩcm. It is tempting to interpret the increase in resistivity with Mn concentration as possibly due to the strong resonant state scattering of the Fermi-electrons with the d-state of Mn. However, for any quantitative interpretation of the data for these systems it is clear that a detailed knowledge of the preparation conditions, morphology of the specific samples being studied, and the temperature dependence of the resistivities are essential. Depending on the preparation conditions, presence of oxide layers in these rapid quenched materials would result in much higher resistivities. On the other hand, presence of any segregated 'pure' Al ($\rho \simeq 2.7$ μΩcm) percolating through the system would create channels that effectively short the resistive networks in the system. Despite, the factors mentioned above, since the electron mean free path in these samples corresponding to the measured resistivities is of the order of atomic distances, it is reasonable to conclude that the major contribution to the observed total resistivity arises from intrinsic resistivity of the icosahedral structure itself. Detailed studies of the role of some of the factors mentioned above will be reported elsewhere.

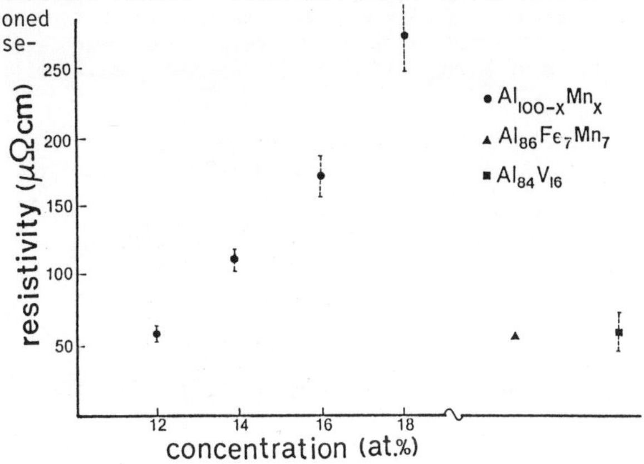


*Fig 3: Room temperature Electrical resistivity for i(Al-Mn), i(Al-V) and i(Al-Mn-Fe).*

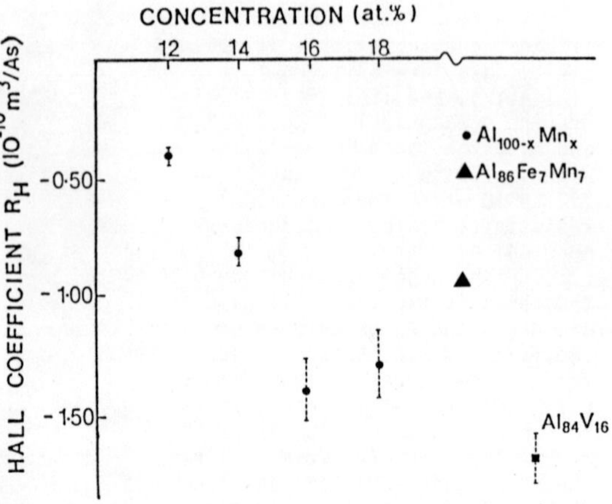


*Fig 4: Room temperature Hall coefficient for i(Al-Mn), i(Al-V) and i(Al-Mn-Fe)*

In fig 4, the Hall coefficient data is presented (also tabulated in Table I). A striking information from these data is that the sign of the Hall coefficient is negative suggesting that charge carriers in these alloys are mainly electronic. Despite the large and increasing resistivity with higher Mn concentrations, the Hall resistivity remains negative and in fact, increases in magnitude by a factor of 3 for a substitution range from 12 to 18 at.% Mn. It is useful to recall that for pure fcc-Al the Hall coefficient $R_H = -0.3 \times 10^{-10}$ $m^3$/As while for simple cubic Mn $R_H = +9.3 \times 10^{-11} m^3$/As. Thus for i(Al-Mn), i(Al-V), and i(Al-Fe-Mn) the 'ordinary' Hall coefficient determined at room temperature is about four times larger ($\simeq 1.2 \times 10^{-10}$ $m^3$/As). Finally, we find, within experimental errors, the ratio $R_H/\rho$ of the Hall and electrical resistivities is essentially a constant, $\approx 0.75$ $cm^2$/Vs, which implies that the carrier mobility is within an order of magnitude a constant for the i(Al-based alloys) studied here. Note that this value is an order of magnitude smaller than that known for pure Al.

*Table I: Room Temperature Hall coefficient, & electrical resistivity of i(Al-alloys).*

| Alloy | Resistivity ($\rho$ - $\mu\Omega$ cm) | Hall coef. ($R_H \cdot 10^{10} m^3$/As) |
|---|---|---|
| $Al_{88}Mn_{12}$ | 55 | -0.42 |
| $Al_{86}Mn_{14}$ | 110 | -o.83 |
| $Al_{84}Mn_{16}$ | 170 | -1.4 |
| $Al_{82}Mn_{18}$ | 270 | -1.3 |
| $Al_{86}Mn_7Fe_7$* | 59 | -1.0 |
| $Al_{84}V_{16}$ | 77 | -1.7 |
| fcc-Al+ | 2.7 | -0.30 |

(*) Data on one sample only; (+) Literature value.

Further studies on the temperature dependence and consequences of crystalization of these and other icosahedral systems will be published elsewhere.

REFERENCES

1 D.R.Nelson, Phys Rev Lett. 50, 982 (1983).
2 S. Sachadev and D.R.Nelson, Phys Rev Lett. 53, 1947 (1984).
3 D.Shechtman, I.Blech, D.Gratias, and J.W.Cahn, Phys Rev Lett. 53, 1951 (1984).
4 M.E.Eberhart, K.H.Johnson, D.Adler, R.C.O'Handley, and McHenry, J. of Non-Crystalline Solids 75, 97 (1985).
5 P. Bak, Phys Rev Lett. 54, 1517 (1985).
6 B. Lundberg and G. Bäckström, Rev. Sci. Instr., 43, 872 (1972).
7 N. Karpe, Thesis, Royal Institute of Technology, Stockholm (1985).
8 K.V.Rao, J. Fidler and H.S.Chen (to be published).
9 see for example: K.V.Rao in Amorphous Metallic Alloys, ed. F.E.Luborsky (Butterworth Co., Ltd,. U.K. 1983) page 401-431.

# FORMATION OF ICOSAHEDRAL Al(Mn) BY PULSED SURFACE MELTING

D. M. FOLLSTAEDT AND J. A. KNAPP
Sandia National Laboratories, Albuquerque, NM 87185

## ABSTRACT

Surface melting of alternating Al/Mn layers on Al and Fe substrates by pulsed electron beams and laser is found by TEM to produce icosahedral Al(Mn). Interpreting the observed microstructures in terms of the calculated temperature histories places limits on the melting point of the phase (660°C < $T_m$ < 960°C), the time needed for its nucleation from the melt at 660°C ($t_n$ < 20 ns), and the rate of quenching from the liquid needed to suppress formation of the competing T phase ($-\dot{T} > \sim 1 \times 10^6$ K/s).

Melt spinning of Al(Mn) alloys containing 14-20 at.% Mn was found last year to produce a phase which exhibits long-range order with icosahedral orientational symmetry [1,2]. This observation attracted immediate interest because this symmetry is inconsistent with lattice translational invariance. The structure of the "icosahedral phase" is still in question [1-3], but many investigators believe that it may be a new type of atomic arrangement called a "quasicrystal". The unique symmetry of this phase makes its formation and properties very much of interest.

We have applied directed energy processes to Al(Mn) surface alloys to study the thermodynamics and formation kinetics of the icosahedral phase [4]. We have demonstrated that the phase can be formed by surface melting techniques; this result is important because the temperature history of alloys quenched with this method can be accurately calculated. The microstructures observed can then be interpreted to place quantitative limits on the melting point of the icosahedral phase and on the kinetic parameters leading to its formation. Our observations on Al(Mn) alloys quenched on both Al and Fe substrates indicate that the melting point of the icosahedral phase exceeds 660°C.

The surface alloys were formed by vapor depositing Al/Mn layers on electropolished substrates of Al or Fe in a vacuum of ~ $5 \times 10^{-7}$ Torr. Typically 16.4 nm of Al and 2.3 nm of Mn were alternately deposited at ~ 0.1 nm/s to produce alloy layers 100-150 nm thick with average concentrations of 16 ± 1 at.% Mn. Nuclear reaction analysis showed that such deposited layers contain at most $1.7 \times 10^{16}$ O/cm$^2$ and $3.5 \times 10^{16}$ C/cm$^2$. To insure that the layers adhered to the substrate, the samples were "ion-beam stitched" by irradiating with $2 \times 10^{15}$ Xe/cm$^2$ at 400 keV. Nomarski optical microscopy of the surfaces also showed features exhibiting flow, which indicates that melting occurred. Samples for transmission electron microscopy (TEM) examination were prepared by jet electropolishing from the substrate side of the sample with a nitric acid solution. Large, free-standing areas of the Al(Mn) surface alloy were exposed and could be examined without interference from the substrate.

Our line source electron-beam annealing (LEBA) treatment focuses a sheet beam of electrons to a line 1 mm x 20 mm, which is incident upon a sample being swept perpendicular to the line. For sweep speeds of 185 cm/s, a point on the sample receives an exposure with a Gaussian time dependence and ~ 570 μs FWHM dwell time. The calculated time dependence of the surface temperature of Al for the measured 75 kw/cm$^2$ peak power used to treat the Al/Mn layers is shown in Fig. 1. The temperature increases to the melting point of the Al substrate and remains at that temperature for ~ 200 μs while the substrate is molten. After the solidification front returns to the alloy layer, the temperature drops at the rate of $-\dot{T} \sim 0.7 \times 10^6$ K/s. Because the alloy layer has high thermal conductivity, is continuous with

Mat. Res. Soc. Symp. Proc. Vol. 58. © 1986 Materials Research Society

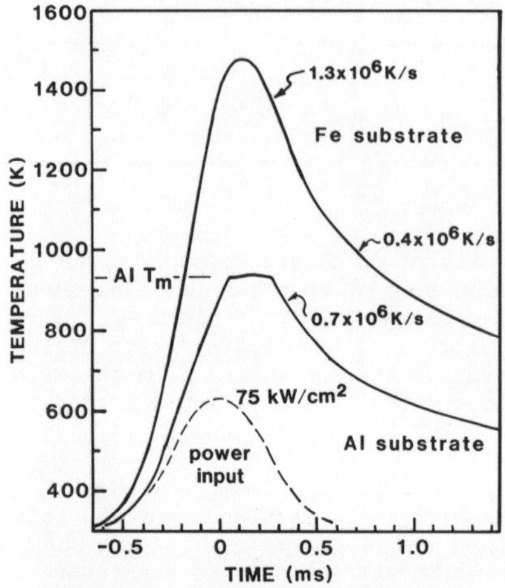


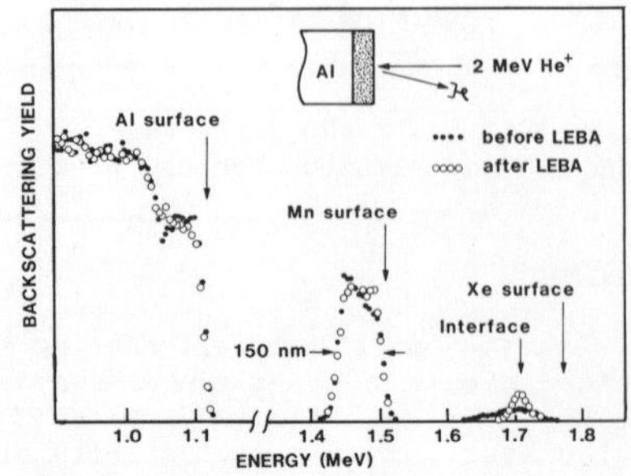


Figure 2. (above) RBS spectra for Al(Mn) alloys on Al substrates before and after LEBA treatment.

Figure 1. (left) Calculated temperature history for Al(Mn) alloys on Al and Fe substrates with LEBA treatment.

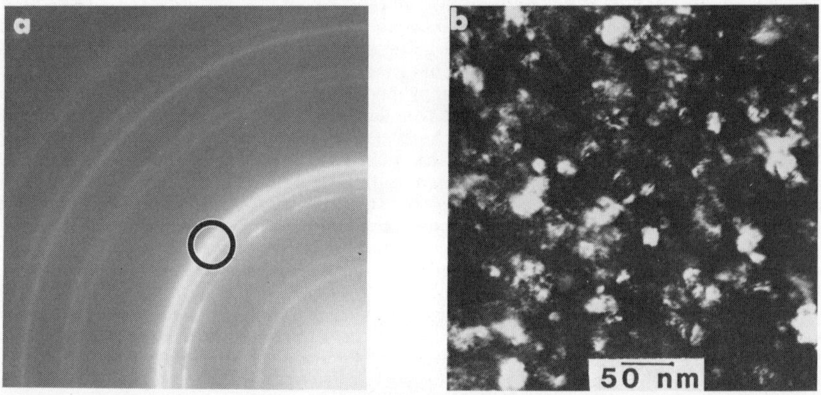


Figure 3. a) Diffraction pattern from Al(Mn) alloy on Al substrate and b) dark-field image of icosahedral grains obtained with indicated rings in a).

the substrate and is thin, its temperature tracks that calculated for the Al immediately beneath the alloy layer.

The depth distributions of Mn in the alloy layer and the implanted Xe were determined with Rutherford backscattering spectroscopy (RBS); spectra before and after LEBA treatment are shown in Fig. 2. The Mn profile was smoothed by the treatment to an essentially constant value of 16 ± 1 at.% across the alloy layer, which is consistent with melting. After implantation, the Xe extended to ~ 0.3 μm, but with the LEBA treatment it became more localized at the alloy/substrate interface 0.15 μm below the surface. This movement indicates that the Xe was zone refined toward the surface by the resolidification front in the Al substrate. The observation that the Xe stopped upon reaching the interface indicates that the alloy was solid when the front reached it. Nomarski optical images of such layers occasionally showed cracks between smooth areas, as might be expected for thin, solid surface layers on liquid substrates.

The microstructure of the alloy layer is also consistent with the formation of solid material ahead of the solidification front. As shown in

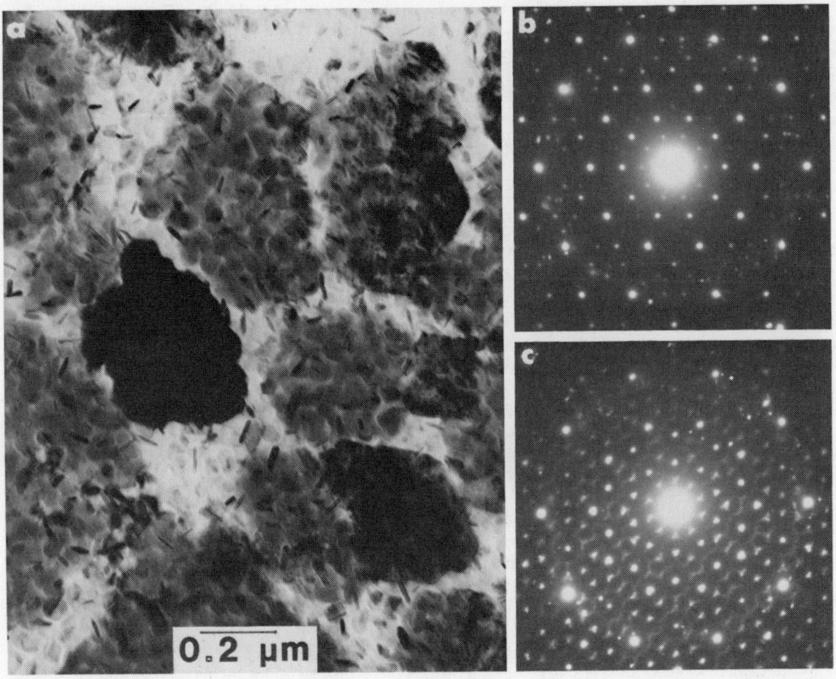


Figure 4. a) Bright-field image of Al(Mn) surface alloy formed on Fe, and diffraction patterns from b) the 5-fold axis of the icosahedral phase, taken from the darkest grain in a), and c) the 10-fold axis of a T-phase grain.

Fig. 3, the alloy layer consists of fine grains of the icosahedral phase plus some Al. The ring electron diffraction pattern in Fig. 3a shows the relatively sharp (111) Al ring (arrowed) plus other rings which index to the strongest rings of icosahedral Al(Mn) observed in x-ray powder patterns [5]. Dark-field imaging with the two brightest rings revealed grains 10-50 nm in diameter, as seen in Fig. 3b. (A small fraction of the largest grains may be Al.) The small size and the random orientation (indicated by the ring pattern) of the icosahedral grains imply that the phase nucleated in a highly undercooled melt. The temperature goes no higher than a few degrees above the melting point of Al (660°C), which thus places a lower limit on the melting point of the icosahedral phase. The liquidus in the equilibrium phase diagram gives an upper limit on the melting point of the icosahedral phase; recent investigations suggest that the central composition of the icosahedral phase is centered at 20 at.% Mn [6], for which the liquidus is at 960°C [7]. Thus we obtain 660°C $< T_m <$ 960°C.

Formation of the icosahedral phase has also been observed to occur within much shorter times by using an electron beam with a shorter pulse (92 ns) [4], and a pulsed laser (22 ns) [8]. These observations place restrictions on the nucleation and growth kinetics of the phase. The time spent below the upper limit on $T_m$ but before the substrate resolidification front reaches the alloy is taken as an upper limit on the time available for nucleation in the melt. Calculation for the pulsed laser shows that the time required for nucleation must satisfy $t_n <$ 20 ns; this time was spent between 660-760°C. The very short nucleation time and fine grain size (~ 2 nm) obtained with pulsed laser irradiation provide stringent tests for structural models of the icosahedral phase.

When the LEBA treatment was applied to Al/Mn layers on an Fe substrate, the resulting microstructure was very different from that for Al substrates as seen in Fig. 4. The surface alloy consisted primarily of equiaxed grains ~ 0.5-1 μm in diameter. Individual grains were identified with electron diffraction; most patterns looked like those of the T phase, a phase with decagonal symmetry that has also been identified in other melt quenched Al(Mn) alloys [6,9,10]. However some of the grains were found to be the icosahedral phase [5]. Diffraction patterns from these two phases are shown in Fig. 4b) and c). The grain size varied between areas corresponding to different Fe substrate grains. Interdiffusion between the alloy and the Fe substrate was not detectable with RBS, as expected for an unmelted substrate.

The calculated temperature history for the Fe substrate is also shown in Fig. 1; the maximum temperature was 1480 K, which is below the melting point of the Fe substrate, 1809 K. The subsequent cooling rate was as high as $1.3 \times 10^6$ K/s, and was $0.4 \times 10^6$ K/s at 660°C. This temperature history resembles that expected for melt spinning, and the resulting grain size is accordingly similar to that obtained with melt spinning. We suggest that in melt-spun and LEBA/Fe alloys, the icosahedral grains nucleate above 660°C, where the phase is less undercooled than on Al and a lower nucleation density is produced. If the icosahedral phase were forming below 660°C on Al, a grain size similar to that obtained with the Fe substrate should have been observed because the cooling rates are similar below 660°C for the Al and Fe substrates. Comparison of these two microstructures thus indicates that the icosahedral phase formed at 660°C on Al, and above 660°C on Fe.

The T phase has been found to become more prevalent with slower cooling rate and higher Mn concentrations [6]. Since both the T and icosahedral phases are observed with the Fe substrate, its quenching conditions (Fig. 1) are not adequate to suppress the T phase. We suggest that rates $\dot{T} > \sim 1 \times 10^6$ K/s are needed to avoid T-phase formation when quenching from above the liquidus with 16 at.% Mn. All the rings in Fig. 2a fit either the icosahedral phase or fcc Al; this sample apparently contains no T-phase material. Thus the icosahedral phase is stable against transformation to the T phase for at least 200 μs at 660°C, as well as to other possible phases such as $Al_6Mn$. The results with the Fe and Al substrates suggest that T-phase formation occurs above 660°C, and thus the limits on the melting point of the icosahedral phase also apply to the T phase.

In conclusion, the observations on both Al and Fe substrates indicate that the icosahedral phase formed at 660°C on the Al substrate. With that interpretation, limits can be placed on the melting point (660-960°C) and time to nucleate (<20 ns) of icosahedral Al(Mn). These restrictions must be met by successful thermodynamic and structural models of this novel phase.

We wish to thank M. Moran and G. Schuh for technical assistance. This work performed at Sandia National Laboratories was supported by the U.S. Department of Energy under contract number DE-AC04-76DP00789.

REFERENCES

1. D. Shechtman, I. Blech, D. Gratias and J. W. Cahn, Phys. Rev. Lett. 53, 1951 (1984).
2. R. D. Field and H. L. Fraser, Mater. Sci. Eng. 68, L17 (1984).
3. D. Levine and P. J. Steinhardt, Phys. Rev. Lett. 53, 2477 (1984).
4. J. A. Knapp and D. M. Follstaedt, Phys. Rev. Lett. 55, 1591 (1985).
5. P. A. Bancel, P. A. Heiney, P. W. Stephens, A. I. Goldman and P. M. Horn, Phys. Rev. Lett. 54, 2422 (1985).
6. R. J. Schaefer, L. A. Bendersky, D. Shechtman, W. J. Boettinger and F. S. Biancaniello, to be published.
7. Metals Handbook (ASM, Metals Park, Ohio, 1973), 8th ed., Vol. 8, p.262.
8. J. A. Knapp and D. M. Follstaedt, MRS Symposium, December 1985.
9. L. Bendersky, R. J. Schaefer, F. S. Biancaniello, W. J. Boettinger, M. J. Kaufman and D. Shechtman, Scripta Met. 19, 909 (1985).
10. L. Bendersky, Phys. Rev. Lett. 55, 1461 (1985).

# QUASICRYSTALS WITH 1-D TRANSLATIONAL PERIODICITY AND A TEN-FOLD ROTATION AXIS

L. BENDERSKY
Center for Materials Research, The Johns Hopkins University, Baltimore, MD and Institute for Materials Science and Engineering, National Bureau of Standards, Gaithersburg, MD

## ABSTRACT

Studies of phase formation in rapidly solidified Al-Mn alloys (composition range 18-22 at% Mn) show that an icosahedral phase is replaced by another noncrystallographic phase, a decagonal phase. The decagonal phase is another example of quasicrystal: It has a noncrystallographic point group (10/m or 10/mmm) together with long-range orientational order and one-dimensional translational symmetry. The decagonal phase is an intermediate phase between an icosahedral phase and a crystal both from the symmetry and from the solidification condition points of view.

## INTRODUCTION

Last year the discovery of an icosahedral Al-Mn phase, which diffracts electrons like a single crystal but has non-crystallographic point group symmetry m35, was announced by Shechtman et al [1]. The icosahedral symmetry is inconsistent with translational periodicity yet the diffraction peaks are sharp. This contradiction can be resolved assuming a structure with quasiperiodic or almost periodic properties. A fourier transform (diffraction pattern) of a quasiperiodic or almost periodic function yields a continuum of true delta functions (Bragg peaks) [2,3]. For a quasiperiodic structure, or quasicrystal, any point symmetry is permitted for the diffraction pattern.

The diffraction properties of the icosahedral phase can be successfully described using the concept of quasiperiodicity. Experimental diffraction patterns and high-resolution images [3,4,5] are in a good agreement with computer simulations, when an icosahedral quasilattice obtained by the Cut and Projection Method [6-8] was used as a model. At the same time, the conventional crystallographic explanation (multiple twinning, large unit cell [9-11]) is unable to explain all existing experimental evidence. It seems that icosahedral phase is a truly quasiperiodic solid. But what about the possibility for an infinite number of non-crystallographic point groups to exist? Recently, two other quasicrystals exhibiting non-crystallographic point groups different from m35 were reported: dodecagonal in the Ni-Cr system [12] and decagonal Al-Mn [13-15]. In this paper I present some details concerning the formation, stability and crystallography of the decagonal phase (also called T-phase). The relationship of this phase to the icosahedral phase is also discussed.

## FORMATION

The conditions for decagonal phase formation were outlined in the paper by Schaefer et al [16]. The phase was observed in rapidly solidified Al-Mn alloys, in some other aluminum-transition metals alloys [17], in ternary Al-Mn-Ge and Al-Mn-Zn [18]. In this paper only the binary Al-Mn will be discussed. In melt-spun ribbons the decagonal phase starts to appear when the alloy composition is higher than ~27 wt% Mn, growing epitaxially on the icosahedral phase dendrites, as shown in Figure 1. A specific orientation relationship is maintained between these two phases. The icosahedral phase

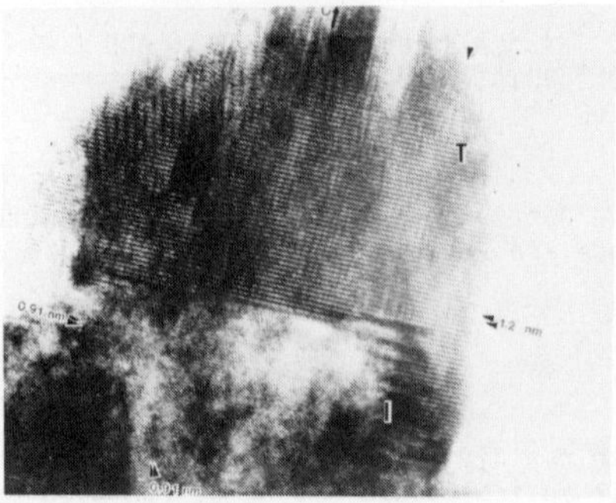

Figure 1. Growth of the decagonal phase (T) on the icosahedral phase dendrite (I). Al-28 wt% Mn melt-spun ribbon.

grows into an undercooled melt [1,16], and a decrease in solidification velocity is expected as the solidified volume increases due to a recalescence effect. Hence, the growth kinetics of the decagonal phase are comparable with the slowest growth of the icosahedral phase. The same conclusions are obtained for an Al-25 wt% Mn alloy, rapidly solidified by electron beam (EB) surface melting [16]. The icosahedral phase growth dominates for the highest EB scan velocities (50 to 200 cm/s), but at slower velocities (10 to 50 cm/s) the decagonal phase gradually replaces the icosahedral phase. Figure 2 shows typical microstructures, observed from 200 cm/s (a) and 10 cm/s (b) EB scans. The decagonal phase (Figure 2b) appears as blocky crystals with striated internal contrast associated with planar defects, but for some orientations it shows a circular appearance. From analysis of different cross-sections and tilting experiments, it was concluded that the decagonal phase has a cylindrical morphology. Most of the crystallites in Figure 2b are parts of a developed dendritic network embedded in an aluminium matrix.

In the composition range from 30 to 37 wt% Mn, a single phase decagonal microstructure is observed. The microstructure consists of fine grains (<1 μm) with wavy grain boundaries. This microstructure probably resulted from growth and impingement of facetted thermal dendrites of the decagonal phase.

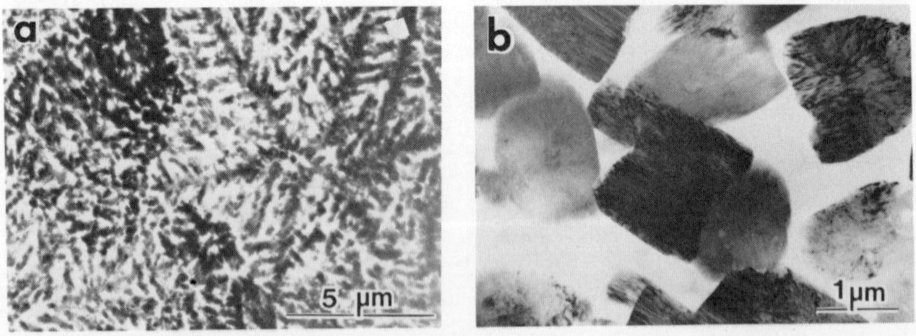


Figure 2. Al-25 wt% Mn, EB surface melting. (a) 200 cm/s, icosahedral phase dendrites; (b) 10 cm/s, decagonal phase.

## STABILITY

Like the icosahedral phase [1], the decagonal Al-Mn phase seems to be truly metastable, but with higher thermal stability than the former. The icosahedral phase transforms to stable $Al_6Mn$ after a 1 hour anneal at 400°C [1]. One hour isothermal annealing of the decagonal phase (37 wt% Mn) shows that crystallization starts at temperature close to 600°C. For 600°C annealing a mixture of the decagonal phase and the so-called $\phi$ phase (hexagonal, isomorphous to $Al_5Co_2$) is observed. For 730°C annealing a mixture of $\phi$ phase and the so-called $Al_4Mn$ phase (hexagonal, a = 2.84 nm, c = 1.24 nm) is observed, where the decagonal phase is completely transformed. Higher temperature annealing (820°C) shows the presence of $\mu$-phase (hexagonal, a = 1.995 nm, c = 2.452 nm). The stability of the icosahedral and decagonal phases has the same hierarchical order as does the non-equilibrium formation of the phases from the undercooled melt.

## CRYSTALLOGRAPHY

Crystallographic analysis of the decagonal phase was presented recently [14]. Based on transmission electron microscopy tilting experiments and convergent beam diffraction, the author concluded that the decagonal phase has a noncrystallographic point group symmetry (10/m or 10/mmm), different from icosahedral. The presence of a unique ten-fold axis implies that planes normal to this axis are either quasiperiodic (like Penrose tiling), or multiply twinned. The second possibility was rejected based on microdiffraction, dark field and high resolution imaging experiments. Figure 3 shows a high resolution image along the ten-fold axis, where no isolated crystalline (periodic) domains can be observed. However, an optical diffractogram (c) taken from selected small regions reproduces a pattern with ten-fold symmetry similar to an electron diffraction pattern (b).

The decagonal phase was found to be quasiperiodic within a plane, but the planes (layers) are stacked with 1.24 nm periodicity, as was shown by diffraction and high resolution imaging with the incident electron beam normal to the ten-fold axis [14]. The conclusion was that the decagonal phase is a quasicrystal with two-dimensional quasiperiodicity and one-dimensional periodicity and long-range decagonal orientational order.

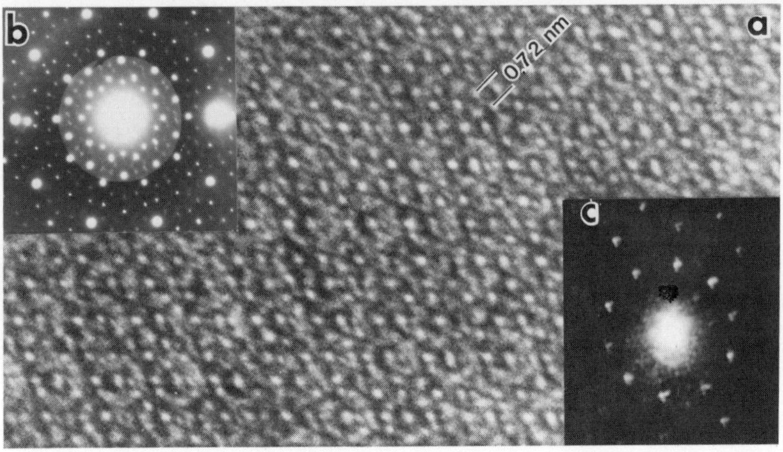


Figure 3. High-resolution image (a) of the decagonal phase taken at ten-fold orientation, SADP (b). (c) is an optical diffractogram from (a).

The cylindrical growth morphology of the decagonal phase is consistent with the cylindrical symmetry 10/m: the ten-fold rotation axis is along the cylinder axis, which is also probably the growth direction. The orientation relationship between the icosahedral and decagonal phases, mentioned

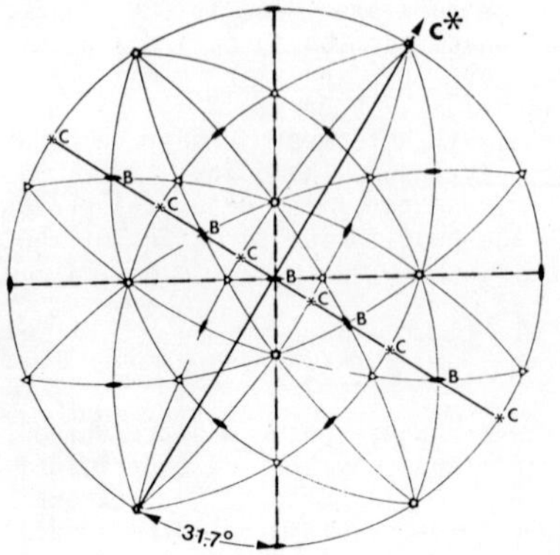


Figure 4. Superimposed stereographic projections of icosahedral and decagonal phases, according to found orientation relationship. B and C are axes of the decagonal phase two-fold patterns, ref. [14]. C* is ten-fold axis.

above, is presented in Figure 4. The ten-fold axis c* of the decagonal phase coincides with one of the six five-fold axes of the icosahedral phase. The two-fold axes of the decagonal phase (zone axis of pattern B, ref. 14) coincides with ten two-fold axes of the icosahedral phase, normal to the five-fold axis, however mirror planes of the two-fold pattern are rotated 31.7°.

ACKNOWLEDGEMENTS

The author thanks DARPA for financial support of this work. My thanks are also due to W. J. Boettinger, J. W. Cahn, D. Gratias, M. J. Kaufman, and R. J. Schaefer for helpful discussions.

REFERENCES

1. D. Shechtman, I. Blech, D. Gratias, and J. W. Cahn, Phys. Rev. Lett., 53, 1951 (1984).
2. H. A. Bohr, "Almost Periodic Functions," Chelsea, NY (1947).
3. J. W. Cahn, D. Shechtman, and D. Gratias, submitted to J. of Mat. Sci.
4. R. Portier, D. Shechtman, D. Gratias, J. W. Cahn, and J. Bigot, New Castle EMAG Conf. Proc. (1985).
5. M. Cornier, R. Portier, and D. Gratias, New Castle EMAG Conf. Proc. (1985).
6. V. Elser, Phys. Rev. Lett., 54, 1730 (1985).
7. M. Duneau and A. Katz, Phys. Rev. Lett., 54, 2688 (1985).
8. P. A. Kalugin, A. Kitaev, and L. Levitov, JETP 41, 119 (1985).
9. R. D. Field and H. L. Fraser, Mat. Sci. Eng., 68, L17 (1984-5).
10. M. D. Ball and D. L. Lloyd, Scr. Met., 19, 1065 (1985).
11. L. Pauling, Nature, 317, 512 (1985).
12. T. Ishimasa, H. V. Nissen, and Y. Fukano, Phys. Rev. Lett., 55, 511 (1985).
13. L. Bendersky, R. J. Schaefer, F. S. Biancaniello, W. J. Boettinger, M. J. Kaufman, and D. Shechtman, Scr. Met., 19, 909 (1985).
14. L. Bendersky, Phys. Rev. Lett., 55, 1461 (1985).
15. K. Chattopadhyay, S. Lele, S. Ranganathan, G. N. Subbanna, and N. Thangaraj, Current Sci., 54, No. 18, 895, 1985.
16. R. J. Schaefer, L. Bendersky, F. S. Biancaniello, W. J. Boettinger and D. Shechtman, submitted to Met. Trans.
17. P. Bancel, private communication.
18. L. Bendersky and F. S. Biancaniello, unpublished results.

# FORMATION AND THERMAL STABILITY OF QUASICRYSTALLINE AND OTHER METASTABLE Al-RICH Al-Mn PHASES*

M.X. Quan and B.C. Giessen, Materials Science Division, Barnett Institute, Northeastern University, Boston, MA 02115

## ABSTRACT

RSP of Al-rich Al-Mn alloys is known to produce several metastable phases, including (with increasing Mn content) supersaturated Al solid solution and two non-crystallographic phases, identified in the literature as an icosahedral phase (I) and a (probably) decagonal phase (T) with translational symmetry in one direction. The formation of these phases in the composition range from 8 to 30 at. pct. Mn has been studied in melt-spun and arc-furnace quenched alloys. The thermal transformation characteristics of all metastable phases were measured by DSC; transformation temperatures, enthalpies and activation energies were determined. At compositions from 17 to 20 at. pct. Mn, two transformations were observed. Noteworthy results include: smeared out (slow-onset) reactions, transformation temperatures dropping sharply toward lower Mn contents, transformation enthalpies rising steeply in the same range, and activation energies ranging from ~40 kcal/gmol for the low-temperature process (at ~763 K) to ~80 kcal/gmol for the high temperature process (~883 K).

## INTRODUCTION

In 1984, Shechtman et al. [1] reported the finding of a metastable phase in rapidly solidified Al-rich Al-Mn alloys that exhibited non-crystallographic symmetry in TEM and was identified as having icosahedral symmetry. Such phases are now designated quasicrystalline and a large body of work has sprang up around them [2-5]. The icosahedral phase was originally observed in $Al_{86}Mn_{14}$ alloys but was subsequently reported [2] to occur in better yields at ~21 at. pct. Mn. This phase is designated I phase in the following.

Subsequently, Bendersky found TEM evidence [6] for a different non-crystallographic phase (designated T phase) in RSP Al-Mn alloys. T phase was reported [6,7] to have a tenfold symmetry axis and to be quasicrystalline in two dimensions while being periodic (crystallographic) in the third dimension.

The conditions under which either of these metastable phases forms are not yet clear and their thermal stabilities have not been reported. We report here the results of a DSC study of the thermal properties of Al-rich Al-Mn alloys produced by RSP under different conditions and include transformation temperatures, transformation enthalpies and activation energies of transformation for the metastable phases found in these alloys.

* Communication No.280 from the Barnett Institute

## EXPERIMENTAL

Arc melted Al-Mn alloys (Al, 99.9+% from Leico and Mn 99.97% from Ventron) were quenched either by the arc-furnace hammer-and-anvil method [8], ("foils") or by melt spinning in inert-gas atmosphere ("ribbons"). Samples were examined by X-ray diffraction (XRD, Mo-Kα radiation), heated at variable $\dot{T}$ (5, 10, 20, 40, 80 $K \cdot s^{-1}$) in the Perkin-Elmer - DSC 2B and reexamined by XRD after terminating the heating process at appropriate temperatures. Activation energies $\Delta E_a$ were derived by the Kissinger method (determination of $T_{trans}$ as a function of $\dot{T}$) [9].

## RESULTS AND DISCUSSION

The structures observed in the as-quenched foils and ribbons, transformation temperatures $T_{trans}$ and transformation enthalpies $\Delta H_{trans}$ are summarized in Table I. $T_{trans}$ and $\Delta H_{trans}$ are plotted against composition in Figures 1 and 2, respectively; all percentages are atomic percent.

The activation energies $\Delta E_a$ for three specific transitions are listed in Table II; their determination by varying the heating rate $\dot{T}$ is shown for one transition in Figure 3. We discuss each group of data.

<u>Structures:</u> In the as-quenched foils and ribbons we have observed the XRD patterns of three phases: aluminum solid solution, I phase and a third phase which is probably T phase. Identification of the I phase was made by comparing it to the well publicized pattern [4] that can be indexed using a 6-vector index set. Only the patterns of I phase with $\leqslant$ 20% Mn could be indexed in this way: at 25 and 30% the pattern of I phase appears distorted, and, although it was not attempted to index these patterns, it appears doubtful that this could be done with the one-parameter 6-vector indexing method. This extension phase is therefore referred to as I'.

The phase designated here as T phase was assigned on the assumption that it is identical with the decagonal T phase reported by Bendersky [6,7]. Although an X-ray powder pattern had not been published for the T phase, a four-peak pattern (of which three form a broad cluster) has been indicated as being characteristic of it [10]. Therefore, when we obtained a pattern of peaks in one particular metastable phase, agreeing in the first two lines with the one described [10], identification of this phase as T-phase was reasonable. The peaks assigned by us to T phase are as given here:

| | d(Å) = | Rel.Int. (%) = |
|---|---|---|
| 1. | 2.247 | 20 |
| 2. | 2.072 | 100 (broad cluster [10]) |
| 3. | 1.454 | 10 |
| 4. | 1.272 | 30 |
| 5. | 1.082 | 15 |

It should be noted that the first two of these peaks are very difficult to differentiate from the first cluster of strong peaks of the equilibrium phases $Al_6Mn$ and other, more Mn-rich phases [11]; primarily for this reason, there is uncertainty attached to the identification of T phase. Further, the first strong peaks of I, Al, and T all fall into the same region of reciprocal space but can be distinguished from each other. Thus, peaks No. 2,4, and 5 of T phase are almost coincident with peaks (110000), (101000) and (200000), respectively, of I phase; however, peaks No. 1 and 3, of T phase, while weaker, do not belong to the I phase set and serve as an identifying mark.

Table I: Compositions, Structures, Transformation Temperatures and Transformation Enthalpies of Rapidly Quenched Al-Mn Alloys (Thermal data taken at $\dot{T}$ = 80 K/s).

| Alloy | Structure After RQ | | $T_{trans}$ (°C)[(a)] | | | $\Delta H_{trans}$ (cal/gmol) | |
|---|---|---|---|---|---|---|---|
| at.% Mn | Thin Foil and Thin Ribbon (35-40 μm) | Thick Ribbon (50-70 μm) | Thin Foil | Thin Ribbon | Thick Ribbon | Thin Foil | Thick Ribbon |
| 8 | Al s.s. [+ Glass (?)] | | 355;430 | | | 666 | |
| 11 | Al s.s. + I | | 370;425 | | | 591 | |
| 14 | I + Al s.s.[(b)] | | 400;475 | 450;480 | | 444 | |
| 17 | I + Al s.s. | T + trace Al s.s. | 1) 475;490<br>2) 590 | 480;495 | 540(T) | 1) 273<br>2) 79 | 160(T) |
| 20 | I (+ trace Al s.s.) | | 1) 450;470<br>2) 605 | 1)~450(v.w.)<br>2) 590;610 | | 1) 246<br>2) 40[(c)] | |
| 22 | --- | T | | | 660(T) | | 187(T) |
| 25 | I' | | 490;510 | | | 247 | |
| 30 | I' | | 470;505 | | | 180 | |

(a) Where two temperatures are listed for a given transition, the first refers to $T_{onset}$ (first detectable deviation from baseline), the second to $T_{trans}$ (by maximum slope method); 1): first exotherm; 2): second exotherm.

(b) Intensity ratio is reversed for ribbon, (Al s.s. + I)

(c) In a different foil, $\Delta H_{trans}$ = 138 and 169 cal/gmol were observed for transition 1) and 2), respectively.

Dependence of Structures on Preparation Method and Composition: Preparation methods found to be capable of producing metastable phases could be divided into three groups: thin (arc-furnace quenched) foils (~35 μ thick); thin ribbon (~40 μ thick), and thick ribbon (50-70 μ thick).

Of these, results for the thin foil and thin ribbons are close to each other with respect to the type of phase retained (see Table I) but differ in the thermal stabilities of these phases (see below). The thick ribbons, by contrast, contain different phases, specifically, there is T phase instead of I at 17 and 22% Mn (the only compositions so studied).

Turning to the effect of composition on the structure of samples prepared by a single method (thin foil), we notice formation of supersaturated Al solid solution (Al s.s.) (and possibly amorphous phase formation) at 8% Mn; I phase appears first at 11% Mn (together with Al s.s.); however, I is not obtained as a single phase for compositions with < 20% Mn. At > 25% Mn, I' forms instead of I.

Transformation Temperatures: Depending on composition and mode of preparation, one or two exotherms were observed in the DSC patterns. In contrast to the shape of exotherms obtained in the crystallization of glassy metals [12], many of the present exotherms had gradual onsets; this complicated the use of the conventional calibration method (steepest-slope tangent extrapolation). Therefore, two transformation temperature values are given for each exotherm showing this feature, where the first value is the earliest observed onset (obtained by drawing a tangent to the "onset tail") and the second value is the one obtained by the conventional method (drawing the steepest-slope tangent).

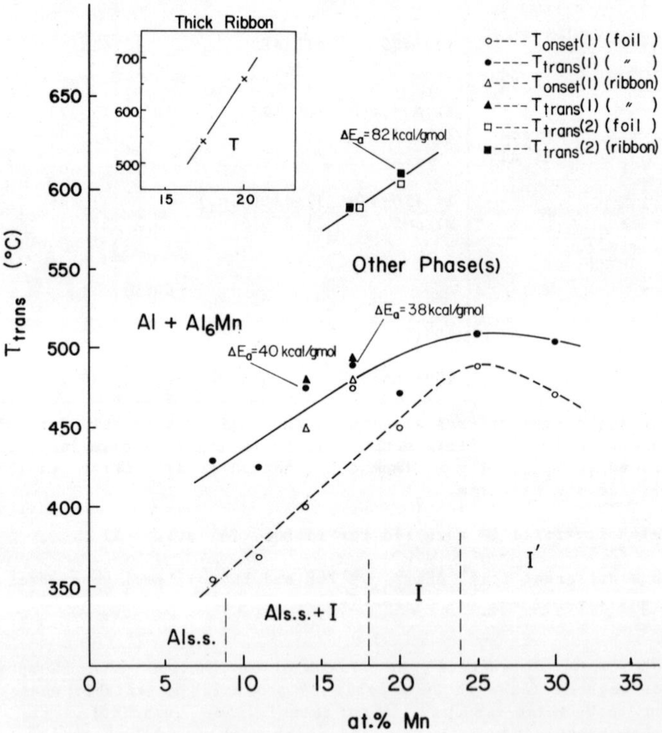


Fig. 1: Onset and transition temperatures of exotherms for Al-Mn alloys prepared by different RSP methods (see text). Alloys contain supersaturated Al s.s., I phase or I' phase. Data are taken at $\dot{T}$ = 80 $K \cdot s^{-1}$. Transformation products are indicated where known. Measured activation energies of transformation are included. Insert shows transformation temperatures for samples containing T phase.

The results listed in Table I and shown in Fig. 1 show two distinctly different sets of exothermic transitions. The first of these rises from values of 355°C (onset) and 430°C (transition) at 8% Mn to ~490°C at 17% Mn; at 20% Mn, the temperature of this transition becomes erratic. Depending on composition, different products form. At ⩾17% Mn, we attribute this transition to the transformation of I phase to undetermined metastable products or product mixtures; at ⩽14% Mn, the transition must be due to I phase and, increasingly, to $Al_6Mn$ forming from supersaturated Al-rich solid solution by transformation (at 14% Mn) or precipitation (below 14% Mn) and possibly, some glassy alloy crystallizing into $Al+Al_6Mn$.

It is not possible to better separate these two exothermic processes (I ⟶ Products and Al s.s. ⟶ Products) which take place in the first exotherm using only the present thermal data. We note that a sharp decrease of $T_{trans}$ towards lower alloying element content concentrations is a common feature in binary glassy alloys, e.g., Fe-B [13] where the effect is explained by assuming that the precipitation of the terminal element (e.g., Fe) can occur progressively more readily.

The second exotherm is found only in an intermediate composition range at 17 and 20% Mn; it is absent below 17% Mn. For samples with > 20% Mn, this transition is also not seen; however, it may occur above the calorimeter operational limit of 630°C. The qualitative labelling of the T-x diagram [Fig.1] is tentative.

Heats of Transition $\Delta H_{trans}$: Despite some uncertainty concerning the transformation onset temperatures, as discussed above, the integrated heats of transition were readily obtained; they have a smooth composition dependence, shown in Fig. 2. It is seen that the enthalpy of the first transition $\Delta H_{trans}$ (1) is strongly composition dependent at ⩽ 17% Mn and

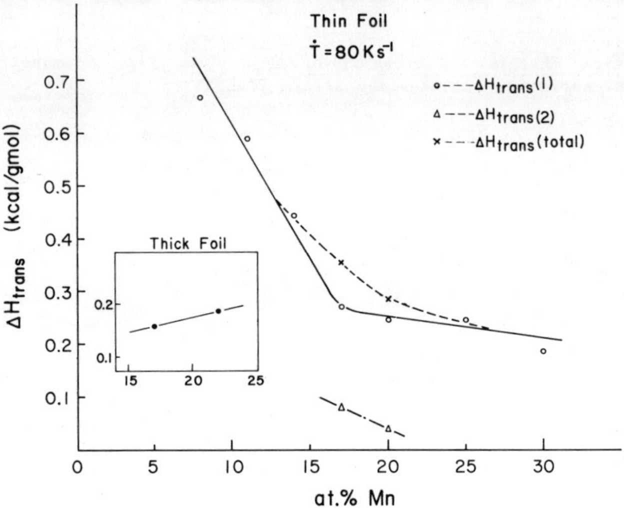


Fig. 2: Enthalpies of transformation for metastable Al-Mn phases (see text and caption, Fig. 1). Insert shows data for T phase.

remains flat at $\geqslant$ 17% Mn, confirming the assignment of this transition to different undetermined processes as a function of composition.

The large values of $\Delta H_{trans}$ (1) at the lowest Mn contents ($\Delta H_{trans}$ = .67 kcal/gmol at 8% Mn) are of the same order as typical values for the heat of crystallization of metal-metal and metal-metalloid alloy glasses ($\Delta H_{cr}$ = 1.0 kcal/gmol for $Zr_{.40}Cu_{.60}$ [14] and $\Delta H_{cr}$ = 1.13 kcal/gmol for $Pd_{.80}Si_{.20}$ [15]). The large value for $\Delta H_{trans}$ is surprising, as this exotherm is ascribed to precipitation from or transformation of a supersaturated solid solution to $Al_6Mn$ (depending on composition); the large value must therefore be due to the establishment of an energetically much more favorable crystalline structure than the terminal solid solution. For transformation of I phase to products, $\Delta H_{trans}$ ~270 cal/gmol are found at $\geqslant$17% Mn.

The value obtained by us for a mixture of I and Al s.s. at 14% Mn ($\Delta H_{trans}$ = 444 cal/gmol) is in fair agreement with a literature value of .5 kcal/gmol reported for a phase mixture including I at the same composition [3].

For the second, unidentified transition, lower values of $\Delta H_{trans}$ ~40-80 cal/gmol are found; however, $\Delta H_{trans}$ for the total exotherms (sum of both transitions) fall on a common curve with the single transition values (Fig. 2). The steep, monotonic increase of $\Delta H_{trans}$ with decreasing % Mn is puzzling, regardless of the nature of the initial and final phase. One observes $T_{trans} \cdot \Delta H_{trans}$ ~const.

The values for $\Delta H_{trans}$ at the intermediate composition Al + 20 at. pct. Mn are apparently sample-dependent in a manner not yet established; however, while the values for $\Delta H_1$ and $\Delta H_2$ differ substantially for two different quenched alloys, the sum $\Delta H_1 + \Delta H_2$ is constant, showing that in different samples there is a redistribution of the heat release between the two exotherms.

<u>Activation Energies of Transition $\Delta E_a$</u>: As Fig.3 demonstrates, the activation energy $\Delta E_a(2)$ for the second transition in $Al_{.80}Mn_{.20}$ ribbon was

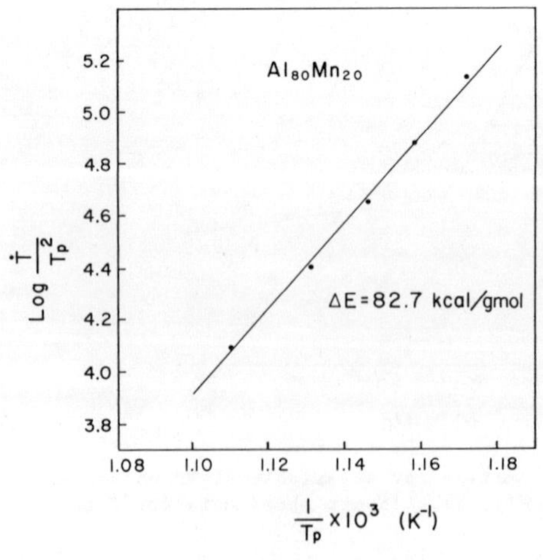


Fig. 3: Activation energy $\Delta E_a$ of second transition [trans(2), Fig. 1], for $Al_{.80}Mn_{.20}$, determined by Kissinger method [9].

Table II: Activation Energies $\Delta E_a$ of Exothermic Transitions(1) and (2) (see Figure 1).

| Composition (at. % Mn) | Transition | Sample Source | $T_{trans}$ ($^oC$) | $\Delta E_a$ (kcal/gmol) |
|---|---|---|---|---|
| 14 | (1) | foil | 475 | 40.1 |
| 17 | (1) | foil | 490 | 37.5 |
| 70 | (2) | ribbon | 610 | 82.7 |

readily obtained by the Kissinger method; further, $\Delta E_a(1)$ for the first transition obtained on foils with 14 and 17% Mn showed closely agreeing values of 40.1 and 37.5 kcal/gmol (see Table II), respectively; the value for the second transition $\Delta E_a(2)$ ~82 kcal/gmol is approximately twice as large as the value for the first transition.

Concluding Remarks: As found for metallic glasses in many alloy systems (e.g., Cu-Zr [14,16]), transformation temperatures and enthalpies are functions of alloy composition only and have no obvious relationships to the compositions of equilibrium phases. In the present system also, transition temperatures and enthalpies are continuous for terminal solid solution and quasicrystalline phases up to ~17% Mn; at higher Mn contents, complex and currently unidentified transition patterns are observed.

The present results appear to be qualitatively similar to results of a similar study reported in another symposium at the same conference [17]. Further research, correlating thermal and electron microstructural data is needed, especially at Mn contents <8% Mn and >20% Mn.

Data on the Young's moduli of quasicrystalline Al-Mn alloys and data on formation and thermal stability of metastable phases in the Al-Cr system will be published elsewhere.

## ACKNOWLEDGMENT

We thank Mr. Y.Z. Lu for numerous assistances and acknowledge support of this work by the Ford Motor Co. and the Barnett Fund. One of us (MXQ) is grateful for personal support by the Institute of Metal Research of the Academia Sinica, PRC.

## REFERENCES

1. D. Shechtman, I. Blech, D. Gratias, and J.W.Cahn, Phys. Rev. Lett. 53, 1951 (1984).

2. R.J. Schaefer and L. Bendersky, MRS Sym. Proc. Vol. 58, edited by B.C. Giessen, D.E. Polk, and A.I. Taub (Materials Research Society, Pittsburgh, PA, 1986).

3. K.F. Kelton and T.W. Wu, Appl. Phys. Lett. 46 (II), I, 1059 (1985).

4. D. Shechtman and I.A. Blech, Metall. Trans. 16A, 1005 (1985).

5. D. Levine and P. J. Steinhardt, Phys. Rev. Lett. 53, 2477 (1984).

6. L. Bendersky, Phys. Rev. Lett. 55, 1461 (1985).

7. L. Bendersky, R.J. Schaefer, F.S. Biancaniello, W.J. Boettinger, M.J. Kaufman, and D. Shechtman, Sc. Metall. 19, 909 (1985).

8. M. Fischer, D.E. Polk, and B.C. Giessen, Rapid Solidification Processing, Principles and Technology, edited by R. Mehrabian, B.H. Kear, and M. Cohen, (Claitor's Publ. Div., Baton Rouge, LA, 1978), p. 140.

9. M.G. Scott and P. Ramachandrarao, Mater. Sci. Eng. 29, 137 (1977).

10. R.J. Schaefer, Private Communication.

11. W.B. Pearson, Handbook of Lattice Spacings and Structure of Metals, Vol. 2, (Pergamon Press Ltd. 1967), p. 574.

12. D.E. Polk, C.E. Dube, and B.C. Giessen, Rapidly Quenched Metals III, Vol. 1, edited by B. Cantor (The Metals Society, London, 1979), p. 220.

13. T. Kémeny, I. Vincze, B. Fogarassy, and S. Arajs, ibid., p. 291 (1979).

14. A.J. Kerns, D.E. Polk, R. Ray, and B.C. Giessen, Mater. Sci. Eng. 38, 49 (1979).

15. F. Spaepen and D. Turnbull, Proc. 2nd. Int. Conf. on Rapidly Quenched Metals, Section I, edited by N.J. Grant and B.C. Giessen (M.I.T. Press, MA, 1976), p. 205.

16. R.V. Raman, Ph.D. Thesis, Northeastern University, 1977.

17. B.G. Bagley and H.S. Chen, MRS Sym. Proc., Vol. 58, edited by B.C. Giessen, D.E. Polk, and A.I. Taub (Materials Research Society, Pittsburgh, PA, 1986).

# FORMATION OF A QUASICRYSTALLINE PHASE FROM THE GLASSY STATE IN Pd-U-Si

A. J. DREHMAN,* S. J. POON,** AND K. R. LAWLESS**
*Ames Laboratory-USDOE, Iowa State University, Ames, IA 50011
**University of Virginia, Charlottesville, 22901

## ABSTRACT

Metallic glasses of composition $Pd_{58.8}U_{20.6}Si_{20.6}$ can be transformed to a metastable single phase with quasicrystalline order by thermal annealing. The icosahedral symmetry of this phase was verified by both electron and x-ray diffraction. Although the metallic glass can be formed over a broad compositional range the homogeneity range of this icosahedral phase is less than 1%. This indicates strong chemical ordering in the quasicrystalline state.

## INTRODUCTION

The discovery by Shectman et al. [1] of an Al-Mn phase exhibiting icosahedral point group symmetry ($m\bar{3}\bar{5}$) and the initial work of Levine and Steinhardt [2] has led to theoretical studies of the quasicrystalline structure [4,5] and its stability [6]. Efforts have also been made to determine the atomic sites [4,7,8]. Even lacking a complete structural model, the observed diffraction patterns can be adequately explained [2,5,9].

Molecular dynamics simulations of undercooled metallic liquids indicate the existence of icosahedral bond-orientational order [10]. This leads to the question of how quasicrystalline phases and metallic glasses are related. Previously we reported [11] the formation of a quasicrystalline phase during the devitrification of $Pd_{60}U_{20}Si_{20}$ metallic glass. Here we report the formation of what appears to be a truly single phase quasicrystalline material.

## EXPERIMENTAL METHODS

Pd-U-Si alloys were prepared by arc-melting the components under an argon atmosphere. Metallic glass ribbons were produced by melt spinning onto a copper wheel under 0.2 atm of helium gas. Ribbons of 15-20 μm thickness were obtained when the wheel surface speed was 50-75 m/s. The ribbons were checked for crystallinity by x-ray diffraction. They were then annealed in evacuated silica tubes which also contained zirconium turnings to act as a getter. The tube section containing the zirconium was preheated to about 700°C.

Annealed ribbons were thinned for transmission electron microscopy using an ion mill with a liquid nitrogen temperature cold stage. Specimens were initially inclined at 15° to the two argon ion beams with the final thinning taking place at about 10°. The operating voltage was 3.5kV. After thinning the very brittle specimens were mounted between folding-grids and secured in a double-tilt stage of the electron microscope. The specimens were examined in the TEM at an accelerated voltage of 120kV. Because of the small grain size, convergent beam electron diffraction was used for all diffraction patterns with a spot size of 40 or 100 nm. A small condenser aperture was used to obtain good angular resolution.

## FORMATION OF THE QUASICRYSTALLINE PHASE

Fully glassy $(Pd_3U)_{1-x}Si_x$ ribbons were formed over the range 16% to 23% silicon. A smaller range of glass formation, 18% to 22% uranium, was found for $Pd_{80-y}U_ySi_{20}$. Glassy ribbons near the composition $Pd_{58.8}U_{20.6}Si_{20.6}$ were transformed to the quasicrystalline phase by annealing between 420 and 490°C. Except for compositions very near the above, small amounts of crystalline phase were also present, as indicated in Fig. 1. The quasicrystalline phase can be formed over a range of less than 1% in contrast to the much larger range of glass formation.

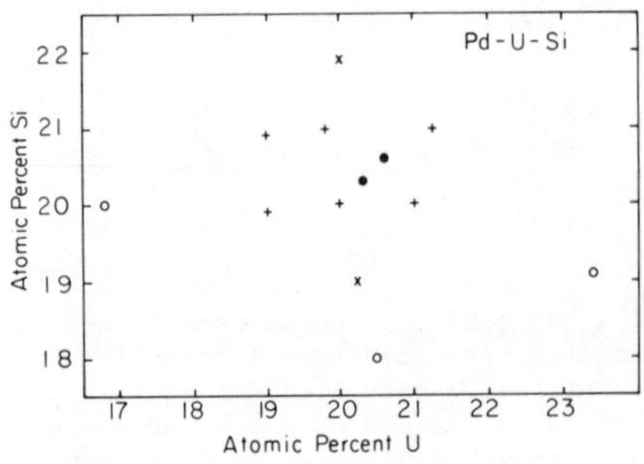


Fig. 1. Structure of devitrified Pd-U-Si glasses:
- ● Quasicrystalline single phase.
- + Quasicrystalline with trace amount of other phases.
- x Quasicrystalline with significant fraction of other phases.
- o Quasicrystalline phase is the minority of the material.

Fig. 2. TEM bright field image of quasicrystalline material.

Glassy ribbons were formed under various melt spinning conditions and this resulted in different annealing conditions required to form the quasicrystalline phase. Typical annealing times were 1.5 to 6 hours. Samples which transformed to the quasicrystalline phase at lower temperatures (as low as 420°C) were formed using a wheel speed of 50 m/s and had a typical grain size of about 50 nm. A faster wheel speed (75 m/s) and more ideal conditions resulted in glassy ribbons which transformed at higher temperatures and had larger grains.

TEM RESULTS

The ribbons which transformed at lower temperatures, although devoid of the glassy phase did contain some crystalline material as evidenced by x-ray diffraction. A TEM micrograph of a typical area is shown in Fig. 2. Grain size varied from about 30 nm to 100 nm with most of the grains being 40-60 nm across. Electron diffraction showed most of the grains to be the quasicrystalline phase, but some grains of a cubic phase were present as well as a very small amount of a heavily faulted phase (which may also be cubic). The faulted phase is visible in the upper-right-hand portion of the micrograph. The more rapidly cooled (75m/s) samples transformed to virtually 100% of the icosahedral phase (no other phases were detected by x-ray diffraction). Larger grains of the icosahedral phase, up to about 180 nm, were found in these specimens.

Convergent beam diffraction patterns from most grains showed symmetry near 2-fold, and only a small amount of tilting was necessary to locate an exact 2-fold symmetry axis, as shown in Fig. 3. This is to be expected since there are more 2-fold axes than 3 or 5-fold. Tilting experiments showed a 5-fold axis, Fig. 3b, approximately 32° from the 2-fold axis and a 3-fold axis, Fig. 3c, about 37° from the 5-fold, in agreement with previous studies on the Al-Mn system. The similarity of these patterns for the Pd-U-Si and Al-Mn systems is amazing.

High resolution microscopy was carried out to see if any evidence for twinning could be found. A typical micrograph with the beam direction near a 2-fold axis is shown in Fig. 4. Although defects were present no evidence for microtwinning was found in images of over 30 different icosahedral grains in three different specimens, varying in size from 30 nm to 180 nm.

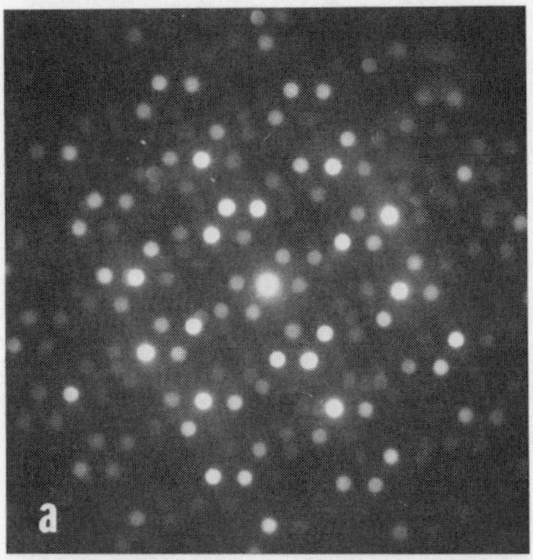


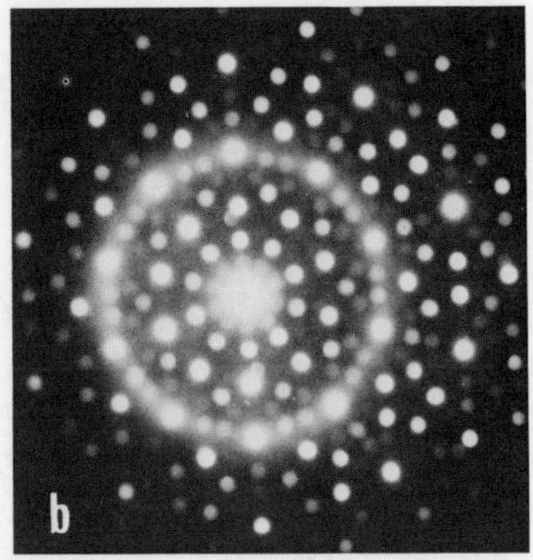


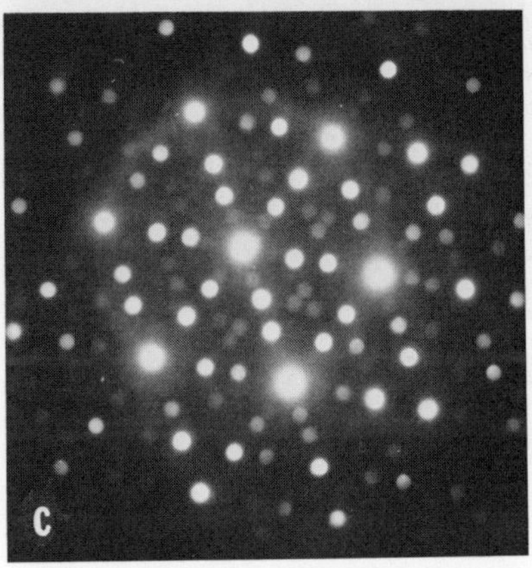


Fig. 3. Convergent beam diffraction patterns showing: (a) 2-fold axis, (b) 5-fold axis, and (c) 3-fold axis.

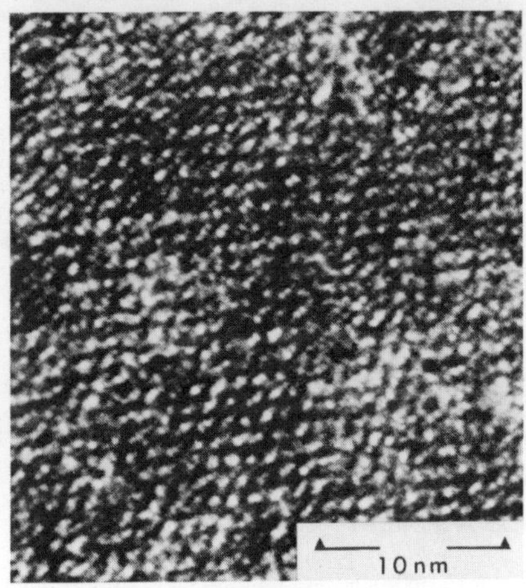


Fig. 4. High resolution image near a 2-fold axis. Two fold symmetry observed most easily at a glancing angle.

X-RAY DIFFRACTION RESULTS

The rapidly cooled (75 m/s wheel speed) $Pd_{58.8}U_{20.6}Si_{20.6}$ metallic glass ribbon was annealed at 490°C for 100 min. This anneal fully devitrified the glass and produced an apparently single phase quasicrystalline sample. All diffraction peaks could be indexed as belonging to the quasicrystalline phase. This is shown in Fig. 5 using the notation of Bancel et al. [9].

Using silicon powder to calibrate each diffraction pattern, the d spacing of the (100000) and (110000) peaks were measured and compared. Single phase $Pd_{58.8}U_{20.6}Si_{20.6}$ patterns had a ratio of 1.0527 ± 0.0005 while a ratio of 1.0555 was obtained for $Pd_{60}U_{20}Si_{20}$ (which contained a small amount of crystalline phase). The ideal ratio, based on a perfect icosahedron, is 1.0515. This indicates two things: (1) The quasicrystalline phase has a small range of homogeneity as the d spacings change with composition. (2) This quasicrystal-

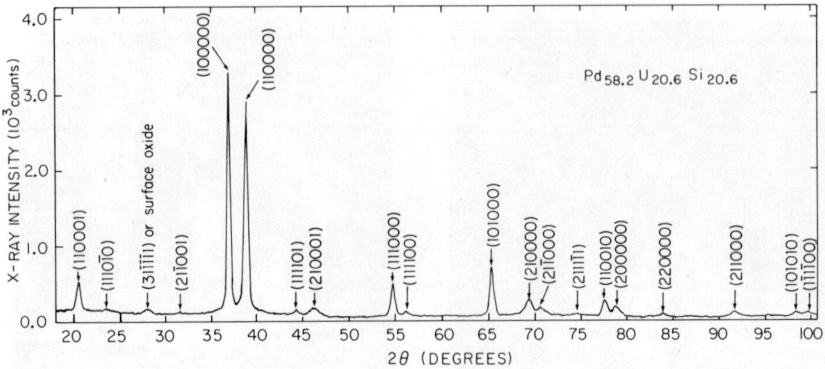


Fig. 5. X-ray diffraction pattern of single phase quasicrystalline material.

line phase can exist without perfect icosahedral symmetry (deviating by less than half a percent). The $Pd_{58.8}U_{20.6}Si_{20.6}$ composition is derived from $Pd_{3+\tau}U_{\tau}Si_{\tau}$ where τ is the golden mean. This is consistent with the quasicrystalline phase being composed of two cells of composition $Pd_3$ and PdUSi, the former being τ times more frequent.

## CONCLUSIONS

A single phase quasicrystalline material has been formed by devitrification of a metallic glass of the same composition. The strong dependence of grain size on the quenching conditions used to form the glass indicates that copious nucleation of the quasicrystalline occurs at high temperatures. This suggests a strong similarity in local order between the glass and the quasicrystalline phase. The small compositional range over which a single phase quasicrystalline material could be formed, however, indicates strong chemical ordering. This in turn implies some form of medium to long range order. The absence of this longer range order may be the primary difference between the glassy and quasicrystalline phases.

## ACKNOWLEDGMENT

This work was supported in part by the Director of Energy Research, Office of Basic Energy Sciences, USDOE, under contract no. W-7405-Eng-82.

## REFERENCES


1. D. Shectman, I. Blech, D. Gratias and J.W. Cahn, Phys. Rev. Lett. 53, 1951 (1984); D. Shectman and I. Blech, Met. Trans. 16A, 1005 (1985).
2. D. Levine and P.J. Steinhardt, Phys. Rev. Lett. 53, 2477 (1984).
3. P. Bak, Phys. Rev. Lett. 54, 1517 (1985).
4. K. Kuriyama, G.G. Long and L. Bendersky, Phys. Rev. Lett. 55, 849 (1985).
5. P.A. Kalugin, A.Yu. Kitaev and L.S. Levitov, Sov. Phys. JETP Lett. 41, 145 (1985).
6. N.D. Mermin and S.M. Trojan, Phys. Rev. Lett. 54, 1524 (1985).
7. K.M. Knowles, A.L. Greer, W.O. Saxton and W.M. Stobbs, Phil. Mag. B52, L31 (1985).
8. P. Guyot and M. Audier, Phil. Mag. B52, L15 (1985).
9. P.A. Bancel, P.A. Heiney, P.W. Stephans, A.I. Goldman and P.M. Horn, Phys. Rev. Lett. 54, 2422 (1985).
10. P.J. Steinhardt, D.R. Nelson and M. Ronchetti, Phys. Rev. B28, 784 (1983).
11. S.J. Poon, A.J. Drehman and K.R. Lawless, Phys. Rev. Lett. 55, 2324 (1984).

# FRACTAL COEFFICIENTS OF AN ICOSAHEDRAL STRUCTURE OF QUASI-CRYSTALS AND AMORPHOUS ALLOYS

J.C.S. LEVY and D. MERCIER
Laboratoire de Magnétisme des Surfaces
Université Paris 7 - 75251 Paris Cédex 05 - France

## ABSTRACT

Different extended icosahedral structures are obtained from energy minimization and local symmetry $Y_h$. The quasi fractal character of one of these is demonstrated from the quasi algorithm of definition, the gamma density of holes and the measured Hausdorff number $\bar{d}$. The spectral dimension $\bar{\bar{d}}$ is measured and dimensions corresponding to higher derivatives are defined.

## INTRODUCTION

Numerous observations of different extended "quasicrystalline" structures satisfying a local icosahedral order have been done from the recent experiment on $Al_{86}Mn_{14}$ [1], and several theoretical models have been proposed [2]. The very definition of the singularity of such a quasi-crystalline structure implies its self-similarity, its unity [3]. But, is it a trivial self similarity of dimension 3 ? The answer is obviously no since icosahedral symmetries are not compatible with translations. However numerous known crystals contain many icosahedral sites in their unit cell [4]. Thus on a large scale a dimension near 3 can be expected.

## ICOSAHEDRAL STRUCTURES AND BROKEN SYMMETRY

The minimization of energy due to pair interactions V( x ) leads to a density $n_0$ of the form

$$n_0(\vec{x}) = \sum_j c_j \exp [i\vec{k}j \cdot \vec{r}] \tag{1}$$

where the $\vec{k}_j$ are the nodes of the integral Fourier transform of the effective pair potential V(x) [5]. A remark on the optimality of the icosahedral cluster for many potentials leads to select the $c_j$'s and $k_j$'s which are compatible with the $Y_h$ group of symmetry [6,7]. Different systems of $c_j$'s lead to dislocation [8] or phasons in the language of incommensurate order [9], while different systems of $k_j$'s lead to different structures [10]. More generally such bifurcations can be expected since the group $Y_h$ is an optimal group, and an extended structure tends towards a high symmetry such as that of the full rotation group for a nearly liquid state [10]. Because of this optimal property, the completion of an extended icosahedral structure necessarily involves some breakings of symmetry which enable us to obtain approached symmetries which are near these of the full rotation group. Different symmetry breakings may occur. For instance the 20 equilateral spherical triangles of the icosahedron may be broken into 80 spherical triangles defined from the 3 apices of this triangle and the 3 middle points of its segments. Obviously other such nearly self similar treatments can be done, dividing this spherical triangle in $n^2$ nearly equal spherical triangles. The previous one will make the fivefold axis a tenfold one as recently observed [11]. Breaking the fivefold symmetry leads to symmetries which are compatible with the crystal. Hence numerous bifurcations may occur.

In this work we refer to a typical structure already studied by us [6,7] with a discrete atomic defined by the maximum values of $n_1(\vec{r})$ larger than 8 and not too close :

$$n_1(\vec{r}) = 2\cos(kz) + 2\sum_{p=1}^{5}\cos\left(k_0 z + k_1 x \cos\frac{2\pi p}{5} + k_1 y \sin\frac{2\pi p}{5}\right) \tag{2}$$

with $k_0 = \frac{\tau}{\sqrt{5\tau-3}} \cdot k$ , $k_1 = 2\sqrt{\frac{\tau-1}{5\tau-3}} \cdot k$ , and $\tau$ the gold mean.

## QUASI-FRACTAL CHARACTERS

Take the direction x, there are 30 such directions in the space since there are 3 such directions in each of the 20 equilateral triangles of the icosahedron. Thus this is nearly an "arbitrary" cut. $n_1(x)$ reads :

$$n_1(x) = 2+2\cos k_1 x - 4\cos\left(k_1 x \frac{\tau}{2}\right) + 4\cos\left[k_1 x \left(\frac{\tau-1}{2}\right)\right] \tag{3}$$

Since for maxima, $k_1 x$ and $k_1 \tau x/2$ must be nearly integer values of $\pi$ these integers belong to a Fibonnacci series $f_n$ such that $f_{n+1}=f_n+f_{n-1}$. Then the maximum condition reads [7] = $k_1 x = 4 f_n \pi$ (4)

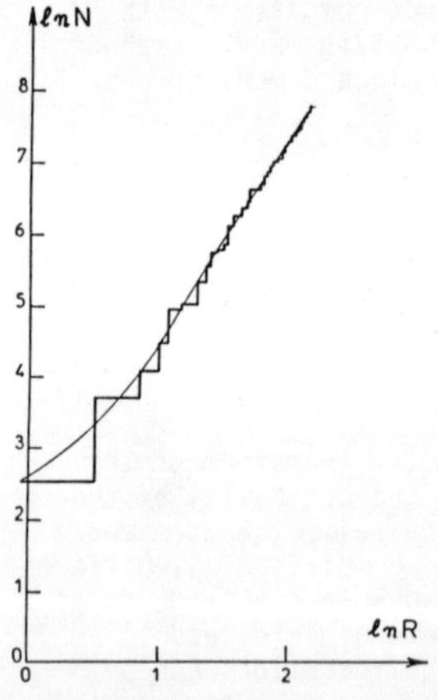


Figure 1 : Ln N (R) vs Ln R for a sphere a radius R. The slope increases from 2.7

Obviously higher n is, higher is the maximum value of $n_1$. Thus the hierarchy of points follows : 0 the origin is the first, then taking N very large defines $x_1 = 4 f_N \pi/k_1$. The second is defined in the interval

$$[0,x_1] \text{ by } x_2=4f_{N-1}\pi/k_1 \approx \frac{x_1}{\tau} \tag{5}$$

This algorithm is obviously recursive and has to be stopped when N-n takes a value of order 5 [7]. This looks like an inexact Cantor process, the approximation $\tau \# 1.5$ leads to an Hausdorff dimension $(d)_1$ = Log 2/Log 3 since in an interval of $2\tau \# 3$, an interval of 2 is selected. This leads for the whole structure to an Hausdorff dimension

$$d = 2 + \text{Log } 2/\text{Log } 3 = 2.63 \tag{6}$$

not very far from the measured Haussdorff dimension in a cluster of 1681 atoms, which is about 2.7 as shown in Figure 1.

## ANOTHER EVIDENCE FOR A QUASI FRACTAL STRUCTURE

Basically, the Hausdorff dimension is a measurement of density, and in the problem of compact structures, quasi crystals or amorphous materials, there is a "free volume" problem due to geometry [12]. In a simple, approximate mannner, the density may be said to satisfy an isotropic equation such as a Laplacian one with a central defect. Thus the density n ($\vec{r}$) is a sum of modified Bessel functions $I_\nu$ ($\sqrt{r}$), the indices $\nu$ being determined by the very nature of the defect [13]. The limiting form of $I_\nu$ for small r defines an exponent which is directly related to $\bar{d}$. Moreover such a modified Bessel function behavior is known to result from a Poisson distribution of gamma densities $f_{\alpha,\nu}$ (x) of increasing indices $\nu$[14] :

$$f_{\alpha,\nu}(x) = \frac{1}{\Gamma(\nu)}\alpha^{\nu}\ x^{\nu-1}\ e^{-\alpha x} \tag{7}$$

And gamma densities are associated with fractals [15]. Such gamma densities have been observed in the distribution of holes in such a structure [16] or in dense randomly packed monoatomic structures [17]. Finally the migration of holes according to a "randomized random walks" leads to such Bessel functions [14] as expected for the density in amorphous and liquid materials.

## OTHER FRACTAL COEFFICIENTS

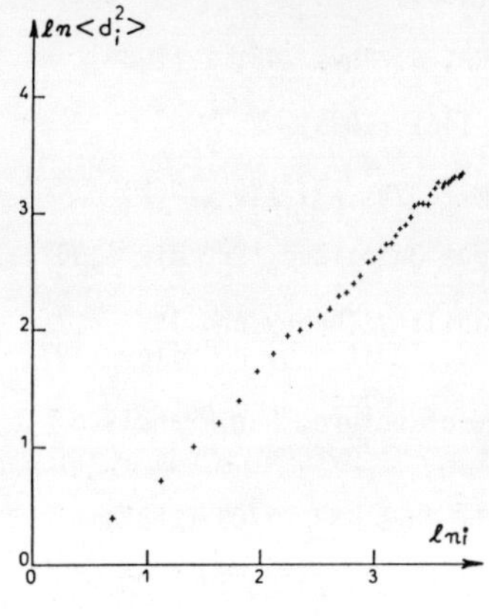


Figure 2 : Ln <$R^2$(t)> vs Ln t

Alexander and Orbach [18] noticed that the dynamic properties such as phonons or magnons are linked with differential operators and that there is a strong connection between Laplacian operator and random walks on nearest neighboring sites. They define a dynamic or spectral dimension $\bar{d}$ by $\bar{d} = 2\nu\ \bar{d}$ (8). where $\nu$ is the random walk exponent on such a structure, with the value 1/2 for a crystalline structure. Figure 2 shows the measurement of $2\nu$ in the same structure. Different slopes $2\nu$ are seen on this figure, they are associated with the quasi-fractal character of this structure, i.e. especially for low **n**, the recursive character of equations (5) is approximate. Comments on the localization effect of such different slope are obvious.

Other random walks can be defined, for instance on next nearest neighbor only, or on neighbors taken at a distance $\rho$ lower than $\rho_M$ and larger than $\rho_m$. Such random walks weight differently the second order derivatives and higher order derivatives for crystalline samples [19]. Thus they define a continuum of fractal coefficients which analyze each part of the long range contribution of correlation effects. And obviously the connection with derivatives means the connection with moments of arbitrary exponents in the Fourier transform, i.e. derivative of continuous index.

The authors acknowledge a fruitfrul discussion with Professor L. MICHEL on the optimal character of 1/h and its representation by functions of $r^2$, $Y_6$'s and $Y_{10}$'s.

REFERENCES

1. D.S. SHECHTMAN, I. BLECH, D. GRATIAS and J.W. CAHN, Phys. Rev. Lett. 53, 1951 (1984)

2. Review by D.R. NELSON and B.I. HALPERIN, Science, 229, 223 (1985)

3. D. LEVINE and P.J. STEINHARDT, Phys. Rev. Lett. 53, 2477 (1984)

4. F.C. FRANCK and J.S. KASPER, Acta Crystallogr. 11, 184 (1958)

5. J.C.S. LEVY, Surf. Sci. 104, 1 (1981)

6. D. MERCIER and J.C.S. LEVY, Phys. Rev. B27, 1292 (1983)

7. J.C.S. LEVY, J. Phys. 46, 215 (1985)

8. D. LEVINE, T.C. LUBENSKY, S. OSTLUND, S. RAMASWAMY and P.J. STEINHARDT, Phys. Rev. Lett. 54, 1520 (1985)

9. P. BAK, Phys. Rev. Lett. 54, 1517 (1985)

10. M.V. JARIC, L. MICHEL and R.T. SHARP, J. Phys. 45, 1 (1984)

11. L. BENDERSKI, Phys. Rev. Lett. 55, 1461 (1985)

12. D. TURNBULL and M.H. COHEN, J. Chem. Phys. 34, 120 (1961)

13. L. LANDAU and E. LIFSCHITZ, Mécanique Quantique, Ed. MIR (1967)

14. W. FELLER, An Introduction to Probability Theory and its Applications, vol. II, p. 59, 2nd Ed., Edited by J. Wiley (1971)

15. B. MANDELBROT, The fractal geometry of nature, San Francisco Freeman editor (1982)

16. J.C.S. LEVY and D. MERCIER, J. Appl. Phys. 53, 7709 (1982)

17. J.L. FINNEY and J. WALLACE, J. Non crystalline Solids, 28, 293 (1978)

18. S. ALEXANDER and R. ORBACH, J. Phys. (Paris) Lett. 43, L625 (1982)

19. Handbook of Mathematical Functions, p. 885, edited by M. ABRAMOWITZ and I.A. STEGUN, Washington (1964)

PART V A

# Microstructure and Properties of Crystalline RSP Alloys— Light Metals: Aluminum, Magnesium and Titanium

# THE STATUS AND POTENTIAL OF RAPID SOLIDIFICATION OF MAGNESIUM ALLOYS

F. HEHMANN AND H. JONES
Department of Metallurgy, University of Sheffield, Sheffield S1 3JD, U.K.

## ABSTRACT

In spite of giving exceptionally low density and excellent machinability at relatively low cost, magnesium alloys have found only restricted application as engineering materials to date because of limitations in formability, corrosion resistance, strength and creep resistance. Although the very earliest work on record to employ rapid solidification as a means to enhance engineering properties was carried out on magnesium-base alloys, subsequent work employing more modern techniques has not, until very recently, been motivated by the need for alloy development. The present study combines a critical survey of relevant published work with an assessment of the potential of magnesium-base alloys for development by the rapid solidification route.

## 1. INTRODUCTION

Major efforts in rapid solidification research over the last decade have been devoted to engineering metals that already find widespread structural application in unalloyed and/or alloyed form i.e. iron, aluminium, nickel and titanium. Arguably the most promising results have been obtained for aluminium, for which restrictions imposed by equilibrium alloying behaviour set definite limits to what can be achieved by conventional ingot metallurgy. Even more serious restrictions in this respect, limit the application of magnesium as a structural material in spite of cost and fusibility similar to aluminium and even lower density (in fact the lowest density of all engineering metals). Some of the limiting characteristics of magnesium and its alloys are:

(i) crystallography: cph with consequent limitations on the multiplicity of slip modes which limits deformability when temperature is not elevated and also the capacity for strengthening by alloying or cold working while retaining adequate ductility. Much of the conventional alloy development has been pursued to improve casting alloys for which there is a wide range and finishing of these is aided by the excellent machinability of magnesium and its alloys (the most machinable of all the engineering metals).

(ii) chemical reactivity: its extreme position in the electrochemical series combined with its inability to readily form a protective self-healing passive surface film (comparable to alumina on aluminium and its alloys) in corrosive environments make magnesium and its alloys particularly vulnerable to galvanic attack when coupled with more noble metals. This itself provides an important application for magnesium as a competitor to aluminium and zinc for sacrificial protection of steel in corrosive environments.

(iii) thermal stability: of properties such as strength and creep resistance, because of limited resistance to coarsening of precipitates resulting from conventional alloying (magnesium seems to be no worse than aluminium in this respect at least).

The capability of rapid solidification to induce the constitutional change and microstructural refinement that could be instrumental in circumventing such limitations has been known for decades but no systematic attention has been given to exploring these possibilities for magnesium alloys since the pioneering studies [1-4] at Dow Chemical Company for extrusions of atomized powder.

The present purpose is to review the limited work reported to date on the effects of rapid solidification on magnesium and its alloys and to assess their potential for future development by this route. The literature comprises: more basic studies concerned primarily with constitutional change, mainly to metallic glass formation and associated electronic properties since magnesium has turned out to be an important glass-former but also some important observations of new or extended crystalline phases together with limited studies of microstructural change; and more applied studies mainly concerned with the mechanical properties of engineering alloy compositions consolidated from rapidly-solidified particulate.

## 2. CONSTITUTIONAL AND MICROSTRUCTURAL EFFECTS

The reported observations [6-55] of constitutional changes produced by rapid solidification of magnesium alloys are summarized in Table 1. These observations comprise solid solubility extension of Al[5], Mn[10], Cu[19], Ga[44,45] and Zr[10] in $\alpha$Mg, retention of a disordered high temperature solid solution to room temperature over the composition range Mg-20 to 80at.%Cd[45], formation of nonequilibrium crystalline phases in Mg-Si[7], Mg-Sn[6,46,47] and Mg-Pb[6,52] and formation of amorphous (noncrystalline) phases in Mg-Ca[8,9], Mg-Fe[11], Mg-Co[12], Mg-Ni[13,14], Mg-Cu[8,13-21], Mg-Zn[22-43], Mg-Ga[44,45] Mg-Sb[48,51], Mg-Au[19] and Mg-Bi[48-50,53-55]. The noncrystalline phases in Mg-Fe, Mg-Co, Mg-Sb, Mg-Au and Mg-Bi were formed as thin films 0.01 to 1$\mu$m in thickness by vapour deposition in high or ultrahigh vacuum on a variety of substrates at very low temperatures. In the other systems the effects were obtained by rapid solidification from the liquid state and, for specific compositions (e.g. $Mg_2Cu$) or composition ranges (e.g. Mg-30 to 35at.%Zn) by sputter [18] or vapour [41] deposition. In most cases some form of splat-quenching to foil > 30$\mu$m thick was used to achieve rapid solidification from the melt, although melt-spinning to ribbon of similar thickness has been used for Mg-Zn and Mg-Cu. The effect of ternary additions on glass formation has been explored in a number of cases, for example, the addition of 10at.%Cu permitting glass formation to be obtained at Ca-contents as low as 20at.% from the 50at.% minimum in the Mg-Ca binary, while the Mg-11at.%Ni, Mg-14.5at.%Cu and Mg-30at.%Zn glasses have each been shown to accommodate substantial additions of ternary additions not known to form glasses with Mg alone [8,14,21, 23,35,39]. The Mg-Mn alloy extended solid solutions show the highest thermal stability reported in Table 1, short-term decomposition occurring at 520 to 570 K, compared with 350 to 450 K for most of the glassy compositions, with the Mg-Bi glasses and nonequilibrium fcc Mg-Sn and Mg-Pb phases decomposing rapidly around or below room temperature. Even the Mg-30at.%Zn glass, with a crystallization temperature $\sim$ 380 K on continuous heating, undergoes decomposition within a period of 6 years at room temperature according to Calka [56]. The Mg-Zn and Ca-Mg glasses have received considerable attention from solid state physicists as prime examples of metallic glass formation between components that are both simple metals, particular attention having been given to their electronic structure and transport properties [20,21,26-41,57-67]. Mechanical properties have received virtually no attention, though Calka et al [22a] reported a Vickers hardness of 220 kg/mm$^2$ and ultimate tensile strength as high as 830 MPa for the Mg-Zn glass, more than twice the values for the strongest commercial Mg-base alloys.

With very few exceptions [22a,43,44a,56], none of these reports of constitutional change included any reference to the associated microstructure. Kattamis [68], however, reported that the solidification microstructure of the Mg-5.3wt.%Zn-0.6wt.%Zr (ZK60) alloy remained nondendritic on splat-quenching, but that the grain size underwent refinement. Laser or electron-beam melted surfaces, in contrast, rapidly solidified with a dendritic microstructure,

TABLE 1 Summary of constitutional effects reported for rapid solidification of magnesium alloys. KEY: A = amorphous solid, C = new crystalline phase, D = sputtered deposit, E = extended solid solution F = vapour deposited film, H = retained disordered solid solution, I = interdendritic quench, R = melt-spun ribbon, S = splat-cooled foil, X = chemical symbol, Z = atomic number.

$C_\alpha$ = terminal solid solubility in cph Mg

$C_{eu}$ = eutectic composition RS = rapidly-solidified

| Addit-ion Z X | Equilibrium limits, at%X $C_\alpha$ | $C_{eu}$ | RS product and thick-ness (μm) | Constitutional effect and its composition range (atomic % X) (ternary add'ns are underlined) | Decomp-osition temp.(K) | Refer-ence |
|---|---|---|---|---|---|---|
| 13 Al | 11.8 | ∿31 | S - | E: to 22.6 Al (x2) | - | 5 |
| 14 Si | 0.003 | 1.34 | S - | C: anti $PbCl_2$ ($Mg_2Si$)[7] | - | 6,7 |
| 20 Ca | 0.5 | 10.5 | S - | A: 50-75Ca*, 20 or 40Ca+10Cu | 415/430 | 8,9 |
| 25 Mn | 1.0 | 0.9† | S 200-300 | E: to 2.5Mn (x2.5) | 520/570 | 10 |
| 26 Fe | 0.0004 | 0.008 | F 0.5 | A: 0.3-0.8 Fe | - | 11 |
| 27 Co | low | 1.9 | F 0.2-1.0 | A: 30-90Co | - | 12 |
| 28 Ni | <0.04 | 11.3 | S 30-50 | A: 10-18Ni* | 390/490 | 13 |
| | | | | A: 11.3Ni+0/30Zn, 0/9Sn or 0/5Sb | 380/440 | 14 |
| 29 Cu | 0.013 | 14.5 | S 30-50 | A: 12-22Cu* | 370/430 | 8,13-15 |
| | | | | A: 14.5Cu+0/11Ni, 0/30Zn, 0/6Sn* 0/18Pb, 0/10Ag; 14.5/20Cu+0/15Sn | 350/380 | 8,14,15 |
| | | | R 25 | A: 19.6, 14.5Cu; 20/40Cu+0/20Ag | - | 16-18 {20b, 21b} |
| | | | D 10 | A: $Mg_2Cu$ | - | 18 |
| | | | F 0.08/0.1 | E: 10Cu, A: 30Cu | - | 19 |
| | | | F 0.2 | A: 20-100(CuAg) | - | 20a |
| 30 Zn | 3.3 | 28.1 | R 30±20 | A: 25-35Zn[22], 28-32Zn[30-4] | 380 | 22,26-34, 38,40,42 |
| | | | R 22-25 | A: 20-40Zn | 350/400 | 36,37 |
| | | | S - | A: 23-40Zn*[23,24]; 22.5-35Zn[33 | 355-430 | 23,24,33 |
| | | | F - | A: 28-35Zn | 350 | 41 |
| | | | I ∿10 | A: 26Zn | - | 43 |
| | | | R 30? | A: 30Zn Subst. 0-8at% 19 add'ns | 390/460 | 23 |
| | | | | A: 30Zn Subst. Cu, In, Sn or 0-20Ga | | 21,35 |
| | | | | A: 30Zn plus 0.5 to 4Gd | | 39 |
| 31 Ga | 3.1 | 19.1 | S - | E: to 10Ga; A, 15-20Ga | <(390) | 44 |
| 40 Zr | <1.04 | 0.161† | S 200-300 | E: to 0.32Zr (?) | - | 10 |
| 48 Cd | 100 | ◇ | S - | H: 20 to 80Cd | - | 45 |
| 50 Sn | 3.4 | 10.7 | F - | C: orthorhombic 7-55Sn | 420 | 46 |
| | | | S - | C: fcc 14-18Sn | room T | 6,47 |
| 51 Sb | <0.04 | 11.5 | F 0.01-0.1 | A: 20-100Sb | 450 | 48-51 |
| 79 Au | 0.1 | 7.0 | F 0.1 | A: 30Au | room T‡ | 19 |
| 82 Pb | 7.8 | 19.1 | S 5 | C: fcc ($Cu_3Au$ type I) 16-23Pb | room T | 6,52 |
| 83 Bi | 1.1 | 14.3 | F 0.01-0.3 | A: 20-80Bi | 150/350 | 48-50, 53-55 |

◇ continuous solid solution †peritectic ‡6 months

* partially amorphous outside these composition ranges

attributable to epitaxial regrowth on the underlying grain structure of the unmelted alloy. Rapid galvanic corrosion occurred on immersion in 3½wt.%NaCl solution between the electropositive αMg dendrite cores and the electronegative interdendritic $Mg_7Zn_3$ phase*, especially for the electron-beam melted material that contained higher porosity. Samples solidified rapidly enough to suppress the formation of such an electropositive interdendritic phase were expected to show improved corrosion resistance, as demonstrated by Kattamis for a splat-cooled foil which had been heat-treated long enough to take the $Mg_7Zn_3$ phase into solid solution. Chill-block melt-spun ribbons of the Mg-4.5 and 6wt.%Zn alloys were reported [69] to exhibit an increasing volume fraction of nonequilibrium second phase both with increasing distance from the chilled side within the thickness of a ribbon and with increasing total thickness of the ribbon. These results indicated that a ribbon thickness below 10μm would be required to achieve completely segregation-free solidification in these alloys by this particular technique of rapid solidification. Melt-spun ribbons 0.2 to 0.3mm thick of a Mg-2wt.%Mn alloy exhibited noncellular regions in which βMn particles were more uniformly distributed throughout the αMg matrix, and cellular regions in which βMn was concentrated intercellularly [70]. The formation of intercellular βMn as a direct result of microsegregation during solidification is not consistent with the peritectic mode of solidification ($k_o>1$) normally experienced by Mg-Mn alloys in this composition range, when equilibrium αMn forms as the second phase. Evidently, however, nonequilibrium βMn forms in the melt-spun ribbons which need not necessarily require the peritectic mode of solidification. Indeed the results are entirely consistent with a eutectic mode ($k_o<1$) in the cellular regions of the specimen. The noncellular regions could have formed massively, following prior formation of the uniform distribution of βMn particles associated with these regions or alternatively followed by solid state precipitation of these particles during recalescence. This explanation is fully consistent with present understanding of the formation of segregation-free and segregated microstructures by rapid solidification although it is in conflict with the explanation proposed [70] by the originator of these microstructural observations. Meschter and O'Neal [71] reported on the microstructures of twin-roller quenched flakes of Mg-9wt.%Li alloy with and without addition of 1wt.%Si or Ce, compared with the as-received chill-cast condition, which showed plates of αMg (cph) 16±4μm thick spaced 43±3μm apart in a β(bcc) matrix. Corresponding values for the flake material were 1.8±0.6μm and 4.4±2.1μm, addition of silicon giving further refinement to 0.7±0.3 and 1.2±0.5μm, and of cerium giving 3.6±1.4 and 3.4±1.4μm. Rapid solidification thus gave microstructural refinement by a factor of 10 compared with chill-casting, while addition of silicon gave further refinement by a factor of 3. The refining effect of silicon was attributed to dispersoids of $Mg_2Si$ typically 50nm in diameter distributed evenly between the α and β phases and this dispersoid was shown to be effective in inhibiting coarsening of the α+β structure during 2-h heat treatments above 473 K.

## 3. ENGINEERING ALLOYS AND PROPERTIES

The various approaches reported to date directed towards development of improved engineering properties in magnesium alloys via rapid solidification processing are summarized in Table 2. These studies comprise early work by the Dow Chemical Co. using first gas-jet atomization [1,4,72,73,77,78] and then rotating disc atomization [2,3,74,75] and related work [80] using rotating-electrode atomized material, and recent work by MIT [81] and by Allied Corporation [82] employing as-spun or pulverized melt-spun ribbon, and by McDonnell-Douglas [83] using chopped twin-roll quenched flake. All of these studies employed extrusion to consolidate the resulting RS particulate into bulk form suitable for mechanical testing and evaluation. Alloys investigated have included both standard wrought alloy compositions and variations from

* The referee advises that the correct stoichiometry of this phase is $Mg_{51}Zn_{20}$ (I. Higashi et al, J Solid State Chem., 1981, 36, 225).

TABLE 2 Summary of reported approaches to improvement of engineering properties in magnesium alloys via rapid solidification. Consolidation of RS particulate was by extrusion in all cases. Key: PES = permissable extrusion speed; I/M = ingot metallurgy; P/M = powder metallurgy; CYS, TYS = compressive, tensile yield strength; RT = room temperature; $\bar{d}$ = mean powder particle size; $d_{max}$ = maximum powder particle size; SCR = stress corrosion resistance.

| RS particulate | Alloy Composition(wt%) | Summary of Results |
|---|---|---|
| Gas-jet atomized ($\bar{d} \sim 0.2$ mm) | AZ31(Mg-3Al-1Zn-0.2Mn)<br>M1(Mg-1.2Mn-0.1Ca)<br>ZK60(Mg-6Zn-0.7Zr)<br>AMZ111(Mg-1Al-1Mn-1Zn) | Tensile properties and CYS comparable with or better than some alloys via I/M [1,72a] |
| | Mg-0.3Zr<br>Mg-0.5Zr-0.2 or 0.5MM<br>(plus 50%Mg-1.2Mn-0.1Ca) | 'Interference hardening' via co-extrusion with blended-in Al or Mg-32Al alloy powders[1,73] |
| | AZ31(Mg-3Al-1Zn-0.2Mn)<br>AZ61(Mg-6Al-1Zn-0.2Mn) | Improved SCR via coextrusion with blended-in M1[1] |
| | A6,AZ61/63/80/92<br>ZK60 + 0.5Ca | PES increased via coextrusion with blended-in M1 or AMZ210 [1] |
| | Mg-1Zn - 0.4 to 1.6Si | Increased TYS and CYS [4] cf same alloy via I/M [76] |
| | Mg-2.1 or 6Si<br>Mg-4.5 or 8.3 Ba | Increased strength at 590/700K cf identically-processed ZK60 [77] |
| | Mg-0.6 to 4.3Zn-0.05 to 0.64 Zr including AK11 (Mg-1Zn-0.6Zr) | Increased strength, ductility and PES via coextrusion with blended-in Al + Bi, Ca, Th, Di or Ag [4,78] |
| Spin-disc atomized ($\bar{d} \sim 0.4$ mm)[74,75] | ZK60B(Mg-6Zn-0.6Zr) | CYS increased to TYS value and indep. section size of I/M ZK60A[3]<br>Less property directionality [2] |
| | ZE62(Mg-6Zn-2MM-0.5Zr) | Increased strength, stability and weldability cf P/M ZK60B [3] |
| | ZK10(Mg-1Zn-0.5Zr) | Increased PES with min. property loss cf via blended-in Al powder[3] |
| Rotating-electrode atomized ($d_{max}$ < 0.7mm)[79] | ZK60A(Mg-6Zn-0.45Zr) | Up to 1.5 x strength and 5x Charpy impact value cf I/M ZK60A [80] |
| Melt-spun ribbon | Mg-2.5 to 6.4Zn;ZK 60 | Tensile properties like [80] but no delam. on fracture faces [81] |
| Pulverized melt-spun ribbon | Mg-5to 8Al-0.2 to 1Zn-0 to 1.5Si-0 or 0.6Mn* | Increased strengths and corr'n res. of I/M ZK60A + AZ91HP/92A [82] |
| Chopped twin-roll quenched flake | Mg-9Li<br>Mg-9Li-2Si<br>Mg-9Li-2Ce | Up to 1.5 x RT strength and 3-4 x TYS at 423K cf I/M Mg-9Li [83] |

* In atomic %

MM = mischmetal

these. Tensile properties and compressive yield strength at ambient temperature have received most attention with resistance to corrosion, stress corrosion, impact and exposure at elevated temperature having been explored in isolated cases. Busk and Leontis [1] investigated the effect of extrusion temperature/speed over the ranges 590 to 700K and 25 to 200 mm/s for three commercial (AZ31, M1 and ZK60) and one experimental (AMZ111) wrought alloy(s) extruded from atomized powder and from ingot material. For the same extrusion conditions, tensile and compressive yield strengths were up to twice as large and ultimate strength up to 30% larger for powder extrusions compared with ingot extrusions. Similar findings were also reported for Mg-1wt.%Zn-1wt.%Si alloy [4]. The $Mg_2Si$ and $Mg_2Ba$ dispersoids generated by atomization of Mg-2 or 6wt.%Si and Mg-4.5 or 8.3wt.%Ba alloys were reported to result in a 3 to 5-fold increase in strengths at 590 K compared with similarly-processed ZK60 alloy [77]. Encouraging results were obtained also by co-extrusion of blends of different alloy compositions in the form of atomized powders. So-called 'interference hardening' [1,4,73,78] was produced by blending Zr-bearing Mg alloy powders with Al or eutectic Mg-Al alloy powders. Precipitation resulting from interdiffusion during extrusion and heat treatment, for example, nearly doubled the yield strength of Mg-0.3wt.%Zr and increased its ultimate strength by 30%. The same approach was used to impart resistance to stress corrosion from M1 alloy into a stress corrosion sensitive alloy such as AZ61, and to increase the permissable extrusion speed (above which hot-cracking occurs) of such alloys by blending-in 50% of M1 alloy that is not susceptible to hot-shortness at high extrusion speeds.

Drawbacks in employing gas-jet atomization for magnesium alloys include nonuniformity of powder particle size including a proportion of highly flammable and adherent fines and the need to filter, clean and recycle large volumes of an inactive atomizing gas [74]. One alternative is to achieve atomization by directing a stream of melt on to a rapidly-rotating heated disc [74,75]. A steel disc was found to be particularly effective for magnesium alloys containing Zn and Zr and led to small-scale industrial production of ZK60B (Mg-6wt.%Zn-0.6wt.%Zr) alloy atomized pellets for extrusion into a variety of sections for structural applications [2]. Pellet extrusions showed a particular advantage over ingot extrusions of similar composition in that compressive yield strength matched tensile values for any section size, whereas it was progressively lower with increase of section size for extrusions from ingot material. Addition of 2wt.% mischmetal to the ZK60B composition resulted in a weldable version of the alloy, designated ZE62, less sensitive to exposure to elevated temperature [3]. Further work on the ZK60 composition employed powder atomized by the rotating-electrode process [79]. Increased strength and impact values resulted from extrusion at temperatures in the range 394 K down to room temperature, compared with 644K employed commercially for ZK60A. The 50% increase in yield strength produced was consistent with a reduction in grain size from 10 to 100μm to 1μm, assuming a Hall-Petch factor of 1 $kg/mm^{3/2}$ . Das and Chang [82] however, suggest that the increase in longitudinal properties obtained by extrusion at room temperature may have been associated with inferior transverse properties since significant delamination was observed on the fracture surfaces. Ductile fractures were obtained however in recent work by Flemings and Mortensen [81] on extrusions of melt-spun ribbon of this alloy extruded at 483 K. The parallel study by Das and Chang [80] employing pulverized melt-spun or planar-flow-cast ribbon was directed towards investigating the effect of changes in alloy chemistry, based on the composition Mg-8at.%Al-1at.%Zn with increasing levels of added silicon which at 1.5at.%Si gave yield strength 50% higher than for I/M-ZK60A-T5 and 65% higher than the same material without the silicon addition. The composition Mg-8.1at.%Al-1.3at.%Si-0.6at.%Mn-0.2at.%Zn showed a particularly good combination of strength and ductility and also corrosion resistance in 3%NaCl solution superior to the alloy containing 3at.% less Al and no Mn. The major effect of the silicon addition was considered to be the formation of a fine dispersion of $Mg_2Si$ that helped to retain a fine grain size of the alloy matrix during hot-compaction and extrusion. Finally, the recent study by Meschter et al [83] using chopped twin-roll flake showed 50 to 60% higher yield and ultimate strengths for Mg-9wt.%Li-2wt.%Si or Ce compared with ingot-processed Mg-9wt.%Li. Half of this increment was shown to arise from rapid solidification in the absence of the Si or Ce addition, the remaining half resulting from the fine dispersion of $Mg_2Si$ or $Mg_9Ce$ generated by rapid solidification when Si or Ce was present. Similarly, yield strength at 423 K was trebled for extrusions of Mg-9wt.%Li flake compared with the ingot extrusion, and quadrupled when cerium was present in the flake alloy.

## 4. POTENTIAL

A major factor limiting the choice and level of wrought alloying additions is the maximum equilibrium solid solubility $C^{eq}_{max}$. Figure 1 shows $C^{eq}_{max}$ in magnesium [84-6] as a function of the atomic number of the alloying addition and identifies some 21 alloy elements known to exhibit potentially useful levels of $C^{eq}_{max}$($\gtrsim$ 1at.%) in Mg i.e. Li(17), Al(11.8), Sc(15), Mn(1), Zn(3.3), Ga(3.1), Y(3.8), Zr(1.0), Ag(3.8), Cd(100), In(19.4), Sn(3.35), Sm(∼1), Tb(4.6), Dy(6), Ho(5.4), Er(6.9), Tm(6.3), Yb(1.2), Lu(8.4), Tl(15.4) Pb(7.75), Bi(1.12) and Pu(3.2). All but 2 (Sc and Ga) of the first 9 feature prominently in conventional alloy development based on magnesium. Magnesium levels of addition are well within $C^{eq}_{max}$ for the binary, as shown in Table 3 with the exceptions of Li in alloy LA141, the higher level being required to stabilize the bcc β-phase of the Mg-Li system as the matrix phase, of Cu in ZCM711 and of lanthanide additions in HK31/32 and QE22 casting alloys. It is significant that all of the solid-soluble additions in Figure 1 showing $C^{eq}_{max}$ $\gtrsim$1at.% show Hume-Rothery atomic size factors within the ±15% limit for extensive solid solubility (most of them being within a ±12% size factor limit) and that the smallest size factors ($\gtrsim$8% for Li,Sc,Ga,Cd,In,Sn,Tl,Pu) tend to give higher solid solubilities ($\gtrsim$10at.%). Other elements within the ±15% size factor limit are He, (Si),(P),(S),Cl,Ti,Ge,As,Se,Br,Nb,Mo,Pd,Sb,Te,Ce,Pr,Nd,Pm,Gd,Hf,Ta,W,Re,Pt,Au, Hg,Th,Pa,U,Np and Am with several of these within a ±8% size factor limit (Si,P,S,Cl,Ge,As,Se,Nb,Ta,Hg,Pa,V,Np and Am). All of these are possible candidates for solid solubility extension by rapid solidification, although this effect has been reported so far for only one less solid-soluble element (Cu, by vapour quenching [19], and this in spite of a size factor of 20% exceeding the 15% limit) and many of these may not fulfil other necessary conditions for solid solubility extension. Experience with aluminium-based alloys, for which solid solubility extension by rapid solidification is obtained for most alloy additions of favourable size factor [87], would encourage the expectation that extension of solid solubility in magnesium will not be limited to Al,Mn,Ga and Cu. Desirable consequences could include:

(i) sufficient decrease of the axial ratio $^c/_a$ of the cph Mg-based solid solution to promote nonbasal slip and so enhance formability and ductility, thereby allowing increased levels of strengthening while retaining useful ductility. Only lithium and silver of the more soluble additions have been reported to decrease the $^c/_a$ ratio of magnesium [88].

(ii) improved corrosion or oxidation resistance as a result of shifting the electrode potential of magnesium to more noble values, thereby reducing the intensity of galvanic action, or by formation of self-healing passive films.

(iii) generation, by suitable heat treatment, of zone precipitation effects giving increased strength and/or thermal stability.

Levels of alloy addition that decrease or do not unduly increase the low density of magnesium and increase the elastic modulus are evidently of particular interest with respect to potential applications.

Further possibilities arise from the tendency to form new or extended alloy phases on rapid solidification. This includes the formation of cubic

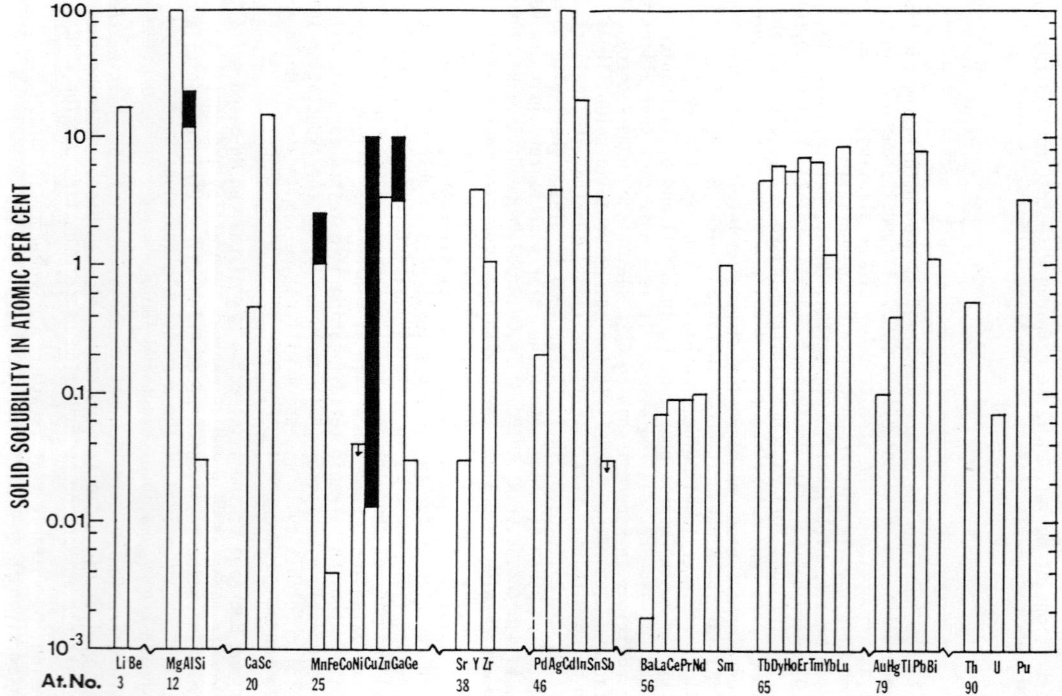


Figure 1. Maximum equilibrium solid solubility in magnesium [84-6] as a function of atomic number of binary solute addition, with filled bars indicating nonequilibrium extensions reported to have been achieved by rapid solidification.

TABLE 3 Maximum level $C_{max}^{com}$ of alloy addition (wt%) used in commercial wrought magnesium alloys compared with maximum equilibrium solid solubility $C_{max}^{eq}$ (wt%) in the binary.

| Alloy element X | | Alloy designation and level of addition (wt%) | | |
|---|---|---|---|---|
| Atomic number | Identity | Designation | $C_{max}^{com}$ | $C_{max}^{eq}$ |
| 3 | Li=L | LA141 | 14 | 5.5 |
| 13 | Al=A | AZ80 | 8.5 | 12.9 |
| 20 | Ca | AZ21A | 0.2 | 0.8 |
| 25 | Mn=M | M1 | 1.5 | 2.2 |
| 29 | Cu=C | ZCM711 | 1.2 | 0.04 |
| 30 | Zn=Z | ZCM711 | 6.5 | 8.4 |
| 39 | Y=W | WE54* | 5.25 | 12.6 |
| 40 | Zr=K | ZK61 | 0.8 | 3.8 |
| 47 | Ag=Q | QE22* | 2.5 | 15.0 |
| 58 | CeMM=E | HK31/32* | 3.2 | 0.5 |
| 60 | NdMM=E | QE22* | 2.5 | 0.6 |
| 90 | Th=H | HK31 | 3.2 | 4.8 |

Alloy Compositions (wt%):

| | |
|---|---|
| AZ21A: | Mg-2.0Al-1.0Zn-0.2Ca-0.1Mn |
| AZ80: | Mg-8.5Al-0.5Zn-0.1Mn |
| *HK31: | Mg-3.2Th-0.7Zr |
| *HK32: | Mg-3.2Th-2.1Zn-0.7Zr |
| M1: | Mg-1.5Mn |
| *QE22: | Mg-2.5Ag-2.0NdRE-0.6Zr |
| *WE54: | Mg-5.25Y-3.5RE-0.5Zr |
| ZCM711: | Mg-6.5Zn-1.2Cu-0.7Mn |
| ZK61: | Mg-6.0Zn-0.8Zr |

* Casting alloys

phases, bcc in the electron concentration (VEC) range 1 to 1.75 and fcc in the range 2.25 to >3 in preference to cph in the range 1.8 to 2.2 [89]. Examples include the equilibrium bcc β-phases in the Mg-Li and Mg-Sc systems [84-6] and the nonequilibrium fcc phases in Mg-Sn [47] and Mg-Pb [52]. Systems vulnerable to formation of a nonequilibrium fcc phase field include Mg-Ni,Mg-Co,Mg-Bi and Mg-Te [90] though competition is expected from terminal solid solubility extension in cph-Mg, from stable compound formation as well as from glass formation. The possibility of extending glass formation to higher Mg-contents than obtained hitherto also merits consideration. The limited thermal stability of such glasses or other nonequilibrium or extended phases need not, of course, be a barrier to potential application since suitably-controlled decomposition can result in a highly refined microstructure with highly desirable properties not readily attainable without first forming the nonequilibrium precursor phase [91-2].

The possibility of dispersing thermally stable insoluble intermetallic compounds in a magnesium alloy matrix via rapid solidification in order to, for example, maintain refined matrix grain size or improve creep resistance, also merits careful consideration. Figure 2 shows melting point of most Mg-rich compound [84-6] as a function of atomic number of binary alloy addition. Compounds such as $Mg_2Si$, $Mg_2Ge$ and $Mg_3Sb_2$ which combine high melting point with low equilibrium solid solubility (Figure 3) are specially notable and $Mg_2Si$, $Mg_2Ba$ and $Mg_9Ce$ have already been dispersed in Mg-based alloys by rapid solidification with encouraging results [77,71,80,83] (see §3), though the benefits of such a strategy remain to be fully evaluated.

Although the full potential of rapid solidification processing of magnesium alloys can be expected to be realized in such essentially novel alloy compositions, it is also important to evaluate any benefits that might accrue to established alloys, wrought or cast, when so processed. A preliminary programme at Sheffield evaluated the effect of rapid solidification via twin-piston splat quenching on the hardening responses of two casting alloys (ZH62, Mg-5.5Zn-1.8Th-0.7Zr and ZE41, Mg-4.2Zn-1.3RE-0.7Zr)* and two wrought alloys (EZ32, Mg-2.7RE-2.2Zn-0.6Zr and LA141, Mg-14Li-1.22Al)*. The results are shown in Figures 4a and 4b. Splats of ZH62 maintained an average hardness of $\sim$ 80 to 90 kg/mm$^2$ in isochronal (1h) treatments up to 673 K and room temperature treatment for up to 500 h with evidence of additional hardening in the temperature range 570 to 650 K and after $\sim$ 100 and 300 h at 295 K. Splats of ZE41 showed more of a hardening response in similar treatments, hardness reaching 105 kg/mm$^2$ after 1h and 573 K or 200 or 500 h at 298 K and retaining the as-quenched value of $\sim$ 80 kg/mm$^2$ after 1h at 673 K. Splats of EZ32 gave an even bigger response, hardness exceeding 130 kg/mm$^2$ after 1h at 373 K and still exceeding 80 kg/mm$^2$ after 1h at 673 K. Likewise splats of LA141, hardened from an initial value of 70 to 110 kg/mm$^2$ in 1h at 373K or 200 h at 295 K, retaining more than 80 kg/mm$^2$ after 1h at 673 K or 500 h at 295 K. This value of 110 kg/mm$^2$ appears to be the highest ever reported for this ultralight alloy (specific gravity 1.35). Alamo and Banchik [93], for example reported a peak value of 90 kg/mm$^2$ after $\sim$ 100 h at room temperature. Further work on novel and related alloy compositions is in progress.

## 5. CONCLUSIONS

(1) Splat quenching and other techniques of rapid solidification result for magnesium alloys in glass formation in 10 binary and related ternary systems, nonequilibrium crystalline phases in 3 systems and extended solid solubility in 5 systems. The glasses and new crystalline phases found so far show limited thermal stability and observations of microstructural effects for nonglassforming alloys have been few.

(2) Extrusions of rapidly solidified particulate of several engineering alloys and some novel compositions show examples of increased static and impact strengths independent of section size with less dependence on sign of stress, direction of testing and temperature.

(3) Opportunities for further development via rapid solidification include introduction of stable dispersoids, further extensions of solid solubility and formation of cubic or glassy phases with higher Mg-contents, as means of enhancing formability, ductility, strength, corrosion resistance and thermal stability, including creep resistance.

## ACKNOWLEDGEMENT

The authors are grateful to Magnesium Elektron Ltd. for supplying samples of alloys ZH62, ZE41, EZ32 and LA141.

---

* compositions in wt.%

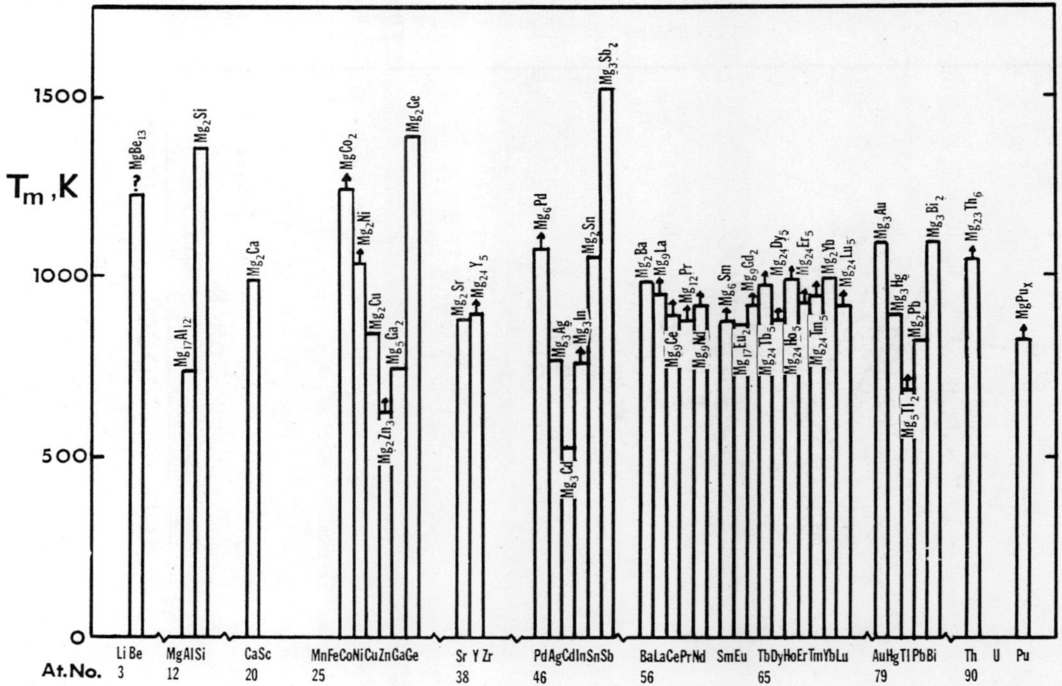


Figure 2. Equilibrium melting temperature of the most magnesium-rich binary intermediate phase [84-6] as a function of atomic number of alloying element.

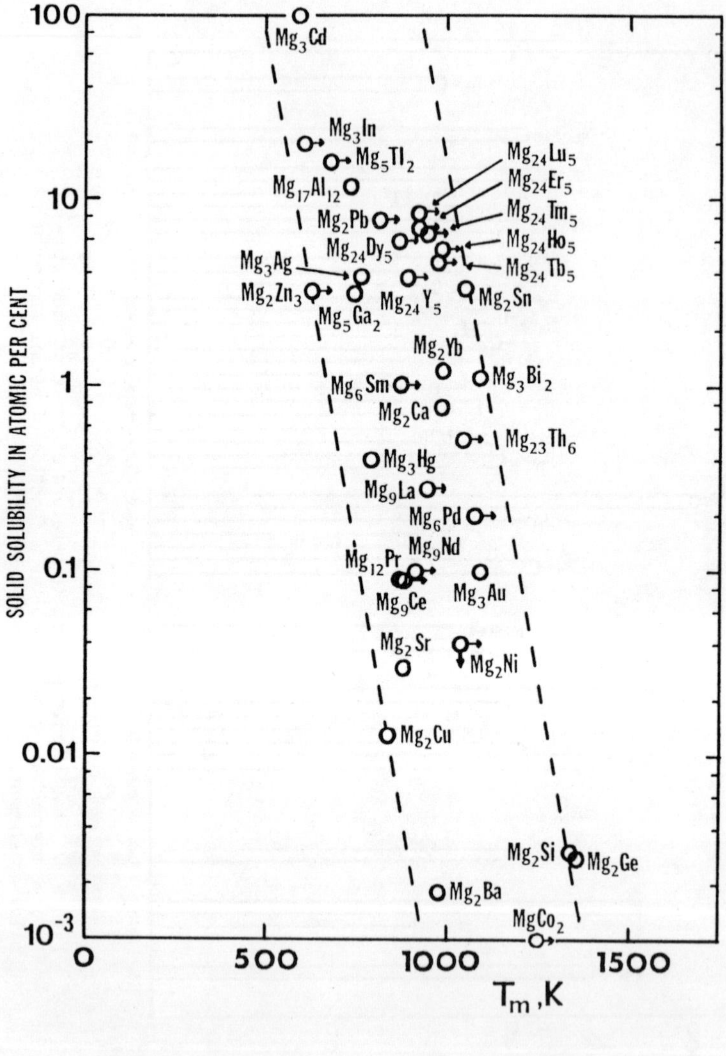


Figure 3. Maximum equilibrium solid solubility in magnesium as a function of melting point $T_M$ of most magnesium-rich binary intermediate phase, extending the plot first given by Carapella [94].

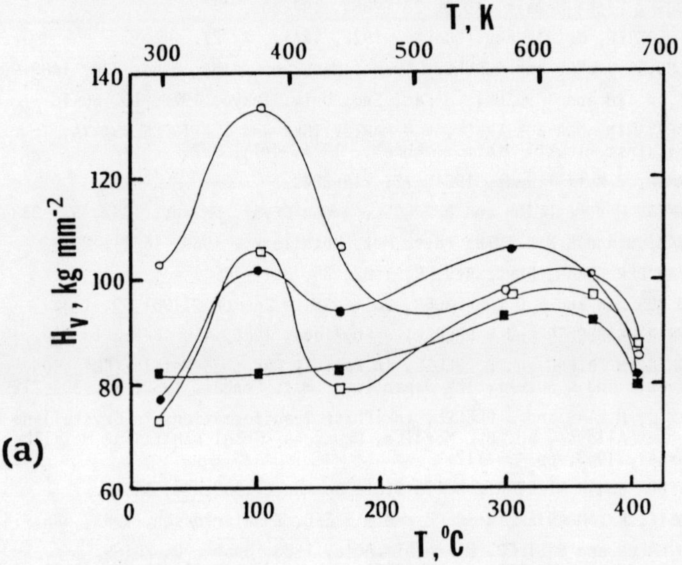


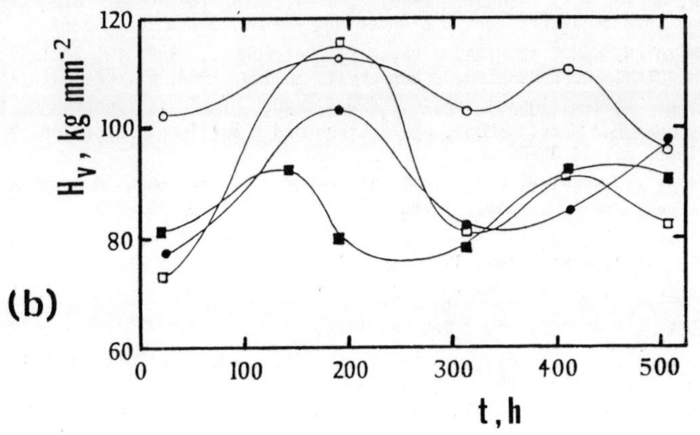


Figure 4. Hardness as a function of (a) temperature T of isochronal (1h) heat treatment and (b) time t at room temperature for four splat-quenched magnesium alloys ■ ZH62 ● ZE41 ○ EZ32 □ LA141.

## REFERENCES


1. R S BUSK and T E LEONTIS, Trans AIME, 1950, 188, 297-306.
2. G S FOERSTER and H A JOHNSON, Product Engineering, 1958, 29(19), 80-81.
3. R S BUSK, Light Metals, 1960, 23 (266), 197-200.
4. G S FOERSTER, Metals Eng. Quart., 1972, 12(1), 22-27.
5. H L LUO, C C CHAO and P DUWEZ, Trans. Met. Soc. AIME, 1964, 230, 1488-90.
6. H ABE, K ITO and T SUZUKI, J Fac. Eng. Univ. Tokyo, 1970, 10, 60-1.
7. A F BELYANIN, N A BUL'ENKOV, V R MARTOVITSKY and T A TOPORENSKAYA, Tr. Moskov. Inst. Toukoi Khim. Tekhnol., 1974, 4(1), 3-7.
8. F SOMMER, Z Metallkunde, 1981, 72, 219-224.
9. F SOMMER, W VOGELBEIN and B PREDEL, J Non-Cryst. Solids, 1982, 51, 333-43.
10. N I VARICH and B N LITVIN, Phys. Met. Metallogr., 1963, 16(4), 29-32.
11. A M VAN DER KRAAN, Phys. Rev. B, 1982, 25, 3311-18.
12. K H J BUSCHOW and P G VAN ENGEN, Solid State Commun., 1981 39, 1-3.
13. F SOMMER, G BUCHER and B PREDEL, J Physique, 1980, 41, C8-563 to 566.
14. F SOMMER, M FRIPAN and B PREDEL, in 'Rapidly Quenched Metals' (RQ4), ed. T Masumoto and K Suzuki, The Japan Inst. Met, Sendai, 1982, pp. 209-212.
15. F SOMMER, H HAAS and B PREDEL, in 'Phase Transformations in Crystalline and Amorphous Alloys', ed. B L Mordike, Deutsche Gesellschaft fur Metallkunde, Oberursel, 1983, pp. 95-112.
16. T MATSUDA and U MIZUTANI, Solid State Commun., 1982, 44, 145-9.
17. E NASSIF, P LAMPARTER, W SPERL and S STEEB, Z Naturforsch, 1983, 38a, 142-8.
18. (a) S FALCH and S STEEB, Mikrochim. Acta, 1983, Suppl. 10, 261-9; (b) S FALCH, P LAMPARTER and S STEEB, Z Naturforsch, 1984, 39a, 1175-83.
19. S MADER, J Vac. Sci. Technol., 1965, 2, 35-41; in 'The Use of Thin Films in Physical Investigations' ed. J C Anderson, Academic, New York, 1966,
20. (a) U MIZUTANI and T YOSHIDA, J Phys. F; Met. Phys., 1982, 12, 2331-48; (b) U MIZUTANI and K YOSHINO, J Non-Cryst. Solids, 1984, 61, 62, 1313-18.
21. U MIZUTANI and T MATSUDA, J Phys. F; Met. Phys., 1984, 14, 2995-3006; in 'Rapidly Quenched Metals' (RQ5), ed. S Steeb and H Warlimont, Elsevier, New York, 1985, pp. 1035-38.
22. A CALKA et al., Scripta Met., 1977, 11, 65-70; in 'Amorphous Magnetism II', ed. R A Levy and R Hasegawa, Plenum, New York, 1977, pp.197-205.
23. A CALKA and H MATYJA, in 'Amorphous Metallic Materials', ed. P Duhaj and P Mrafko, VEDA, Bratislava, 1980, pp. 71-82.
24. P G BOSWELL, Mater. Sci. Eng., 1978, 34, 1-5.
25. J J HAUSER and J TAUC, Phys. Rev. B, 1978, 17, 3371-80.
26. M LIARD et al, in 'Rapidly Quenched Metals III', ed. B Cantor, The Metals Society, London, 1978, Vol. 2, pp. 29-35.
27. J HAFNER, E GRATZ and H J GÜNTHERODT, J Physique, 1980, 41, C8-512 to 515.
28. J B SUCK et al, J Phys. C: Solid State Phys., 1980, 13, L1045-51; ibid., 1981, 14, 2305-17; as ref. 14 (1982) pp. 407-410; Phys. Rev. Lett., 1983, 50, 49-52; J Non-Cryst. Solids, 1984, 61, 62, 295-302.
29. H RUDIN, S JOST and H J GÜNTHERODT, J Non-Cryst. Solids, 1984, 61, 62, 291-4.
30. T MIZOGUCHI et al, J Physique, 1980, 41, C8-183 to 185; J Non-Cryst. Solids, 1984, 61, 62, 285-290.
31. U MIZUTANI and T MIZOGUCHI, J Phys. F: Metal Phys., 1981, 11, 1385-91.
32. N SHIOTANI et al, J Phys. Soc. Japan, 1981, 50, 828-834; as ref. 14, pp.667-70.

33. T MATSUDA and U MIZUTANI, J Phys. F; Metal Phys., 1982, 12, 1877-90; and as ref. 14, pp. 1315-18.

34. M ITO et al, Phil. Mag., 1984, A49, L9-L12; J Non-Cryst. Solids, 61, 62, 303-308.

35. T MATSUDA, N SHIOTANI and U MIZUTANI, J Phys. F: Met. Phys., 1984, 14, 1193-1204.

36. Z ALTOUNIAN, T GUO-HUA and J O STROM-OLSEN, J Mater. Sci., 1982, 17, 3268-74

37. M N BAIBICH, W B MUIR, Z ALTOUNIAN and T GUO-HUA, Phys. Rev. B, 1982, 26, 2963-6.

38. G FRITSCH, J WILLER, A WILDERMUTH and E LÜSCHER, J Phys. F: Metal Phys., 1982, 12, 2965-74.

39. S J POON, P L DUNN and L M SMITH, J Phys. F: Met. Phys., 1982, 12, L101-6; P L DUNN and S J POON, ibid.pp. L273-8.

40. R VAN DEN BERG, S GRONDEY, J KÄSTNER and H V LÖHNEYSON, Solid State Commun., 1983, 47, 137-140.

41. N NGUYEN VAN, S FISSON and M L THEYE, J Non-Cryst. Solids,1984,61/62,1325-9; M-O THEYE, V NGUYEN-VAN and S FISSON, Phys. Rev. B, 1985, 31, 6447-55.

42. A SADOC, R KRISHNAN and P ROUGIER, J Phys. F: Metal Phys., 1985, 15, 241-7.

43. Y V S S PRASAD, S L AGARWAL, K CHATTOPADHYAY and P RAMACHANDRARAO, as ref. 21,pp.39-42 .

44. (a) B PREDEL and K HULSE, (a) J Less Common Metals, 1979, 63, 45-56, 245-56; (b) B PREDEL, Physica, 1981, 103B, 113-122.

45. A C MOTA, P BREWSTER and R WANG, Phys. Lett., 1972, 41A, 99-101.

46. R SUGANUMA, J Phys. Soc. Japan, 1959, 14, 685-6.

47. H ABE, K ITO and T SUZUKI, Trans. Jap. Inst. Met., 1970, 11, 368-370.

48. R P FERRIER and D J HERRELL, J Non-Cryst. Solids, 1970, 2, 278-283;1970, 4, 338-344.

49. J H SLOWIK and F C BROWN, Phys. Rev. Lett., 1972, 29, 934-937.

50. J H SLOWIK, Phys. Rev. B, 1974, 10, 416-431.

51. R D BRINGANS and C M SUTTON, Solid State Commun., 1976, 19, 277-9.

52. H ABE, K ITO and T SUZUKI, Acta Met., 1970, 18, 991-994.

53. R P FERRIER and D J HERRELL, Phil. Mag., 1969, 19, 853-868.

54. M J SIK and R P FERRIER, Phil. Mag., 1974, 29, 877-894.

55. C M SUTTON, Solid State Commun., 1975, 16, 327-330.

56. A CALKA, unpublished preprint of contribution submitted to RQ5, Wurzburg, September 1984.

57. L VON HEIMENDAHL, J Phys. F: Met. Phys., 1979, 9, 161-9; J HAFNER and L VON HEIMENDAHL, Phys. Rev. Lett., 1979, 42, 386-9.

58. J HAFNER, Phys. Rev., B, 1980, 21, 406-25; 1983, 28, 1734-9; in 'Amorphous Materials-Modelling of Structure and Properties', Met. Soc. AIME, Warrendale, Pa., 1983, pp 201-9; in 'Structure of Non-Crystalline Materials', Taylor and Francis, New York, 1983, pp. 539-49; Phys. Rev. B, 1983, 27, 678-95; J Phys. C:Solid State, 1983, 16, 5773-92; J Non-Cryst. Solids, 1984, 61, 62, 523-8; ibid., 1985, 69, 325-45.

59. R W COCHRANE et al., Physica, 1981, 107B, 131-2.

60. J B SUCK, H RUDIN, H J GÜNTHERODT and H BECK, J Phys. F: Met. Phys., 1981, 11, 1375-83; J B SUCK and H RUDIN, in 'Glassy Metals II', ed. H Beck and H J Güntherodt, Springer-Verlag, Berlin, 1983, pp. 217-59.

61. M SEN and S CHATTERJEE, Phys. Stat. Sol. (b), 1982, 111, 353-9.

62. S M MUJIBAR RAHMANN, Z Phys. B - Cond. Mat ., 1982, 45, 307-13.

63. (a) L V MEISEL and P J COTE, Phys. Rev. B, 1983, 27, 4617-27, (b) P J COTE and L V MEISEL, J Non-Cryst. Solids, 1984, 61, 62, 1167-72.

64. J LAAKKONEN and R M NIEMINEN, J Phys. F: Met. Phys., 1983, 13, 2265-80.

65. R HARRIS and B G MULIMANI, Phys. Rev. B, 1983, 27, 1382-5.

66. J HAFNER and A PHILIP, J Phys. F: Met. Phys., 1984, 14, 1685-91.

67. T MATSUDA, U MIZUTANI, W B MUIR and M FROM, J Phys. F: Met. Phys., 1984, 14, L21-3.

68. T Z KATTAMIS, in 'Lasers in Metallurgy', ed. K Mukherjee and J Mazumdar, The Met. Soc. of AIME, Warrendale, Pa., 1981, pp. 1-10.

69. L J MASUR, J T BURKE, T Z KATTAMIS and M C FLEMINGS, in 'Rapidly Solidified Amorphous and Crystalline Alloys', ed. B H Kear et al, Elsevier, New York, 1982, pp. 185-9.

70. P SKJERPE, Mater. Sci. and Technol., 1985, 1, 316-20.

71. P J MESCHTER and J E O'NEAL, Met. Trans. A, 1984, 15A, 237-40.

72. (a) D S CHISHOLM and R S BUSK, U S Pat. 2630623 (1953), Brit. Pat. 662 312 (1951); (b) D S CHISHOLM, U S Pat. 2676359 (1954); (c) G F HERSHEY and N R COLBRY, U S Pat. 2934787 (1960).

73. T E LEONTIS and R S BUSK, U S Pat. 2659131 (1953), Brit. Pat. 690783 (1953).

74. N R COLBY and G F HERSHEY, U S Pat. 2699 576 (1955), Brit. Pat. 746301 (1956)

75. D S CHISHOLM and G F HERSHEY, U S Pat. 2752196 (1956), Brit. Pat. 783685 (1957).

76. G S FOERSTER, U S Pat. 3067028 (1962).

77. G S FOERSTER, U S Pat. 3182390 (1965).

78. G S FOERSTER, U S Pat. 3219490 (1965).

79. A R KAUFMANN, U S Pat. 3099041 (1963).

80. S ISSEROW and F J RIZZITANO, Internat. J Powd. Met., 1974, 10, 217-27.

81. M C FLEMINGS and A MORTENSEN, AD-A150 270, September 1984.

82. S K DAS and C F CHANG, in 'Rapidly-Solidified Crystalline Alloys', ed. S K Das et al., The Met. Soc. AIME, Warrendale, Pa., in press.

83. P J MESCHTER, R J LEDERICH and J E O'NEAL, Met. Trans. A, in press.

84. M HANSEN, Constitution of Binary Alloys, McGraw Hill, New York, 1958, and Supplements by R P ELLIOTT, 1965 and F A SHUNK, 1969.

85. W G MOFFATT, The Handbook of Binary Phase Diagrams, Genium Publ. Corp., Schenectady, NY, 1978 and updates.

86. Bull.of Alloy Phase Diagrams, 1980, 1, 108-9; 1982, 3, 60-74; 1984, 5, 23-30, 36-48, 348-74; 454-76, 579-92; 1985, 6, 37-42, 59-66, 149-67, 235-50 and to be published.

87. H JONES, in 'Rapidly Solidified Metastable Materials', ed. B H Kear and B C Giessen, Elsevier, New York, 1984, pp. 303-315.

88. R S BUSK, Trans. AIME, 1950, 188, 1460-64.

89. N ENGEL, Acta Met., 1967, 15, 565-7.

90. F HEHMANN, M.Met. Dissertation, Sheffield, 1984.

91. R RAY, Metal Progress, 1982, 121 (7), 29-31.

92. S K DAS, L A DAVIS, J R Y WANG and D KAPOOR, in 'Rapid Solidification Processing: Principles and Technologies III', ed. R Mehrabian, NBS, Gaitherburg, Md., 1982, pp. 559-564.

93. A ALAMO and A D BANCHIK, J Mater. Sci., 1980, 15, 222-9.

94. L A CARAPELLA, Metal Progress, 1945, 48, 297-307.

# MECHANICAL PROPERTIES OF RAPIDLY SOLIDIFIED MG-ZN ALLOYS

D. THEMINES*, W. RIEHEMANN, W. HENNING AND B.L. MORDIKE
Institut für Werkstoffkunde und Werkstofftechnik, Technische Universität, Agricolastr. 2, 3392 Clausthal-Zellerfeld, FRG
* Equipe Matériaux-Mikrostructure, L.A. 251, ISMRa Université Caen, Rue d' Edimbourg, 14032 Caen, France

## INTRODUCTION

Rapidly quenched alloys differ from those that are conventionally solidified in the degree of supersaturation, the microstructure and the grain size. The quenching rate can thus be used to modify the mechanical properties. A fine grain size increases the hardness and ductility. The precipitation behaviour is modified by the high density of heterogeneous nuclei or high density of quenched in defects in that the rate of precipitation is increased. Discontinuous precipitation should make a significant contribution to the increase in strength as a result of the fine grain size.

The aim of this work was to investigate the effect of rapid quenching on the mechanical properties of Mg-Zn alloys. Fig. 1 shows the equilibrium diagram. The concentration of the alloys investigated are shown as vertical lines. The Mg-Zn system is particularly suited for an investigation of the effects of rapid quenching for the following reasons:

(1) There is a very narrow range of solid solution which should be expanded by rapid solidification leading to increased solid solution hardening.
(2) The deep eutectic indicates the possibility of glass formation, as was pointed out by Calka [1,2].
(3) Microcrystalline microstructures can be prepared by rapidly quenching from the melt and also by crystallization of amorphous ribbons.

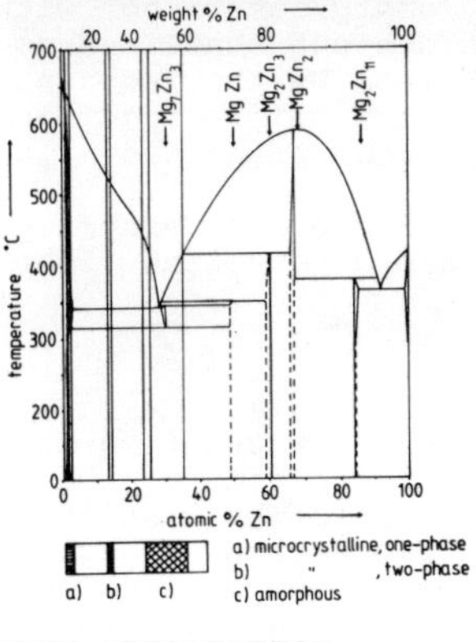


Fig. 1
Equilibrium diagram Mg-Zn according to Elliott [3]. The vertical lines indicate the compositions of the metallic tapes

## SPECIMEN PREPARATION

The alloys were melted under argon in a high frequency furnace using 99.5% pure Mg and 99.98% pure Zn. They were cast in the form of 6 mm diameter cylindrical rods. Metal ribbons were prepared by the melt spinning technique.

The alloys were melted under vacuum (0.4 Pa) in a boron nitride crucible and forced through a 0.5 mm diameter nozzle onto the rotating copper wheel. The peripheral speed of the copper wheel was 16 m/s and the nozzle inclined at 16° to the wheel normal. The compositions of the ribbons were determined by atomic absorption spectroscopy.

## EXPERIMENTAL RESULTS

Metallographic and X-ray investigations showed that alloys in the concentration range up to 2.2 at% Zn solidify as a single crystalline phase.

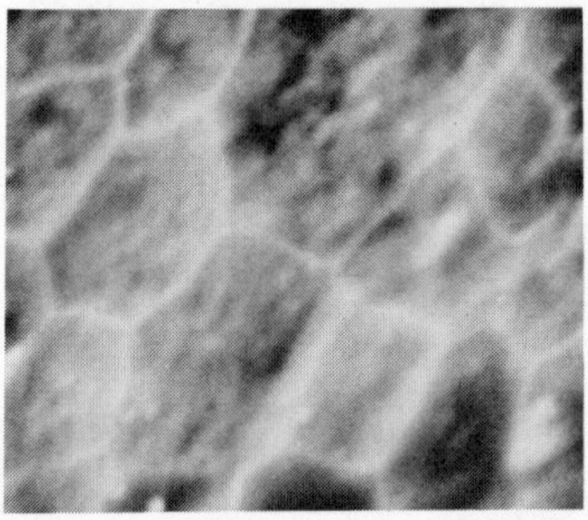

Fig. 2
Scanning electron microscope picture of surface crystals in a pure Mg ribbon

Fig. 3
Scanning electron microscope picture of fracture surface of amorphous ribbon with a few quenched in crystals

Fig. 2 shows the microstructure of a rapidly quenched pure Mg ribbon. The ribbon was microcrystalline as was the case for other single phase specimens. Fig. 3 shows a scanning electron microscope picture of a predominantly amorphous ribbon with 23.2 at% Zn. The alloy with the highest Zn content contained no crystals. The quenching rate as determined by the secondary dendrite arm separation in the partly crystalline specimen was $\sim 10^6$ K/s. This agrees well with the values predicted by heat transfer calculations [4].

The lattice parameters were determined using X-ray diffraction for the single and two phase crystalline ribbons. The results are shown in Fig. 4.

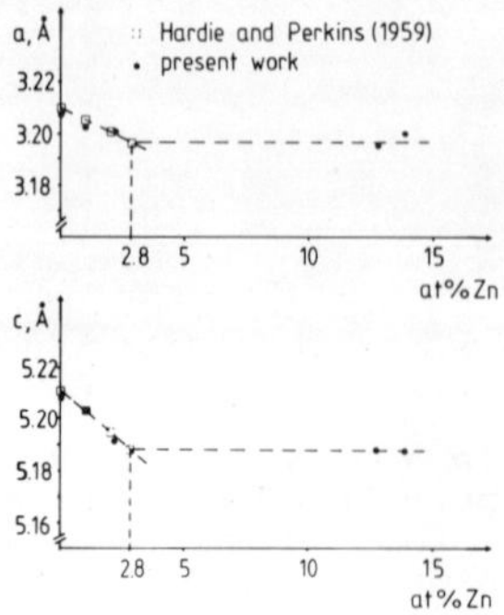


Fig. 4:
Lattice constants of Mg-Zn alloys as function of the zinc content. The results of Hardie and Perkins [5] were obtained on conventionally solidified material where the present work concentrated solely on rapidly quenched ribbons

The lattice parameters obtained agree well within the limits of accuracy with those of Hardie and Perkins [5]. It is clear that at quenching rates of $10^6$ K/s the range of solid solubility is not extended.

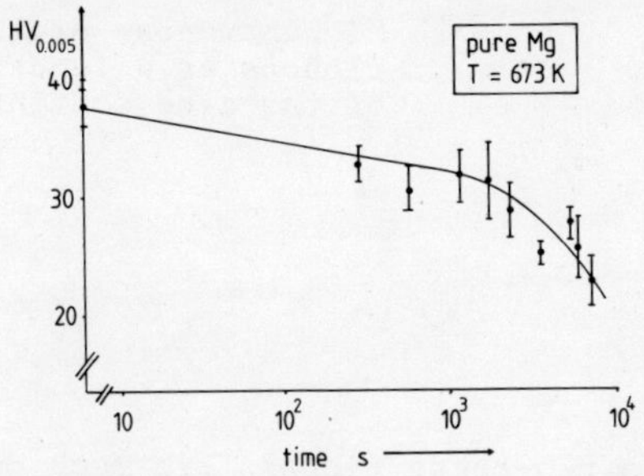


Fig. 5
Microhardness of Mg ribbons annealed at 673 K as a function of annealing time

In order to investigate the influence of grain size on the strength rapidly quenched pure Mg specimens were annealed for various times at 673 K. Fig. 5 shows a decrease in the hardness with increasing ageing time.

The grain size of some annealed specimens could be determined using electron microscopy. Fig. 6 shows the hardness of these specimens as a function of the grain size.

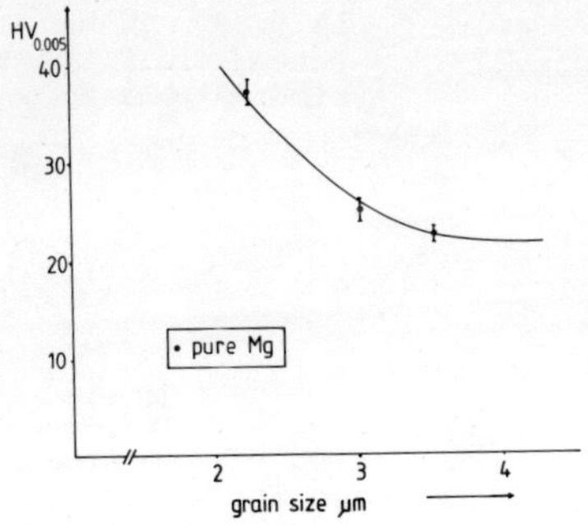


Fig. 6
Microhardness of Mg ribbons annealed at 673 K as a function of the grain size

The monotonic decrease in hardness with the ageing time is not necessarily a result of grain coarsening; it may be due to annealing out of quenched in defects.

It can be seen in Fig. 7 that in the region of solid solution hardening there is the typical increase in hardness with zinc content. It is apparent that an extension of the range of solid solubility by higher quenching rates would not lead to a significant increase in hardness.

The hardness of the two phase alloys is about a factor of three higher than in the range of solid solution, which can be explained by the high volume fraction of the second phase. The second phase is not, as X-ray measurements have shown, the intermetallic compound MgZn but $Mg_2Zn_3$.

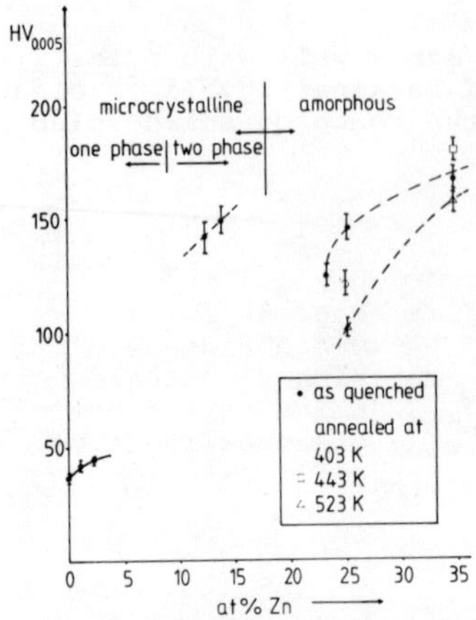


Fig. 7
Microhardness of ribbons as a function of the zinc content.

The hardness values of the two-phase ribbons are much higher than for conventionally cast material of the same composition. This is due to the refinement of the microstructure and the associated increase in interfacial surface area.

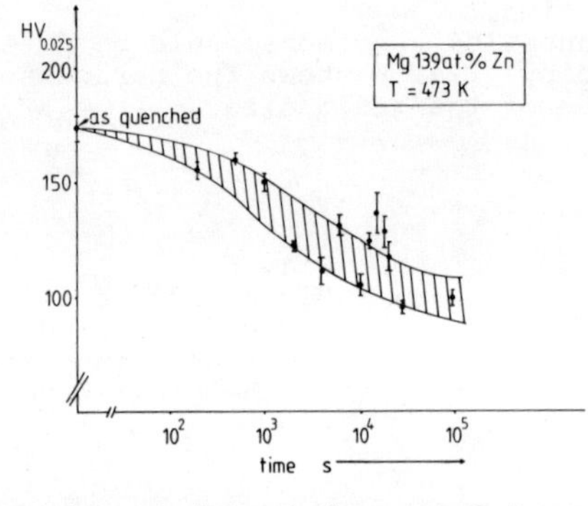


Fig. 8
Microhardness of two-phase Mg alloys with 13.9 at% Zn as a function of the ageing time at 473 K

During ageing at increased temperatures the hardness of two-phase alloys decreases due to coarsening of the constituent phases. This is clear in Fig. 8 which shows the microhardness data for a Mg-13.9 at% Zn specimen, annealed at 473 K. The hardness started to fall after about 1000 s. This relatively short time may be due to the high volume of grain and phase boundary assisted diffusion and the high density of quenched in defects.

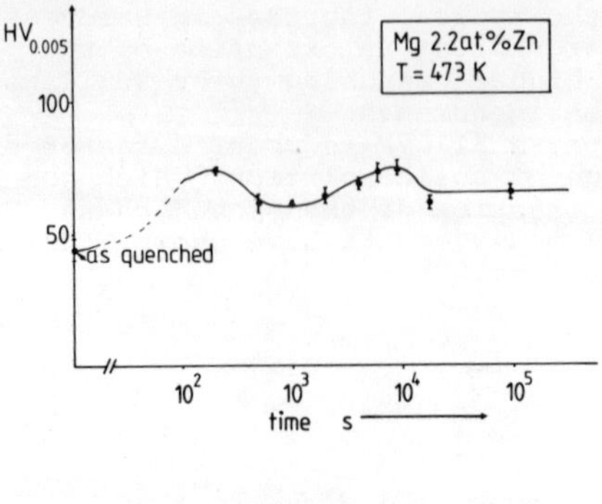


Fig. 9
Isothermal annealing of a single phase ribbon of Mg-2.2 at% Zn at 473 K

During annealing of a Mg-2.2 at% Zn-specimen the maximum hardness is reached within the first 100 s. The hardening phase is $Mg_2Zn_3$. The reason for a secondary hardening maximum may be the formation of the Mg rich phase MgZn. This phase has not yet been identified. The rapid rate of precipitation was reconfirmed by lattice parameter measurements as a function of the ageing time. As Fig. 10 shows, the excess Zn in solid solution precipitates completely within the first 100 s.

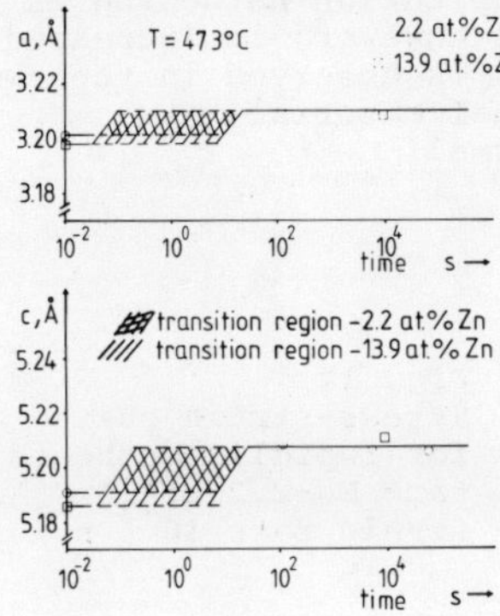


Fig. 10
Isothermal annealing of ribbons Mg-2.2 at% Zn and Mg-13.9 at% Zn at 473 K

The hardness of amorphous material is usually very high. The Mg-Zn amorphous ribbons Mg-34.7 at% Zn were the hardest of all alloys investigated (Fig. 7). It is surprising that the hardness can be increased by crystallization. The microstructural effects must be stronger in this case.

Calorimetric measurements (Fig. 11) have shown that the metastable phases which form on crystallization transform to stable phases at two different temperatures.

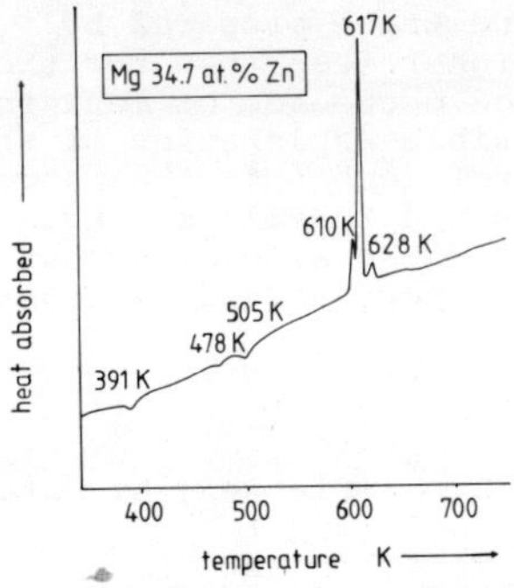


Fig. 11
Crystallization of completely amorphous ribbon Mg-34.7 at% Zn for a heating rate of 0.167 K/s

The amorphous state crystallizes initially at 391 K. The phases which form are α-solid solution and $Mg_7Zn_3$ as was demonstrated by X-ray measurements. The breadth of the diffraction peak points out fine constituents and a high degree of lattice distortion. Both effects favour a high hardness and could possibly explain the increase in hardness after crystallization. During crystallization at 478 K the phase $Mg_2Zn_3$ forms first and finally at 508 K, MgZn. The hardness is then lower than that of the amorphous state. Recovery, coarsening and the decrease in volume fraction of hardening phase all play a role.

SUMMARY

The results can be summarized as follows:
- the quenching rate achieved was ∿ $10^6$ K/s
- the grain size of the crystalline ribbon was in the micron range
- an extension of the α-solid solution range is not achieved at this quenching rate
- solid solution hardening and precipitation hardening is normal, the rate of precipitation, however, is increased
- a significant increase in hardness is observed in the two phase range compared to conventional material
- hardening by grain refinement is small

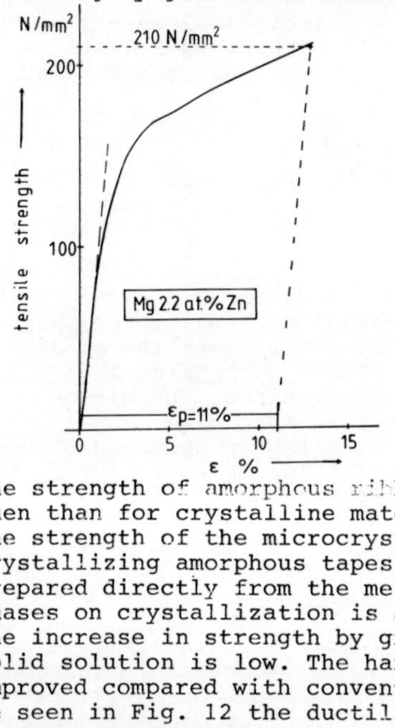


Fig. 12
Stress-strain curve for rapidliy quenched tape Mg-2.2 at% Zn, strain rate $10^{-3}$ $s^{-1}$

- the strength of amorphous ribbons is, as expected, higher then than for crystalline materials
- the strength of the microcrystalline tapes prepared by crystallizing amorphous tapes is higher then than for those prepared directly from the melt. The occurence of metastable phases on crystallization is a possible explanation of this
- the increase in strength by grain refinement in the range of solid solution is low. The hardness and tensile strength are improved compared with conventional Mg-Zn alloys. As can be seen in Fig. 12 the ductility is increased significantly.

## Literature

1. A. Calka, M. Madhaoa, D. E. Polk, B. C. Giessen, H. Matyja and J. von der Samde, Ser. Mat. 11, 65 (1977).
2. A. Calka and H. Matyja, Amorphous Metallic Materials, ed. P. Juhaj and P. Mrafka (VEDA, Bratislava, 1980)
3. R. P. Elliott, Constituion of Binary Alloys, (McGraw-Hill, 1965)
4. H. Jones, Rapid Solidification of Metals and Alloys, (The Institution of Metallurgists, London, 1982)
5. D. Hardie and R. N. Perkins, Phil. Mag. 4, 815 (1959)

# PLASTIC DEFORMATION OF A RAPIDLY SOLIDIFIED Al-Li-Zr ALLOY

N. J. Kim and S. K. Das
Allied Corporation, Metals and Ceramics Laboratory, Corporate R&D
P.O. Box 1021R, Morristown, New Jersey 07960

## ABSTRACT

The deformation behavior of a rapidly solidified Al-3.7%Li-0.5%Zr alloy has been investigated using transmission electron microscopy. Aging of the alloy in a conventional manner produces a homogeneous precipitation of δ' with small amount of heterogeneously distributed composite precipitates. The deformation behavior of the alloy in such aging conditions is dominated by the readily shearable δ', resulting in a pronounced planar slip. A homogeneous distrubution of composite precipitates promoted by double-aging treatment, however, alters the deformation mode to wavy slip by inducing the by-passing of dislocations around the shear resistant composite precipitates.

## INTRODUCTION

There is a growing need for structural alloys with improved specific strength to achieve substantial weight savings in aerospace applications. Aluminum-lithium alloys offer the potential of meeting the weight savings due to the pronounced effects of lithium on increasing the elastic modulus and decreasing the density of aluminum alloys. Moreover, heat treatment of alloys results in the precipitation of a coherent, metastable phase, δ' ($Al_3Li$) which offers considerable strengthening. Nevertheless, development and wide spread application of the Al-Li alloy system has been impeded mainly due to its inherent brittleness. The poor toughness of Al-Li alloys is mainly due to the pronounced planar slip. Additions of Cu and Mg to Al-Li alloys have been shown to promote slip dispersion through the precipitation of Al-Cu-Li, Al-Cu-Mg and/or Al-Li-Mg intermetallics (See e.g., Ref. 1-3).

Recently, it has been suggested that the development of Zr modified δ' precipitates might modify the deformation behavior of Al-Li alloys (4). It has been shown that, in Al-2.34%Li-1.07%Zr alloy, solution treatment produced a fine distribution of $Al_3Zr$ precipitates, which acted as nucleation sites for δ' during subsequent aging, forming composite $Al_3(Li,Zr)$ precipitates.

In the present paper, a new approach to promote the precipitation of composite phase is presented, emphasizing the effect of resultant composite precipitates on the deformation behavior of a rapidly solidified Al-Li-Zr alloy.

## EXPERIMENTAL PROCEDURE

The material used in this investigation was a ternary Al-3.7%Li-0.5%Zr alloy. This alloy was rapidly quenched from the melt into continuous ribbon using the jet casting process. The ribbons were mechanically comminuted to -60 mesh powder and then consolidated into bulk compacts by vacuum hot pressing followed by hot extrusion to a final rectangular shape of 63.5 mm by 3.8 mm (18:1 extrusion ratio, 6:1 aspect ratio).

The extruded bar was solutionized at 540°C, quenched in cold water, and was given various aging treatments. Slices 0.3 mm thick were taken from the broken tensile specimens, and thin foils for transmission eletron microscopy were made by jet polishing with the electrolyte of 33% $HNO_3$ and 67% methanol.

 

## RESULTS AND DISCUSSION

### 1. Microstructure

Aging of the alloy at 180°C produces a homogeneous precipitation of δ' within the matrix. Precipitation of δ' is also observed around the dark $Al_3Zr$ cores, forming composite $Al_3(Li,Zr)$ precipitates (Figure 1a). The increase in aging temperature and/or time results in a coarsening of δ' within the matrix and also around $Al_3Zr$ cores. However, the volume fraction and distribution of composite precipitates do not change. In all aging conditions, the distribution of composite precipitates is not uniform and their volume fraction is small as compared to the observation of Gayle and Vander Sande (4) who reported the homogeneous distribution of large volume fraction of composite precipitates in 1% Zr containing alloy. The nucleation of composite precipitates in single-aged alloy is mainly a result of heterogeneous precipitation of δ' upon pre-existing $Al_3Zr$ particles (4-7). The observed large differences in the volume fraction and distribution of composite precipitates between two studies might well be due to the difference in the Zr content (0.5% vs. 1%) and, more importantly, to the differences in the distribution and volume fraction of $Al_3Zr$ dispersoids formed during and/or prior to solution treatment, which are strongly affected by the prior processing of the alloy. In the present alloy, it has been found that only a small volume fraction of $Al_3Zr$ dispersoids are heterogeneously distributed in as-solutionized condition and most of Zr is in supersaturated solid solution (7).

It is this supersaturated Zr which gives rise to the characteristic morphology of composite precipitates observed in double-aged alloy. Figure 1b is a typical microstructure of double-aged alloy, showing the homogeneous distribution of large volume fraction of composite precipitates. Most of the δ' are associated with composite precipitates. Mechanism of composite precipitation during double-aging is not yet clear, but it has been recently suggested that vacancy/Zr interaction during low temperature aging is responsible for the phenomenon (7).

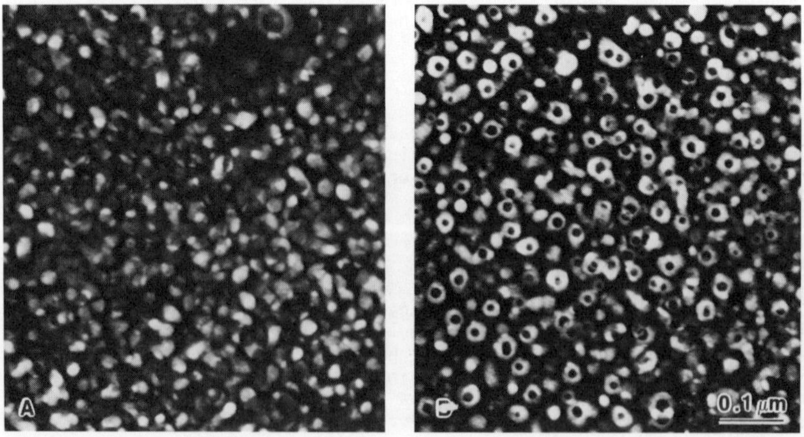


Fig. 1. Superlattice dark field micrographs; (a) single-aged (180°C/16 hrs.) and (b) double-aged (160°C/4 hrs. + 180°C/16 hrs.). There is a homogeneous precipitation of composite phase in double-aged alloy.

2. Deformation Mode

The above described changes in the distribution and volume fraction of composite precipitates from single-aged alloy to double-aged alloy significantly affect the deformation behavior of the alloy. Figure 2a shows the bright field image of deformed single-aged alloy. It clearly shows that there is a pronounced planar slip during deformation of single-aged alloy. The planar slip observed in single-aged alloy is mainly due to the shearable nature of δ' precipitates which result in decreased resistance to dislocation slip on planes containing the sheared δ' precipitates. The existence of heterogeneously distributed composite precipitates in single-aged alloy is not expected to modify the deformation behavior. It is affected more by the higher volume fraction of homogeneously distributed δ' precipitates.

In contrast to the single-aged alloy, double-aged alloy shows the wavy and homogeneous mode of deformation as shown in Figure 2b. As has been discussed previously, the only notable microstructural difference between single-aged alloy and double-aged alloy is the volume fraction and distribution of composite precipitates. Hence, it is apparent that the drastic change in the deformation mode from planar slip in single-aged alloy to wavy slip in double-aged alloy is due to the homogeneous distribution of composite precipitates in double-aged alloy. The shear-resistant nature of composite precipitate can be understood when one considers that the metastable $Al_3Zr$ particle in the Al-Zr alloys is highly resistant to dislocation shear, although $Al_3Zr$ is of the same cyrstal structure ($Li_2$) as δ'. The effect of composite precipitates on modifying the deformation mode can be more readily seen in weak beam dark field image (Figure 3). Superlattice dislocation pairs, which are the characteristic of sheared precipitates, are not observed and single dislocations move in a wavy mode. Many dislocation loops can also be seen. By comparing δ' superlattice dark field image and weak beam dark field image of the same area, these dislocation loops were identified to surround the δ'/$Al_3Zr$ or δ'/matrix interface of composite precipitates (7). This means that dislocations by-pass the composite precipitates leaving the dislocation loops around them, indicating the high resistance of composite precipitates to dislocation shear.

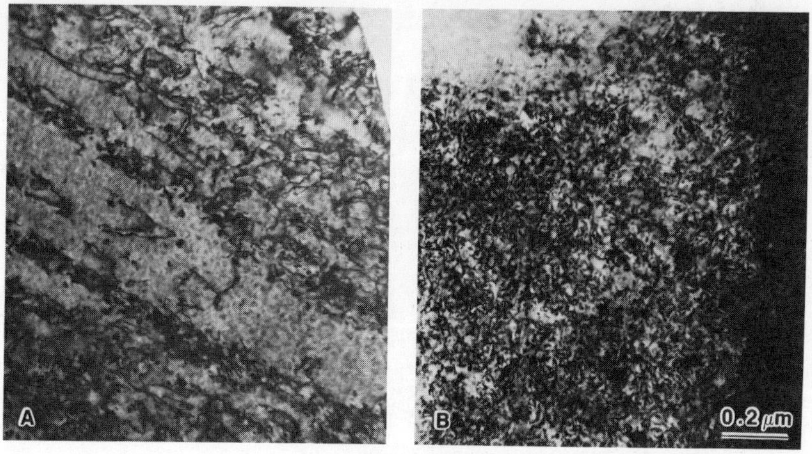


Fig. 2. Deformation substructures; (a) single-aged alloy showing planar slip and (b) double-aged alloy showing wavy slip.

## SUMMARY

1. Double-aging of a rapidly solidified Al-Li-Zr alloy produces a homogeneous distribution of composite $Al_3(Li,Zr)$ precipitates.
2. Deformation mode of double-aged alloy is quite homogeneous, while conventionally aged alloy deforms by the pronounced planar slip.
3. Composite precipitates have been shown to be quite resistant to dislocation shear. Dislocations by-pass the composite precipitates leaving dislocation loops around them.

## ACKNOWLEDGEMENTS

The authors would like to thank Drs. D. J. Skinner, C. M. Adam and Mr. R. L. Bye for helpful discussions.

## REFERENCES

1. T. H. Sanders, Jr. and E. A. Starke, Jr., eds., Aluminum-Lithium Alloys I, TMS-AIME, Warrendale, PA, 1981.

2. T. H. Sanders, Jr. and E. A. Starke, Jr., eds., Aluminum-Lithium Alloys II, TMS-AIME, Warrendale, PA, 1984.

3. Proceedings of 3rd International Aluminum-Lithium Conference, Oxford, England, 1985, Institute of Metals, to be published.

4. R. W. Gayle and J. B. Vander Sande, Scripta Metall., 18, p. 473 (1984).

5. P. J. Gregson and H. M. Flower, J. Mat. Sci. Letters, 3, p. 829 (1984).

6. P. L. Makin and B. Ralph, J. Mat. Sci., 19, p. 3835 (1984).

7. N. J. Kim, R. L. Bye, D. J. Skinner and C. M. Adam, submitted to International Conference on Rapidly Solidified Materials, San Diego, CA, 1986.

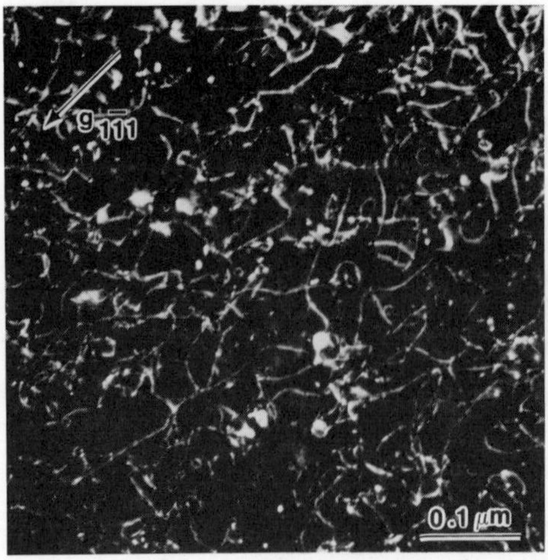


Fig. 3. Weak beam dark field micrograph of deformed double-aged alloy.

# Effect of Rolling Parameters on the Properties of Al-Fe-V Sheet Rolled on an Experimental Mill

A. Brown and D. Raybould
Metals and Ceramic Lab
Allied Corporation
P.O. Box 1021R
Morristown, New Jersey 07960

## ABSTRACT

In recent years, interest in high temperature aluminum alloys has increased. However, nearly all the data available is for simple extrusions. This paper looks at the properties of sheet made from a rapidly solidified Al-10Fe-2.5V-2Si alloy. The sheet is made by direct forging followed by hot rolling, this is readily scalable, so allowing the production of large sheet. The room temperature strength and fracture toughness of the sheet are comparable to those of 2014-T6. The high temperature strength, specific stiffness and corrosion resistance are excellent. Recently, improved thermomechanical processing and new alloys have allowed higher strengths and fracture toughness values to be obtained.

## INTRODUCTION

Several developments have increased the interest of aerospace designers in both light weight, high temperature and high stiffness materials. For high strength structural applications, titanium and aluminum alloys are often the only materials available. Improvements are being made to both of these [1]; however, it is the advanced aluminum alloys which have shown the most significant improvement in properties [2]. In particular, rapidly solidified Al-Fe type alloys with good high temperature strengths have been reported. The potential of these alloys was shown in the early 70's by workers at Tube Investment/British Aluminum [3] [4]. More recently, using the Al-Fe system as a base, improved alloys with additions such as Ce,Mo,V, and Zr have been made by several organizations [5] [6]. Vanadium has one of the lowest diffusivities in aluminum and Al-Fe-V alloys have been shown to have excellent high temperature properties [2] [7].

Until now the designers have not had available data for complex shapes or for sheet; the latter is one of the most important potential applications for these types of materials. This paper will deal with the initial work on the production of sheet from Al-Fe-V type alloys. The work is experimental and, therefore, deals with relatively small sheet, but the design techniques used can be scaled-up. Techniques like extrusion followed by rolling were not used; instead, direct powder rolling type techniques were employed, even though these are known to be difficult and can result in some loss in mechanical properties. In fact, as experience was gained, properties equal or superior to those of extrusion were obtained.

## EXPERIMENTAL

An Al-10Fe-2.5V-2Si (452-S) was used for this program. Properties of similar extruded alloys have been previously reported [7]. The alloy was jet cast to a thin ribbon with a predominantly zone 'A' microstructure. This may be described as a microeututic which is unresolvable by optical microscopy. Some zone 'B' microstructure was also present; although this was resolvable by optical microscopy its structure was still very fine. The ribbon was commuted into powder of less than 60 mesh. It was degassed and then forged into plates. The plates were hot rolled on Allied's Stannet rolling mill, using 6" diameter rolls. The starting plates were 1 3/4" thick and were rolled by repeat passes to 0.24" to 0.06" thick sheet.

All the sheet produced was tensile tested using standard specimens. Some sheet was tested with the as rolled surface, some had all surfaces milled and polished. No difference was observed. Specimens were produced by three different machine shops. Some specimens were tensile tested at temperature, 30 minutes after exposure. Fracture toughness testing used compact tension specimens. The pre-cracking and final tests were conducted on a computer-controlled hydraulic tester. The tests were carried out to ASTM standard E-399. For the sheet it was impossible to comply with the thickness requirement. Therefore, the effect of sample thickness is investigated. As a comparison, data for commercial alloys is also given. The corrosion resistance was determined by exposure to a salt fog, according to ASTM B117. The elastic modulus was determined dynamically, tests were carried out up to 700°F.

## RESULTS AND DISCUSSION

The forged plates were approximately 15" x 20" x 1 3/4", [Figure (1)]. The plate was cut up by a band saw to 6" by 4" slabs which were multiple pass hot rolled to sheet. Cross rolling was not required. Edge cracking was present in some sheet to a maximum depth of 1/16" but did not occur for most of the sheets. It should be mentioned that, in subsequent large scale work, significant edge cracking did occur; however, this was eventually avoided or limited to insignificant proportions. The presence or absence of edge cracking is not solely related to the alloys.

Typical tensile properties are given in Table I. The properties shown are for specimens taken along the rolling direction; transverse properties are very similar, but the Y.S. and U.T.S. may be approximately 1 Kpsi lower.

TABLE I: Tensile Strengths of the As Rolled Sheet

| Thickness [Inch] | Y.S. [Kpsi] | UTS [Kpsi] | El % |
|---|---|---|---|
| 0.22 | 64 ±0.5 | 72 ±0.5 | 14 ±0.5 |
| 0.095 | 58 ±0.3 | 65 ±0.5 | 14 ±0.5 |
| 0.060 | 50 ±0.3 | 57 ±0.5 | 16 ±2.0 |

A strength of 72 Kpsi with 14% elongation is acceptable for most applications. However, this is for the thicker sheet, and Table 1 shows a decrease in strength as the thickness is reduced. This was found to be caused by the longer exposure to temperature, which degraded the microstructure. Improved processing, based on considerations such as the deformation

Figure 1 Forged Plates of Al -10Fe-2.5V-2Si

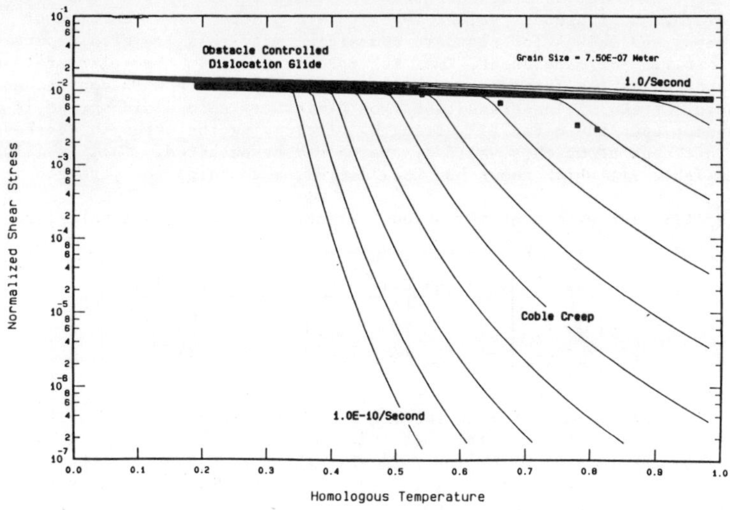


Figure 2 Deformation map for Al-10Fe-2.5V-2Si.

map, Figure (2), allowed high strengths to be obtained. In fact, 0.08" thick sheets were eventually made with strengths over 85 Kpsi.

The difficulty in processing the alloy can be seen from the deformation map, Figure (2), especially when it is realized that exposure to the processing temperature can deteriorate the alloy microstructure and properties. In contrast, tests carried out at their potential operating temperature show no changes after over 100 hours of exposure. Short time exposure (1 hr) to high temperatures followed by a slow cool was investigated as a means of removing residual stress remaining from the rolling, Table II. The properties are unaffected demonstrating that annealing is unnecessary.

TABLE II: The Effect on the Mechanical Properties of Annealing for 1 Hour Followed by Slow Cooling

| | RT | 350°C | 400°C | 450°C |
|---|---|---|---|---|
| YS(Kpsi) | 62.5 | 61.5 | 62 | 61.6 |
| UTS(Kpsi) | 69.5 | 70.0 | 69.5 | 70.0 |
| El(%) | 14.0 | 13.5 | 11.5 | 12.0 |

The elevated temperature strengths of the sheet, Figure (3), include data for a sheet with a low room temperature strength (62 Kpsi) to show that even after some coarsening of the microstructure due to processing, the elevated temperature strength is still good. Recently processed sheet has strengths equal to the higher properties of the extrusion, Figure (3).

The sheets all have excellent elongations of over 10%, Table 1. This does not guarantee a good fracture toughness. However, the tensile test is simple and valid, while fracture toughness testing of thin sheet is difficult and open to questions of validity. For such thin specimens fracture toughness values should be treated with caution and perhaps used only as an indicator of relative toughness between alloys tested by the same technique. Therefore, the toughness values reported are plotted as a function of thickness and values for standard commercial materials are also reported, Figure (4). For many sheet, Kmax is approximately two times larger than the Kq reported as the fracture toughness. For thick plate Kmax approaches the value of $K_{IC}$(Kq). The transition from plain strain to plain stress is also dependent upon the materials strength. Increasing the strength decreases the thickness at which a valid $K_{IC}$ value may be obtained. This is illustraed in Table III which shows how Kmax decreases to Kq as the strength increases.

Table III: The effect of the sheets strength on Kmax, sheet thickness constant

| UTS (Kpsi) | Kq (Ksi$\sqrt{in}$) | Kmax Ksi$\sqrt{in}$ |
|---|---|---|
| 60 | 16 | 35 |
| 65 | 20 | 25 |
| 70 | 22 | 24 |

From data such as Figure (4) the Kq of the sheet of alloy 452-S appears to be around 18±5 ksi$\sqrt{in}$, the fact that some specimens were still slightly bent from the rolling might indicate that the high values are the more accurate, but further work is required to confirm this. Repeat tests on several sheet showed that Kq and Kmax are consistantly 2 Ksi$\sqrt{in}$ higher in the

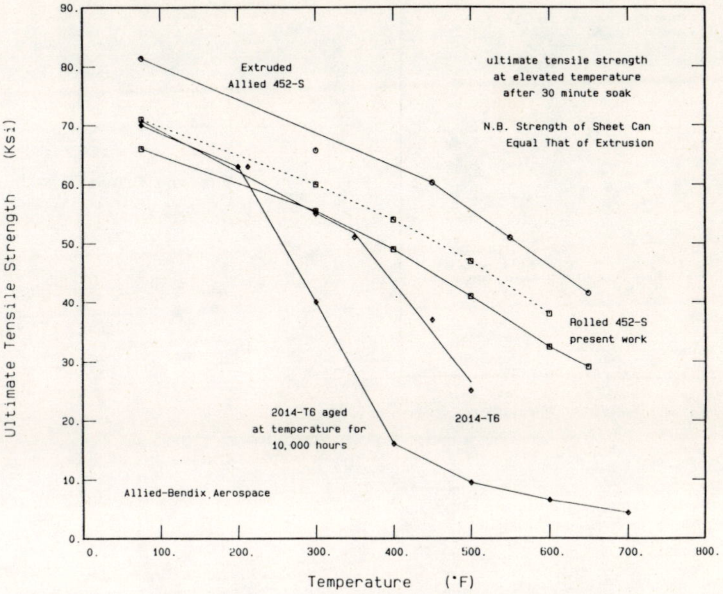


Figure 3 Elevated Temperature Strength for Three Different Room Temperature Strength Sheet.

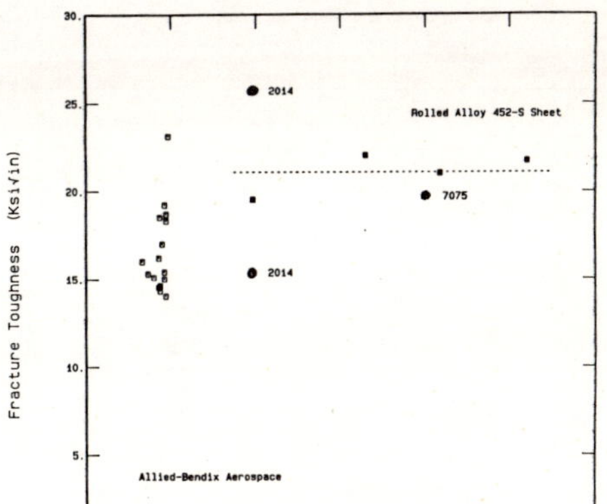


Figure 4 The fracture Toughness (Kq) as a Function of Thickness.

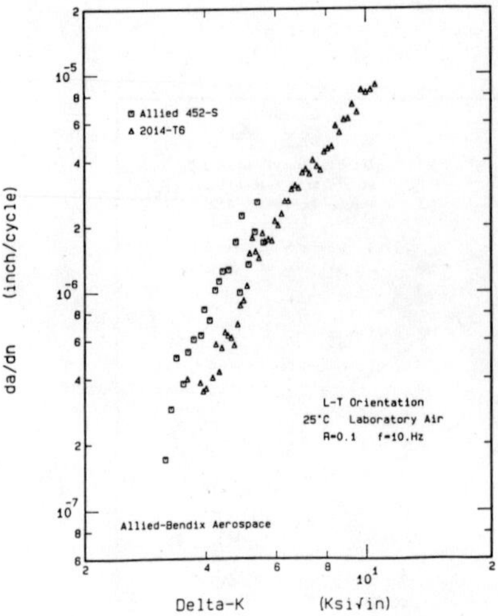


Figure 5 Fatigue Crack Growth Rate.

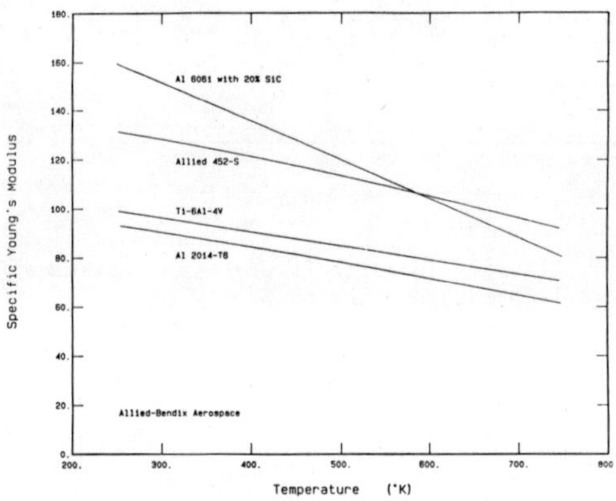


Figure 6 Specific Elastic Modulus as a Function of Temperature.

Figure 7 Salt Fog Corrosion Resistance.

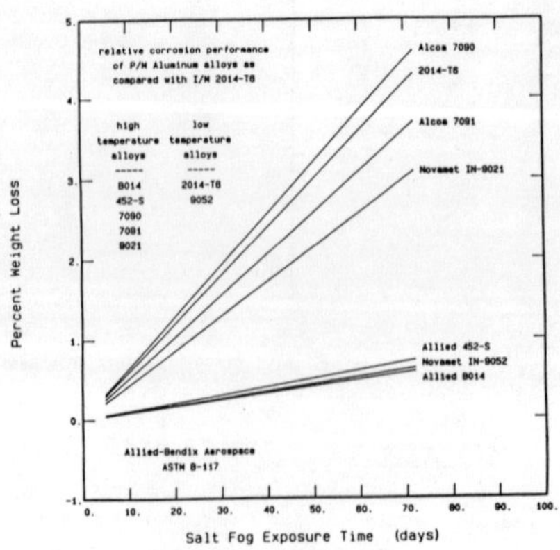

TL compared to the LT direction. The Kq we measure for the sheet of alloy 452-S approaches that measured for 2014-T6. The crack growth rate is also comparable to that measured for 2014-T6, [Figure (5)].

This plus the good elongations, indicates that the sheet does have an acceptable toughness. A concurrent alloy development program has, in any case, developed alloys with similar elevated temperature properties and increased toughness [8]; rolling of these alloys is being investigated.

The dynamic elastic modulus proved relatively insensitive to thermomechanical processing. The values measured are nearly 40% higher than for present commercial alloys; 14.0 Msi at R.T. and 11.3 Msi at 600°F. These values are even more impressive when the specific Young's modulus is determined, Figure (6).

Although not intended as a corrosion resistant alloy, the salt fog test, Figure (7) does indicate that the sheet has excellent corrosion resistance. Data is also shown for 2014 and other P.M. alloys. Again, the actual thermomechanical processing appears unimportant. This perhaps indicates that in all cases the microstructure is homogeneous, although factors such as a modified oxide skin may be responsible for the good corrosion resistance rather than the fine microstructure. However, chemical composition is important, as reducing the alloying addition improves the corrosion resistance, as shown by alloy 14, Figure (7).

## CONCLUSIONS

1. The technique of forge and hot rolling allows elevated temperature aluminum alloys to be produced in sheet form with good stengths and ductilities. This technique is readily scalable to allow the production of large sheets.
2. The room temperature strength and fracture toughness of sheet of alloy 452-S are comparable to values measured for conventional 2014-T6. Improved thermomechanical processing and new alloys have recently allowed higher strengths and fracture toughness values to be obtained. For instance, YS 80 kpsi, UTS 85 Kpsi , El 9%.
3. The Al-Fe-V-Si sheet has exceptional specific stiffness and corrosion resistance.

## ACKNOWLEDGEMENTS

Important contributions to this work were made by C. M. Adam, R. Bye, S. K. Das, "Benny" Kronus, D. Skinner and D. Timan. The dynamic modulus was measured by Texas A&M. The corrosion tests were carried out by Bendix.

## REFERENCES

1. F. H. Froes and J. R. Pickens, J. of Metals, Jan. 14, 1984.
2. C. M. Adam and R. E. Lewis in Rapidly Solidified Crystalline Alloys, editors S. K. Das, B. H. Kear and C. M. Adam, (TMS-AIME Warrendale, PA 1986).
3. H. Jones, Mat. Sci. & Eng. 5 1 (1969).
4. T. Sheppard and P. Chare, Powder Metal, 16, (32), 437 (1973).
5. C. M. Adam, British Patent 2088409.
6. G. W. Hilderman, U.S. Patent 4,379,719.
7. D. J. Skinner and K. Okazaki, Scripta Met. 18 905 (1984).
8. D. J. Skinner, R. Bye, D. Raybould and A. Brown, Scripta Met. to be published.

# MICROSTRUCTURES AND THERMAL PROPERTIES OF RAPIDLY SOLIDIFIED ALUMINUM-RARE EARTH ALLOYS

D. ELIEZER*, S. J. SAVAGE**, Y. R. MAHAJAN, and F. H. FROES
Air Force Wright Aeronautical Laboratories, Materials Laboratory, AFWAL/MLLS, Wright-Patterson Air Force Base, OH 45433
*Sabbatical leave from the Ben-Gurion University of the Negev, Beer Sheva, Israel
**Now with the Swedish Institute for Metals Research, S-114 28 Stockholm, Sweden

## ABSTRACT

Rapidly solidified (RS) Al-Er, Al-Nd, and Al-Gd ribbons were prepared by pendant drop melt extraction (PDME). The microstructures were studied in detail by optical and transmission electron microscopy. A wide range of microstructures are observed, including "conventional" RS type zones, a cellular structure surrounded by a degenerate eutectic and coupled (lamellar) eutectic structures. A bimodal precipitate distribution is seen in many areas of the ribbons. A series of isochronal heat treatments was conducted to determine the relative thermal stability of these microstructures, which were evaluated by Knoop microhardness measurements.

## INTRODUCTION

Aluminum alloys containing additions of dispersoid forming elements can show attractive mechanical properties at elevated temperatures if the dispersed phase is stable and is present as a fine, homogeneous distribution in the matrix [1]. Examination of the equilibrium phase diagrams of Al-lanthanide alloy systems [2] indicates that such a dispersion may be produced by rapid solidification followed by an appropriate heat treatment. For example, all the phase diagrams show a high liquid solubility (10-30 wt%) and a very low solid solubility (≦0.05 wt%) of the rare earth in aluminum. The rare earths also have low duffusivity and form stable, high melting intermetallic compounds.

In this paper some preliminary results of microstructural characterization and thermal stability of Al-10.3%Gd, Al-10.8%Er, Al-9.8%Ce, and Al-10.7%Nd alloys are presented; the work is part of a continuing program at the authors laboratory to characterize the Al-lanthanide alloys.

## EXPERIMENTAL

Rapidly solidified ribbons of this alloy were prepared by the pendant drop melt extraction (PDME) method [3, 4]. Ribbons produced were typically 80 μm thick and 2 mm wide.

Samples were characterized by optical and transmission electron microscopy (TEM) in the as-solidified state and after heat treatment for 1 hr at temperatures between 100°C and 500°C. TEM foils were prepared by electropolishing from the top surface of the ribbon in order to study the chill zone at the wheel side. The microhardness data were obtained at room temperature using a Knoop indenter with a 5 gm load and 15 second dwell time. Measurements were restricted to the chill zone. As a standard, the microhardness of a sample of as-cast 99.999% Al was measured.

## RESULTS AND DISCUSSION

Figure 1a shows a representative longitudinal section optical micrograph of the as-cast Al-10.8Er PDME ribbon. Three zones are evident. Beginning with the side of the ribbon which solidified in contact with the casting wheel: a featureless chill zone about 70 μm wide which changes to a zone showing columnar grains about 10 μm wide, followed by an equiaxed region C (about 10 μm wide) which extends to the top of the ribbon. Within the featureless zone are barely discernable grain boundaries which indicate a grain size of about 5-10 μm. Figure 1b shows part of the chill zone and the columnar and equiaxed zones at higher magnification. This microstructure is typical of all the Al-RE binary alloys.

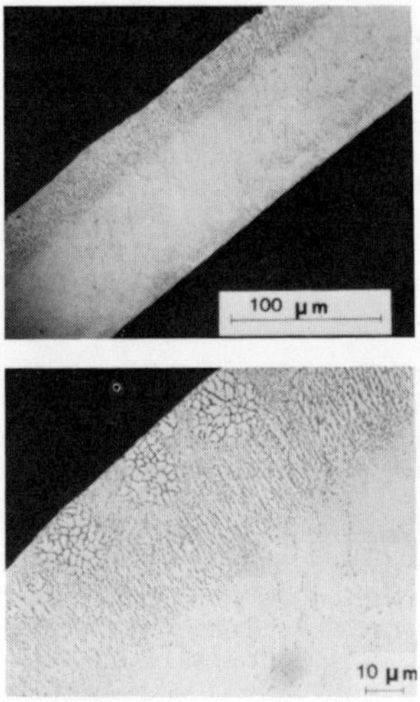


Figure 1. Longitudinal cross section of rapidly solidified Al-10.8Er ribbon.

Figures 2a, 2b, and 2c show typical TEM micrographs of the Al-10.8Er ribbons as cast. Figure 2a shows a region close to the grain boundary where a discontinuous eutectic is visible, with a solute free zone between the grain boundary and a series of banded precipitates within the grain. The latter (Figures 2b and 2c) are seen to have a bimodal distribution, where the larger, angular precipitates (about 100-120 nm diameter) are arranged in regularly spaced bands about 500 nm apart, and a homogeneous dispersion of smaller (20-60 nm diameter) spherical precipitates.

Figures 3a, 3b, and 3c show TEM micrographs of as-cast Al-10.7Nd, Al-9.8Ce, and Al-10.3Gd, respectively. All three figures show an equiaxed cellular microstructure, although the size of the cells varies from about 0.5-1 μm for Al-10.3Gd to 2-2.5 μm for Al-9.8Ce. In the Al-Ce alloy there is a well developed intercellular eutectic phase which is almost completely suppressed in the other two systems.

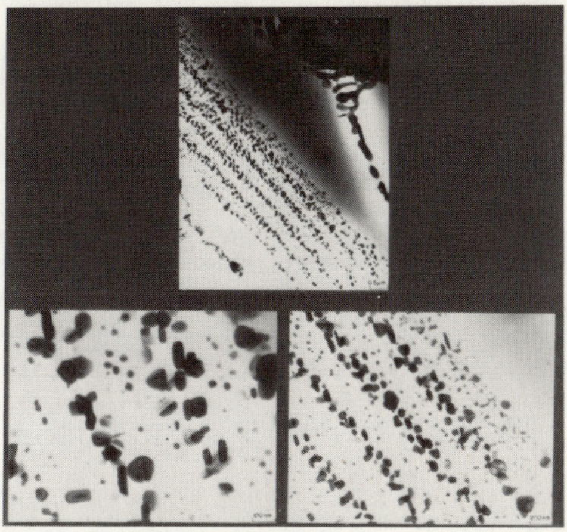

Figure 2. As-cast microstructure of Al-10.8Er PDME ribbon.

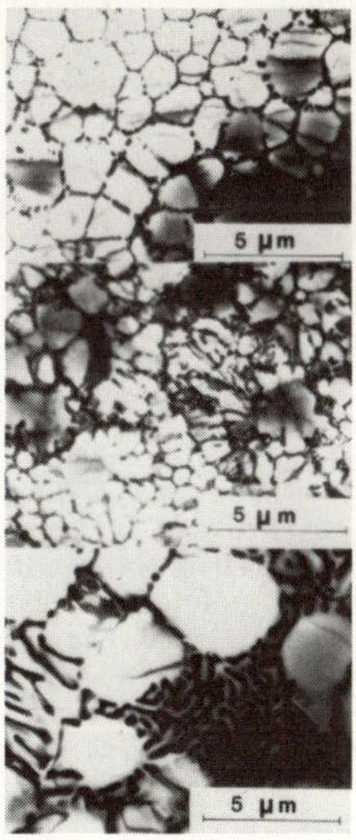


Figure 3. TEM micrographs of as-cast PDME ribbons (Al-10.3Gd, Al-10.7Nd, and Al-9.8Ce).

The thermal stability of these alloys from room temperature to 500°C was monitored by microhardness measurements and TEM on samples aged for 1 hr. Figure 4 shows the isochronal annealing curves for three systems, Al-10.7Nd, Al-9.8Ce, and Al-10.3Gd. The results obtained for the Al-Er alloy showed that the hardness of the heat treated alloy decreased to a level only slightly greater than that of pure aluminum (about 45 KHN) after 1 hr at only 100°C, and remained at that level. The cause of this is unknown. In Figure 4 it can be seen that the Al-10.7Nd alloy exhibits the highest hardness (about 120 kg/mm$^2$) after 1 hr at 300°C, whereas Al-Gd shows a peak at 150°C (115 kg/mm$^2$), and Al-Ce has greatest hardness (110 kg/mm$^2$) after heat treatment at 350°C. After 450°C all the curves are rapidly dropping, and appear to be about to meet at around 60 kg/mm$^2$ at 500-550°C.

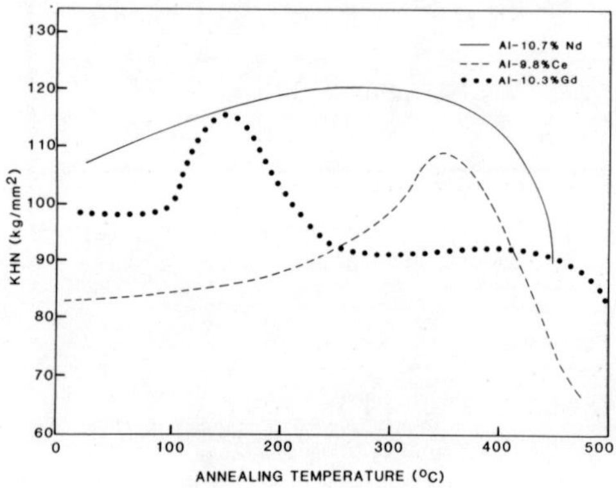


Figure 4. Comparison between isochronal annealing curves of Al-Nd, Al-Ce, and Al-Gd alloys.

Figure 5 shows, for Al-10.3Gd the isothermal annealing curve as a function of time up to 100 hr at 300°C, where no change in microhardness from the as-cast value of 100 kg/mm$^2$ is observed. Figure 6 shows TEM micrographs of Al-Er ribbons isochronal aged between 100°C and 500°C. Figures 6a and 6c (100°C and 500°C, respectively) show a grain boundary region with a precipitate free zone and a few discontinuous second phase particles in the boundary. Within the grains there is a dense array of small (0.06 μm) precipitates. After 1 hr at 500°C these precipitates have coarsened slightly (to about 0.1 μm), as have the grain boundary precipitates. The precipitate free zone has also broadened to about twice its original width. Also seen in this system are a number of small, angular precipitates (Figure 6c) about 0.01 μm, which show Moiré fringe patterns, indicating coherency with the matrix lattice on at least one plane.

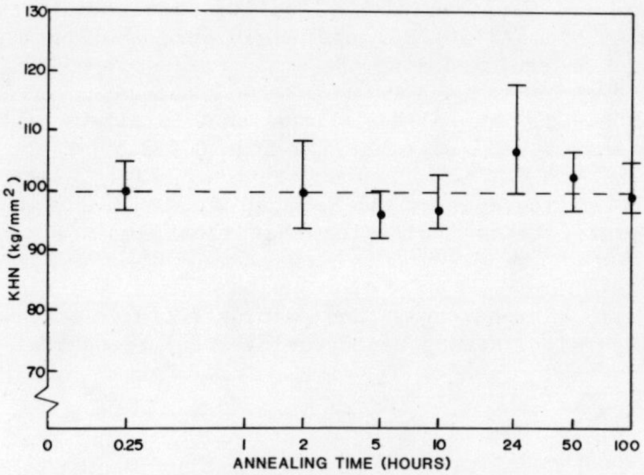


Figure 5. Al-10.3Gd isothermal anneal at 300°C (570°F). Knoop microhardness at 5 gm load.

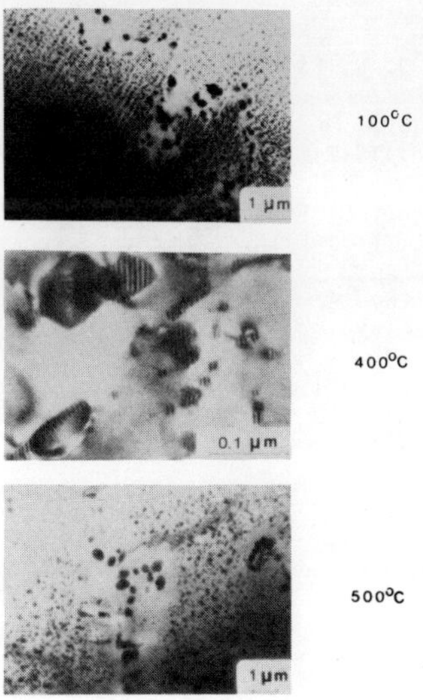


Figure 6. TEM micrographs of Al-Er ribbons isochronal aged.

## CONCLUSIONS

1. Conventional type rapidly solidified microstructures are observed in Al-10.3Gd, Al-9.8Ce, Al-10.7Nd, and Al-10.8Er, showing a chill zone, a columnar zone, and an equiaxed zone.

2. Al-Ce, Al-Gd, and Al-Nd alloys show similar equiaxed cellular microstructures, with a cell size varying from 0.5-2.5 μm.

3. The TEM micrographs of Al-Er aged at 400°C for 1 hr show Moiré fringes in the precipitates indicating that they remain at least partially coherent.

4. Based on microhardness and microstructures the Al-RE systems offers good dispersion strengthened alloys for potential use at high temperatures.

## ACKNOWLEDGMENTS

D. Eliezer and Y. R. Mahajan acknowledge the National Research Council for provision of Resident Research Associateships, S. J. Savage acknowledges the sponsorship of a Visiting Scientist position by U.S. Air Force Contract No. F33615-84-C-5116.

## REFERENCES

1. F. H. Froes and J. R. Pickens, J. Metals 36 (1), 14-28 (1984).

2. K. A. Gschneidner, Jr. and F. W. Calderwood, Rare Earth Information Center Report No. IS-RIC-PR-5, 1983.

3. S. J. Savage and F. H. Froes, J. Metals 36 (4), 20-33 (1984).

4. R. E. Maringer and C. E. Mobley, in Rapidly Quenched Metals III, Vol. I, edited by B. Cantor (The Metals Society, London, 1978), pp. 49-56.

# METASTABLE EXTENSIONS OF INTERMEDIATE PHASES IN SOME ALUMINUM - RARE-EARTH METAL SYSTEMS*

M.X. Quan, P. Haldar, J. Werth and B.C. Giessen,
Materials Science Division, Barnett Institute and Department of Chemistry, Northeastern University, Boston, MA

## ABSTRACT

Using the arc-furnace hammer-and-anvil method, metastable alloys in the Al-rich portions of four Al-RE systems (10-30 at. pct. RE; Re = Gd, Ho, Er and Y) were prepared and examined by XRD. The principal results are: all equilibrium alloy phases [$Al_4RE$ (T phase), $Al_3RE$ (CN12 close-packed phases) and $Al_2RE$ (Laves phase)] have substantial metastable solid solubilities for excess Al; e.g., in the Al-Y system, L phase is extended from 66.7 up to 86 at. pct. Al, suppressing any formation of the $Al_3RE$ phase in that system.

For early RE elements up to and including Gd, T phase is favored; for alloy chemically intermediate RE elements such as Y, L phase is dominant and for late RE elements (after Ho), the CP phases are extended most. DSC work indicates the formation of additional metastable phases occurring during equilibration.

## INTRODUCTION

Recently, the alloy chemistry of metastable phases rich in Al has attracted research interest because of the finding of quasicrystalline alloy phases in rapidly quenched Al-rich Al-Mn alloys and other metastable alloys of Al with transition metals [1]. In the course of ongoing research on other Al-based alloy systems that might contain such phases, a survey of metastable, Al-rich Al-RE alloys produced by RSP was undertaken (where RE is a rare earth metal or Y). An additional motivation for this work arises from the current interest in Al alloys strengthened by metastable solid solutions or precipitation products from such solutions [2].

The present paper gives preliminary results on the metastable alloy chemistry of the systems Al-Gd, Al-Y, Al-Ho and Al-Er in the regions with 10-30% RE (all percentage data are in atomic percent), reporting the structures of arc-furnace hammer-and-anvil quenched samples examined in the as-quenched state.

### Alloy Chemistry of Al-rich Al-RE Alloys:

The Al-rich portions of all Al-RE systems are similar in the major features [3] which are briefly reviewed in the following. Starting from 100% Al, one finds the following alloy phases and phase diagram characteristics:

1. The solid solubility of RE elements in Al is very low (0.01% for Al(Ce) solid solution), as is readily explained by the substantial size difference [4].

* Communication No. 281 from the Barnett Institute

2. There are phases of stoichiometry $Al_4RE$ or $Al_{11}RE_3$ with related, comparatively complex structure [3]; their occurrence is limited to light RE metals up to RE = Tb.

3. All systems contain phases $Al_3RE$ with 12-coordinated, close-packed (CN12-CP) structures of the $Ni_3Sn$, $BaPb_3$, $HoAl_3$ and $AuCu_3$ types. These $Al_3RE$ structures can be described as consisting of hexagonal $Al_3RE$ layers stacked such that minority atoms never are first nearest neighbors; the stacking sequence varies with the RE element as shown in Table I. This Table lists the structure type, the number of layers in the repeat unit in the hexagonal representation (the 9 and 15 layer structures are rhombohedral and the 3-layer structure is of the cubic $AuCu_3$ type) and the percentage of cubic close packing (ccp) polyhedra in the structure [5]. Where they are reported [3], both high and low temperature phases are given. It is seen that the seemingly complex variation of the stacking sequences in $Al_3RE$ phases as a function of the respective RE element can be interpreted simply on the basis of some straightforward structural principles:

   (a) The stacking order changes from 0% ccp stacking (= 100% hcp stacking) for light RE phases (up to Gd) to 100% ccp stacking for heavy RE's (starting with Tm).

   (b) Between these regions, there is a range of transition structures having 33.3, 50, and 60% ccp stacking arranged such that the percentage of ccp stacking increases on going from Gd to Tm.

   (c) High temperature $Al_3RE$ phases occur only in this transition region and have stacking sequences higher in ccp stacking than the corresponding low temperature phases, confirming the well known trend to form cubic phases or, in some cases here, "more cubic-like" phases at higher temperatures.

Table I: The CN12 Close-Packed (CN12-CP) Structures of $Al_3RE$ Phases

| | | Gd[a] | Y[b] | Tb | Dy | Ho | Er | Tm[c] |
|---|---|---|---|---|---|---|---|---|
| Structure Type: | HT | -- | $BaPb_3$ | -- | $HoAl_3$ | $AuCu_3$ | $AuCu_3$ | -- |
| | LT | $Ni_3Sn$ | $Ni_3Sn$ | $BaPb_3$ | $TiNi_3$ | $HoAl_3$ | $HoAl_3$ | $AuCu_3$ |
| NL[d] | HT | -- | 9 | -- | 15 | 3 | 3 | -- |
| | LT | 2 | 2 | 9 | 4 | 15 | 15 | 3 |
| % ccp[e] | HT | -- | 33 | -- | 60 | 100 | 100 | -- |
| | LT | 0 | 0 | 33 | 50 | 60 | 60 | 100 |

a) Elements between Ce and Gd omitted; for Pr, Nd and Sm, the $Al_3RE$ phases are isostructural with $Al_3Gd$.

b) Y is placed in this location due to its alloy chemical characteristics.

c) Elements after Tm omitted; for Yb and Lu, the $Al_3RE$ phases are isostructural with $Al_3Tm$.

d) NL = number of c.p. $Al_3RE$ layers in hexagonal repeat unit (structures with NL = 9 and 15 are rhombohedral and the structure with NL = 3 is cubic).

e) % ccp = percent cubic packing.

(d) While Y is not an RE element, it is appropriately located with respect to the structure of $Al_3Y$ between Gd and Tb.

The morphotropism of these $Al_3RE$ phases is the alloy chemically richest structural feature of the Al-RE systems. It may be noted that the transition from 0 to 100% ccp-like stacking in $Al_3RE$ runs in the opposite direction from the structure type change observed for the RE metal elements themselves which also have close-packed structures but change from 100 to 0% ccp on passing from light to heavy RE's [6].

4. The phase diagrams are dominated by $Al_2RE$, the thermally most stable Al-RE phase. In all systems $Al_2RE$ is a Laves phase (L phase) with the fcc $MgCu_2$ type.

## EXPERIMENTAL

Alloys were arc-melted from high purity elements and were quenched by the hammer-and-anvil method [7] into foils of 30-70 μm thickness and studied by RXD using Mo-Kα radiation.

## RESULTS AND DISCUSSION

Fig.1 represents and Table II summarizes the results obtained on the four systems studied in this survey. The "quenched-phase plot" [8,9] given for each system in Fig. 1 contains the equilibrium phases (as line compounds) with their designations given below the diagram; the compositions of the alloys studied and their after-quenching constitution (single or multiple phase) are given by the ovals in the diagrams; the metastable phases found after quenching are marked above the diagrams, suggesting the metastable composition range (phase field) in which they fall. Table II is arranged by stoichiometric compound composition and describes the suppression of phases by RSP or, alternatively, the composition ranges in which they are formed as metastable phases. We review the results briefly, first by system, then by alloy phase type.

### Alloy Systems:

Al-Gd: RSP extends the composition range of the L phase approximately up to the stoichiometric composition of the T phase, suppressing formation of the $Al_3RE$ phase. T phase is extended metastably to higher Al contents.

Al-Y: Here also a metastable extension of the L phase suppresses formation of the $Al_3RE$ phase. Since there is no T phase in this system, L extends without competition to quite high Al contents (up to >86% Al).

Al-Ho: In this system as in the preceding ones, L phase is extended to higher Al contents, up to and slightly beyond the $Al_3RE$ stoichiometry (to about 77% Al); however, its range is extended not as far as in the Al-Gd and Al-Y systems. Instead, the CN12-CP $Al_3RE$ phase gains in stability relative to the L phase, possibly as a result of the decreasing size of the RE atom; $Al_3RE$ appears off-stoichiometrically between 78 and ~88% Al and has the $AuCu_3$ type structure. However, despite the close structural relationship of terminal fcc Al and $Al_3Ho$, a continuous metastable transition phase between both of these phases could not be produced.

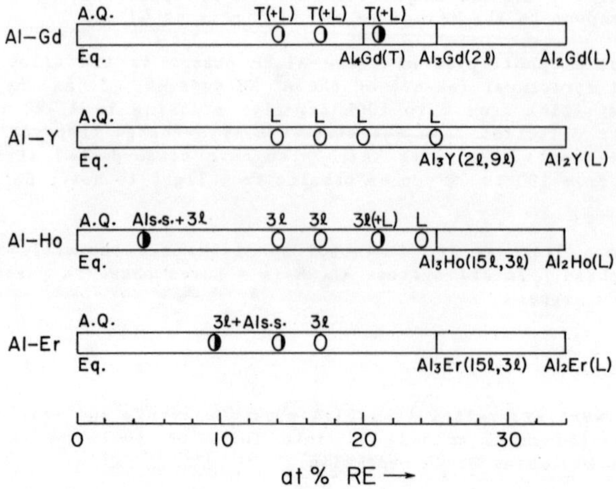


Fig. 1: Equlibrium phases (below diagrams), constitution after RSP (in diagrams) and metastable phases observed after RSP (above diagrams) in Al-rich Al-Gd, Al-Y, Al-Ho and Al-Er alloys ("Quenched-Phase Plots" [8,9]).

Table II: Results of Rapid Solidification Experiments on Intermetallics in Four Al-RE Systems.

| | $Al_2RE$ | $Al_3RE$ | $Al_4RE$ |
|---|---|---|---|
| Al-Gd | metastable extension to ~78% Al | Formation suppressed by formation of metastable extension of C15 Laves phase | metastable extension from 80 to >~86% Al |
| Al-Y | metastable extension to >86% Al | Formation suppressed by formation of metastable extension of C15 Laves phase | Not formed in equilibrium or as metastable phase |
| Al-Ho | metastable extension to ~78% Al | Formation of $AuCu_3$ type phase as metastable phase over composition range ~80% Al ->88% Al | Not formed in equilibrium or as metastable phase |
| Al-Er | metastable extension not determined; structure of alternate structure | Metastable extension of $AuCu_3$ type phase from 75% Al to >88% Al | Not formed in equilibrium or as metastable phase |

It may be pointed out here that on heating, metastable off-stoichiometric $Al_3(Ho_{1-x}\ Al_x)$ transforms to the equilibrium phases via other CN12 close-packed phases not found in the equilibrium phase diagram [10].

Al-Er: For this system there are only tentative results. First, there is some evidence that the L phase is partially replaced by an $Al_2RE$ phase of different structure. Second, for Al-Er, as for Al-Ho, the CN12 close-packed phase ($AuCu_3$ type) is extended to higher Al contents.

Metastable Phase Fields:

The metastable phase fields indicating the alloy phase type occurring during RSP of Al-RE alloys as a function of composition are indicated in Fig. 2 for T, L and CP.

T Phase: For RE elements up to Gd, the T phase forms and extends to higher Al contents, as seen for Al-Gd.

L Phase: Metastable extension of L has a maximum for Al-Y, as shown; in systems with earlier RE metals L phase is restricted by the stability of the T phase, while in systems with later RE metals, L is restricted by the phase field of CN12-CP phases. XRD intensity studies [10] indicate that off-stoichiometric L phases accommodate the excess Al atoms in RE atom sites by partial, random substitution.

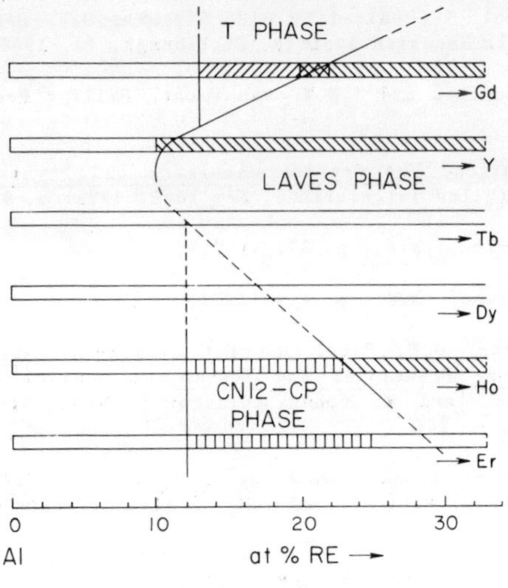


Fig. 2: Phase fields of occurrence of stable and metastable phase extensions in Al-RE alloys showing T phase ($BaAl_4$ type), Laves phase ($MgCu_2$ type) and CN12-CP phases (see text).

CP Phase: Starting with Al-Tb, CN12-CP phases appear, at first off-stoichiometrically shifted from $Al_3RE$ to higher Al contents, then, with decreasing stability of L, also at lower Al concentrations. The $AuCu_3$ type was the only $Al_3RE$ phase type found in the as-quenched alloys in the systems reported here.

## CONCLUDING REMARKS

Structural studies, lattice parameters, thermal stabilities, and phases appearing on thermal treatment will be reported subsequently [10]. Exploratory work in the Al-rich portions of Al-RE systems with light RE metals (up to Gd) was also done but showed only few metastable phases or phase extensions. By contrast, the Al-RE systems starting with RE = Gd reported here have a fairly rich and interesting metastable alloy chemistry, characterized primarily by substantial metastable solid solubility of Al in the equilibrium phases.

## ACKNOWLEDGMENT

We are pleased to acknowledge support of work on metastable alloy phases by the Ford Motor Company.

## REFERENCES

1. D. Shectman, I. Blech, D. Gratias, and J.W. Cahn, Phys. Rev. Lett. 53, 1951 (1984).

2. S.J. Savage, D. Eliezer, Y.R. Mahajan, and F.H. Froes, MRS Sym. Proc., Vol. 58, edited by B.C. Giessen, D.F. Polk, and A.I. Taub (Materials Research Society, Pittsburgh, PA, 1986).

3. K.H.J. Buschow and J.H.W. van Vucht, Philips Res. Repts. 22, 233 (1967).

4. W.B. Pearson, The Crystal Chemistry and Physics of Metals and Alloys, (Wiley Interscience, New York, 1972) p. 151.

5. W.B. Pearson, ibid., p. 321 (1972).

6. W.B. Pearson, ibid., p. 114 (1972).

7. M. Fischer, D.E. Polk, and B.C. Giessen, Rapid Solidification Processing, Principles and Technology, edited by R. Mehrabian, B.H. Kear, and M. Cohen (Claitor's Publ. Div., Baton Rouge, LA,1978), p. 140.

8. B.C. Giessen in Adv. in X-Ray Analysis, Vol. 12, edited by C.S. Barrett, G.R. Mallett, and J.B. Newkirk, (Plenum Press, N.Y.1969), p. 23.

9. W.B. Pearson, ibid., p. 288 (1972).

10. P. Haldar, M.X. Quan, and B.C. Giessen, to be published.

# PROPERTIES OF ALUMINUM THIN FILMS PREPARED BY TARGETS FACING TYPE OF SPUTTERING SYSTEM

T. HIRATA, S. KAKU* AND M. NAOE
Tokyo Institute of Technology, Fac. of Engineering, Dept. of Electrical and Electronic Engineering, 2-12-1 O-okayama, Meguro-ku, Tokyo 152, Japan
*Osaka Vacuum, Ltd., 3-6 Kitahama, Higashi-ku, Osaka 541, Japan

## ABSTRACT

The Targets Facing Type of Sputtering (TFTS) system can deposit very dense films on substrates at a high rate without undesirable influences of the plasma. Using this system, Al films 0.7 ∿ 1.5 μm thick were deposited in an argon gas pressure $P_{Ar}$ of $9.7 \times 10^{-2}$ Pa.

The grains with irregular shape grew as $P_{Ar}$ became high. However, the number of grains per unit area decreased and their size became small when the bias voltage to the substrate Vb was -40 V. Consequently, the film surfaces became smooth. The Al films became hardest at Vb of about -40 V and were twice as hard as ones prepared without the bias voltage.

The step coverage of these Al films was very satisfactory to the wiring of the integrated circuit (IC) by adjusting Vb and $P_{Ar}$ at the proper values.

## INTRODUCTION

Nowadays, aluminum thin films play important roles in various fields of electronic engineering. They also are very useful as reflecting layers in optical and magneto-optical recording media.

Concerning the production of the integrated circuit (IC), 6-inch Si wafers have been advanced in practical use and 8-inch wafers will be used in the near future. Usually, Al thin films for electrodes and wires in IC are prepared by electron beam evaporation and DC magnetron sputtering methods. The DC magnetron sputtering method is more convenient for in line production and produces better quality films than electron beam evaporation. However, the film quality is still not satisfactory since hot particles such as γ-electrons and negative ions attack and damage the growing films. Especially, in the case of the Al sputtering, the dicharge current concentrates on the narrow area of the target surface and melts material in the area due to the locally extreme heating. Consequently, many splinters may be ejected from the target and hillocks may deposit on the growing film.

On the other hand, the TFTS system hardly exposes the growing films to the plasma[1]. In addition, the TFTS system can effectively apply the bias voltage to the substrate for control of the structure and properties of films, since the substrates are free from the plasma and are not attacked by energetic particles such as γ-electrons and negative ions.

In the range of $9.7 \times 10^{-2}$ ∿ 2.0 Pa of $P_{Ar}$, the surface appearances of the films on the lower side of $P_{Ar}$ are very smooth and lustrous, while ones on the higher side of $P_{Ar}$ are rough with the formation of larger grains.

In grains of the bias sputtered films, the number of the grains per unit area decreased and the film surfaces became smooth. In addition, those became hardest at Vb of -40 V. The TFTS is also excellent for step coverage because it has two facing targets and no dead regions exist in the groove of the wafer.

EXPERIMENT

The TFTS system used in this study is shown in Fig.1. It can deposit the film on the substrate at a high rate above 1 μm/min. Furthermore, the substrate temperature does not rise extremely since the growing film on it would not be exposed to the plasma between two facing targets.

This system has two targets of diameter of 100 mm. The argon plasma can be confined between these targets with the magnetic field of 100 ∿ 200 Gauss perpendicular to them. The applied voltage was about 500 V and the discharge current was about 0.3 A. The substrate holder is electrically floating. The bias voltage to the substrate Vb was applied by using a radio frequency (rf) power source at a frequency of 13.56 MHz. Vb was in the range of 0 ∿ -150 V. The Al films 0.7 ∿ 1.5 μm thick were deposited at $P_{Ar}$ in the range of 9.7 x $10^{-2}$ ∿ 2.0 Pa.

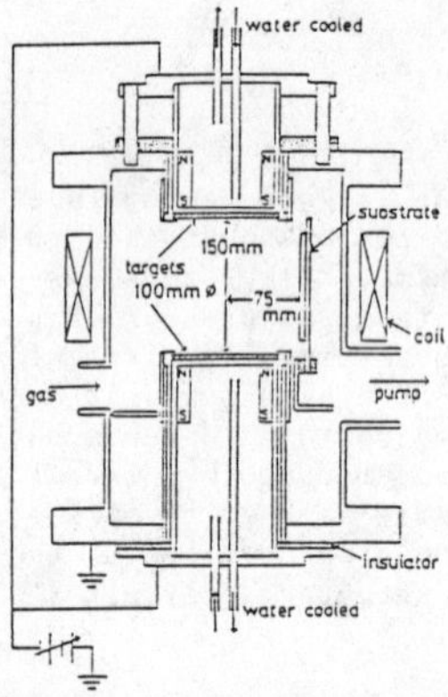


Fig.1 Targets Facing Type of Sputtering System

RESULTS AND DICUSSION

Figure 2 shows the surface appearance on various $P_{Ar}$. The film surfaces became rougher as $P_{Ar}$ became higher. When $P_{Ar}$ was in the order of $10^{-2}$ Pa, the film surface appeared to be very smooth and lustrous. The grains with irregular shape grew as $P_{Ar}$ became high. This change of the surface may be

(A) $P_{Ar}$ = 9.7 x $10^{-2}$ Pa (B) $P_{Ar}$ = 7.7 x $10^{-1}$ Pa (C) $P_{Ar}$ = 2.0 Pa

Fig.2 Surface Appearance of Film on Various Pressure $P_{Ar}$

due to the difference of the energy of the sputtered particles from the targets. When $P_{Ar}$ is low, then, the mean free path of the sputtered particles is long and they can reach the substrate with a large initial energy. As $P_{Ar}$ becomes higher, they come into collision with the argon particles more often. Their energy decreases as they repeat the collision. Consequently, the excelent Al films with smooth and lustrous surface appearance were able to be prepared on the lower side of $P_{Ar}$.

Figure 3 shows the surface appearance of Al thin film prepared by magnetron. The hillock can be seen as mentioned above.

Figure 4 shows the surface appearance of films on the various bias voltage at $P_{Ar}$ of 0.2 Pa. The number of grains per unit area decreased and the film surfaces became smooth when Vb was -40 V. The surface became rougher when Vb became lower or higher than -40 V. Figure 5 shows the dependence of the Knoop hardness $H_K$ on Vb. The films prepared at Vb of -40 V were hardest and were twice as hard as ones prepared without the bias voltage. The Al films prepared by the TFTS system were 3 times harder than the ones prepared by the magnetron sputtering method.

The step was covered with the uniform Al film by adjusting Vb and $P_{Ar}$. It has been confirmed as shown in Fig.6 that the step coverage of these Al films by means of TFTS technique was satisfactory to the IC wiring.

The TFTS system has original merits for step coverage compared to other sputtering systems and electron beam evaporation. As it has two facing targets, there is no dead region in the groove of the wafer. In other sputtering systems, since the growing films are highly exposed to the plasma, the structure and properties of the films cannot be effectively improved by applying Vb to the substrate. On the other hand, in the TFTS system the plasma between two facing targets can be confined with the magnetic field applied perpendicular to them. Therefore, the growing films are scarcely exposed to the plasma or hot particles and so the effect of the bias application can be remarkable.

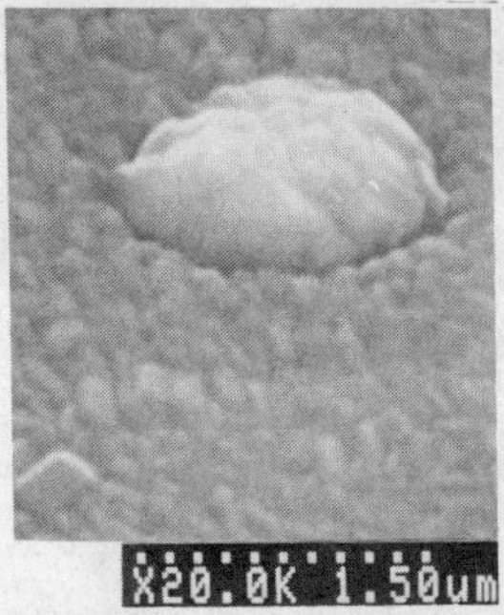


**Fig.3 Hillocks by Magnetron Sputtering**

(A) Vb = 0 V (B) Vb = −40 V (C) Vb = −100 V

**Fig.4 Surface Appearance of Film on Various Bias Voltage Vb**

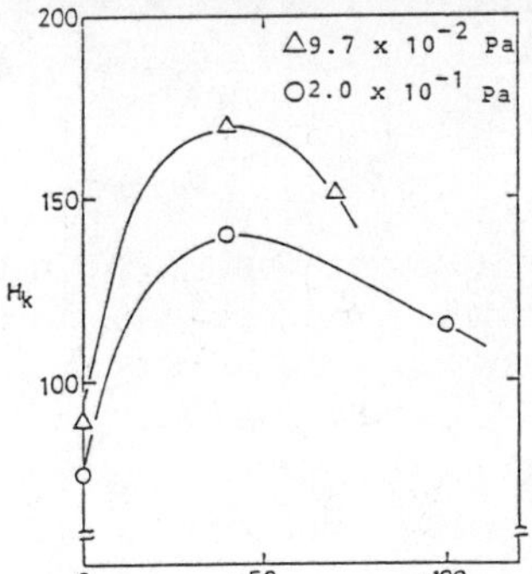


**Fig.5 Vb Dependence of Knoop Hardness $H_K$**

CONCLUSION

The TFTS system has a high capability for improving the quality of thin Al films for IC, because the substrate is almost free from the plasma.

The Al thin films prepared by the TFTS system were fine-grained and were harder than ones prepared by other sputtering systems.

By applying Vb of -40 V to the substrate, the Al films became still harder and had a very smooth, lustrous and fine- grained surface appearance. The step coverage of the Al thin films was very satisfactory for the

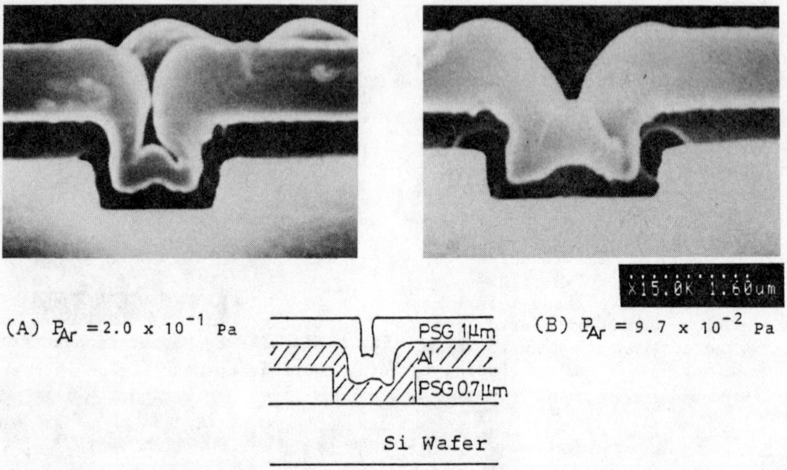


Fig.6 Cross Section View of Step Coverage by TFTS System

production of the electrodes and wires in IC. Therefore, such hard Al thin film composed of fine-grains may be very suitable for other fine processing in semiconductor devices.

REFERENCE

(1) M. Naoe, S. Yamanaka and Y. Hoshi: IEEE Trans. Mag., MAG-16(1980)646.

# RAPIDLY SOLIDIFIED TITANIUM

F. H. FROES[a] AND R. G. ROWE[b]

[a]Air Force Wright Aeronautical Laboratories, Materials Laboratory, AFWAL/MLLS, Wright-Patterson AFB, OH 45433

[b]General Electric Company, Corporate Research and Development Center, P.O. Box 8, Bldg. K-1, 263 MB, Schenectady, NY 12301

## ABSTRACT

The attributes and requirements for rapid solidification (RS) are discussed and related to the characteristics of the titanium system. The extreme reactivity of liquid titanium requires the development of RS production methods tailored to this system. Compaction and processing techniques are presented followed by a discussion of the various alloy classes which have been produced by the RS approach: conventional alloys such as Ti-6Al-4V, rare earths, metalloids, eutectoid formers, beta alloys, amorphous materials, high aluminum alloys, aluminides, and other minor classes. Finally some thoughts on the future of this technology are presented.

## INTRODUCTION

### Scope

Powder metallurgy (PM) can conveniently be separated into two general areas, the traditional press-and-sinter approach, where alloy elements are blended together, and the newer prealloyed technique [1]. For the titanium system a second sub-division of the prealloyed method can be made [2]. In the first category the PM approach is used as a low cost method for production of conventional alloy parts at essentially the same mechanical behavior levels as cast and wrought ingot metallurgy (IM) material. The second involves RS leading to production of unique alloys with enhanced mechanical behavior levels or unique microstructures in conventional alloys again with the same effect. It is this subject of unique alloys/ microstructures which will be presented in the present paper.

The technology which surrounds solidification methods such as the Plasma Rotating Electrode Process (PREP) of conventional titanium alloys like Ti-6Al-4V will not be covered here. Thus the shape making techniques such as the Ceramic Mold Process will not be discussed in this paper. Also compaction or processing methods which relate generally to conventional titanium will not be covered. The interested reader is referred to a number of comprehensive review articles on conventional titanium PM, which cover both press-and-sinter (blended elemental) and prealloyed approaches, for further details [2-4].

The present paper will cover both crystalline and amorphous materials, but the emphasis will be on applications with structural requirements rather than physical requirements. A general discussion on the attributes of RS will be followed by specific characteristics relating to the titanium system. The potential for application of RS to titanium alloys will be presented. A discussion of production methods for RS titanium will be followed by compaction techniques and processing methods. The alloy systems investigated to date will be discussed and finally some thoughts on where this technology may go in the future will be presented.

## Rapid Solidification

### Attributes

The potential attributes of RS have been well documented [5] and will only be presented in summary form here. The advantages which can be gained can most conveniently be divided into constitutional and microstructural effects. RS allows large departures from equilibrium constitution (the identity and compositional ranges of phases formed) resulting in large extensions of solid solubility, formation of non-equilibrium or metastable crystalline phases and metallic glasses, and retention of disordered crystalline structures in normally ordered materials such as intermetallic compounds. Additionally it leads to changes in the morphology and dimensions of the microstructure (the size and location of the phases present). This change is in the direction of a more uniform and finer microstructure with a large reduction in solute segregation effects. The microstructural features which are refined include the grain size, dendrite spacing, coarse constituent particles, and precipitates and dispersoids. To date there has been little evidence of increased dislocation density due to RS, but there is strong evidence for significant increases in vacancy concentration. Other advantages of RS are increased tolerance to tramp elements, a big gain in days of recycling. Since RS is also a PM technique other attributes include circumvention of working problems by production of net or near net shapes, and elimination of highly textured products especially in hexagonal close packed materials such as titanium.

### Requirements

RS usually refers to rapid solidification of the molten metal to the solid state, involving rapid cooling through the solidification temperature range. This is achieved with two basic requirements, a small cross-sectional dimension of molten metal which must be in good contact with an effective heat sink. The methods developed to achieve these requirements have been categorized in a number of ways, though most conveniently as spray (atomization), chill (ribbon), and weld (surface) methods [5, 6]. Only the first two will be considered in detail as production methods in the present paper, the weld method will be discussed only as a technique to survey or screen alloy compositions.

### Titanium Characteristics

The extreme reactivity of titanium in the molten state sets it apart from most other metals when considering RS techniques. Thus containment must be of major concern in developing RS methods for titanium. Contamination can take place from the mold, guide tubes (tundish) or nozzle involved with either the spray or chill methods. Thus the RS of titanium alloys has generally involved the use of containerless processes (e.g., local melting and centrifugal atomization) which however can lead to problems because of the absence of a large mass of molten metal to give chemical homogeneity and the lack of super-heat. Specific production methods used will be discussed in more detail in the section on Powder Production.

### Total Cooling Path

Although the technology being discussed is referred to as rapid "solidification" it is also critical that subsequent solid state cooling must be rapid, especially in the high solid state temperature range where diffusion is most rapid [6]. This requirement for a fast solid state quench is particularly necessary because the metastability of extended solid solutions or new phases, and the fine microstructure make them extremely vulnerable to solid state decomposition [5]. This effect is of major concern in techniques such as melt-spinning and melt-extraction in which cooling can be much less rapid following release from the substrate (conduction dominated) to the surrounding environment (convection/radiation dominated). Separation of the microstructural changes due to the solidification event (generally dendritic in nature) and high temperature solid state decomposition are often not easy, but are important for optimization of an alloy microstructure and therefore the bottom line mechanical properties.

## Titanium Phase Diagrams and Potential for RS

### Titanium Phase Diagrams

A number of authors have attempted to categorize titanium alloy phase diagrams [7, 8], all agreeing that there are two major divisions -- alpha stabilized and beta stabilized systems. Of these perhaps the most convenient is that developed by Molchanova [8], Figure 1. Here the alpha stabilizers are divided into those having complete stability, in which the alpha phase can coexist with the liquid (e.g., Ti-O and Ti-N) and there is a simple peritectic reaction, and those which have limited alpha stability in which decomposition of alpha takes place by a peritectoid reaction into beta plus a compound (beta peritectoid). Examples of the latter type of system are Ti-B, Ti-C, and Ti-Al. Molchanova also divides the beta stabilizers into two categories beta isomorphous and beta eutectoid. In the former system an extreme beta solubility range exists with only a restricted alpha solubility range. Examples are Ti-Mo, Ti-Ta, Ti-V, with elements such as Zr and Hf occupying an intermediate position since they have complete mutual solubility in both the alpha and beta phases. For the beta eutectoid systems the beta phase has a limited solubility range and decomposes into alpha and a compound (e.g., Ti-Cr and Ti-Cu). This class can also be further sub-divided depending on whether the beta transformation is fast (e.g., Ti-Cu, Ti-Ni, and Ti-Sn) or sluggish (e.g., Ti-Cr, Ti-Mn, and Ti-Fe).

### Solubility Limits

A summary of the solubility extensions determined in titanium alloy systems to date is shown in Table I. Here the maximum equilibrium solubility is the solubility in the high temperature beta phase. The extended solubility level indicated is the value judged by the present authors to be the level to which the matrix in question did not exhibit second phase particles, involving the solute, in the as-RS condition. Estimates of the solubility extensions for beta isomorphous and eutectoid formers have been made [9], however the criterion used is not the same as used by the present authors.

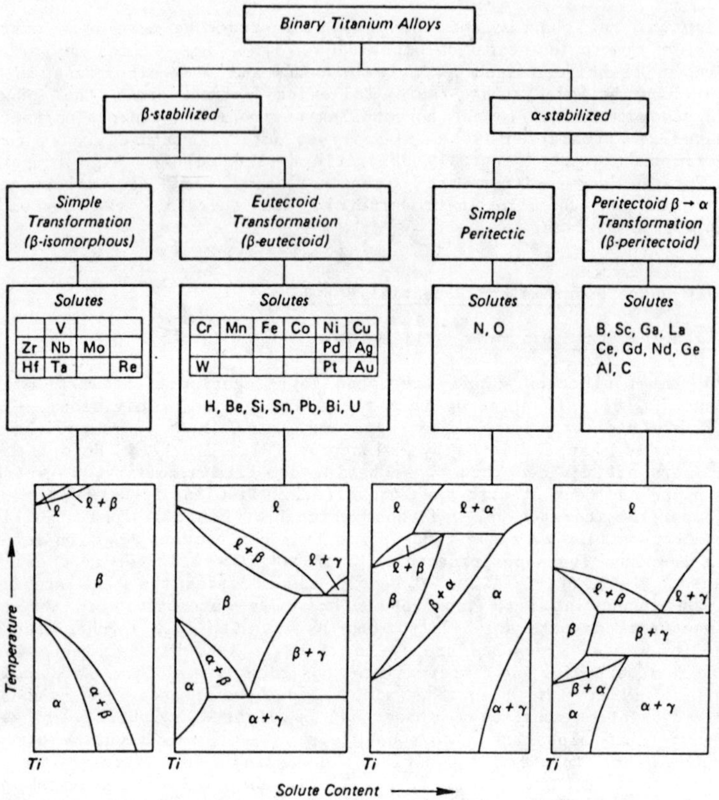


Figure 1. Categories of titanium alloy phase diagrams [8].

TABLE I. SOLUBILITY LIMITS OF SOME ALLOYING ELEMENTS UNDER EQUILIBRIUM AND RAPID SOLIDIFICATION CONDITIONS

| Element | Matrix | Maximum Equilibrium Solubility[a], at% | Extended Solubility[b], at% | Technique | Reference |
|---|---|---|---|---|---|
| B | Binary | 0.5 | 6.0 | Splat | 10, 11 |
| C | Binary | 3.1 | 10.0 | Splat | 10, 11 |
| Si | -10at%Zr | 5.0 | 6.0 | Splat | 10, 12, 13 |
| Ge | -5Al-2.5Sn | 8.2 | >5.0 | EBSQ[c] | 14 |
| Nd | Binary | ∿0.3 | >1.0 | EBSQ | 15 |
| Y | -8Al | ∿0.5 | >2.0 | Laser | 16 |
| Gd | Binary | ∿0.3 | >0.5 | EBSQ | 15 |
| La | -5Al, -5Sn | ∿1.5 | ≧1.5 | Splat | 11, 17 |
| Ce | Binary | ∿1.5 | >1.0 | Splat | 11, 18 |
| Er | -5Al | ∿0.3 | ≧1.5 | Splat | 11 |
| Dy | Binary | ∿0.3 | >0.6 | EBSQ | 15 |

[a]In beta titanium. [b]As-rapidly solidified product.

[c]Electron Beam Splat Quench.

## RS Potential for Titanium Alloys

As a transition metal titanium is characterized by its ability to provide greater than 1 atomic percentage (at%) solid state solubility for more than 50 elements [2]. Thus the potential for extended alloying is not as great as for the aluminum system where the corresponding number of elements is only eight [5].

The classes of elements for which the alloy potential in the titanium system using RS appears greatest, can be summarized as:

(a) the rare earths (including Y) which normally exhibit low solubility

(b) the metalloids (B, C, Si) which exhibit low solubility

(c) the eutectoid formers, which although they exhibit quite high solubility, are very segregation prone.

# TITANIUM RAPID SOLIDIFICATION

## Powder Production

The production of RS titanium alloys is complicated by the fact that metal containment is difficult. Even the most chemically stable ceramic compounds are dissolved in liquid titanium [19] and to date no scheme to minimize the reaction has become commercially viable. Thus the melt processing for this system is different than that which has been used in other metal systems [6].

The accepted technique for melting titanium and titanium alloys is cold hearth arc melting [20]. A solid layer (skull) of titanium serves as the melt containment surface. The advantage of this process is that there is virtually no contamination during melting. The disadvantage is that superheat in the liquid is only local, near the arc (or plasma) source, and that both solid and liquid co-exist in the melt pool.

An alternate melting technique is localized melting with an intense localized heat source (electron beam, laser, arc) so that melt is contained by the remaining solidified part of the ingot. The characteristics of local melt-based processes are the same as for skull melting, but the possibility of local composition fluctuation due to heterogeneity in the ingot stock is present. Most RS processes for titanium alloys rely upon either cold hearth or local melting schemes for melt containment.

### Spray or Atomization

Titanium powder is produced on a semi-commercial basis by the Plasma Rotating Electrode Process (PREP) [2-4]. The powder is contamination-free, but is relatively coarse (average size about 150 μm) on the scale of most RS processes. Its cooling rate has been estimated to be from $10^3$°C/sec to $10^5$°C/sec [21]. Heterogeneity of the powder due to local fluctuation of the ingot is occasionally observed [22]. The issue of whether PREP powder has a sufficient solidification rate for the production of fine dispersions of rare earth oxides without the presence of heavily segregated interdendritic regions is still being resolved (see Rare Earth Additions section).

A process similar to PREP in concept is Laser Spin Atomization (LSA) [23, 24]. Powder produced by this process is finer than PREP, and more rapidly solidified. It has been suggested that the cooling rate for 100 micron diameter powder produced by this technique is $10^5$°C/sec, but a well defined dendritic structure is still observed at powder surfaces. LSA powder of an alloy of Ti-8Al-1Mo-1V-1.5Er (wt%) has been examined by transmission electron microscopy [25], and found to exhibit a supersaturated solid solution of Er in the alloy after solidification.

Other local melting/centrifugal atomization techniques have been developed for conventional titanium alloys such as the French PSV (Pulvérisation Sous Vide) electron beam process [2, 3, 6]. But these have generally not yet been applied to RS titanium systems. A possible exception is the open stream technique used in combination with a spinning disc reportedly in use in the Soviet Union [26].

Gas atomization of RS titanium alloys has been reported by two entirely different techniques. In the first of these, titanium powder was produced by ultrasonic gas atomization (USGA) utilizing induction melting in a coated ceramic crucible [27]. However it is doubtful that this is a production process since significant pickup of the ceramic in the form of the oxide is likely [19]. The powder produced had a size range of 10-100 micron diameter and an estimated cooling rate, based upon dendrite arm spacing, of $10^4$-$10^6$°C/sec.

A system for titanium gas atomization utilizing cold hearth melting and bottom ejection through a refractory metal-skull nozzle has also been reported [28, 29]. Open stream inert gas atomization produced titanium alloy powders with virtually no contamination, and an average powder size for Ti-6Al-4V of 190 microns. This system with a 3 lb powder capacity, approaches a prototype production scale. The distribution of dispersoids in titanium-rare earth alloys produced by this technique is presently being evaluated.

Chill or Substrate-Quenching

Titanium RS processes based upon substrate quenching have also been developed. Pendant drop melt extraction (PDME) has become a widely utilized process for the production of laboratory-scale quantities of titanium and other refractory and reactive metal alloys [6], although chemical homogeneity may be of some concern with this method. For alloys such as Ti-6Al-4V, filament of thickness from 50-125 microns shows no evidence of dendritic segregation. Strongly segregation prone elements such as boron produce a two-zone microstructure which is columnar in the bottom half of the ribbon, and equiaxed dendritic in the top half [30]. Filament of titanium alloys has been produced in quantities of 50-100 gm per run. Melt extracted filament has exhibited fine dispersoids of rare earth oxides after consolidation and heat treatment suggesting that during solidification, a metastable solid solution containing up to 1 at%Er has been maintained. However, bimodal particle distributions are typical in vacuum melt extracted alloys suggesting some solid state decomposition during cooling [30-32].

Another localized melting technique which has been developed is the electron beam splat quench (EBSQ) process [6, 27] which produces horseshoe-shaped splat foils. Flake thickness is similar to filament produced by the melt extraction process at 75-150 micron. As-rapidly solidified flake of titanium alloys containing rare-earth additions solidified with little evidence of second phase precipitation, demonstrating retention of a metastable solid solution in this system [15].

Prototype scale cold hearth arc melting systems have been developed in order to melt spin titanium alloys [6, 30, 33, 34]. Ribbon is produced which is somewhat thinner (30-50 μm) than that which is achieved in melt extraction processes because of the higher wheel speeds that can be tolerated in melt spinning [35]. Alloys containing boron or other strongly segregating elements solidify in the same two-zone microstructure that has been observed in melt extracted alloys of the same composition, but the scale of the microstructure is finer [30].

Other Techniques

Titanium alloys have been rapidly solidified for microstructural studies by the process of laser or electron beam surface melting [6]. Heat is extracted from the melt by the adjacent solid alloy, and hence very intimate melt-substrate contact and high solidification rates from $10^3$-$10^6$°C/sec are achieved [36].

Other techniques for direct titanium powder production may also be applicable to RS processing based on the fineness of the microstructure observed (Hurd and Goldschmidt Processes [37]) and the specifics of the technique involved (Albany Titanium Process, ALTi [37]). An example of the fine beta grain size which can be produced by the Hurd process is shown in Figure 2. All these techniques are still in the early stage of development.

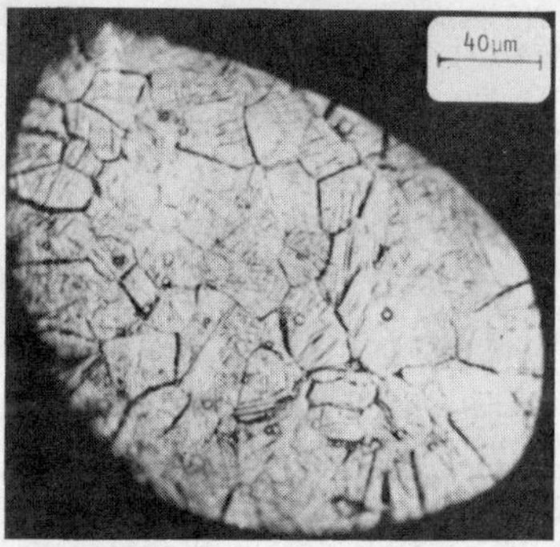


Figure 2. Microstructure of Ti-6Al-4V powder produced by the Hurd Process [37]. Note fine beta grain size and martensitic structure indicating fast cooling.

## Compaction and Processing

### Thermomechanical Processing (TMP)

To date mainly conventional compaction techniques such as extrusion, vacuum hot pressing, and hot isostatic pressing have been used with RS titanium alloys [2-4]. It has been found that RS titanium alloys can be successfully consolidated by HIP and extrusion at a temperature of 840°C (1545°F), but that HIP alone does not produce sufficient bonding for high mechanical integrity compacts [22]. Dispersoid-bearing alpha titanium alloys have been consolidated at this temperature with little dispersoid coarsening. Extrusions exhibit both fine dispersions and ultrafine grain size [30, 31]. Compacts of dispersion-modified titanium alloys have also been successfully consolidated by a similar process of vacuum hot pressing at 820°C (1510°F) and isothermal forging at 800°C (1470°F) [27]. A small amount of work using high strain rate compaction techniques such as gun compaction [38] and explosive methods [39-41] has been carried out but the applicability to RS alloys is still to be defined.

### Thermochemical Processing (TCP)

Hydrogen can easily be introduced into titanium powder in a hydrogen rich atmosphere, and equally easily removed by a vacuum anneal leading to development of refined microstructures of various types in the Ti-6Al-4V and similar alloys [2-4, 42].

In an attempt to use TCP to control the microstructure of eutectoid forming elements, hydrogen was introduced into a Ti-Cu alloy [43]. This hydrogenated material and baseline Ti-Cu alloy were taken through a simulated compaction cycle. It was found that after this cycle, and a subsequent dehydrogenation treatment, that the material processed with hydrogen in it exhibited a considerably finer microstructure than the non-hydrogenated material.

Recent work [44] on dispersion strengthened high temperature RS titanium alloys indicated that the most advantageous microstructure for creep behavior was the "traditional" Widmanstatten/basketweave alpha structure in combination with a second phase dispersion. However, the challenge was to obtain the desired alpha morphology (which generally necessitates going above the beta transus temperature) while not unduly coarsening the dispersion. It has been demonstrated that hydrogen dramatically reduces the beta transus. By adding hydrogen to the dispersoid forming alloys and going above the modified beta transus temperature it was hoped that the desired alpha morphology would be obtained along with a fine dispersoid. However it was found that the increased diffusion rate in the beta phase field resulted in excessive coarsening.

## Alloy Systems

### Matrix Microstructure Development

The matrix microstructure of as-produced RS titanium alloys can be significantly more complicated than alloys such as aluminum because of the allotropic change and the occurrence of martensite. The present section will only deal with the microstructure of the matrix, subsequent sections will be concerned with how alloying elements modify and enhance these microstructures.

In richly beta stabilized alloys RS material quenches to a beta matrix which is relatively uncomplicated [45]. In leaner alloys however the RS material passes through the beta region and is unstable enough that on quenching to room temperature a hexagonal martensite structure occurs [21]. On subsequent heating, the matrix can then evolve in two ways depending on the beta grain size. If the beta grain size is relatively large (≥5 μm) lenticular alpha forms from the martensite lathes [21] in a manner analagous to that which occurs in ingot metallurgy material. If, however, the beta grain size is small (≤5 μm), a quite different equiaxed alpha morphology develops. This alpha grows to consume the martensitic structure in a similar manner to that of the normal lenticular alpha. Examples of this alpha morphology are shown in Figure 3 [14]. A more detailed discussion of the effect of cooling rate on beta grain size and hence alpha morphology is presented below for the Ti-6Al-4V alloy.

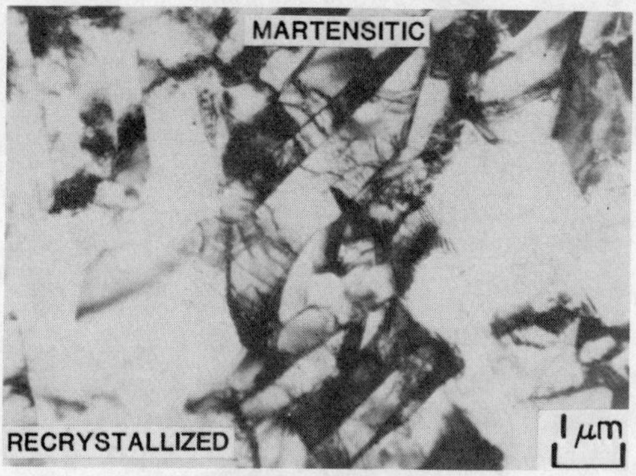


Figure 3. Equiaxed alpha growing from martensite plates in a rapidly solidified Ti-5Al-2.5Sn-0.5Si alloy aged for 5 hr at 535°C (995°F) [14].

Conventional Alloys

The most comprehensive study done to date on RS conventional titanium alloys has been on the Ti-6Al-4V alloy [2-4, 21] using cooling rates varying from $10^4$-$10^7$°C/sec. From this work two significant effects were noted: the beta grain size decreased as the cooling rate was increased (Figure 4), and at the higher cooling rate studied, equiaxed alpha was produced on subsequent high temperature annealing (Figure 5). This latter effect is not observed on annealing either conventional powder or unworked IM material, and may result from either the short time available for strain redistribution during the extremely high cooling rate or as a direct consequence of the fine beta grain size. This equiaxed alpha is likely to have attractive fatigue initiation characteristics.

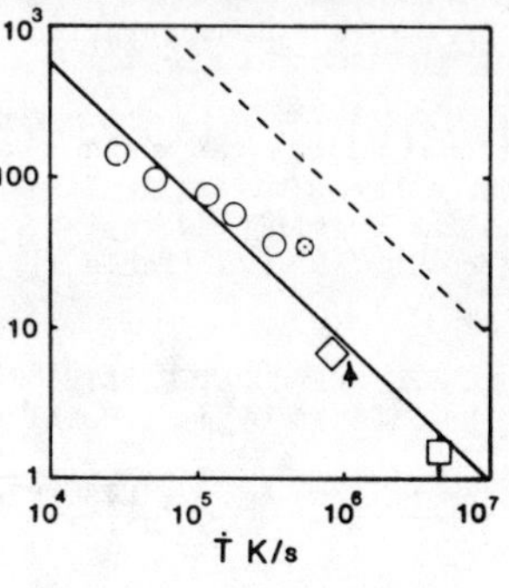


Figure 4. Variation of beta grain size in Ti-6Al-4V with cooling rate [21]. Dashed line is predicted for aluminum [46].

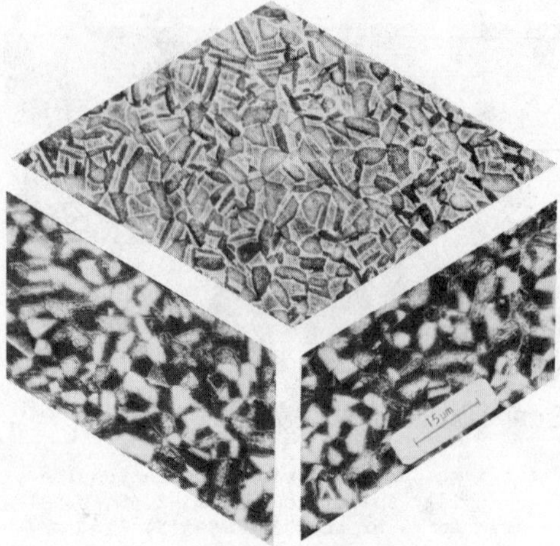


Figure 5. Equiaxed alpha morphology in Ti-6Al-4V produced after annealing fine grained rapidly solidified material [21].

The result of plotting beta grain size (L) as a function of estimated cooling rate ($\dot{T}$) (Figure 4) gives the single power relation:

$$L = A_2 \dot{T}^{-n_2}$$

where $n_2 = 0.93 \pm 0.2$ and $A_2 = 3.1 \times 10^6$ μm $(K/s)^{n_2}$

The exponent $n_2$ in particular does not differ significantly from that predicted for aluminum [46], though absolute values of L for Ti-6Al-4V are an order of magnitude less than that expected for aluminum from this model. This work represented the first clear demonstration that the predicted power law [46] can apply.

Other work on conventional alloys has shown the advantage to be gained from RS processing. In an alloy similar to Ti-6Al-4V, but containing Mo instead of V, both strength and ductility were increased as the cooling rate was increased from $10^3/10^4$ to $10^6/10^7$ K/s (Table II) [47].

TABLE II. EFFECT OF COOLING RATE ON TENSILE PROPERTIES OF A Ti-Mo-Al ALLOY [47]

| Production Method | UTS, MPa | RA, % |
|---|---|---|
| Casting | 895 | 12 |
| Cast + Hot Worked | 1070 | 22 |
| $10^3/10^4$ Cooling + HIP | 1070 | 22 |
| Flake $10^6/10^7$ Cooling + HIP | 1135 | 30 |

A study of the Beta III alloy (Ti-11.5Mo-6Zr-4.5Sn) indicated that the beta grain size decreased as the cooling rate increased (∿40 μm in PREP, 2-4 μm in EBSQ) [45]. These grain sizes were significantly smaller than in corresponding Ti-6Al-4V product (∿135 μm and ∿7 μm, respectively), indicating the effect of increasing solute content in decreasing grain size for a given cooling rate [5]. A further observation in the Beta III was that upon aging alpha precipitated interdendritically in PREP material and at grain boundaries in the EBSQ alloy.

For the Ti-15V-3Al-3Sn-3Cr alloy a relationship was developed between the calculated cooling rate ($\dot{T}$, K/s) and the secondary dendrite arm spacing (d, mm) using a heat transfer model [36] of:

$$d = 80\dot{T}^{-0.34}$$

Rare Earth Additions

The RS of titanium alloys containing a rare earth element produced an ultrafine dispersion of stable rare earth oxides with particle sizes much finer than that produced in earlier work on IM material [48-50], even when mechanical alloying was used [51]. Both $Er_2O_3$ [15, 52-55] and $Y_2O_3$ [15, 16] were found to produce a fine dispersoid in rapidly solidified titanium alloys. Virtually no dispersoid was present in the RS condition [15, 25, 55, 56] provided solid state cooling was sufficiently fast. Aging at various temperatures produced extremely fine dispersoids, with aging at 500°C (930°F) producing the finest dispersoid [55]. Fabrication of components requires exposure to consolidation cycles at temperatures up to 950°C (1740°F). The stability of the dispersoid is a critical issue to temperatures considerably higher than the service temperature.

A comparison of the dispersoid distribution in a series of ternary titanium- and rare earth-oxygen alloys indicated that alloys containing La, Nd, Dy, and Er had the highest resistance to coarsening at 800°C (1470°F) [15]. In these alloys, dispersoids had an average diameter range from 650-980Å with interparticle spacings from 0.3-0.8 micron. This translated to an Orowan strengthening increase of up to 85 MPa (12 ksi).

The stability of the $Er_2O_3$ dispersoid has been shown to be very good in alpha titanium-aluminum alloys as long as the alloy remains in the alpha titanium phase field. The alloy Ti-6Al-2Sn-4Zr-2Er (wt%), rapidly solidified by melt extraction, was annealed for 1 hr at 950°C [30, 31]. A dispersoid 200-400Å diameter, with an interparticle spacing of ∿0.3-0.4 micron, was maintained in the grain interiors of this alloy.

Dispersoids containing aluminum and tin as well as rare earth elements have been suggested in Ti-Al-Ce and Ti-Sn-La alloys [17, 18, 57]. However other studies postulate that the dispersoid in the former of these alloys are $Ce_2O_3$ [58]. The stability of rare earth aluminide or stannide dispersoids would appear to be less than that of the rare earth sesquioxide. Therefore it is possible that the aluminide or stannide dispersoids may act as precursors to the formation of the oxides. The stability of dispersoids based upon Ce and La in Ti-Al and Ti-Sn alloys have not shown the same degree of resistance to coarsening which has been observed in dispersoid systems based upon the sesquioxide $Er_2O_3$ [10, 18, 59]. Annealing the alloys at 800°C (1470°F) results in coarsening of the dispersoid to particle diameters of on the order of 1000-2000Å, while $Er_2O_3$ dispersoids exhibit 200-400Å size after annealing at 950°C (1740°F) in alpha titanium alloys containing Al and Sn [17, 30, 31, 55].

The rare earth sulfides and oxysulfides form a fine dispersion in alpha titanium alloys and exhibit good resistance to coarsening even after exposure to temperatures as high as 1000°C (1830°F) [31, 32, 60].

Dispersoid coarsening has been observed to be significantly higher near grain boundaries for both $Er_2O_3$ and CeS dispersoids after elevated temperature exposure [30–32, 60, 61], with the width of the depleted region increasing with annealing temperature. Figure 6 [61] shows the depletion and coarsening in $Ti_3Al$-2Er which had been consolidated by HIP at a temperature of 850°C (1560°F) [61]. The $Er_2O_3$ dispersoid was relatively unaffected in grain interiors, yet heavily depleted and coarsened near grain boundaries. In this case this is probably a result of enhanced diffusion when dispersoids are intercepted by the grain boundaries [61]. In other instances it has been shown that selective grain boundary beta transformation during annealing also produced dispersoid coarsening [31].

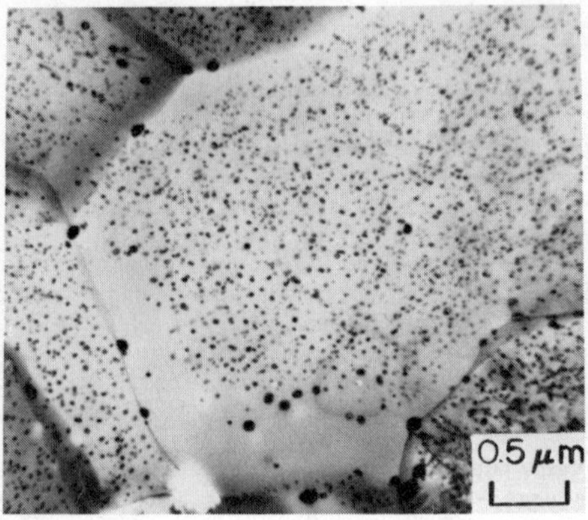


Figure 6. Depleted grain boundary region and particle coarsening in a $Ti_3Al$-2Er alloy HIP'd at 850°C (1560°F) [61].

The rare earth dispersoid in titanium alloys appears to effect the behavior of the material in a number of ways. First, the dispersoid produces hardening due to the Orowan effect [62]. There are also indirect effects due to a grain size reduction [15, 31, 32, 60, 63] and scavenging of interstitial elements thereby reducing interstitial strengthening. The net result in ternary Ti-RE-O alloys was a strengthening on the order of 40 MPa (6 ksi) at 700°C (1290°F) [63] and a significant creep rate reduction.

Preliminary work on a Ti-6Al-2Sn-4Zr-2Mo-0.1Si-2Er alloy produced using the PREP process [64] indicates that even after compaction, by extrusion at 760°C (1400°F), a relatively fine dispersion of rare earth oxide is present (maximum size 500Å, majority of the particles <100Å) (Figure 7). This suggests that the PREP process may be useful for production of alloys containing rare earth oxides and that in fact a cooling rate of $\sim 10^4$ K/s [21] may be more realistic for this production method than the often quoted $10^2$ K/s.

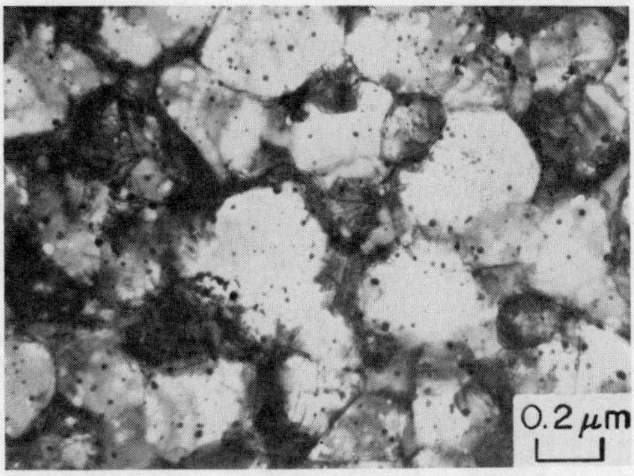


Figure 7. Fine dispersion of $Er_2O_3$ particles in PREP Ti-6Al-2Sn-4Zr-2Mo-0.1Si-2Er alloy as-extruded at 760°C (1400°F). Note majority of particles are <100Å [64].

Metalloid Additions (C, B, Si, Ge)

The metalloid elements C, B, Si, and Ge have the potential for good chemical stability exceeded only by the rare earths. However, the stability of these elements is only moderate in titanium alloys, and their useful temperature range is below that of rare earth oxide dispersion strengthened alloys [57], i.e., below 600°C (1110°F).

RS titanium alloys containing boron have been produced by laser surface melting, splat quenching, melt extraction, and melt spinning [30, 48, 57, 65, 66]. In virtually all studies with the possible exception of hammer and anvil splat quenching, boron decorates the solidification interfaces. Boron-containing alloys exhibit a two-zone microstructure of columnar growth changing to equiaxed dendritic structure near the top of melt extracted and melt spun ribbons, contrasting with the segregation free columnar solidification of alloys containing rare earth additions. Because of this segregation, boride-containing alloys exhibit stronger resistance to grain coarsening of annealed ribbon than alloys containing rare earth additions [30]. Fine, needle-like TiB [30] precipitates develop at temperatures above 500°C (930°F) (Figure 8) [11, 57, 65, 66] and coarsen dramatically at 800°C (1470°F) [11, 30, 66].

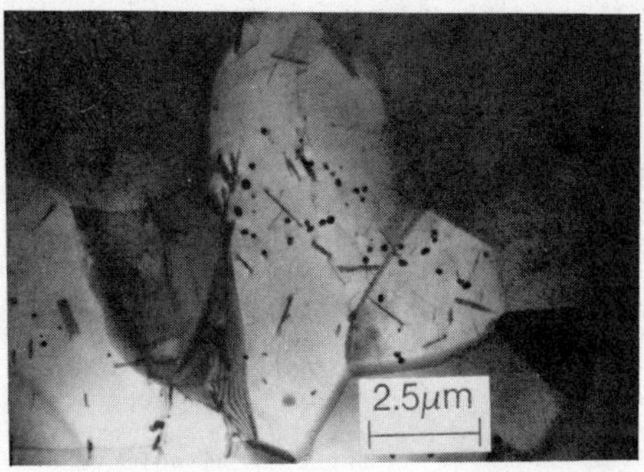


Figure 8. Fine needle-like TiB precipitates formed by exposing material at 925°C (1700°F) [30].

Silicon and germanium do not segregate as strongly as boron in RS alloys. Both elements have been quenched to metastable solid solutions as-quenched for levels of silicon or boron to 1.5 wt% [57, 67-69]. However, additions of 5Si showed a strongly segregated solidification structure [67]. Annealing these RS alloys at intermediate temperatures (550-700°C [1020-1290°F]) produced fine generally spherical precipitates [13, 57, 59, 67, 70] although at the higher temperatures, the particles were relatively coarse [59], particularly in alpha-beta alloys [68]. Annealing at 900-950°C (1650-1740°F) produced particle coarsening to ∿½ micron diameter. The dispersed particles were identified as $Ti_5(Si,Ge)_3$ in all cases [67, 68].

Rapidly solidified titanium alloys containing carbon have been shown to have good strength and ductility as-quenched. However, they lose both strength and ductility after aging at 700°C (1290°F) due to precipitation and rapid coarsening of titanium carbide [66]. Annealing carbon-bearing alloys at 900°C (1650°F) produced ½ micron diameter carbides.

The inherent instability of the metalloid compounds in titanium alloys limits the upper temperature bound of their use to temperatures on the order of 500-600°C (930-1110°F). However RS is an effective means of introducing relatively high volume fractions of fine titanium-metalloid compounds. The potential for alloys based upon metalloid additions is yet to be fully identified.

## Eutectoid Formers

The eutectoid formers (e.g., Ni, Cu, Co, Fe, Cr, Si, and W) are characterized by a phase diagram at the titanium rich end, which involves both the eutectoid reaction ($\beta \rightarrow \alpha$ + compound) and often at higher temperatures and solute content, a eutectic decomposition ($L \rightarrow \beta$ + compound). This class of alloy has been studied in detail using ingot metallurgy [71]. It is normally possible to obtain relatively high levels of solute in the beta phase and lower though still significant levels in solution in the alpha phase. A problem which arises however is that the eutectoid formers normally show a high partitioning coefficient between the liquid and the solid phase and a large temperature difference between the liquidus and solidus meaning that they are very segregation prone. Thus though there have been a number of attempts to develop titanium alloys containing eutectoid formers using conventional ingot metallurgy only limited success has been achieved due to ingot segregation and also poor workability [72]. Alloys with attractive combinations of strength and ductility have been developed [73, 74] but not commercialized. In one comprehensive study a few percentage of Be, Cu, Co, Fe, Ni, and Si were added to stable beta base compositions, but with a large adverse effect on hot workability and rapid aging kinetics in the case of the Ni and Si containing alloys [75]. From this study ultimately developed the commercial alloy Ti-10V-2Fe-3Al which is now seeing use in high strength applications. The addition of Si is used to enhance high temperature behavior of commercial alloys, and 2.5Cu is used in the British moderate strength alloy IMI-230. Apart from these isolated instances use of the eutectoid formers has not been realized.

Because of the relatively large solid state solubility of the eutectoid formers the major advances which should be possible for this alloying class are avoidance of segregation, refinement of the microstructure, and possible formation of metastable crystalline phases and a change in the kinetics of decomposition. Amorphous materials are discussed separately below.

The great advantage in avoiding segregation which can be achieved using RS is shown in Figure 9 for a Ti-22Fe alloy [76]. The conventionally produced ingot material shows extensive segregation of the iron rich eutectic constituent. In contrast the RS material exhibits a very homogeneous microstructure with a fine grain size (0.2-0.3 μm).

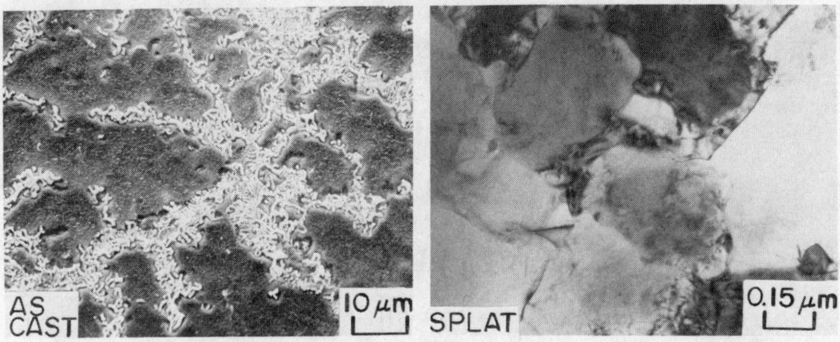


Figure 9. Ti-22Fe alloy showing difference in segregated eutectic between ingot material and RS product [76].

Early work on the addition of Fe and Cu (to levels of 8 wt% and 5 wt%, respectively) to the Ti-6Al-4V alloy produced by the melt extraction technique showed calculated yield strengths in excess of 1725 MPa (250 ksi) after consolidation [77].

More recent studies by Sastry and co-workers [12, 65, 66, 78, 79] investigated Ni levels up to 7 wt% with and without additions of Al to 10 wt%, produced by laser glazing, splat flake, and electron beam splat quenching (EBSQ). As-quenched microhardness more than twice as high as in a pure titanium alloy decreased rapidly on annealing in the range 600-800°C (1110-1470°F). The best combination of properties was judged to occur in a Ti-3Ni-6Al alloy where a 910 MPa (130 ksi) YS and 1010 MPa (145 ksi) UTS occurred in conjunction with 6% elongation.

By far the most comprehensive study of the RS characteristics of titanium alloyed with eutectoid formers has been carried out by Krishnamurthy and co-workers. This study included evaluation of Ti-Ni [80-82], Ti-Co [83], Ti-Fe [84, 85], Ti-W [85, 86], Ti-W-Ni, Ti-W-Ni-Si [87], Ti-Cr-Al [88], Ti-Si [85], and Ti-Cr [85] alloys.

In Ti-Ni material, both RS and conventional IM material showed a hardness decrease with aging time due to $Ti_2Ni$ precipitation and/or coarsening. However at 500°C (930°F) the RS material exhibited a significantly better hardness retention with time, which was attributed to the finer scale precipitation in the RS material. A detailed study of the precipitation morphology in hypoeutectoid (5.5 wt%) and eutectoid (7 wt%) nickel alloys revealed that a bainitic [71] non-lamellar decomposition occurred in the leaner alloy while a pearlitic [71] lamellar decomposition took place at the eutectoid composition, in both cases during the slow solid state cooling following melt spinning or electron beam splat quenching. Careful characterization of an EBSQ Ti-7Ni alloy (Figure 10) indicated a eutectoid lamellar structure of alpha titanium and $Ti_2Ni$ occurred in a shorter time than the 2 seconds predicted for the elapsed time for this material assuming radiative cooling after separation from the wheel [89]. In hammer and anvil splat quenched material a fully martensitic structure was observed. The formation of the previously unobserved lamellar structure was rationalized in terms of the fine grain size which occurs in RS material.

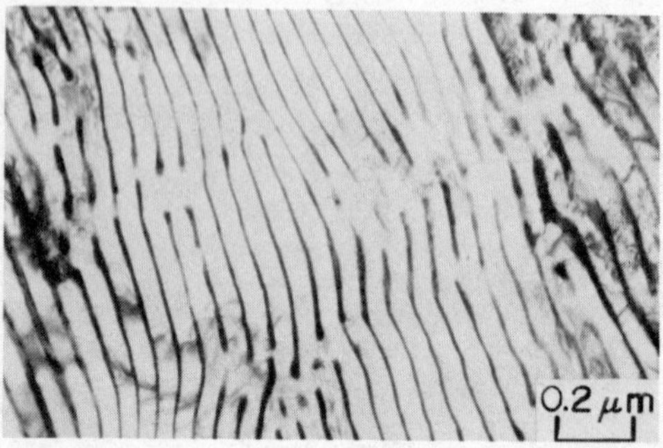


Figure 10. Eutectoid lamellar structure produced in Ti-7Ni alloy after separation from disc in EBSQ material [89].

A study of RS hypoeutectoid Ti-9Co alloy [83] showed that this material exhibited a finer grain size and more rapid and uniform precipitation than in conventional material. Formation of alpha was followed at longer times by precipitation of $Ti_2Co$, and at higher temperatures (575°C [1070°F]) by a higher hardness in RS material.

A detailed evaluation of the effect of cooling rate on the as-quenched microstructure of Fe [84, 85], W [85, 86], Si [85], and Cr [85] containing alloys indicated that a variety of structures were observed ranging from predendritic, to cellular, to equiaxed dendritic [5]. The beta grain size (L) exhibited a relationship of the form:

$$L = B_o \dot{T}^{-n}$$

where ($\dot{T}$) is the cooling rate, and the constants $B_o$ and n varied between $3 \times 10^4$ and $3 \times 10^5$ $(°K\ sec^{-1})^n$, and 0.62 and 0.79, respectively. However re-evaluation of the data suggests that the slope may have been over estimated and that a value closer to that predicted for Al [46] and determined for Ti-6Al-4V [21] of 0.9 may be more appropriate. For laser melted material the dendrite arm spacing (d) was also related to cooling rate:

$$d = B_1 \dot{T}^{-n_1}$$

with $B_1 = 45$ and $n_1 = 0.31$, respectively.

Rapidly solidified W containing alloys and ternary Ti-W-Ni and quaternary Ti-W-Ni-Si alloys exhibited significant hardening peaks, due in the early stages to omega formation [87]. The Ni addition broadened (in time) the hardening peak, with however some decrease in level, and appeared to suppress omega formation.

The eutectoid-former system for which RS showed the most significant advantage was the -Cr system with the addition of 4 wt% Al [88]. Here a 15 wt% Cr alloy with the Al addition was evaluated under RS (chill block melt spun) and conventional conditions. The RS material showed a much finer beta grain size than the conventional material (30 μm compared to 300 μm) and more rapid hardening which was maintained for a considerably longer time. This enhanced behavior was related to the finer grain size and much more uniform precipitation, the latter effect perhaps due to the higher vacancy concentration.

Preliminary results on a Ti-8V-5Fe-1Al PREP powder product indicate that strengths in excess of 1380 MPa (200 ksi) can be obtained with ductility levels close to 10% elongation [90].

Thus it can be concluded that RS processing does offer an advantage for the Ti-eutectoid system. Specifically a more homogeneous, more thermally stable precipitation reaction occurs in the RS material. However little has yet been done to optimize compositions to obtain the best mechanical property combination.

## Beta Alloys

Except for the conventional beta alloys such as Beta III and Ti-15V-3Al-3Sn-3Cr discussed above there has been little done on beta alloys using the RS technique. Three exceptions are a study of the microstructure and creep behavior of a Ti-6Al-15V-2Er alloy [91] and microstructural studies of a Ti-25V-4Ce-0.6S [92] and a heavily stabilized Ti-24V-10Cr alloy [93, 94]. The Ti-25V-4Ce-0.6S alloy showed a very fine dispersoid leading to a predicted strength in excess of 1725 MPa (250 ksi). The Ti-6Al-15V-2Er alloy showed no dispersoid coarsening to temperatures as high as 760°C (1400°F) (beta transus estimated at 830°C [1525°F]). Second phase particles occurred in relatively large grains (~15 μm) in the Ti-24V-10Cr alloy (Figure 11) suggesting that this system may have potential for high strength applications.

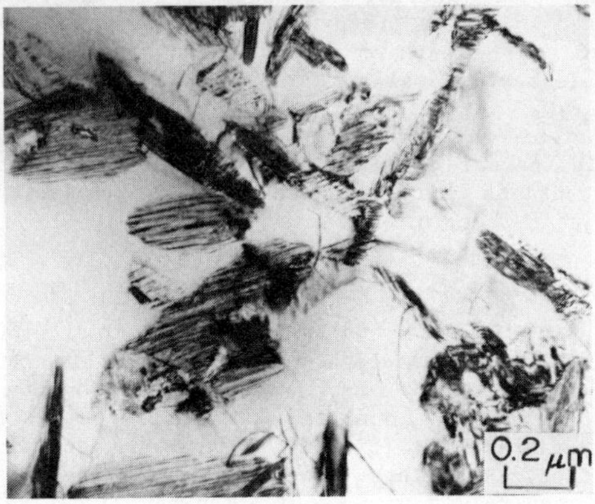


Figure 11. Precipitates present in Ti-24V-10Cr alloy after rapid solidification and aging for 65 hr at 480°C (900°F), indicating potential for high strength applications [93].

## Amorphous Alloys

Metallic glasses or amorphous alloys have been classified in two major groups, metal-metal and metal-metalloid systems [95]. Compositions which readily form glasses are often found at or near deep eutectics where the lower free energy values for the liquid, compared to competing crystalline phases, favors the formation of the amorphous material. This type of feature is quite often observed in titanium phase diagrams therefore this system exhibits a propensity for glass formation. Since the major theme of the present paper is towards the use of RS titanium alloys for structural applications, the present section will be restricted to this topic. However the interested reader is referred to a number of other articles for further details of titanium base glassy materials for enhanced physical properties such as electrical and magnetic behavior [96-101] and general discussions on metallic glasses in this system [57, 102-108].

The strength of Ti-25Zr-10Be (Metglas 2204, $Ti_{50}Be_{40}Zr_{10}$) was estimated from hardness values to be 2345 MPa (340 ksi) UTS [77]. Even higher values have been obtained in a series of amorphous Ti-M-Si (M = Mn, Fe, Co, Ni, Cu) alloys where strengths close to 2750 MPa (400 ksi) were recorded [102]. In other work [103] Ti-Zr-Ni-B and Ti-Zr-Ni-Si glasses were formed using the hammer and anvil splat quenching technique, and subsequently devitrified by aging. Hardness values corresponding to strengths in excess of 2070 MPa (300 ksi) UTS were obtained with grain sizes less than 1 μm even after exposures as long as 2 hr at 950°C (1740°F). High hardness were also found in Ti-Si, Ti-M-Si, and Ti-M-B alloys [107].

Much further work is needed in this area, particularly on the use of amorphous precursors for usable crystalline structural alloys [108].

### High Aluminum, Alpha Titanium Alloys

High aluminum alloys offer the potential for good high temperature behavior due to a dispersion of $Ti_3Al$ precipitates, however ductility is generally low. Work on highly alloyed alpha titanium alloys indicated that consolidated alloys containing a dispersoid are significantly finer in grain size than the same alloy without a dispersoid [31, 32]. However this fine grain size resulted in a fast creep rate which probably could be improved by an excursion into the beta phase field to grow the grains. The beta heat treated dispersoid-bearing alloys, which had a dispersion of coarse 0.2-0.4 micron $Er_2O_3$ particles, had a post-creep ductility of 2%, in contrast to dispersoid-free material which had nil ductility. It appears that the coarse dispersoid reduces the tendency to planar slip and allows higher aluminum contents which, without dispersoid modification, would have had unacceptable post-creep brittleness [31].

### Titanium Aluminides

The titanium aluminides because of their ordered structure are potentially useful alloys for high temperature applications. However the planarity of slip results in low ductility at room temperature. Rapid solidification offers the potential for improved ductility in these intermetallics by disordering [5] grain refinement, and development of fine dispersoid particles. When these fine dispersoids are oxides an additional gain in ductility can result from deoxidation of the matrix.

Early work on RS of this alloy class utilized TiAl-W alloys atomized by the PREP process. This material exhibited good creep resistance due to a fine precipitate of tungsten rich beta phase in the alpha-2 + gamma lath structure of the TiAl composition [110].

Rapid solidification of alloys with a base composition of $Ti_3Al$ with the addition of 0.4 at%Er has been shown to produce a fine dispersoid of $Er_2O_3$, Figure 12 [61, 111]. This dispersoid does not coarsen in grain interiors to temperatures as high as 1000°C (1830°F) [111]. Under HIP conditions $Er_2O_3$ dispersoid coarsening in a $Ti_3Al$-Nb alloy was comparable to that observed during an anneal at the same temperature. However extrusion caused rapid coarsening; which was contrasted with cerium sulfide or cerium oxysulfide containing alloys where virtually no coarsening occurred [60, 61]. The presence of a dispersoid in $Ti_3Al$-Nb alloys also produced a refinement in grain size. The ductility and fracture toughness of dispersoid containing alloys was dependent on the test used. Hardness indentation ductility was improved by grain refinement but fracture toughness was not [60]. Much further work is required in this generic system to optimize the microstructure and hence mechanical properties.

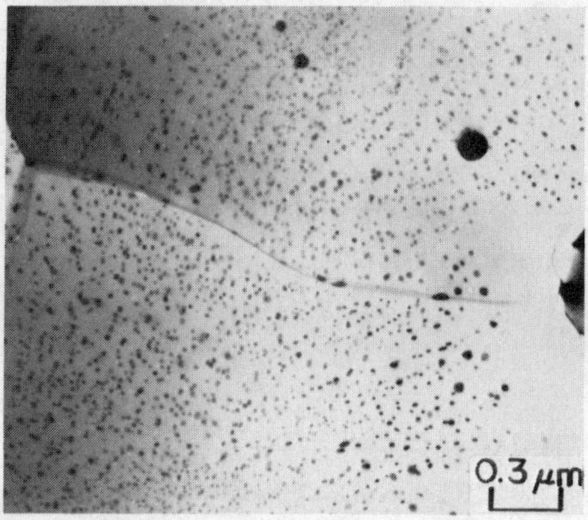


Figure 12. Fine stable dispersion of $Er_2O_3$ particles in $Ti_3Al$ aged 750°C (1380°F)/1 hr [60].

Other Systems

Other alloy elements which could be added to titanium to give attractive characteristics are the low density elements Li, Mg (Al and Si have a similar effect). The challenge here is to successfully add these elements which have a boiling point below the melting point of titanium. However work is in progress in a number of laboratories to achieve these alloying possibilities by innovative melting techniques [112, 113], use of the ALTi powder making method [37], blended elemental approaches [114], mechanical alloying [115], and direct production from the vapor phase [116].

## CONCLUSIONS AND FUTURE THOUGHTS

The study of RS titanium alloys has got off to a later start than in some other systems such as aluminum primarily because of problems such as the reactivity of molten titanium. However there is obviously much to be gained from this approach. Conventional alloys can be produced with refined and more desirable microstructures. The addition of the generally catastropic rare earths and metalloids can be greatly refined and made useful via RS. Segregation in alloys containing eutectoid formers can be controlled much better using the RS technique making this alloy class particularly attractive for high strength applications. Preliminary work suggests that new and improved beta alloys can be synthesized by RS. The surface has only just been scratched in using amorphous glassy materials as a precursor to production of useful crystalline alloys. Some of the inherent problems associated with the ordered aluminides may be circumvented using much faster cooling rates. Finally, development of lower density titanium alloys may be possible with RS.

The technology of RS titanium has effectively been explored for only about a half decade. Yet already a myriad of very attractive avenues for exploration have been defined. In concert with coating developments for high temperature applications, and useful fiber reinforcement, this technology should see many applications within the next decade.

## ACKNOWLEDGMENTS

The authors would like to acknowledge useful discussions and comments from their colleagues: H. I. Aaronson, H. B. Bomberger, R. R. Boyer, T. F. Broderick, D. Eylon, H. L. Fraser, S. Fujishiro, M. F. X. Gigliotti, F. J. Hehmann, J. P. Hirth, A. G. Jackson, H. Jones, Y-W. Kim, E. F. Koch, D. G. Konitzer, S. Krishnamurthy, H. A. Lipsitt, Y. R. Mahajan, I. A. Martorell, R. E. Omlor, S. M. L. Sastry, S. J. Savage, P. R. Smith, D. B. Snow, J. A. Sutliff, K. R. Teal, R. G. Vogt, I. Weiss, S. H. Whang, J. C. Williams, and C. F. Yolton. In addition we would like to recognize the high level of efficiency of Ms. Karen A. Sitzman in typing the text.

## REFERENCES

1. R. M. German, Powder Metallurgy Science (MPIF, Princeton, NJ, 1984).

2. F. H. Froes and J. R. Pickens, J. of Metals 36 (1), 14-28 (1984).

3. F. H. Froes and D. Eylon, in Titanium, Science and Technology, Vol. 1, edited by G. Lutjering, U. Zwicker, and W. Bunk (DGM, Oberursel, West Germany, 1985), pp. 267-286.

4. Titanium Technology: Present Status and Future Trends, edited by F. H. Froes, D. Eylon, and H. B. Bomberger (The Titanium Development Association, Dayton, OH, 1985).

5. H. Jones, Rapid Solidification of Metals and Alloys, Monograph No. 8, (Institution of Metallurgists, London, 1982).

6. S. J. Savage and F. H. Froes, J. of Metals 36 (4), 20-33 (1984).

7. H. Margolin and J. P. Nielson, in Modern Materials, Advances in Development and Application, Vol. 2, edited by H. H. Hauser (Academic Press, 1960), pp. 225-325.

8. E. K. Molchanova, Phase Diagrams of Titanium Alloys, (Israel Program for Scientific Translations, Jerusalem, 1965).

9. S. H. Whang and C. S. Chi, presented at the MRS Fall Meeting, Boston, MA, 1985 (and to be published).

10. C. S. Chi and S. H. Whang (private communication, 1985).

11. S. H. Whang (private communication, 1985).

12. J. E. O'Neal, S. M. L. Sastry, T. C. Peng, and J. F. Tesson, Micro. Sci. 11, 143 (1983).

13. S. H. Whang, Y. Z. Lu, and Y-W. Kim, J. Mat. Sci. Letts. 4, 883 (1985).

14. A. G. Jackson, PhD thesis, University of Cincinnati, 1983.

15. S. M. L. Sastry, P. J. Meschter, and J. E. O'Neal, Met. Trans. A 15A, 1451 (1984).

16. D. G. Konitzer, B. C. Muddle, and H. L. Fraser, Met. Trans. A 14A, 1979 (1983).

17. Y. Z. Lu, C. S. Chi, and S. H. Whang, in Rapidly Quenched Metals, Vol. 1, edited by S. Steeb and H. Warlimont (Elsevier, Amsterdam, 1985), pp. 949-952.

18. S. H. Whang, U.S. Patent No. 4 512 826 (23 April 1985).

19. C. E. Holcombe and T. R. Serandos, Met. Trans. B 14B, 497 (1983).

20. S. J. Noesen, in International Transactions Vacuum Metallurgy Conference 1967, edited by E. L. Foster (Am. Vac. Soc., New York, 1967), p. 503.

21. T. F. Broderick, A. G. Jackson, H. Jones, and F. H. Froes, Met. Trans. A, 16A, 1951 (1985).

22. C. H. Smith, General Electric Company (Schenectady, NY) Report No. 84CRD071, 1984.

23. D. G. Konitzer, K. W. Walters, E. L. Heiser, and H. L. Fraser, Met. Trans. B 15B, 149 (1984).

24. T. C. Peng, S. M. L. Sastry, and J. E. O'Neal, Met. Trans. A 16A, 1897 (1985).

25. D. G. Konitzer and H. L. Fraser, Proceedings of the Powder Metallurgy Conference, Toronto, Ontario, Canada, 1984 (to be published).

26. N. F. Anoshkin (private communication, 1980).

27. S. M. L. Sastry, T. C. Peng, P. J. Meschter, and J. E. O'Neal, J. of Metals 35 (9), 21-28 (1983).

28. C. F. Yolton and J. H. Moll, U.S. Patent No. 4 544 404 (1 October 1985).

29. C. F. Yolton and J. H. Moll, presented at the First International ASM Conference on Rapidly Solidified Materials, San Diego, CA, 1986 (and to be published).

30. R. G. Rowe, T. F. Broderick, E. F. Koch, and F. H. Froes, presented at the First International ASM Conference on Rapidly Solidified Materials, San Diego, CA, 1986 (and to be published).

31. M. F. X. Gigliotti, R. G. Rowe, G. E. Wasielewski, G. K. Scarr, and J. C. Williams, presented at the MRS Fall Meeting, Boston, MA, 1985 (and to be published).

32. R. G. Rowe and E. F. Koch, presented at the First International ASM Conference on Rapidly Solidified Materials, San Diego, CA, 1986 (and to be published).

33. R. Ray, U.S. Patent No. 4 471 831 (18 September 1984).

34. S. H. Whang, C. S. Chi, and Y. Z. Lu, in Rapidly Quenched Metals, Vol. 1, edited by S. Steeb and H. Warlimont (Elsevier, Amsterdam, 1985), pp. 115-118.

35. R. B. Pond and J. M. Winter, Mat. Sci. Engr. 23, 87 (1976).

36. T. C. Peng, S. M. L. Sastry, J. E. O'Neal, and J. F. Tesson, in Lasers in Materials Processing, edited by E. A. Metzbower (ASM, Metals Park, OH, 1983), p. 241.

37. F. H. Hayes, H. B. Bomberger, F. H. Froes, L. Kaufman, and H. M. Burte, J. of Metals 36 (6), 70-76 (1984).

38. D. Raybould, in Progress in Powder Metallurgy 1982, Vol. 38, edited by J. G. Bewley and S. W. McGee (MPIF, Princeton, NJ, 1983), pp. 575-592.

39. O. V. Roman and V. G. Gorobtsov, in Shock Waves and High-Strain-Rate Phenomena in Metals, edited by M. A. Meyers and L. E. Murr (Plenum Press, New York, 1980).

40. V. D. Linse, in Sagamore Army Materials Research Conference Proceedings, edited by Gordon Bruggeman and Volker Weiss (1983), pp. 381-404.

41. O. V. Roman, Y. V. Boiko, V. G. Gorobtsov, and V. I. Shelegov, in Titanium, Science and Technology, Vol. 1, edited by G. Lutjering, U. Zwicker, and W. Bunk (DGM, Oberursel, West Germany, 1985), pp. 333-337.

42. B. A. Kolachev and V. K. Nosov, in Titanium, Science and Technology, Vol. 1, edited by G. Lutjering, U. Zwicker, and W. Bunk (DGM, Oberursel, West Germany, 1985), pp. 625-631.

43. K. R. Teal, A. G. Jackson, and F. H. Froes (work in progress, 1984-1986).

44. M. F. X. Gigliotti (private communication, 1985).

45. T. F. Broderick, F. H. Froes, and A. G. Jackson, in Rapidly Solidified Metastable Materials, Vol. 28, edited by B. H. Kear and B. C. Giessen (Elsevier Science Publishers, New York, 1984), p. 345.

46. P. G. Boswell and G. A. Chadwick, Scripta Met., 2, 459 (1977).

47. A. F. Belov and I. S. Polkin, "Modern Trends in Titanium Production and Processing," Germany Metallurgical Society Workshop, University of Nuremberg, Erlangen, July 1982.

48. T. C. Peng, S. M. L. Sastry, and J. E. O'Neal, in Lasers in Metallurgy, edited by K. Mukherjee and J. Mazumder (TMS Publications, New York, 1981), pp. 279-292.

49. J. E. O'Neal, T. C. Peng, and S. M. L. Sastry, in Proceedings of 39th Annual EMSA Meeting, edited by G. W. Bailey (Claitor's Publishing Division, Baton Rouge, LA, 1981), pp. 66-67.

50. B. C. Muddle, D. G. Konitzer, and H. L. Fraser, in Strength of Metals and Alloys, Vol. 1, edited by R. C. Gifkins (Pergamon Press, Oxford, 1983), pp. 313-318.

51. I. G. Wright and B. A. Wilcox, Battelle Columbus Laboratories Report No. AD-781-133, 1974.

52. D. B. Snow, in Laser Processing of Materials, edited by K. Mukherjee and J. Mazumder (TMS Publications, Warrendale, PA, 1984), pp. 83-98.

53. D. G. Konitzer, B. C. Muddle, R. Kirchheim, and H. L. Fraser, in Rapidly Quenched Metals, Vol. 1, edited by S. Steeb and H. Warlimont (Elsevier, Amsterdam, 1985) pp. 953-956.

54. D. G. Konitzer, B. C. Muddle, and H. L. Fraser, Scripta Met. 17, 963 (1983).

55. D. G. Konitzer, B. C. Muddle, H. L. Fraser, and R. Kirchheim, in Titanium, Science and Technology, Vol. 1, edited by G. Lutjering, U. Zwicker, and W. Bunk (DGM, Oberursel, West Germany, 1985), pp. 405-410.

56. D. G. Konitzer (private communication, 1985).

57. S. H. Whang, J. of Metals 36 (4), 34-40 (1984).

58. H. L. Fraser (private communication, 1985).

59. S. H. Whang, "RS Titanium Alloys for High Temperature Applications," (to be published in J. Mat. Sci.).

60. R. G. Rowe, J. A. Sutliff, and E. F. Koch, presented at the 1985 MRS Fall Meeting, Boston, MA, 1985 (and to be published).

61. J. A. Sutliff and R. G. Rowe, presented at the 1985 MRS Fall Meeting, Boston, MA, 1985 (and to be published).

62. L. M. Brown and R. K. Ham, in Strengthening Methods in Crystals, edited by A. Kelly and R. B. Nicholson (Elsevier, New York, 1971), p. 9.

63. S. M. L. Sastry, T. C. Peng, and L. P. Beckerman, Met. Trans. A 15A, 1465 (1984).

64. R. G. Vogt, D. Eylon, and F. H. Froes (work in progress, 1983-1986).

65. S. M. L. Sastry, T. C. Peng, P. J. Meschter, and J. E. O'Neal, J. of Metals 35 (9), 21-28 (1983).

66. S. M. L. Sastry, T. C. Peng, and J. E. O'Neal, in Titanium, Science and Technology, Vol. 1, edited by G. Lutjering, U. Zwicker, and W. Bunk (DGM, Oberursel, West Germany, 1985), pp. 397-404.

67. A. G. Jackson and F. H. Froes, in Titanium, Science and Technology, Vol. 1, edited by G. Lutjering, U. Zwicker, and W. Bunk (DGM, Oberursel, West Germany, 1985), pp. 381-387.

68. C. S. Chi and S. H. Whang, in Rapidly Solidified Metastable Materials, Vol. 28, edited by B. H. Kear and B. C. Giessen (Elsevier Science Publishers, New York, 1984), pp. 353-360.

69. P. R. Frausto, A. G. Jackson, A. H. Clauer, and J. L. McCall, in Microstructural Science, Vol. 10, edited by White, Richardson, and McCall (Elsevier Science Publishers, New York, 1982), pp. 103-112.

70. S. H. Whang, Y. Z. Lu, and B. C. Giessen, in Rapidly Solidified Metastable Materials, Vol. 28, edited by B. H. Kear and B. C. Giessen (Elsevier Science Publishers, New York, 1984), pp. 367-373.

71. G. W. Franti, J. C. Williams, and H. I. Aaronson, Met. Trans. A 9A (11), 1641 (1978).

72. H. B. Bomberger and F. H. Froes, AFWAL Report No. TR-84-4164, 1985.

73. J. Gross, Mallory-Sharon Titanium Corporation Internal Report No. 1000R164, 1956.

74. P. A. Farrar and H. Margolin, Army Materials and Mechanics Research Center Report No. WAL TR 401/303-13, 1967.

75. D. B. Hunter, AFWAL Report TR-405/2-15, Part II, 1966.

76. S. Krishnamurthy, R. G. Vogt, D. Eylon, and F. H. Froes, in Progress in Powder Metallurgy, Vol. 39, edited by H. S. Nayar, S. M. Kaufman, and K. E. Meiners (MPIF, Princeton, NJ, 1984), pp. 603-623.

77. E. W. Collings, R. E. Maringer, and C. E. Mobley, AFML Report No. TR-78-70, 1978.

78. S. M. L. Sastry, T. C. Peng, and J. E. O'Neal, presented at the 1984 MPIF/APMI International PM Conference, Toronto, Ontario, Canada (to be published).

79. T. C. Peng, S. M. L. Sastry, and J. E. O'Neal, in Rapid Solidification Technology III, edited by R. Mehrabian (NBS, Gaithersburg, MD, 1982), pp. 452-457.

80. W. A. Baeslack III, S. Krishnamurthy, and F. H. Froes, in Strength of Metals and Alloys, Vol. 2, edited by H. J. McQueen, J-P. Bailon, J. I. Dickson, J. J. Jonas, and M. G. Akben (Pergamon Press, Oxford, 1985), pp. 1633-1638.

81. S. Krishnamurthy, A. G. Jackson, H. Jones, and F. H. Froes, "Beta-Eutectoid Decomposition in Rapidly Solidified Titanium-Nickel Alloys," (submitted to Met. Trans., 1984).

82. S. Krishnamurthy, A. G. Jackson, D. Eylon, R. R. Boyer, and F. H. Froes, in Rapidly Quenched Metals, Vol. 1, edited by S. Steeb and H. Warlimont (Elsevier Science Publishers, Amsterdam, 1985), pp. 945-948.

83. S. Krishnamurthy, I. Weiss, D. Eylon, and F. H. Froes, in Strength of Metals and Alloys, Vol. 2, edited by H. J. McQueen, J-P. Bailon, J. I. Dickson, J. J. Jonas, and M. G. Akben (Pergamon Press, Oxford, 1985), pp. 1627-1632.

84. W. A. Baeslack III, L. Weeter, S. Krishnamurthy, P. R. Smith, and F. H. Froes, in Rapidly Solidified Metastable Materials, Vol. 28, edited by B. H. Kear and B. C. Giessen (Elsevier Science Publishers, NY, 1984), pp. 375-379.

85. S. Krishnamurthy, R. G. Vogt, D. Eylon, and F. H. Froes, in Rapidly Solidified Metastable Materials, Vol. 28, edited by B. H. Kear and B. C. Giessen (Elsevier Science Publishers, NY, 1984), pp. 361-366.

86. W. A. Baeslack III, S. Krishnamurthy, and F. H. Froes, "A Study of Rapidly-Quenched Microstructures in a Hypereutectoid Ti-40 wt%W Alloy," (accepted for publication in J. of Material Science, 1985).

87. S. Krishnamurthy, A. G. Jackson, I. Weiss, and F. H. Froes, "Aging Response of Rapidly Solidified Titanium-Tungsten Alloys with Nickel and Silicon Additions," presented at the First International ASM Conference on Rapidly Solidified Materials, San Diego, CA, February 1986 (and to be published).

88. S. Krishnamurthy, D. Eylon, I. Weiss, and F. H. Froes, "Age Hardening Behavior of a Rapidly Solidified Ti-15Cr-4Al Alloy," presented at the TMS-AIME Conference on Mechanical Behavior of RS Material, New York City, February 1985 (and to be published).

89. S. Krishnamurthy and F. H. Froes (work in progress, 1985).

90. R. G. Vogt, P. R. Smith, D. Eylon, and F. H. Froes (work in progress, 1983-1986).

91. T. F. Broderick, F. H. Froes, and J. A. Snide (work in progress, 1985-1986).

92. T. F. Broderick and F. H. Froes (work in progress, 1984-1986).

93. F. H. Froes and P. R. Smith (unpublished work, 1985).

94. M. Gutierrez, I. A. Martorell, and F. H. Froes (work in progress, 1986).

95. D. E. Polk and B. C. Giessen, Metallic Glasses, (ASM, Metals Park, OH, 1978), pp. 1-35.

96. R. S. Carbonara, T. A. Gaspar, R. V. Raman, R. E. Maringer, and J. L. McCall, MCIC Report No. 81-45, 1981.

97. Yu. K. Kovneristy, E. K. Osipov, E. A. Trofimova, A. A. Baikov, N. D. Tomashov, and I. B. Skvortsova, in Titanium, Science and Technology, Vol. 1, edited by G. Lutjering, U. Zwicker, and W. Bunk (DGM, Oberursel, West Germany, 1985), pp. 203-208.

98. L. E. Tanner and R. Ray, Scripta Met. 11, 783 (1977).

99. L. E. Tanner, Scripta Met. 12, 703 (1978).

100. L. E. Tanner and Ranjan Ray, Scripta Met. 14, 657 (1980).

101. L. E. Tanner and R. Ray, Acta Met. 27, 1727 (1979).

102. A. Inoue et al., in Proceedings of the Industrial Conference on Titanium, (1980), pp. 1936-1944.

103. S. H. Whang and Y. Z. Lu, in Proceedings of RSP, Principles and Technologies, III, edited by R. Mehrabian (NBS, Gaithersburg, MD, 1983), pp. 286-290.

104. S. H. Whang, Mat. Sci. Eng. 57, 87 (1983).

105. S. H. Whang, J. Non-Cryst. Sol. 61 and 62, 841 (1984).

106. S. H. Whang, Scripta Met. 18, 309 (1984).

107. A. Inoue, H. Kimura, S. Sakai, and T. Masumoto, in Titanium '80, Science and Technology, Vol. 2, edited by H. Kimura and O. Izumi (TMS Publications, Warrendale, PA, 1980), pp. 1137-1145.

108. C. Suryanarayana, A. Inoue, and T. Masumoto, in Titanium '80, Science and Technology, Vol. 1, edited by H. Kimura and O. Izumi (TMS Publications, Warrendale, PA, 1980), pp. 699-707.

109. Y. Q. Gao and S. H. Whang, J. Non-Cryst. Sol. 70, 85 (1985).

110. P. L. Martin, M. G. Mendiratta, and H. A. Lipsitt, Met. Trans. A 14A, 2170 (1983).

111. D. G. Konitzer and H. L. Fraser, presented at the 1984 MRS Fall Meeting, Boston, MA, 1984 (and to be published).

112. H. B. Bomberger and F. H. Froes (work in progress, 1983-1986).

113. C. F. Yolton, P. R. Smith, H. B. Bomberger, and F. H. Froes (work in progress, 1985-1986).

114. S. M. L. Sastry (work in progress, 1985-1986).

115. Y. R. Mahajan, F. H. Froes, H. B. Bomberger (work in progress, 1984-1986).

116. C. Dekock, F. H. Froes, and R. G. Vogt (work in progress, 1985).

# SURFACE TREATMENT OF TITANIUM USING LASERS

B.L. MORDIKE AND H.W. BERGMANN
Institut für Werkstoffkunde und Werkstofftechnik, Technische Universität, Agricolastr. 2, 3392 Clausthal-Zellerfeld, FRG

## ABSTRACT

Thermochemical treatments of titanium surfaces by laser are presented and the beneficial effects attainable eg wear resistance, corrosion resistance etc., discussed. It is shown that it is possible to produce titanium surfaces with load bearing capability.

## INTRODUCTION

Titanium is renowned for its high strength/weight ratio and has been used, therefore, for high duty aircraft applications. It exhibits good corrosion resistance and is non-magnetic and thus suitable for marine and submarine applications. There is an ever increasing amount of recycled, secondary titanium available which cannot be employed in the aircraft industry because of the stringent and conservative specifications but which nevertheless offers considerable advantages in other fields over current materials - ferrous and nickel base for example. The present work shows how the disadvantages of titanium, namely high friction and poor wear resistance can be overcome by laser thermal chemical treatment.

## PRINCIPLES

Titanium can be hardened by solid solution hardening, or precipitation/dispersion hardening. It is not possible to harden titanium by martensitic hardening and interstitial hardening usually causes embrittlement.

Solid solutions can be produced in the surface layer by melting-in a pre-coated layer (e.g. D-gun or electroplated Ni, Fe, Cr, etc.). Precipitation hardened surfaces can be produced by either injecting particles into the melt [1] or by alloying the laser heated surface to react with the atmosphere to produce precipitates. The reaction with the atmosphere can be described by the law of mass action [2]. It is thus possible by fixing the activity (fugacity) of the reactant gas to determine the degree of reaction and direction of the reaction.

For instance, $$2Ti + N_2 \rightleftarrows 2TiN$$

If the surface is melted the volume of material precipitating in the reaction is much larger, as convection ensures a homogenization of the molten layer. Convection has the advantage that the enrichment takes place over the whole melted layer, typically a few millimeters compared with microns in CVD or other techniques [3,4,5,6].

## EXPERIMENTAL DETAILS

Various titanium alloys were produced in the surface by laser surface treatment. In thermochemical treatment an appropriate gas atmosphere is produced which permits the heated surface to react in a controlled manner. The essential parameter is the partial pressure of the reactant gas. Smooth crack free surfaces could be produced by ensuring lamellar flow of the reactant gas across the heated surface and the inclination and azimuthal angles. The following properties were determined: mechanical properties (tensile, hardness and fatigue tests), wear, corrosion and component behaviour. The surfaces were investigated for crystallographic structure, microstructure, chemical composition and topography.

## RESULTS

Fig. 1 shows the equilibrium diagram for TiN and TiC. In both cases a hard and high melting point compound exists and can be formed on the surface by heating in the appropriate atmosphere.

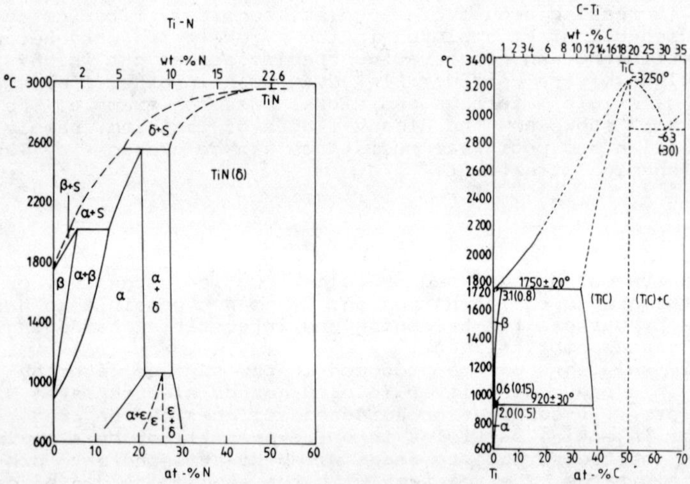


Fig. 1 Equilibrium diagrams for TiN and TiC

Fig. 2 shows the structure of a TiN layer from the surface to the substrate with the corresponding hardness profile. The layer consists of TiN dendrites, the amount of which decreases with increasing distance from the surface and this explains the decrease in hardness.

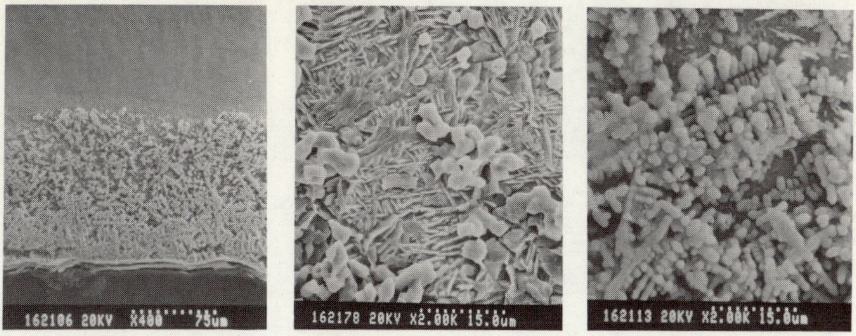


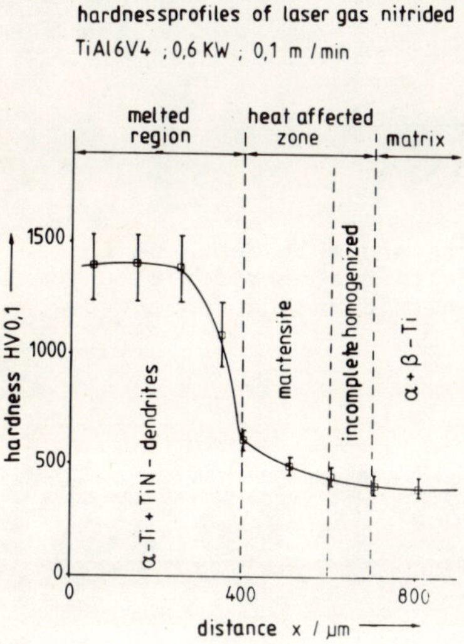


Fig. 2:
Structure of a TiN layer
a) general view,
b) dendritic region,
c) transition region,
d) hardness profile

Fig. 3 shows the difference in structure between TiN and TiC layers. In the first case the layer consists of TiN dendrites and substrate whereas in the second TiC also precipitates from solution.

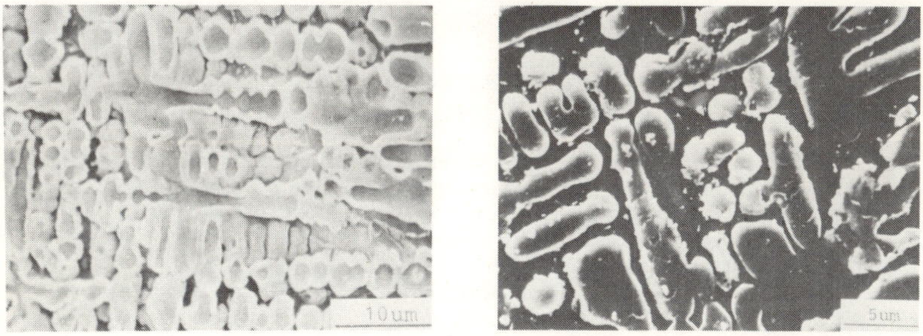


Fig. 3 Difference in structure between a) TiN and b) TiC

Fig. 4 shows the influence of different treatment gases, preheating and feed rate on the melted depth for a 500 W laser.

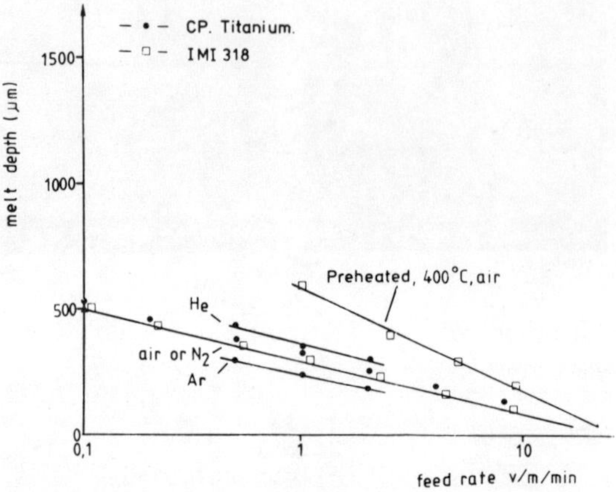


Fig. 4:
Influence of different treatment gases, preheating and feed rate on the melted depth

The dependence of microstructure and melt depth on the laser power is demonstrated in Fig. 5. Melting a nitrided titanium specimen in an inert gas causes denitriding and hence a decrease in the hardness, Fig. 6.

Fig. 5: Effect of different laser power
a) $CO_2$, 500 W, CW, b) $CO_2$, 5 kW, CW

The corrosion behaviour of nitrided titanium is compared with that of the untreated material in Fig. 7. The nitrided specimen passivates at much lower current densities (10 - 100 times less). Subsequent denitriding restores the behaviour of the untreated specimen.

The gas alloyed specimens exhibit in general better wear resistance and lower coefficients of friction (3). Detailed examinations are still in progress.

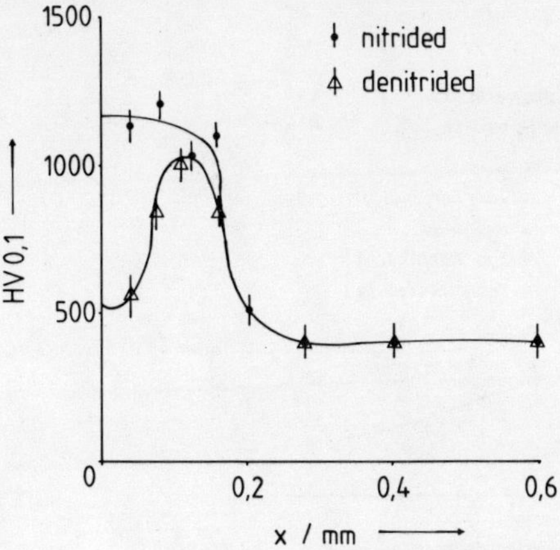


Fig. 6:
Effect of the nitrogen partial pressure

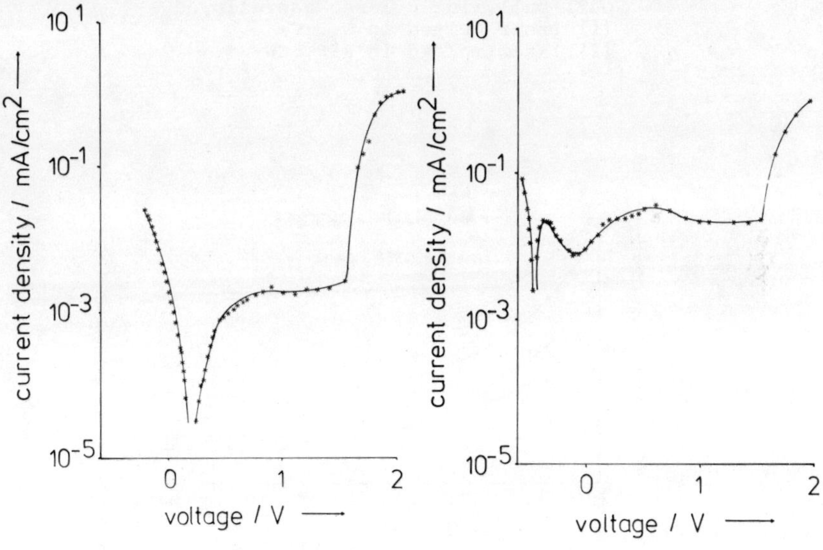


Fig. 7 Corrosion behaviour of treated and untreated titanium in 0.1 N HCl
a) nitrided material b) initial material

Laser surface melting with and without gas alloying reduces the fatigue resistance. Fig. 8 shows the influence on the fatigue in pull-pull and rotating bending.

Application of the technique is illustrated in Fig. 9 by rolls and gear box synchronous rings.

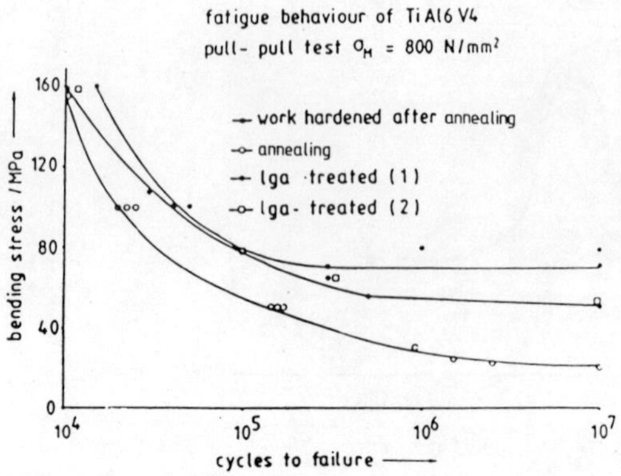


Fig. 8a: Influence of laser melting on the fatigue pull-pull, lga = laser gas alloyed
(1) laser melted in $N_2$, CW
(2) laser melted in air, CW

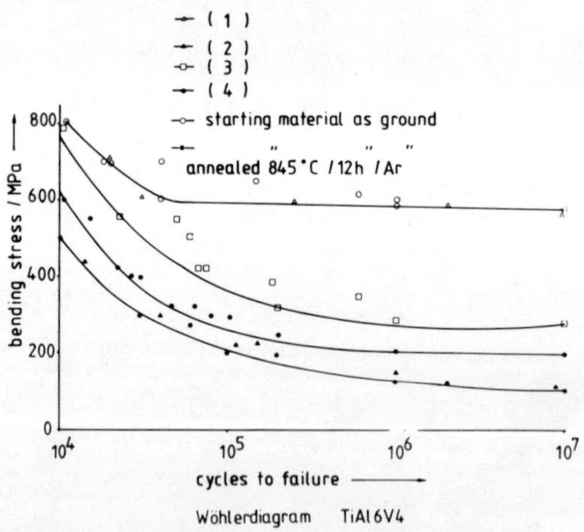


Fig. 8b: Influence of laser melting on the fatigue rotating bending
(1) laser melted in air, CW
(2) laser melted in $N_2$, CW
(3) laser melted in He, CW
(4) laser melted in $N_2$, pulsed

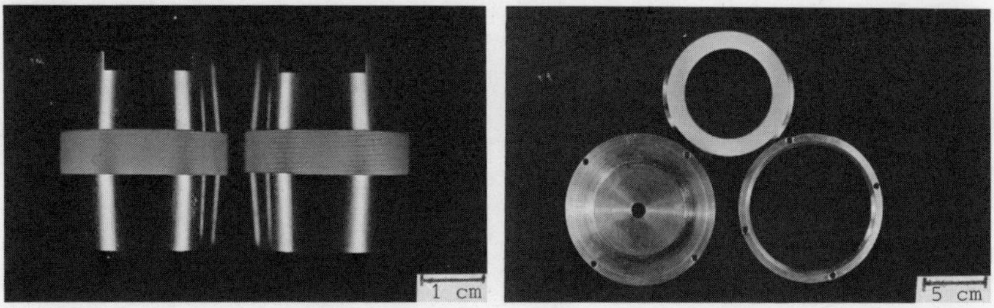


Fig. 9: Examples for application

## CONCLUSIONS

The nitriding or carburizing treatment offers distinct advantages. It must be decided in each case what depth and composition is required for the particular application. Methods must be devised whereby the deterioration in fatigue properties can be overcome or minimised.

## ACKNOWLEDGEMENT

The authors are indebted to the government of Lower Saxony for financial assistance and also BMFT and the Volkswagen Foundation. Mr. Lee is be thanked for carrying out some of the experiments.

## REFERENCES

1. K.P. Cooper and J.D. Ayers, Surface Eng. Vol. 1 No. 4, 263 (1985)
2. B.L. Mordike, Conf. on Laser Surface Treatment of Metals, San Miniato, Italien, 1985, to be published
3. T. Bell and H.W. Bergmann, Heat Treatment 1985, London, to be published
4. H.W. Bergmann, Z. Werkstofftechnik, in print
5. H.W. Bergmann, T. Bell and S.Z. Lee, Proc. 7th Int. Cong. and Int. Trade Fair with 2nd Int. Nd-YAG Laser Conf., München 1985, to be published
6. W. Steen, Surface Eng. Vol. 1 No. 1, 23 (1985)

# TENSILE AND CREEP PROPERTIES OF RAPIDLY SOLIDIFIED TITANIUM ALLOYS CONTAINING COMPLEX MATRICES AND FINE DISPERSOIDS

M.F.X.Gigliotti*, R.G.Rowe*, G.E.Wasielewski**, G.K.Scarr**, and J.C.Williams***

* General Electric Corporate Research and Development, Schenectady, NY
** General Electric Aircraft Engine Business Group, Evendale, OH
*** Carnegie-Mellon University, Pittsburgh, PA

## ABSTRACT

Titanium alloys containing alpha matrices with various rare earth compound dispersions were prepared by melt extraction. The melt extracted material was HIPped, extruded, and thermally exposed to yield various microstructures including transformed beta and equiaxed alpha. Tensile and creep tests were conducted on material in the different microstructural conditions. Trends in mechanical behavior as a function of alloy chemistry and process history are discussed. The alloys with a transformed beta microstructure had superior creep resistance. Alloys containing dispersoids had better tensile ductilities compared with those which did not contain a dispersoid.

Portions of this work were carried out under Air Force Contract F33615-83-C-5034.

## INTRODUCTION

Rapid solidification and deformation processing of titanium alloys containing rare earth additions have been used to produce a fine dispersion of the rare earth phase in titanium alloys [1-3]. Various binary and ternary titanium alloys have been evaluated to assess the role of rare earth compound dispersoids on strengthening [4].

Elements of the lanthanide series, or rare earth elements, have low solubilities in alpha titanium, and appear to be good dispersion strengthening candidates. For those rare earth elements whose phase diagrams with titanium are known, there appear two distinct alloy classes. If the rare earth's melting point is less than 1100 C, the binary diagram with titanium is a monotectic. If the rare earth's melting point is above 1300 C, the binary system is a simple eutectic. Since there is little solubility of rare earths in titanium, there will be liquid rare earth phase above the minimum melting point of the alloy.

Many rare earths have too low a melting point to be useful dispersiods in titanium, since those molten below 1000 C would coarsen during consolidation or thermomechanical processing. However, in the presence of oxygen, stability is likely to be enhanced by oxide formation. This should allow low melting point rare earth elements to form oxides and exhibit stability above their melting point. Those rare earths most promising would include Gd and Er, and by extension -- Tb, Dy, Y, Ho, Tm and Lu.

Some interest has been shown in rare earth additions using conventional ingot metallurgy. Kaschuk and Svetlov [5] found considerable grain refining in a Ti - 5 Al alloy with minor additions of rare earths, with an increase in tensile strength and oxidation resistance.

Sastry et al., using rapid solidification (r.s.), have shown that higher levels of rare earth additions produce dispersoids which act to strengthen the alloy [6]. Further, Sastry et al. found that Er was a preferred addition to titanium, because of a lower rate of coarsening than other rare earths [7], and that the dispersoids can be formed in Ti - Al alloys as well [8].

Most recently, Konitzer et al. have reported the dispersoid formed by erbium metal additions to titanium to be erbium sesquioxide [9].

It is the purpose of this work to trace the structure development of rare earth-containing multielement titanium alloys, and the mechanical properties of these alloys after different thermal treatments.

## EXPERIMENTAL PROCEDURE

The alloys for evaluation are listed in Table I. There are two series of alloys each beginning with an essentially alpha titanium base at two aluminum levels. The alloy bases are called TE and UT.

The first series base, TE, has the composition Ti- 9 a/o Al-1.5 a/o Zr-1 a/o Sn. This base was utilized to characterize microstructure. Alloys TY and TZ permit comparison of rare earth additions at concentrations below and above that necessary to tie up all the oxygen as a sesquioxide. Alloy UQ is a 1 a/o Y addition to the base allowing a comparison between Y and Er. UK and UR enable an evaluation of refractory elements.

The second base, UT, has the nominal composition Ti- 12.2 a/o Al-1.3 a/o Zr- 1.2 a/o Sn- 0.5 a/o Cb-0.1 a/o Mo. Alloy UV is a UT modification containing Si. Alloy UW contains an Er addition, and alloy UX contains Ce plus S.

The alloys were prepared by non-consumable arc melting and drop casting 1.2 cm diameter by 5 cm long rods into copper chill molds. Next, the alloys were rapidly solidified by pendant drop melt extraction in vacuo. The melt extraction was accomplished by melting the top of a vertically oriented alloy rod with an electron beam and extracting filament with a chill wheel rotating in the horizontal plane.

The rapidly solidified alloys were HIPped in steel cans and extruded in steel jackets. The experimental conditions are listed in Table II. After extrusion, the steel was machined off the titanium alloys, and samples of the alloys encapsulated for thermal exposures. Thermal exposure samples were placed in titanium tubes welded closed in an argon glove box. Each titanium tube was then sealed in an evacuated fused silica tube.

Table I
Titanium Alloy Compositions -- atom %

| Alloy | Al | Zr | Sn | Cb | Mo | Other |
|---|---|---|---|---|---|---|
| TE | 9.0 | 1.5 | 1.0 | | | |
| TY | 9.0 | 1.5 | 1.0 | | | 0.3 Er |
| TZ | 8.9 | 1.5 | 1.0 | | | 1.0 Er |
| UA | 9.0 | 1.5 | 1.0 | | | 0.3 Er, 0.1 B |
| UK | 8.9 | 1.5 | 1.0 | 1.0 | | 0.7 Er |
| UQ | 8.9 | 1.5 | 1.0 | | | 1.0 Y |
| UR | 8.9 | 1.5 | 1.0 | | | 1.5 Ta, 1.0 Y |
| UT | 12.2 | 1.3 | 1.2 | 0.5 | 0.1 | |
| UV | 12.2 | 1.3 | 1.2 | 0.5 | 0.1 | 0.3 Si |
| UW | 12.1 | 1.3 | 1.2 | 0.5 | 0.1 | 0.6 Er |
| UX | 12.1 | 1.3 | 1.2 | 0.5 | 0.1 | 0.3 Ce, 0.2 S |

Table II
Process Details

| Alloy Name | HIP Conditions | Extrusion Conditions |
|---|---|---|
| TY,TZ,UA, | 840 C/207 MPa | 840 C/8:1 |
| UK,UK,UR, | 3 hr | |
| UT,UV,UW, | 1550 F/30 ksi | 1550 F/8:1 |
| UX | 3 hr | |
| UT,UV,UW, | 900 C/207 MPa | 900 C/8:1 |
| UX | 3 hr | |
| | 1650 F/30 ksi | 1650 F/8:1 |

Tensile and creep tests were conducted using a specimen with a 1.3 cm (0.5 in.) gage length and 0.20 cm (0.80 in.) gage diameter. Final machining of gage and threaded ends was done after thermal exposure.

Transmission electron microscopy was used to characterize the microstructure of the consolidated materials.

## RESULTS

Chemical Analysis. Melt extracted filaments or consolidated pieces of each of the alloys were analyzed by wet chemistry. The results are listed in Table III. There was a consistent loss in Al between the amount added and that analyzed in the r.s. product. Similarly there was a loss in the rare earth. The input chemistry of alloys TY, TZ, UA, UQ, and UR was the same as the aim. The input chemistry of alloys UT, UV, UW, and UX had excess Al added to compensate for losses.

Table III
Analyzed R.S. Alloy Compositions -- atom %

| Alloy | Al | Zr | Sn | Cb | Mo | Other |
|---|---|---|---|---|---|---|
| TYadd | 8.97 | 1.50 | 1.00 | | | 0.30Er |
| anal. | 7.81 | 1.27 | 0.98 | | | 0.17Er |
| TZadd | 8.91 | 1.49 | 0.99 | | | 1.00Er |
| anal. | 8.33 | 1.29 | 0.99 | | | 0.66Er |
| UAadd | 8.96 | 1.49 | 1.00 | | | 0.30Er,0.10B |
| anal. | 7.96 | 1.32 | 1.00 | | | 0.19Er,0.13B |
| UKadd | 8.94 | 1.49 | 0.99 | 0.99 | | 0.67Er |
| anal. | 8.51 | 1.40 | 0.98 | 1.05 | | 0.30Er |
| UQadd | 8.91 | 1.49 | 0.99 | | | 1.00Y |
| anal. | 8.15 | 1.32 | 0.99 | | | 0.83Y |
| URadd | 8.94 | 1.49 | 0.99 | | | 1.49Ta,0.67Y |
| anal. | 8.30 | 1.38 | 0.91 | | | 1.52Ta,0.55Y |
| UTadd | 13.66 | 1.28 | 1.18 | 0.49 | 0.10 | |
| anal. | 10.63 | 0.98 | 1.20 | 0.44 | 0.08 | |
| UVadd | 13.63 | 1.28 | 1.18 | 0.49 | 0.10 | 0.25Si |
| anal. | 9.90 | 1.01 | 1.18 | 0.40 | 0.09 | 0.27Si |
| UWadd | 13.57 | 1.27 | 1.17 | 0.49 | 0.10 | 0.59Er |
| anal. | 11.14 | 1.02 | 1.19 | 0.44 | 0.09 | 0.39Er |
| UXadd | 13.58 | 1.27 | 1.17 | 0.49 | 0.10 | 0.34Ce,0.23S |
| anal. | 10.87 | 1.06 | 1.21 | 0.42 | 0.08 | 0.29Ce,0.22S |

Chemical analysis of the arc melted feedstock indicated no Al or rare earth losses during non-consumable arc melting. The Al loss appears due to the melt extraction process being conducted in vacuo. The loss of rare earth metal seems due to oxidation of the rare earth during melt extraction, since a slag-like film is seen on the edges of the pool.

Microstructures of r.s. ribbons. The as solidified microstructures of the two alloys without a rare earth addition, UT and UV, have equiaxed grains as solidified, with several grains across a filament thickness. Figure 1a shows the cross section of a melt extracted filament of alloy UV.

In contrast, the alloys with a rare earth addition displayed a fine, columnar grain structure extending across the filament cross section. The structure within the grain appears martensitic. No second phase particles are apparent. Figure 1b shows the as solidified structure of alloy UW.

The alloy UA, containing Er plus B, had the columnar microstructure seen in the other alloys containing a rare earth. Additionally, the filament contained arrays of fine particles in the boundaries of the columnar structure. It is possible that these are borides rejected at the solid liquid interface during solidification. This microstructure is shown in Figure 1c.

Microstructure after HIP. The microstructures of the alloys after HIP were dependent on the presence of a rare earth dispersoid. The alloys UT and UV which do not contain a rare earth display an equiaxed microstructure, with grains about 20 microns across, Figure 2a. The prior filament boundaries can be identified by particle arrays that may be related to

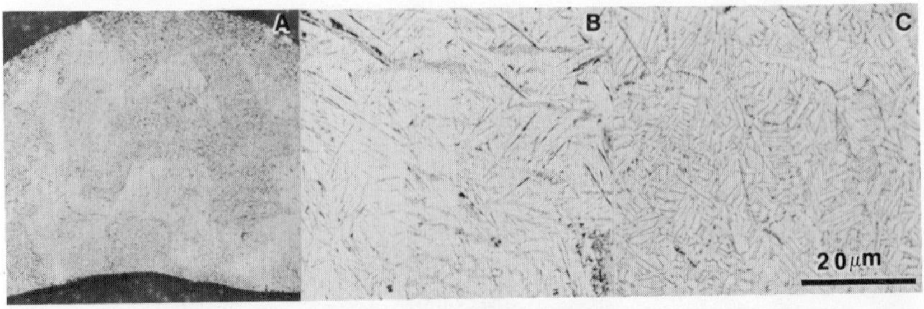


Figure 1. Microstructure of melt extracted alloys. a) UV, Ti-12.1 Al-1.3 Zr-1.2 Sn-0.5 Cb-0.1 Mo-0.25 Si. b) UW, Ti-12.1 Al-1.3 Zr-1.2 Sn-0.5 Cb-0.1 Mo-0.6 Er. c) UA, Ti-9.0 Al-1.5 Zr-1.0 Sn-0.3 Er-0.1 B.

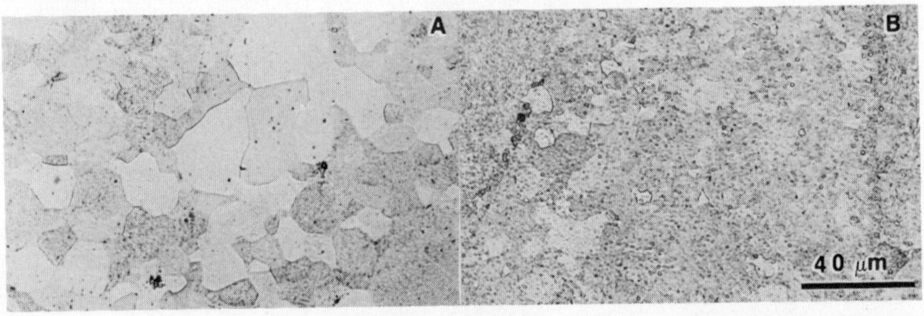


Figure 2. Microstructure of as-HIP at 840 C. a) UT, Ti-12.2 Al-1.3 Zr-1.2 Sn-0.5 Cb-0.1 Mo-0.25 Si. b) UW, Ti-12.1 Al-1.3 Zr-1.2 Sn-0.5 Cb-0.1 Mo-0.6 Er.

surface oxides on the filaments. The alloys containing rare earth additions exhibit a much finer grain size as-HIP, Figure 2b. Additionally, the prior filament boundaries are much more pronounced in these alloys, which could be related to rare earth oxides forming on the filament surface during melt extraction.

The as-HIP microstructure of the rare earth alloys exhibited a bimodal distribution of dispersoid sizes. The microstructure contained large particles from the solidification process and also a dispersion of much finer particles produced by subsequent precipitation from the supersaturated solid solution produced by R.S.

Alloys UT, UV, UW, and UX were hipped at two conditions-- 840 C/207 MPa/3 hrs and 900 C/207 MPa/3 hrs (1544 F/30 ksi and 1652 F/30 ksi). No microstructural differences were identified between alloys HIPped at the two different conditions.

Microstructure as extruded and thermally treated. The alloys were evaluated as-extruded and after a variety of thermal exposures, listed in Table IV.

The level of rare earth influenced the as-extruded microstructure. Figure 3 illustrates the as-extruded microstructures of alloys TY (0.3 a/o Er) and TZ (1.0 a/o Er). As-extruded TY exhibits an equiaxed microstructure, with low dislocation density typical of a recrystallized structure, while TZ retains a worked microstructure, with narrow elongated grains bounded by regions of high defect density.

Table IV - Thermal Treatments of R.S. Ti Alloys

| Alloy | Thermal Exposure | Alloy | Thermal Exposure |
|---|---|---|---|
| TY,TZ,UA, | 750 C/2 h | UT,UV,UW,UX | 750 C/2 h//600 C/48 h |
| UQ,UR | 750 C/2 h | UT,UV,UW,UX | 875 C/2 h//600 C/48 h |
| TY,TZ,UA | 900 C/2 h | UT,UV,UW,UX | 1000 C/2 h//600 C/48 h |
| TY,TZ,UA | 1000 C/2 h//750 C/ 2 h | UT,UV,UW,UX | 1200 C/2 h//600 C/48 h |

The level of a rare earth addition significantly effects the size distribution of dispersoid. Figure 4 displays alloys TY (0.3 a/o Er), UT (0.6 a/o Er), and TZ (1.0 a/o Er) after heat treatments of 2 hrs at 750 C (1382 F). The amount of dispersoid clearly increases with Er level. All Er levels have both fine and coarse particles, with no major difference in size distribution among the three alloys.

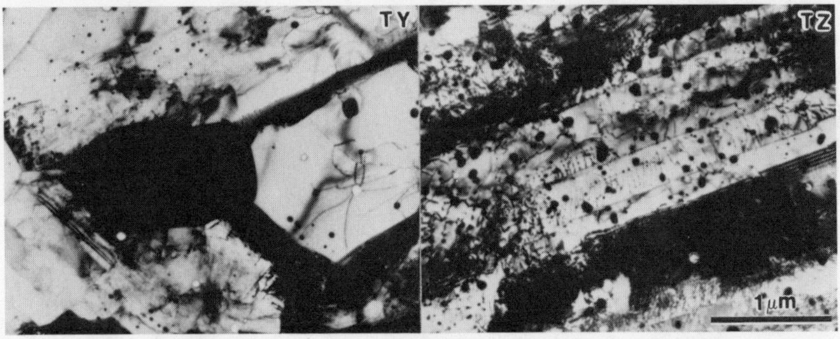


Figure 3. As extruded microstructures of alloys TY (0.3 a/o Er) and TZ (1.0 a/o Er).

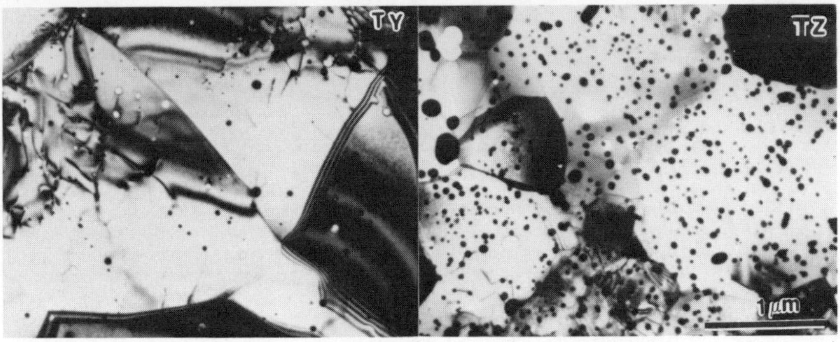


Figure 4. Microstructure after 750 C exposure of alloys TY (0.3 a/o Er) and TZ (1.0 a/o Er).

Both the Er and Ce plus S dispersoids have good stability (i.e. coarsening resistance) up to 1000 C (1832 F). Figure 5 displays Alloy UW (0.6 a/o Er) after 2 hour exposures at 875 C, 1000 C, and 1200 C (1407 F, 1832 F, 2192 F). Figure 6 displays alloy UX (0.3 Ce, 0.2 S) after similar exposures. In both alloys the dispersoid distribution is fine and uniform after the 875 C exposure. After the 1000 C exposure, the particles in the middle of the grains are fine, while those near the edges of the grain coarsen. The 1200 C exposure resulted in fully transforming the alloy to

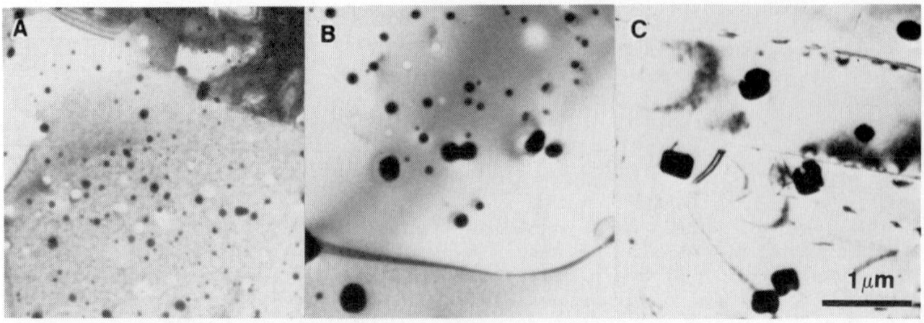


Figure 5. Microstructure after thermal exposure of alloy UW, Ti-12.1 Al-1.3 Zr-1.2 Sn-0.5 Cb-0.1 Mo-0.6 Er. a) 875 C/2hrs//600 C/48 hrs. b) 1000 C/2hrs//600 C/48 hrs. c) 1200 C/2hrs//600 C/48 hrs.

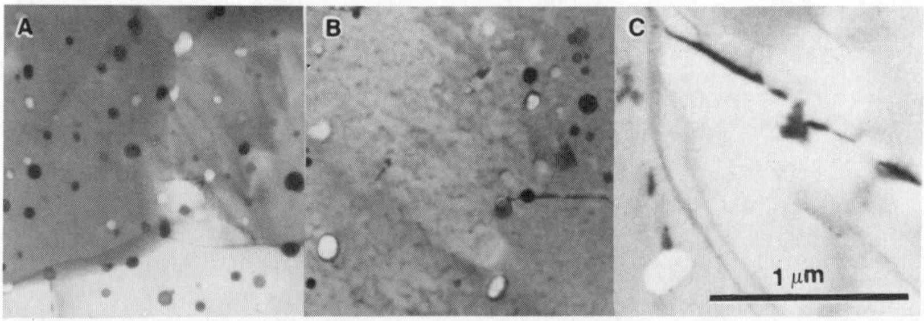


Figure 6. Microstructure after thermal exposure of alloy UX, Ti-12.1 Al-1.3 Zr-1.2 Sn-0.5 Cb-0.1 Mo-0.3 Ce-0.2 S. a) 875 C/2hrs//600 C/48 hrs. b) 1000 C/2hrs//600 C/48 hrs. c) 1200 C/2hrs//600 C/48 hrs.

beta. After the 1200 C exposure, the Er dispersoids have severely coarsened, while the Ce plus S dispersoids have dissolved and reprecipitated on cooling as coarse lenticular particles on the alpha titanium laths.

The stability of the dispersoids at 875 C and the coarsening or dissolving of the alloys on exposure to beta phase at 1200 C are both reasonable results based on relative diffusion rates in alpha and beta titanium. The coarsening of dispersoids in grain boundaries at 1000 C could be due to small amounts of beta phase present at 1000 C or to grain boundary diffusion at this temperature. A sample of alloy UW (0.6 a/o Er) was heated to 1000 C for 20 minutes and water quenched. TEM of this alloy revealed transformed beta phase present at the alpha grain boundaries and coarsened particles within the beta region. This would imply that the coarsened particles in the alpha grain boundaries were due to the presence of beta phase at 1000 C. However, there was also an absence of fine particles near alpha-alpha boundaries. It is possible that both grain boundary diffusion as well as the high bulk diffusion rates of beta phase can give rise to coarsening at 1000 C.

Tensile and creep tests. Creep tests were conducted on alloys UT, UV, UW, and UX after two thermal treatment conditions -- 1000 C/2 hrs//600 C/48 hrs and 1200 C/2 hrs//600 C/48 hrs (1832 F//1112 F and 2192 F//1112 F). The first thermal exposure resulted in an equiaxed alpha microstructure, the second in a transformed beta microstructure. The creep conditions were 650 C/138 MPa (1202 F/20 ksi). The equiaxed alpha tests were conducted in air, the transformed beta tests were conducted in argon. For each condition, a specimen of Ti6242S, heat treated 1040 C/1 hr//600 C/8 hrs (1904 F//1112 F) was also tested in creep.

The creep curves are graphically displayed in Figures 7 and 8. The best creep resistance at these conditions occurs for the transformed beta alloys. Further, among these alloys, the best creep resistance occurs in the Si bearing alloy UV, which contains no dispersoid, and has the largest grain size. Tensile tests were performed at room temperature on the alloys after creep exposure. The tensile data are listed in Table V. The best ductilities are obtained for the alloys containing dispersoids.

The matrix grain size is the dominant factor in creep resistance of the alloys tested. For the two thermal exposures studied, the lowest creep rate occurred in the dispersoid free alloy, the alloy with the largest grain size. However, room temperature tensile tests of precrept alloys indicates the dispersoid may have the beneficial effect of interrupting tendencies to planar slip in grains, thus yielding improved ductility. This is probably due to the dispersoid itself as well as the resultant fine grain size which also ameliorates planar slip by reducing slip length. Figure 9 shows that the fracture surface of the dispersoid containing alloy UW is much finer and rougher than the fracture surface of alloy UT. UT fracture surface is characterized by large facets about the size of individual grains, while the fracture surface of UW does not show crystallographic features.

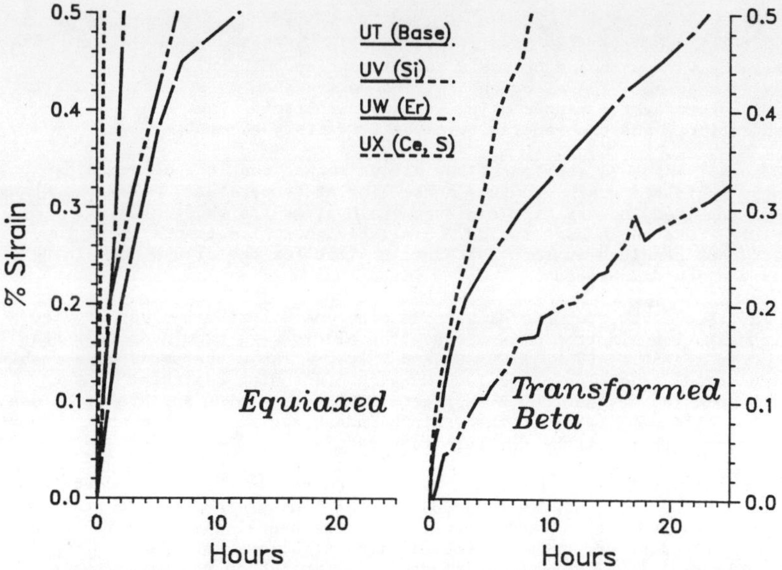


Figure 7. Creep curves of alloys UT, UV, UW, and UX having an equiaxed alpha microstructure. Test conditions 650 C, 138 MPa, air.

Figure 8. Creep curves of alloys UT, UV, UW, and UX having a transformed beta microstructure. Test conditions 650 C, 138 MPa, argon.

Table V - RT Tensile Tests of Precrept Alloys

| Alloy | Structure | .2%YS ksi | UTS ksi | %El | %RA | Alloy | Structure | .2%YS ksi | UTS ksi | %El | %RA |
|---|---|---|---|---|---|---|---|---|---|---|---|
| UT | alpha | ---- | 100.0 | 0.7 | 3.0 | UW | alpha | 144.3 | 153.9 | 1.4 | 5.2 |
| UT | beta | ---- | 92.7 | - | 6.3 | UW | beta | 134.5 | 140.7 | 2.0 | 4.0 |
| UV | alpha | 141.7 | 148.2 | 1.0 | 5.4 | UX | alpha | ----- | 81.1 | 0.2 | 1.1 |
| UV | beta | ---- | 122.7 | - | 6.2 | UX | beta | 143.7 | 144.3 | 0.4 | 4.0 |

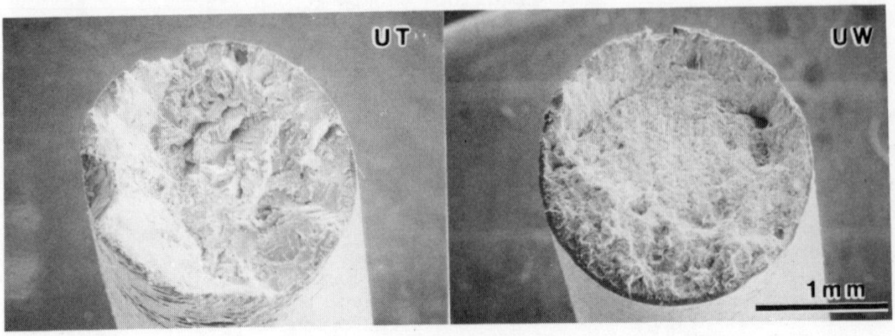


Figure 9. Tensile fracture surfaces of r.s. titanium alloys UT and UW.

SUMMARY

Multielement titanium base alloys containing rare earth compound dispersoids were produced by rapid solidification, HIPped, extruded, and thermally exposed. The alloy chemistries were chosen to evaluate two alloy bases in a systematic manner using various dispersoid types and concentrations, and the general roles of refractory elements.

Thermal exposure studies of the alloys showed that the dispersoids strongly inhibited grain growth. Annealling at temperatures below the alpha transus resulted in very fine equiaxed grain sizes for those alloys containing rare earth dispersoids. Similarly, heat treatments above the beta transus resulted in finer beta grain sizes for the alloys containing the rare earth dispersoids.

The rare earth compound dispersoid size and distribution was studied via transmission electron microscopy. The alloys were examined after heat treatments of 750 C, 875 C, 1000 C, and 1200 C. The dispersoids were stable in size and distribution up to 1000 C, with coarsening beginning in the grain boundaries at the 1000 C exposure. This coarsening appears to be due to higher diffusion rates in the grain boundary region due to a small amount of beta phase there at the exposure temperature.

Preliminary creep and tensile tests conducted on the alloys reveal that very high creep rates are associated with the fine grained alloys produced by alpha anneals. Creep rates were lower after annealling above the beta transus. The post creep tensile ductilities of alloys containing a dispersoid are higher than those of the alloys which do not have dispersoids.

ACKNOWLEDGEMENTS

Lee Perocchi was responsible for overseeing scheduling of the experiments and participated in most phases of material production, testing and characterization. Dave Burnham prepared the arc melted titanium alloys; Sid Young made the melt extracted filaments; Rick Nagy HIPped the alloys; John Hughes and Dick Laing extruded the alloys. Chris Canestraro and Paul Dupree conducted the mechanical tests. Curt Rodd did the optical metallography. Their care and diligence is gratefully acknowledged.

REFERENCES

[1] Vordahl, Milton B.: Dispersoid Titanium and Titanium-Base Alloys. U.S.Patent 3,622,406, November 23, 1971.

[2] Rath, B.B.; O'Neal, J.E.; and Lederich, R.J.: Grain Refinement in Titanium-Erbium Alloys. In: Proceedings Thirty Second Annual Meeting Electron Microscopy Society of America, ed.: C.J. Arceneaux. Claitor's Publishing Division, Baton Rouge, 1974, pp. 522-523.

[3] O'Neal, J.E.; Sastry, S.M.L.; and Lederich, R.J.: The Effect of Erbium and Yttrium Additions on Deformation Twinning in Polycrystalline Titanium. Microstructural Science, vol. 7, 1979, pp. 157-164.

[4] Sastry, S.M.L.; Peng, T.C.; and O'Neal, J.E.: Superposition of Solid-Solution-, Precipitation-, Grain-Size-, and Dispersion- Strengthening in Ti-Al-X Alloys. Titanium Science and Technology, G. Lutjering, U. Zwicker, and W. Bunk, eds., Deutsche Gesellschaft fur Metallkunde E.V., 1984, pp. 1811-1818.

[5] Kaschuk, V.A. and Svetlov, M.B.: Investigation of some Properties of VT5L Alloy with Rare Earth Additions. Titanium and Titanium Alloys, J.C. Williams and A.F. Belov, eds., Plenum Press, 1982, pp. 2201-2208.

[6] Sastry, S.M.L.; O'Neal, J.E.; Lederich, R.J.; and Rath, B.B.: The effect of yttrium and erbium dispersoids on the deformation behavior of titanium. J. Materials Science, vol. 14, 1979, pp. 179-183.

[7] Sastry, S.M.L.; Peng, T.C.; Meschter, P.J.; and O'Neal, J.E.: Dispersion Strengthened Powder-Metallurgy Titanium Alloys. AFWAL TR-83-4092, 1982.

[8] Sastry, S.M.L.; Peng, T.C.; Meschter, P.J.; and O'Neal, J.E.: Rapid Solidification Processing of Titanium Alloys. J. Metals, vol. 35, no. 9, September 1983, pp. 21-28.

[9] Konitzer, D.G.; Muddle, B.C.; Fraser, H.L.; and Kirchheim, R.: Refined Dispersions of Rare Earth Oxides in Ti-Alloys Produced by Rapid Solidification Processing. Titanium Science and Technology, G. Lutjering, U. Zwicker, and W. Bunk, eds., Deutsche Gesellschaft fur Metallkunde E.V., 1984, pp. 405-410.

# FORMATION OF METASTABLE PHASES IN RAPIDLY QUENCHED BINARY Ti ALLOYS

S.H. WHANG AND C.S. CHI
Department of Metallurgy & Materials Science
Polytechnic University
333 Jay Street
Brooklyn, New York 11201

## ABSTRACT

Rapid quenching of binary Ti alloys from the melt results in various metastable phases. A systematic study has been conducted in order to elucidate principles associated with the formation of metastable phases in binary Ti alloys resulting from rapid quenching. These metastable phases that include α', α" phases, metastable β phase, and ω phase are discussed with regard to their occurrence and the extension of α phase as a function of cooling rate. Effect of cooling rate and mechanical stress applied during cooling on metastable phase formation was investigated.

## INTRODUCTION

When metals and alloys are subjected to rapid quenching from the molten state, various metastable states are introduced in alloy phases and microstructures as well. These metastabilities in rapidly quenched alloys and materials have been well documented in recent years [1].

Metastable phases in binary Ti alloys known as α', α", metastable β phase and ω phase have been studied in the past [2,3]. All these phases are competing each other in a sense and overlapped compositionally in a narrow range [4,5]. In particular, martensite transformation and occurrence of ω phase can't be separable compositionally while ω phase formation appears to be associated with electronic concentration per atom [6,7].

In recent years, Ti alloys have been routinely processed by rapid solidification techniques at the cooling range of $10^5$-$10^7$ K/sec. The resulting alloy phases increase their meta-stability with increasing cooling rate. It is of interest to understand how further increased cooling rate affects solid solubility of solute, martensite transformation temperature and ω phase formation.

## EXPERIMENTS

Splat foils (20um thick uniform) of binary Ti alloys were produced by the hammer-and-anvil technique. Solubility extension, occurrence of ω phase and martensite transformation were studied by varying solute concentration.

In addition, foils of different cooling rates and mechanical stress levels were prepared by varying 1) the hammer pressure; 2) the size of splat alloy piece while alloy composition was kept constant: $Ti_{97.2}Fe_{2.8}$.

These foils were electropolished into thin films for TEM examinations. Alloy phase identification was carried out by bright field micrograph as well as by selected area electron diffraction patterns.

 

## RESULTS AND DISCUSSION

### a) Solubility Extension of Solute

Titanium rich terminal compositions of binary Ti alloy systems that can be classified into four different types of phase diagram were investigated as to solid solubility extension in the α phase. The four types of phase diagrams are 1) β-isomorphous; 2) eutectoid type (at β→α transformation); 3) peritectoid type (at β→α transformation); 4) monotectic - peritectic type.

Figure 1 shows TEM micrograph of splat quenched Ti-Mo alloys. These three micrographs reveal three different stages of transformations as a function of concentration. Martensite structure (α', α") are present in both $Ti\text{-}Mo_{2.5}$ (Fig. 1a) and $Ti\text{-}Mo_5$ (Fig. 1b). In addition, diffraction patterns of $Ti\text{-}Mo_5$ alloy (Fig. 1b) show the existence of ω and β phases. In contrast, $Ti\text{-}Mo_{10}$ alloy reveals both ω and β phases without α', α". These results demonstrate that occurrence of ω phase and martensite transformation compositionally overlap each other.

Solubility extension of α phase in a number of binary Ti alloys at the cooling rate $10^6$ K/sec is tabulated in Table 1. Moderate increase in

Table 1**

| Type | Alloy System | Max. Eq. Solu. of α | Ext. Sol. of α | Invariant Reaction Pt. |
|---|---|---|---|---|
| β-Isomorphous | Ti-V | 3.5 | 6 | |
| | -Nb | 2.5 | 5 | |
| | -Ta | 2.2 | 7.5 | |
| | -Mo | 0.4 | 2.0 | |
| Eutectoid | Ti-Cr | 0.5 | < 5 | 13.5 |
| | -Mn | 0.35 | 2.5 | 14.5 |
| | -Fe | 0.44 | 2.5 | 15 |
| | -Co | 0.8 | 2.5 | 6 |
| | -Ni | | 5 | |
| | -Cu | 1.2 | 7.5 | 5.5 |
| | -Si | 0.8 | 6 | 1.1 |
| Peritectoid | Ti-C | ~ 0.55 | 10 | |
| | -B | 0.43 | 10 | |
| | -Al | 1.5 | | |
| | -Ge | 2.7 | 2.5 | |
| | -Sn | 9 | 10 | |
| Monotectic-peritectic | Ti-Y | 0.1 | ~ 1 | 0.2 |
| | -La | <1 | ~ 0.3 | ~1 |
| | -Ce | <1 | ~ 0.6 | ~1 |

** All atomic percent

the solubility is observed in β-isomorphous type system, significant increase in eutectoid type systems and no appreciable increase situation in monotectic-peritectic type systems. There are two different trends in peritectoid type systems, i.e., very large increase in Ti-metalloid solutes (C,B); no increase in Ti-Ge and -Sn systems. Furthermore, lattice parameter increases with increasing boron concentration in Ti-B systems [8], which indicates interstitial solid solution of boron in Ti lattace. Therefore, it seems obvious that such unusual solubility extension can't be explained by resorting to equilibrium phase diagrams: Ti-B and Ti-C where substitutional solid solution is dominant. In β-isomorphous system, rapid quenching tends to stabilize β phase while supressing α+β decomposition. Hence, the solubility extension appears to occur at the expense of α+β field i.e., α+β field is reduced to a narrow compositional range. Solid solubility extension in monotectic-peritectic systems is not significant, probably due to the large atomic size of rare earth metals and the positive heat of formation.

b) Martensite Transformation and ω Phase Formation

Previous report indicates that martensite transformation temperature in binary Ti alloys does not change with the cooling rate exceeding the critical cooling rate: 32°C/s[9]. In order to elucidate the discontinuity effect of cooling rate on martensite transformation, specimen foils of different cooling rates were prepared and studied by TEM, which are shown in Fig. 2a, b, and c. The alloy produced by the highest cooling rate: $10^6$K/sec does not show any martensite transformation(Fig. 2a), but the alloy produced at a half of the highest cooling rate by increasing the alloy thickness to 40 um shows fine martensite structure(Fig. 2b). Furthermore, when the cooling rate is reduced to one third of the highest cooling rate, the martensite structure has fully grown into long rod shape(Fig. 2c). It is interesting to point out that again martensite and ω phase co-exist in Fig. 2b and c. Also, it should be noted that ω phase shown in Fig. 2b is much finer than that in Fig. 2c. In the similar manner, ω phase in Fig. 2a is difficult to identify in contrast to those in Fig. 2b and c. From these observations, it becomes clear that both martensite and ω phase are not truly athermal in Ti-Fe alloy system. Also, the $M_s$ line and lower boundary of ω phase formation can not be separated compositionally as shown in these micrographs.

This circumstance can be visualized in a schematic diagram shown in Fig. 3, in which $M_s$ line may be suppressed by increasing cooling rate from $\sim 10^3$ to $\sim 10^6$. The resulting e/a value for ω phase as well as $M_s$ is shifted to a low value as indicated by a hatched box. This is consistent with non-athermal behavior in Ti-Nb[10]. In fact, various decompositions: $\alpha' \rightarrow \alpha + \beta$; $\alpha'' \rightarrow \alpha''_{lean} + \alpha''_{rich} \rightarrow \alpha+\beta$; metastable $\beta \rightarrow \omega+\beta$ have been observed in Ti-Nb alloy [10,11].

It was reported that ω formation in Ti-10V-2Fe-3Al is triggered by mechanical stress during solidification [12]. It is interesting to see whether ω phase formation is affected by increasing applied pressure during cooling in Ti-Fe alloy, i.e., by increasing the splat hammer pressure. The alloys in Fig. 4a and 4b were produced under different mechanical pressures: 62.5 psi and 200 psi, respectively. Nevertheless, no difference in density and size of ω phase can be found between them. Based on this evidence, ω phase formation in this alloy is not affected by the applied stress of this range.

SUMMARY

1. For a given cooling rate, the solid solubility increase in binary Ti alloys is closely related to the equilibrium phase diagram features. An exception, however, is found in the peritectoid type systems, where the increasing tendency between metal and metalloid solutes is clearly different.

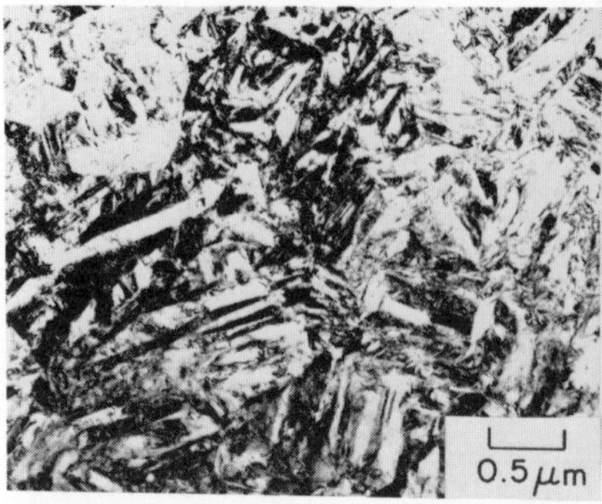


Fig. 1a As-quenched Ti-$Mo_{2.5}$ foil: 20μm thick

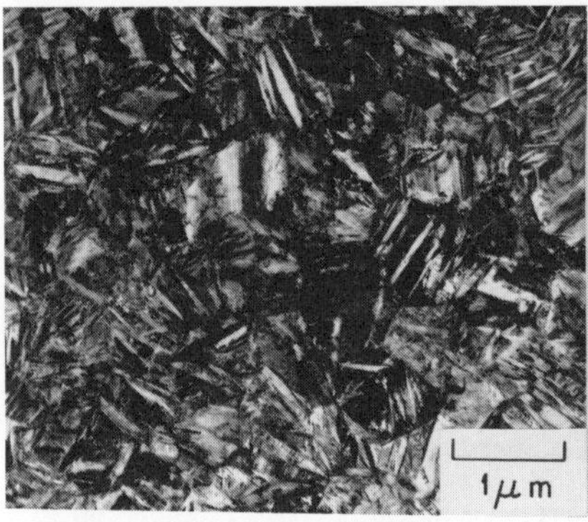


Fig. 1b As-quenched Ti-$Mo_5$ foil: 20μm thick

Fig. 1c As-quenched Ti-$Mo_{10}$ foil: 20μm thick

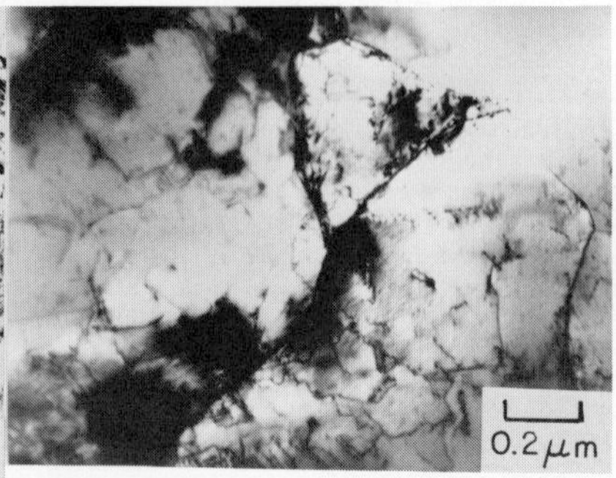


Fig. 2a As-quenched Ti-$Fe_{2.8}$ foil: 20μm thick

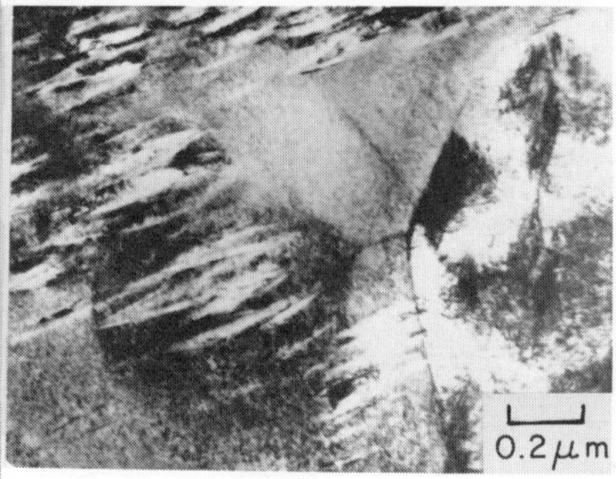


Fig. 2b As-quenched Ti-$Fe_{2.8}$ foil: 40μm thick

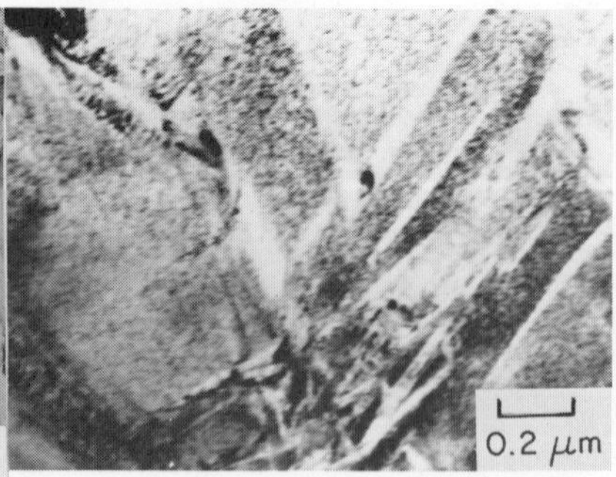


Fig. 2c As-quenched Ti-$Fe_{2.8}$ foil: 65μm thick

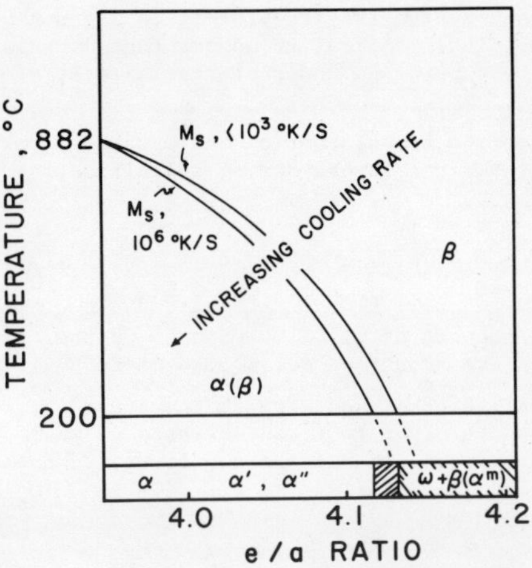


Fig. 3 Schematic Diagram of $M_S$ Temperature as a Function of e/a Ratio.

Fig. 4a As-quenched Ti-$Fe_{2.8}$ foil: 30μm thick and 62.5psi Ar pressure for the hammer.

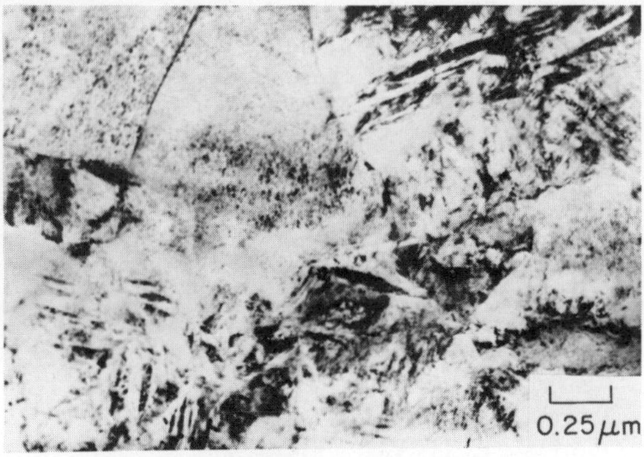

Fig. 4b As-quenched Ti-$Fe_{2.8}$ foil: 32μm thick and 200psi Ar pressure for the hammer.

evidenced by lattice parameter increase in Ti-B alloy system.

2. Both martensite transformation and ω phase growth can be suppressed by rapid quenching. This indicates that martensite transformation may not be athermal in a rigorous sense. Also, it is logical to assume that ω phase formation in binary Ti alloys may be suppressed under cluster size by a sufficient cooling rate.

3. No identifiable effect on martensite transformation and ω phase formation is found in the increase in mechanical stress during solidification and cooling.

## ACKNOWLEDGEMENT

We gratefully acknowledge the support of Ti alloy program by the Office of Naval Research, Arlington, VA (Contract No. N00014-85-K-0787).

The authors thank Professor Bill Giessen and Barnett Institute, Northeastern University for their support given to titanium research while the authors were members of the Institute. Thanks are due to Mr. Young-Woon Kim for his TEM study on this work.

## REFERENCES

1. Proc. 2nd Int. Conf. on Rapidly Quenched Metals, N.J. Grant and Giessen, Eds., MIT Press 1976; Proc. 3rd Int. Conf. on ROM, B. Cantor Ed., The Inst. of Metals, London 1978; Proc. 4th Int. Conf. on ROM, T. Masumoto and K. Suzuki, The Japan Institute of Metals, Sendai, Japan 1981; Proc. 5th Int. Conf. on ROM, S. Steeb and H. Warlimont, Wurzbur, Germany, 1984.

2. J.C. Williams, Titanium Science and Technology, R.I. Jaffee and H.M. Burte, Eds., Pellus Press, New York, 1973, p. 1494.

3. E.W. Collings, The Physical Metallurgy of Titanium Alloys, American Society of Metals, Metals Park, OH 44073, 1984, p. 75.

4. B.S. Hickman, Trans. TMS-AIME, 345, p. 1329-35, 1969.

5. S.L. Sass, J. Less-Common Metals, 28, p. 157-173, 1972.

6. Yu.A. Bagariatskii, G.I. Nosova, and T.V. Tagunova, Transl. Dok. Akad. Nauk., SSSR, 122, p. 593-96, 1958.

7. B.S. Hickman, Met. Trans, 245, p. 1329-1336, 1962.

8. Y.Z. Lu and S.H. Whang, unpublished work.

9. K.S. Jepson, A.R.G. Brown, and J.A. Gray, The Science , Technology and Application of Titanium, R.I. Jaffee and N.E. Promisel, Eds., Pergamon Press, 1970, p. 677-690.

10. H.M. Flower, R. Davis, and D.R.F. West, Titanium and Ti Alloys, Scientific and Technological Aspects, Eds., J.C. Williams and A.F. Belov, Plenum Press, 1982, p. 1703-15.

11. R. Davis, H.M. Flower, and D.R.F. West, Acta Metall, 27, p. 1041-52, 1979.

12. T.W. Duerig, R.M. Middleton, G.T. Terlinde and J.C. Williams, Titanium '80 Science and Technology, H. Kimura and O. Izumi, Eds., The Met. Society of AIME, 1980, p. 1503-12.

# DISPERSOID MODIFICATION OF $Ti_3Al$-Nb ALLOYS

R. G. Rowe, J. A. Sutliff, and E. F. Koch
General Electric Corporate Research and Development
Schenectady, NY 12301

## ABSTRACT

Titanium aluminide alloys with matrix compositions of essentially $Ti_3Al$ plus 0, 5, 7.5, and 10 a/o Nb and with and without rare earth elements for dispersoid formation were prepared. The alloys were rapidly solidified by melt spinning. Ribbon was consolidated by HIP and extrusion at temperatures below the beta transus temperatures of the alloys. The effects of processing conditions and dispersoid additions on room temperature ductility and fracture toughness were studied.

## INTRODUCTION

The ordered compound $Ti_3Al$ has good strength and creep resistance at high temperatures, but nil ductility below 600°C [1-3]. $Ti_3Al$-Nb alloys at niobium levels of 5 atomic percent and above exhibit some ductility at low temperatures [4]. Refinement of slip length has been thought to be a means of modifying the deformability of $Ti_3Al$ [5-7]. Reduction of grain size reduces slip localization, improving ductility [7]. The introduction of a dispersed phase may also refine the slip length [2], and has been shown to improve post-creep ductility in alpha + alpha-2 titanium alloys [8]. We have evaluated the effect of rapid solidification and the introduction of rare earth dispersoids on the microstructure and properties of $Ti_3Al$-Nb alloys. Alloys based upon $Ti_3Al$ with 0, 5, and 7.5a/o niobium were rapidly solidified as base alloys and with alloy additions to form either $Er_2O_3$ [9,10], or cerium sulfide and oxysulfide dispersoids [11].

## EXPERIMENTAL

The alloy compositions tabulated in Table I were rapidly solidified by arc melt spinning [12,13]. Melt spun ribbon was from 30 to 50 micrometers thick. Two major compositional variations were observed. First, some of the alloys had significant aluminum losses during arc melting of elemental constituents. In addition, all alloys also exhibited some tungsten contamination from the non-consumable arc melting electrode. This may have contributed to variability in the mechanical property data, but not to the trends which we observed.

Ribbon was consolidated by cold compaction, hot isostatic pressing (HIP) and extrusion. Alloys were HIP'ped at a temperature of 900°C and pressure of 30 ksi for 3 hours. Extrusion billets of all $Ti_3Al$-Nb alloys were heated to 925°C and extruded at a reduction ratio of from 6 to 7:1. Alloys without Nb were extruded at a temperature of 1108°C.

HIP at 900°C produced little change in the size or density of $Er_2O_3$ or Ce-S dispersoids which were formed in ribbon aged at 750°C [14]. Extrusion at 925°C resulted in refinement of the grain size of dispersoid-bearing alloys and grain coarsening of alloys without dispersoids [14]. The grain size of extruded alloys is shown in Table II. Extrusion also produced severe coarsening of the $Er_2O_3$ dispersoid to particle diameters from 0.1

to 1.0 micrometers. The microstructure of as-extruded alloy 124 ($Ti_3Al$-5Nb-0.5Er) is shown in Figure 1(a). Extrusion under identical conditions produced little coarsening of the Ce-S based dispersoid. Figure 1(b) is a TEM micrograph of Alloy 106 ($Ti_3Al$-5Nb-0.6Ce-0.2S) as-extruded. The dispersoid was extremely fine and closely spaced. The dispersoid diameter was from 5 to 25 nm. The dispersoid in Figure 1(b) was identified as $Ce_2S_3$ (body centered cubic, a=0.8636 nm) [15].

Table I

Alloy Compositions (Atom %)

| Alloy Number | Ti | Al | Nb | Er | O | Ce | S | W |
|---|---|---|---|---|---|---|---|---|
| 216 | Bal. | 25.0 | | | 0.05 | | | 0.16 |
| 213 | Bal. | 24.8 | | 0.6 | 0.2 | | | 0.08 |
| 111 | Bal. | 23.6 | | | 0.6 | 0.5 | 0.2 | 0.05 |
| 108 | Bal. | 24.6 | 4.7 | | | | | 0.16 |
| 124 | Bal. | 20.0 | 4.7 | 0.5 | 0.4 | | | 0.19 |
| 106 | Bal. | 23.5 | 4.9 | | 0.6 | 0.6 | 0.2 | 0.04 |
| 126 | Bal. | 23.3 | 7.7 | | 0.3 | | | 0.13 |
| 215 | Bal. | 24.8 | 7.5 | 0.6 | 0.4 | | | 0.10 |
| 139 | Bal. | 21.75 | 7.6 | | 0.5 | 0.7 | 0.15 | 0.09 |

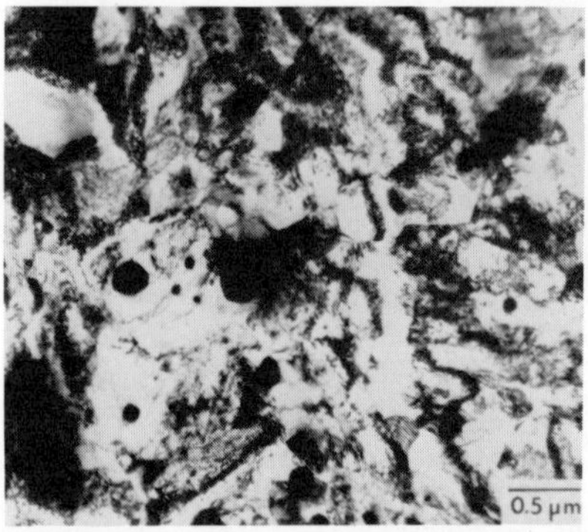


(a)

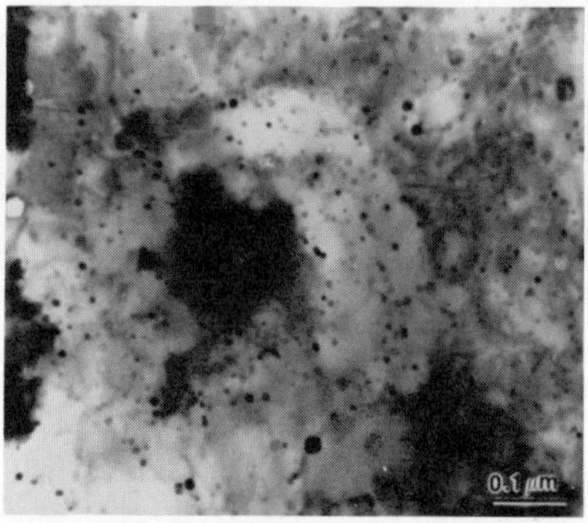


(b)

Fig. 1(a). TEM micrograph of HIP'ped and extruded alloys (a) Alloy 124 (Ti-20.0Al-4.7Nb-0.5Er) and (b) Alloy 106 (Ti-23.5Al-4.9Nb-0.6Ce-0.2S)

Table II

Grain Size of Extruded $Ti_3Al$-Nb Alloys

| | 0 Nb | 5 Nb | 7.5 Nb |
|---|---|---|---|
| Base Alloy | 2μm +10-20μm | 3 - 7μm | 4 - 8μm |
| $Er_2O_3$ Dispersoid | 1 - 2μm +6-10μm | 1/2 - 3μm | 1/3 - 3μm |
| Ce-S Dispersoid | 1/2 - 3μm | 1/3 - 3μm | |

## Ductility and Fracture Toughness

The results of bend tests on as-melt-spun ribbon are tabulated in Table III(a). All ribbons were tested with the wheel contact side of the ribbon facing the outside of the bend. The tabulated results show the calculated outer fiber strain at the first detectable crack in a ribbon [16]. It can be seen that as-melt-spun ribbon of some of the alloys was fully ductile, capable of being bent back upon itself without cracking. The crystal structure of these alloys was ordered body centered cubic (B2) structure, rather than ordered alpha-2 [14]. Ribbon was aged at 750°C. This produced fully ordered alpha-2 and, in the alloys with dispersoid formers, additional precipitation of an ultrafine dispersoid. The bend test results on aged ribbon are shown in Table III(b). The bend ductility of dispersoid-containing alloys was generally lower than the same alloy without a dispersoid.

Table III. Ribbon Bend Ductility (%)

(a) As-Melt-Spun Ribbon

| | 0 Nb | 5 Nb | 7.5 Nb |
|---|---|---|---|
| Base Alloy | 1.2 | 100* | 100* |
| $Er_2O_3$ Dispersoid | 1.0 | 4.3* | 8.3* |
| Ce-S Dispersoid | 1.1 | 4.1 | 2.0 |

* Confirmed ordered beta crystal structure

(b) Aged Ribbon

| | 0 Nb | 5 Nb | 7.5 Nb |
|---|---|---|---|
| Base Alloy | 1.0 | 1.3 | 3.5 |
| $Er_2O_3$ Dispersoid | 1.0 | 1.5 | 1.8 |
| Ce-S Dispersoid | 1.3 | 1.2 | 1.8 |

The effect of the reduction of grain size and the presence of a dispersoid on the ductility of extruded alloys was evaluated by indenting transverse metallographic sections with a Rockwell 'C' hardness indentation. The results, tabulated in Table IV, show that alloys without a dispersoid all had some detectable cracking around the indentation. None of the alloys containing an $Er_2O_3$ dispersoid cracked, in spite of the fact that the dispersoid had coarsened dramatically during extrusion. All alloys containing the Ce-S dispersoid cracked at inclusions which appeared to be unmelted particles carried into the melt spun ribbon. Metallographic sections of the remnant skulls revealed larger particles with the same color distributed throughout the melt. In some cases, nozzle blockage occurred due to these unmelted particles.

Three-point bending fracture bars were EDM machined from extruded rod with dimensions: 2.8 mm x 2.8 mm x 25.4 mm. They were center notched to a depth of 1.5 mm. Because of the small size of the fracture bars, the bars were not fatigue pre-cracked, although the base of the EDM notch was found to be pre-cracked or embrittled to a depth of 2-3 micrometers. Fracture toughness values [17], $K_Q$ for $Ti_3Al$-Nb alloys which had been aged at 787°C after extrusion are tabulated in Table V. It can be seen that the combined effect of the coarse dispersoid and fine grain size did not improve the fracture toughness of the alloy, rather, it appeared to reduce it.

Table IV

Hardness Indentation Test on Extruded Alloys

| | 0 Nb | 5 Nb | 7.5 Nb |
|---|---|---|---|
| Base Alloy | - | Slight to Moderate | Slight Cracking |
| $Er_2O_3$ Dispersoid | No Cracking | No Cracking | No Cracking |
| Ce-S Dispersoid | Severe Cracking | Severe Cracking | Moderate Cracking |

Table V

Fracture Toughness, $K_Q$* of Extruded and Aged Alloys

| | 5 Nb | 7.5 Nb |
|---|---|---|
| Base Alloy | 13.6 (12.2) | 12.3 (11.0) |
| $Er_2O_3$ Dispersoid | 11.7 (10.5) | 10.3 (9.2) |

*$MPa \cdot m^{1/2}$ ($ksi \cdot in^{1/2}$)

## DISCUSSION OF RESULTS

The indentation test did not crack any of the alloys containing an $Er_2O_3$ dispersoid, yet cracked all alloys without a dispersoid. The distinguishing difference between alloys with and without $Er_2O_3$ addition, was grain size and the presence of the dispersoid. Since the dispersoid was coarsened, the grain size difference may be the biggest factor in the apparent ductility of dispersoid-bearing alloys. Grain refinement may have suppressed cracking, or else kept the crack length below the resolution of optical detection. Either result is a positive indication of a beneficial secondary effect of the dispersoid addition.

The refinement of the grain size of dispersoid-bearing alloys during extrusion may be the result of localized concentration of deformation around the dispersoid particles. This would increase the number of recrystallization nucleation sites, reducing grain size [18]. The similar extruded grain sizes of the Er-bearing alloys which had a coarse dispersoid, and the Ce-S bearing alloys which had a fine, high density dispersoid, suggest that grain growth inhibition may not be as important as the interaction of the dispersoid with the deforming matrix in causing grain refinement.

Fracture toughness measurements indicated that grain refinement in combination with the coarsened dispersoid had no toughening effect. The contrast between indentation and toughness results suggests that the two tests may be sensitive to different aspects of the alloy microstructure.

Bend tests of as-melt-spun and aged ribbon of these alloys were, in contrast to the fracture toughness and indenter tests, conducted on alloy samples with an ultrafine dispersoid [14]. The crystal structure of as-melt-spun alloys containing niobium was ordered beta-2 rather than alpha or ordered alpha-2 [14]. Ribbon with a beta-2 matrix and no dispersoid, was completely ductile (able to bend back upon itself without cracking). With the dispersoid, the alloys were ductile, but less so. Upon transformation to alpha-2, the ductility of all alloys was reduced. Thus, the alpha-2 structure appeared to have less ductility than the ordered beta-2 structure.

The bend ductility of the alloys with and without a dispersoid increased with increasing niobium content as has been previously reported in ingot metallurgy studies [1,2,4]. Because of this, it appeared that the niobium content of the matrix remained the major factor contributing to ductility in the ribbon bend tests, and that addition of a fine dispersoid did not materially change this dependence.

Dispersoid modification has been shown to have good potential for alloy exploration in $Ti_3Al$-Nb alloys based upon the creation of new microstructures. The critical test of the effect of a fine dispersoid in $Ti_3Al$-Nb alloys remains the confirmation of ribbon bend test results in consolidated alloys, however. Two problem areas must first be overcome. First, further development of consolidation techniques are needed in order to maintain a fine $Er_2O_3$ dispersoid after consolidation. Second, melting technique development will be required in order to eliminate unmelted inclusions and permit the evaluation of the properties of alloys containing a fine cerium-sulfide dispersion.

## CONCLUSIONS

- As-melt-spun $Ti_3Al$-Nb alloys containing 5 and 7.5 a/o Nb which were ordered beta phase of the B2 crystal structure, had significantly more ductility than the alloy after transformation to the ordered alpha-2 structure.

- Extrusion at 925°C caused severe coarsening of the $Er_2O_3$ dispersoid, but did not coarsen a cerium-sulfide and oxysulfide dispersion.

- Grain refinement in extruded alloys with coarse dispersoids had a beneficial effect in reducing cracking in the indenter ductility test, but provided no improvement in fracture toughness.

- Ribbon bend ductility of $Ti_3Al$-Nb alloys containing a high density of fine dispersoids was related to the ductility of the matrix phase, and was less than that of the corresponding alloy without a dispersoid.

## ACKNOWLEDGMENTS

The authors would like to acknowledge the contributions of Mr. Robert O. Auer to this study. He assisted in the development of the titanium melt spinning system, and the melt spinning of all of the alloys of this study. In addition, he conducted all of the ribbon bend tests. The authors would also like to acknowledge helpful technical discussions with Dr. L. A. Johnson.

Part of this work was supported by the General Electric Company Aircraft Engine Group under the Independent Research and Development Program.

REFERENCES


[1] S.M.L. Sastry and H.A. Lipsitt, Met Trans A, 8A, p. 1543, (1977).

[2] J.C. Williams and M.J. Blackburn, In: Ordered Alloys Proc. 3rd Bolton Landing Conf., Sept. 1969, ed. B.H. Kear, C.T. Sims, N.S. Stoloff, and J.H. Westbrook, Claitors, 1970.

[3] S.M.L. Sastry and H.A. Lipsitt, Titanium '80, Vol. 2, Proc. Int. Conf. on Titanium, Kyoto, Japan, May 1980, p. 1231, TMS/AIME, Warrendale, PA, 1980.

[4] M.J. Blackburn and M.P. Smith, U.S. Patent #4,292,077, Sept. 29, 1981.

[5] P.L. Martin, H.A. Lipsitt, N.T. Nuhfer, and J.C. Williams, Titanium '80, Vol. 2, Proc. 4th Int Conf on Titanium, Kyoto, Japan, Ed: K. Kimura and O. Izumi, Met Soc. AIME, Warrendale, Pa., 1980, p. 1245.

[6] J.C. Williams, in: Titanium and Titanium Alloys III, Proc. Third Int. Conf on Titanium, Moscow, USSR, 1976, p. 1477.

[7] M.G. Mendiratta, S.M.L. Sastry and J.V. Smith, J. Matls Sci, 11, p. 1835 (1976).

[8] M. F. X. Gigliotti, R. G. Rowe, and G. E. Wasielewski, This conference proceedings.

[9] S M L Sastry, P J Meschter, and J E O'Neal, Met Trans A, 15A, p. 1451 (1984)

[10] D. G. Konitzer and H.L. Fraser, Mat. Res. Soc. Symp. Proc., 39, p. 437 (1985).

[11] R. G. Rowe and E. F. Koch To Be Published, Proc., 1986 ASM Conf on Rapidly Solidified Materials, San Diego, Ca., Feb, 1986, ASM, Metals Park, Ohio

[12] H. Tutzauer, P Esquinazi, M E De La Cruz and F De La Cruz, Rev. Sci. Instr., 51(4) p.546 (1980)

[13] S. H. Whang, C. S. Chi, and Y. Z. Lu in: Rapidly Quenched Metals, Proc. Fifth Int Conf. on Rapidly Quenched Metals, Wurzburg, FRG, Sept, 1984, North Holland, Amsterdam, 1985.

[14] J. A. Sutliff and R. G. Rowe, This conference proceedings

[15] Selected Powder Diffraction Data for Metals and Alloys, JCPDS Int. Center for Diffraction Data, Swarthmore, Pa., 1978, index # 27-104, p. 834

[16] G. E. Dieter, Mechanical Metallurgy Mc Graw-Hill, New York, 1961, p. 559

[17] Annual Book of ASTM Standards, 1981, Amer Soc for Testing and Mat'ls., Phila., Pa., 1981, Part 10, p. 588

[18] A. R. Jones in: Grain Boundary Structure and Kinetics, Proc. ASM Materials Science Seminar, Sept 1979, Milwaukee, Wis, Amer Soc for Metals, 1980, p. 379.

# MICROSTRUCTURE OF RAPIDLY SOLIDIFIED $Ti_3Al$ ALLOYS

A. G. JACKSON[a], K. R. TEAL[b], D. EYLON[c], F. H. FROES[b], AND S. J. SAVAGE[b]

[a]Systems Research Laboratories, Inc., 2800 Indian Ripple Road, Dayton, OH 45440

[b]Air Force Wright Aeronautical Laboratories, Materials Laboratory, AFWAL/MLLS, Wright-Patterson Air Force Base, OH 45433

[c]Metcut-Materials Research Group, P.O. Box 33511, Wright-Patterson Air Force Base, OH 45433

## ABSTRACT

The microstructures of ordered $Ti_3Al$-1Zr (alpha$_2$+Zr) and Ti-Al-Nb alloys produced by ingot metallurgy (IM) and by rapid solidification (RS) are compared. The RS $Ti_3Al$-1Zr (alpha$_2$ structure) alloy displayed small antiphase domains (APD) of 10-20 nm in the as-produced condition, but large grain size. The latter observation is rationalized in terms of the cooling history used to produce the material. The RS Ti-Al-Nb alloy exhibited an equiaxed microstructure. The IM Ti-Al-Nb material displays a complete change in microstructure between 1010°C and 1035°C, indicating a narrow alpha+beta phase field.

## INTRODUCTION

Studies of cast and wrought ingot metallurgy (IM) $Ti_3Al$ show the alloy possesses high strength but low ductility at room temperature due to an ordered structure [1]. An investigation of rapidly solidified (RS) $Ti_3Al$ and Ti-Al-Nb alloys was initiated to determine the effects of RS on microstructure and properties, since RS allows both microstructural and constitutional benefits in a number of systems. The RS approach was investigated in this work in an effort to improve the room temperature ductility while maintaining the relatively high strength level. In this paper comparison of RS $Ti_3Al$-1Zr addition is made with IM material and some results for RS and IM alloys of a Ti-Al-Nb ternary are presented.

## EXPERIMENTAL

$Ti_3Al$-1Zr and Ti-Al-Nb billets were both extruded and swaged at 990°C to form bars. The alloys were heat treated at 990°C/2 hr/FC (furnace cool) to produce an equiaxed microstructure. Bars of the $Ti_3Al$-1Zr were converted to RS ribbon in vacuum by the pendant drop melt extraction (PDME) [2] process. Microstructures of each were examined by scanning and transmission electron microscopy on JEOL 100CX and 2000FX microscopes.

## RESULTS

### $Ti_3Al$-1Zr IM and RS

Figure 1 is a TEM micrograph of heat treated $Ti_3Al$-1Zr IM material, showing a lenticular grain morphology and contrast associated with antiphase domain boundaries (APD) [$(11\bar{2}0)$ beam operative]. This structure is characteristic of $Ti_3Al$ which has been subjected to heat treatment at temperatures between 700°C and 1000°C [1]. APD size is ∿600-1500 nm, which is comparable to the size for $Ti_3Al$ annealed at 800°C for 20 minutes.

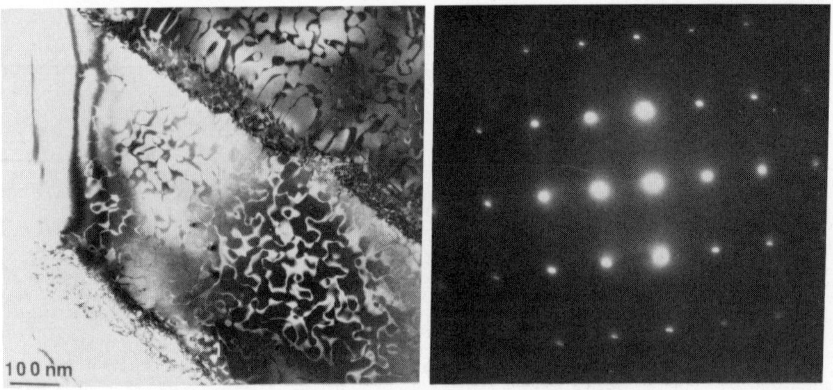


Figure 1. (a) TEM micrograph of IM $Ti_3Al$ and (b) diffraction pattern from (a).

RS $Ti_3Al$-1Zr produced by the PDME process exhibits the microstructure shown in the optical micrograph in Figure 2. The microstructure consists of grains within which are the APDs. The APDs in this structure, shown in Figure 3a (bright field) and Figure 3b [dark field $(11\bar{2}0)$], are 10-20 nm in size, and are smaller than those reported in $Ti_3Al$ ingot material quenched from 1150°C [1], which has an APD size of 25-50 nm.

Figure 2. Optical micrograph of top surface of $alpha_2$ ribbon.

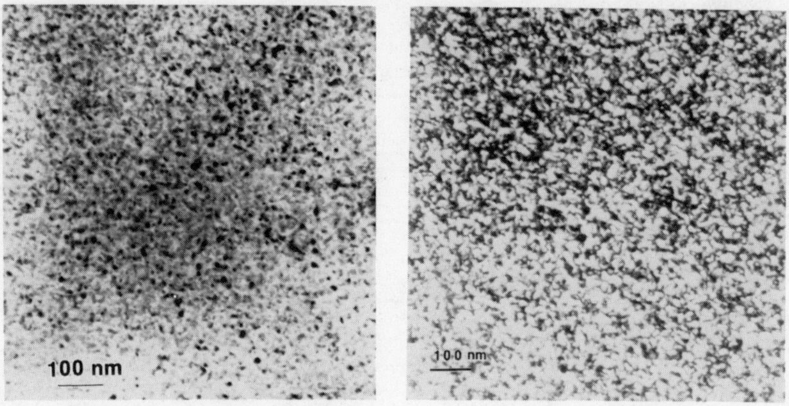


Figure 3. (a) TEM bright field micrograph of RS $Ti_3Al$ and (b) dark field $(11\bar{2}0)$ micrograph highlighting the APDs present.

Figure 4 is a low camera length diffraction pattern which includes the zero order Laue zone (Zolz) and the first order Laue zone (folz) ring. Upon analysis of the Zolz, the zone was found to be $[01\bar{1}1]$ pyramidal zone associated with HCP $Ti_3Al$. As a check on the orientation, the folz ring pattern for $Ti_3Al$ $[01\bar{1}1]$ was calculated [3]. The zone axis reciprocal lattice layer spacing, H, was calculated, using a = 0.579 nm c = 0.466 nm, to be 0.90 $nm^{-1}$. Measured radius of the folz was G = 26.3 $nm^{-1}$, giving an experimental H of 0.87 $nm^{-1}$. Thus the two values are in good agreement, confirming the pattern analysis and zone identification.

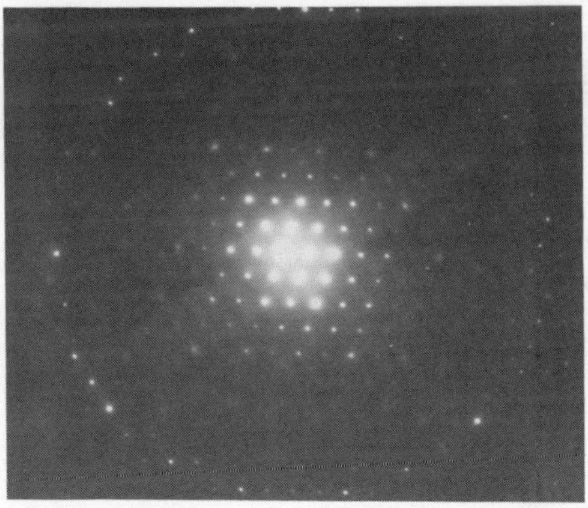

Figure 4. Diffraction pattern of Zolz and folz of $[01\bar{1}1]$ RS $Ti_3Al$ zone.

## $Ti_3Al$-Nb IM and RS

Specimens of IM Ti-Al-Nb alloy were swaged at 990°C followed by 990°C/2 hr/FC heat treatment to produce an equiaxed structure. They were then annealed at 1010°C, 1035°C, and 1095°C temperatures, and water quenched to produce variations in microstructure. Figure 5a shows TEM micrograph of the alloy after solution treatment at 1010°C, the equiaxed alpha (α) morphology is similar to that observed in conventional titanium alloys, and is verified in the diffraction pattern (Figure 5b) with $(10\bar{1}0)$ reflections present to be alpha titanium. The zone is $[\bar{1}2\bar{1}6]$ HCP-Ti.

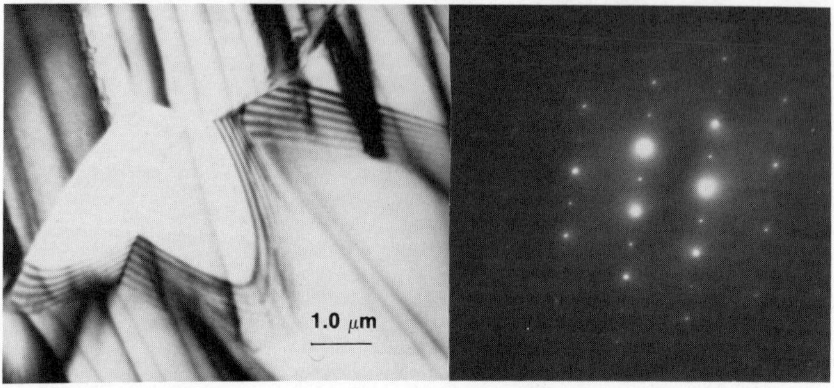


Figure 5. (a) IM Ti-Al-Nb and heat treated at 1010°C and (b) $(\bar{1}2\bar{1}6)$ zone of alpha-Ti.

On heating to 1035°C and quenching, the microstructure displayed a high dislocation density indicating a hexagonal martensitic structure (α') and suggesting that the material had been annealed predominantly in the beta phase field prior to quenching (Figure 6a). Diffraction pattern from this area, Figure 6b, was analyzed to be $[\bar{1}2\bar{1}3]$ HCP-Ti.

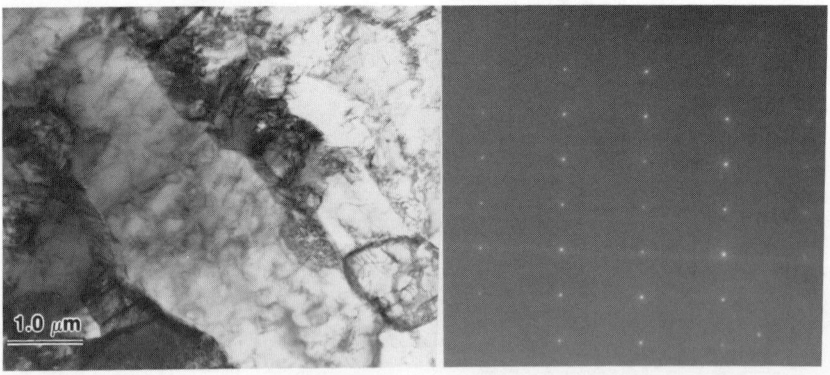


Figure 6. (a) IM Ti-Al-Nb solution treated at 1035°C and (b) diffraction pattern from (a) zone $[1\bar{2}1\bar{3}]$ HCP-Ti.

After heating to 1095°C and quenching, the microstructure consisted of hexagonal martensitic (α') platelets (Figure 7a), the diffraction pattern (Figure 7b) being due to a HCP $[2\bar{1}\bar{1}0]$ zone.

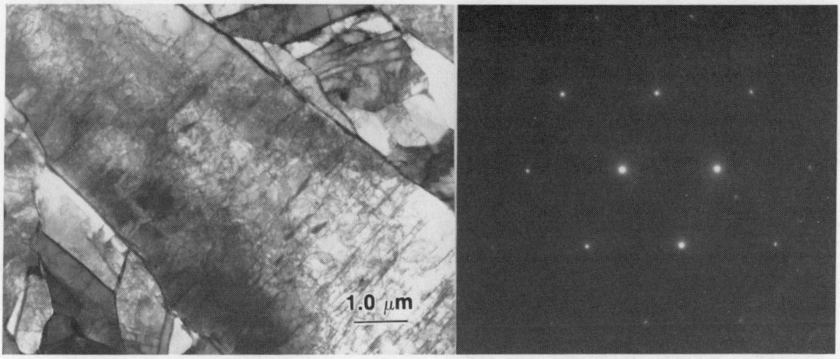


Figure 7. (a) IM Ti-Al-Nb solution treated at 1095°C and (b) diffraction pattern $[2\bar{1}\bar{1}0]$ HCP-Ti.

RS ribbon microstructure, Figure 8, consists of equiaxed grains which are probably beta phase. Analysis of the D.P. from the grains indicated $alpha_2$ was also present.

Figure 8. Microstructure of as-produced RS ribbon of Ti-Al-Nb alloy.

DISCUSSION

The results for the $Ti_3Al$-1Zr RS ribbon indicate that the effect of RS is a noticeable decrease in the size of $alpha_2$ APD compared with IM alloy. The rapid quench from the beta phase field to the $alpha_2$ region causes this decrease. The large grains present in the RS ribbon were somewhat surprising and may have resulted from several effects. If after solidification the cooling rate in the ribbon was relatively slow, then the time spent in the beta phase may have been long enough for beta grain growth to occur. However, a slow cooling rate after solidification is not consistent with the presence of fine APDs. Alternatively, if the ribbon, after initial cooling, was heated to the alpha phase field by contact with other as-produced hot ribbon in the PDME vacuum collection box and then cooled relatively quickly, fine $alpha_2$ would be generated. The result of such an occurrence would be the relatively large beta grains, together with the fine $alpha_2$ observed.

In the $Ti_3Al$ containing Nb the results of this work clearly show that by quenching from 1010°C or 1035°C, a significant change in microstructure occurs. A fully equiaxed alpha titanium is present after the lower temperature anneal while a martensitic lenticular structure is present after the higher temperature anneal. Thus we have gone from fully alpha phase at 1010°C to beta phase at 1035°C. This implies that there is a very narrow alpha+beta phase field between 1010°C and 1035°C.

The RS as-produced ribbon of this alloy group exhibits an equiaxed morphology, which was shown by diffraction analysis to contain both beta and $alpha_2$.

The present work has indicated that the RS approach can result in modifications in the microstructure of the $Ti_3Al$ based material compared to conventional IM processing, and is thus worthy of further exploration. However, caution must be exercised in interpreting microstructures because of the many phase fields and thermal excursions possible in the ribbon during RS processing.

## ACKNOWLEDGMENTS

This work was partially supported by AFWAL Materials Laboratory Contract Nos. F33615-83-C-5073 and F33615-82-C-5078. The authors would like to thank Westinghouse for preparation of the button melts, J. G. Paine of SRL for the conversion of the alloys to ribbon, and to Dr. H. A. Lipsitt for helpful discussions.

## REFERENCES

1. S. M. L. Sastry and H. A. Lipsitt, Met. Trans. A, 8A, 1543 (1977).

2. S. J. Savage and F. H. Froes, J. Metals, 36 (4), 20-33 (1984).

3. D. B. Williams, Practical Analytical Electron Microscopy in Materials Science (Philips Electronic Instruments, Inc., Mahwah, NJ, 1984), p. 128.

# RARE EARTH OXIDE DISPERSOID STABILITY AND MICROSTRUCTURAL EFFECTS IN RAPIDLY SOLIDIFIED $Ti_3Al$ AND $Ti_3Al$-Nb

J. A. Sutliff and R. G. Rowe
General Electric Corporate Research and Development
Schenectady, NY 12301

## ABSTRACT

The microstructures of titanium aluminide alloys containing a rare earth oxide dispersion have been characterized using analytical electron microscopy. The alloys, based on $Ti_3Al$ (alpha-2), contained 0 to 10.7 atom% Nb and 0.5 atom% Er. Alloys were rapidly solidified by melt spinning and were subsequently consolidated by HIP and extrusion. The microstructure of each alloy was examined in the as-cast, as-HIP'ed, and as-extruded conditions. A fine dispersoid spaced less than 100 nm apart was observed in ribbon aged at 750°C. The effects of processing conditions on the dispersoid distribution as a function of matrix chemistry were studied. Hot deformation was also examined to investigate the nature of the interaction between the dispersoids and the matrix during deformation.

Table I

Alloy Compositions (Atomic %)

| Alloy Number | Ti | Al | Nb | Er | O | W |
|---|---|---|---|---|---|---|
| 213 | Bal. | 24.8 | | 0.6 | 0.2 | 0.08 |
| 214 | Bal. | 25.4 | 4.9 | 0.6 | 0.3 | 0.18 |
| 205 | Bal. | 24.6 | 6.0 | 0.5 | 0.4 | 0.04 |
| 215 | Bal. | 24.8 | 7.5 | 0.6 | 0.4 | 0.10 |
| 204 | Bal. | 25.3 | 10.7 | 0.5 | 0.3 | 0.06 |

## INTRODUCTION

Alloys of nominal composition $Ti_3Al$- 0, 5, 6, 7.5, or 10.7 atom% Nb with 0.5 atom% Er were rapidly solidified by melt spinning [1-3]. Ribbons were from 30 to 50 micrometers thick. The alloy compositions, as analyzed by the flame emission method, are tabulated in Table I. Tungsten contamination from the non-consumable arc melting electrode occurred in all alloys. Melt spun ribbons were consolidated by hot isostatic pressing (HIP) and extrusion. The ribbons were cold pressed into low carbon steel cans, and HIP'ed at 850°C, 30 ksi, for three hours and extruded 7:1 at 870°C [4].

## EXPERIMENTAL RESULTS

### As-Melt-Spun and Aged Ribbon

As-melt-spun ribbon had a columnar structure from the bottom of the ribbon to the top. Selected area electron diffraction (SAD) was used to analyze the crystal structure of the as-melt-spun ribbons. The $Ti_3Al$ alloy with no Nb addition was analyzed as the ordered hcp $DO_{19}$ structure, alpha-2 [5]. The alloys containing niobium were analyzed as the ordered bcc phase, beta-2 [6]. In addition to the ordered beta phase, diffraction patterns of as-melt-spun $Ti_3Al$-Nb alloys contained streaks and diffuse intensity maxima, indicative of the precipitation of omega phase. This structure has been

Mat. Res. Soc. Symp. Proc. Vol. 58. 

observed previously in solid state quenched $Ti_3Al$-Nb alloys containing 10 or more atomic percent Nb [7].

Figure 1 is a TEM micrograph of a $Ti_3Al$ ribbon viewed in the columnar growth direction. All as-melt-spun ribbons contained a dispersion of fine (10-50 nm) particles. X-ray spectrometry confirmed that analyzed particles contained erbium. Electron diffraction obtained from extracted particles indicated that the dispersoid was $Er_2O_3$ [8,9]. The oxide dispersions in as-melt-spun ribbons were similar for all the alloys. The average dimension of a dispersoid was between 20 and 40 nm.

We have observed that heat treatment of ribbons for one hour at 750°C resulted in further precipitation of $Er_2O_3$ dispersoid, Figure 2(a) and (b).

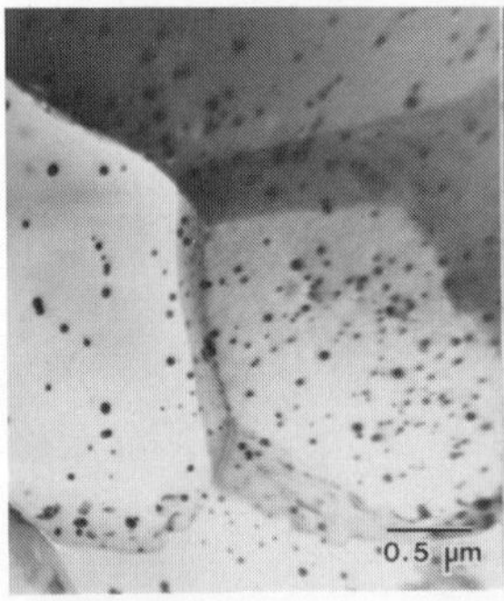


Figure 1. TEM micrograph of as-melt-spun ribbon of $Ti_3Al$-0.5 Er.

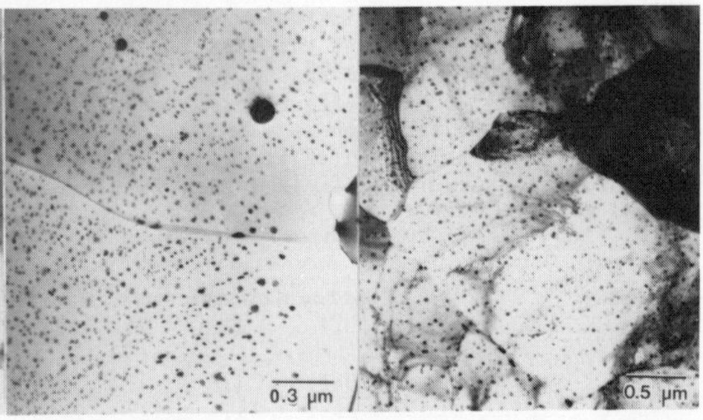


a b

Figure 2. TEM micrographs of dispersoid distributions in 750°C aged ribbon of (a) Alloy 213, ($Ti_3Al$-0.5Er) and (b) Alloy 205, ($Ti_3Al$-7.5Nb-0.5Er)

Dispersoid particles which precipitated during aging were approximately one fifth to one third the diameter of those found in as-melt-spun ribbon and occurred at a much higher density. Aging at 750°C transformed all matrix crystal structures to the ordered hcp, alpha-2. There was no evidence of beta phase in aged ribbon regardless of niobium content.

Ribbons of alloy 205 ($Ti_3Al$-6Nb-0.5 Er) were annealed at 900°C for one hour, Figure 3. The erbium oxide had good thermal stability in grain interiors at this temperature. Particles were coarsened on or near the grain boundaries during the anneal [10,11]. A zone depleted of particles was formed adjacent to the grain boundary. Comparison of alloys with different niobium levels showed that the degree of dispersoid coarsening at grain boundaries decreased with increasing niobium content.

Consolidated Ribbon

The microstructure of compacts which were consolidated by HIP was similar to that of annealed ribbon of the same composition. The grain boundary coarsening in HIP'ed compacts decreased with increasing niobium content. The effect of the niobium level on grain boundary coarsening was more pronounced at higher temperatures. Figure 4(a) and (b) are TEM micrographs of alloys with 6 and 11 atom% Nb respectively, as HIP'ed at 950°C.

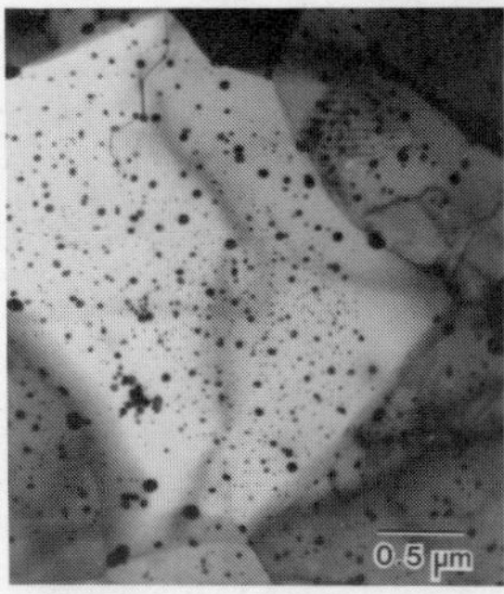

a b

Figure 3. TEM micrograph of alloy 205 ($Ti_3Al$-6 Nb-0.5 Er) aged at 900°C for 1 hour.

Figure 4. TEM micrographs of dispersoid distributions in 950°C HIP'ed alloys (a) $Ti_3Al$-6Nb-0.5Er and (b) $Ti_3Al$-10.7Nb-0.5Er.

Some grain boundaries were observed with dispersoid coarsening on one side and not the other, Figure 5. This observation suggested that coarsening was greatest when the dispersoid particles were intercepted by a moving grain boundary.

Extruded material had a microstructure different from that of aged or HIP'ed material. The grain size was mixed; most grains were equiaxed and less than one micron, others were irregularly shaped and from two to three microns. The fine grains indicated that recrystallization had occured. Figure 6 shows alloy 214 ($Ti_3Al$-5Nb-0.6Er) as extruded at 875°C. Most dispersoids were located on or near grain boundaries with the average dispersoid diameter greater than or equal to 100 nm for all extrusions. Some regions of the extrusion did not exhibit dispersoid coarsening. The large irregular grains retained the fine dispersion observed in HIP samples. This suggests that recrystallization played an important role in the particle coarsening. There was no evidence that the level of niobium significantly affected the degree of coarsening during extrusion.

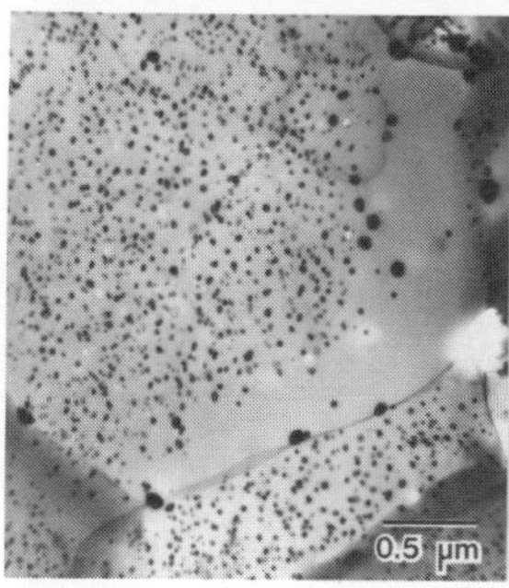


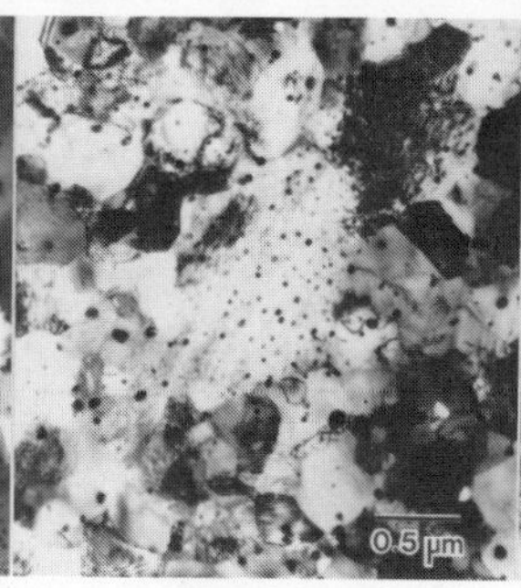


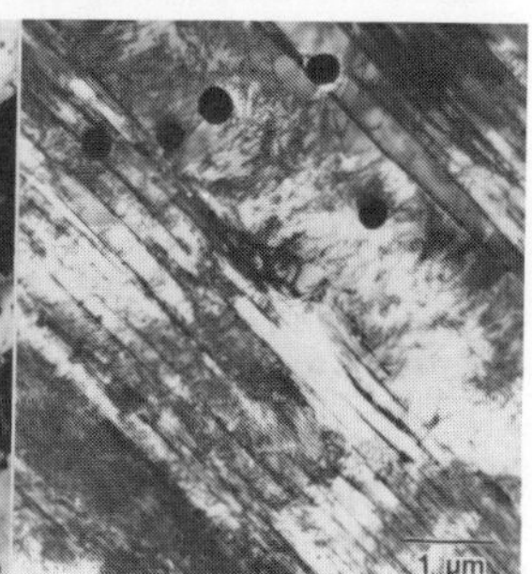


Figure 5. TEM micrograph of alloy 213 ($Ti_3Al$-0.5Er) as HIP'ed at 850°C.

Figure 6. TEM micrograph of alloy 214 ($Ti_3Al$-5Nb-0.5Er) as extruded at 870°C.

Figure 7. TEM micrograph of beta heat treated alloy Ti-22.5Al-11.0Nb-0.5 Er.

Figure 7 shows the microstructure of a Ti-22.5Al-11Nb-0.5Er alloy extruded at 920°C then heat treated in the beta field (1150°C) for 1/2 hr. The dispersoid size was approximately 400 nm. The microstructure was characteristic of the transformation from beta to alpha. Using this microstructure as a reference, we infer that the low temperature extrusions probably did not experience a major excursion into the beta phase.

## DISCUSSION OF RESULTS

The dispersoid distributions observed in as-melt-spun and annealed ribbons can be understood in the context of the ribbons' solidification and precipitation events. All of the alloys solidified as the bcc beta phase. The $Er_2O_3$ dispersion in as-melt-spun ribbons suggested that precipitation began in the beta phase. In $Ti_3Al$-0.6Er, cooling led to the transformation of beta to the ordered hcp phase, alpha-2. As-melt-spun, niobium-bearing alloys did not transform to alpha-2. These alloys had an ordered bcc phase at room temperature. Upon heating, the metastable beta phase transformed to alpha-2. At the annealing temperature of 750°C, there was additional precipitation of $Er_2O_3$ in the alpha phase. The very different conditions for the growth of these oxides, most importantly temperature and matrix structure, resulted in the reduced size of these particles in comparison to those in as-melt-spun ribbons.

### Selective Grain Boundary Dispersoid Coarsening

Exposure to temperatures above 750°C caused dispersoid depletion and coarsening near the grain boundaries of all alloys. Beta transformation has been found to selectively occur at grain boundaries during high temperature annealing and produce severe dispersoid coarsening [10]. Tungsten contamination may affect dispersoid coarsening by causing beta formation. Tungsten is a beta stabilizer and, in conjunction with niobium, may lower the temperature at which beta phase forms. Comparison of dispersoid coarsening in alloys with low tungsten levels indicated that the niobium level determined the degree of dispersoid coarsening regardless of the level of tungsten contamination. Furthermore, the degree of grain boundary dispersoid depletion was inversely related to the total amount of beta stabilizing elements in the alloys. The alloy with the most niobium showed the least amount of coarsening in both HIP'ed and heat treated ribbons. Because of this, it is likely that enhanced mass transport due to grain boundary diffusion dominated the coarsening process rather than selective transformation of the boundary region to the beta phase.

Microstructural features, such as those found in Figures 3 and 5, indicate that particle coarsening and the formation of particle depleted zones were influenced by grain boundary motion. It appeared that as a grain boundary moved through an area containing dispersoids particles intercepted by the boundary were eliminated and larger particles carried along with the boundary. The micrograph in Figure 6 shows large particles which appear to be carried along with a grain boundary.

A comparison of the dispersoid in grain interiors, Figures 5(a) and 5(b), shows that, in these regions, there was virtually no difference in dispersoid coarsening between these alloys. This suggested that the effect of niobium in reducing grain boundary dispersoid coarsening was not a bulk effect, but was itself specific to grain boundary diffusion or migration. It has been established that niobium affects the onset of cross-slip and twinning in $Ti_3Al$ alloys [12]. This suggests that it affects stacking fault or antiphase boundary energies, and we infer that it may also affect grain boundary energies.

## Dispersoid Coarsening During Extrusion

Extrusion at 870°C resulted in extensive dispersoid coarsening. HIP consolidation at 900°C did not result in significant dispersoid coarsening in grain interiors, and HIP at 950°C produced less dispersoid coarsening than the extrusion at 870°C. One explanation is that the alloys may have been adiabatically heated into the beta phase during extrusion. Other explanations for coarsening enhancement by deformation involve pipe diffusion due to the creation of a high dislocation density, stress enhancement, or particle coalescence due to physical motion of particles [13]. We have not been able to discriminate between any of these possibilities for enhanced dispersoid coarsening during extrusion, except to note that the post-extrusion microstructures were all characteristic of a recrystallized single phase rather that an acicular beta transformed structure. Furthermore, the retention of fine dispersoids in some large, apparently unrecrystallized, grains of extruded alloys suggests that dispersoid coarsening during extrusion may be related to recrystallization.

## CONCLUSIONS

- Heat treatment of as-melt-spun ribbon at 750°C resulted in the precipitation of a high density of $Er_2O_3$ particles with a particle size between 5 and 20 nm.
- Annealing or HIP at 900°C for one hour resulted in dispersoid depletion and coarsening near grain boundaries. The degree of coarsening decreased with increasing niobium content.
- Extrusion at temperatures of 870°C and above resulted in extensive coarsening of the dispersoid.
- All $Ti_3Al$-Nb alloys in the as-melt-spun condition had crystal structures which were identified as the ordered bcc B2 structure. All alloys transformed to alpha-2 when annealed at 750°C

## REFERENCES


[1] H. Tutzauer, P Esquinazi, M E De La Cruz and F De La Cruz, Rev. Sci. Instr., v.51(4) pp546-547 Apr(1980)

[2] S. H. Whang, C. S. Chi, and Y. Z. Lu in: Rapidly Quenched Metals, Proc. Fifth Int Conf. on Rapidly Quenched Metals, Wurzburg, West Germany, Sept, 1984, North Holland, Amsterdam, 1985.

[3] R. G. Rowe, T. F. Broderick, E. F. Koch and F. H. Froes, To Be Published, Proc., First Int Conf on Rapidly Solidified Materials, San Diego, Ca., Feb, 1986, ASM, Metals Park, Ohio 1986

[4] R. G. Rowe, J. A. Sutliff, and E. F. Koch, This Conference Proceedings, To be published.

[5] "Selected Powder Diffraction Data for Metals and Alloys", JCPDS Int. Center for Diffraction Data, Swarthmore, Pa., 1978, Index # 14-451

[6] T. Hamajima, G Leutjering and S. Weissmann, Met. Trans., v.3, pp 2805-2810, 1972.

[7] R. Strychor and J. C. Williams, in: Proc. Int. Conf. on Solid-Solid Phase Transformations, Pittsburgh, Pa., 1981, Met soc. AIME, Warrendale, Pa., 1982, pp 249-253.

[8] "Selected Powder Diffraction Data for Metals and Alloys", JCPDS Int. Center for Diffraction Data, Swarthmore, Pa., 1978, Index # 8-50.

[9] D. G. Konitzer and H. L. Fraser, Mat. Res Soc Symposium Proceedings, Dec., 1984, To be Published.

[10] M. F. X. Gigliotti, R. G. Rowe, G. E. Wasielewski, G. K. Scarr, and J. C. Williams, This Conference Proceedings, To be published.

[11] R. G. Rowe and E. F. Koch, To Be Published, Proc., First Int Conf on Rapidly Solidified Materials, San Diego, Ca., Feb, 1986, ASM, Metals Park, Ohio 1986

[12] S. M. L. Sastry and H. A. Lipsitt, in: Titanium '80, Vol. 2, Proc. Int. Conf. on Titanium, Kyoto, Japan, May 1980, pp. 1231-1243, Pub. TMS/AIME, Warrendale, PA, 1980.

[13] T. H. Courtney, in: Proc. Int. Conf. on Solid-Solid Phase Transformations, Pittsburgh, Pa., 1981, Met soc. AIME, Warrendale, Pa., 1982, pp 1057-1076

# PARTICLE COARSENING OF DISPERSOIDS IN RAPIDLY SOLIDIFIED Ti-5Sn-3Y*

Y.Z. Lu and B.C. Giessen, Materials Science Division, Barnett Institute, Northeastern University, Boston, MA 02115,
S.H. Whang, Department of Metallurgy and Materials Science, Polytechnic University, Brooklyn, NY 11201

## ABSTRACT

Rapidly solidified Ti alloys containing rare earth metal dispersoids, such as La or Er, have excellent resistance to coarsening at elevated temperature (700-900°C). In particular, the coarsening of the dispersoid particles in RSP Ti-5Sn-3Y was studied; these particles were found to be $Y_5Sn_3$. The measured growth kinetics were found to be compatible with the modified LSW model; it was concluded from the observed parameters that in this alloy the diffusion rate of Y is the rate determining factor, providing a measurement of the diffusion coefficient of Y in a Ti matrix and yielding its activation energy Q = 272 kJ/mole.

## INTRODUCTION

To date, the design of Ti alloys has been based on the utilization of solution strengthening and basic microstructural manipulation. This design approach has led, e.g., to the development of Ti-6Al-4V [1].

In recent years, new approaches going significantly beyond the conventional design practice have been taken to raise the high temperature capability of Ti alloys further, to 700-1000°C. In one of these approaches, new coarsening-resistant dispersions are introduced into the Ti matrix using rapid solidification processing [2-5]. It was found by trial and error that rare earth metal dispersoids, such as La or Er, have excellent coarsening resistance in Ti alloys at elevated temperatures (700-900°C) [6-7]; by contrast, silicide particles were observed to coarsen rapidly [8]. A preliminary study of the Ostwald ripening behavior of La dispersoids had been reported [6]; however, there is no similar study for other rare earth dispersoids or rare earth compounds in Ti and the particle coarsening mechanism for rare earth dispersoids in Ti is not yet understood.

In this paper, the coarsening of yttrium-containing dispersoid particles in Ti-5Sn-3Y alloys is reported. (Here and in the following, percentages are in weight percent).

## EXPERIMENTAL PROCEDURE

For the present experiments, alloy buttons were prepared from Ti(99.99%) and Y(99.9%) by repeated melting in an arc furnace under argon gas atmosphere. The oxygen level of the Ti used is < 150 ppm. Small alloy pieces from the buttons were splat quenched into thin foils by the hammer and anvil technique. Disk shaped specimens of 20 μm thickness were obtained from the central section of the foil. For heat treatment, samples were wrapped in Ta foil and sealed in quartz tubing with additional Ti foil as getter material under a vacuum of $10^{-5}$ torr. The tubes were annealed isothermally for the desired temperatures and times. In order to identify the chemical composition and crystal structures of the dispersoids in the heat treated alloys, carbon extraction replicas prepared from the annealed foils were studied by STEM (HB5) and electron microprobe. The particle size and distribution were studied in the TEM micrographs using an image analyzer.

## RESULTS AND DISCUSSION

### a) Microstructures

As shown in Fig. 1, the microstructure of the as-quenched Ti-5Sn-3Y alloy shows first, strong grain refinement due to the high cooling rate (~$10^6$ deg/sec) and second, particles of ~5 nm dia. distributed fairly uniformly in the matrix, which are probably attributable to Y solute exceeding the equilibrium solubility limit; extension of solid solubility by rapid solidification is not observed in systems of Ti and rare earth metals [9]. This fine microstructure which includes fine subgrains disappears in the early stage of isothermal annealing and is replaced by equiaxed grains which contain round (spherical) particles, e.g., in Fig. 2 which shows such particles in a Ti-5Sn-3Y alloy annealed at 800°C for 100 hr.

The STEM-EDX spectra of the extraction replicas of the annealed Ti-5Sn-3Y alloy foil indicate that the particles contain Sn and Y (Fig. 3); the small Cu peak is attributed to the copper grid supporting the sample. The weight and atomic ratio of Sn and Y were calculated from the peak ratio using the Cliff-Lorimer equation [10]. The calculated ratio is 39%Sn and 61%Y (±10%) which is compatible with the ratio 44.5%Sn and 55.5%Y in $Y_5Sn_3$; furthermore, the selected area

**Fig. 1: Ti-5Sn-3Y Foil, As-quenched (Bright Field).**

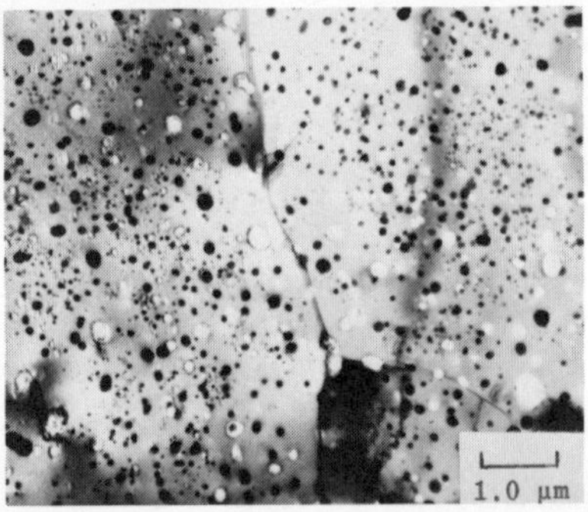


**Fig. 2: Ti-5Sn-3Y Foil, Annealed at 800°C for 100 hr.**

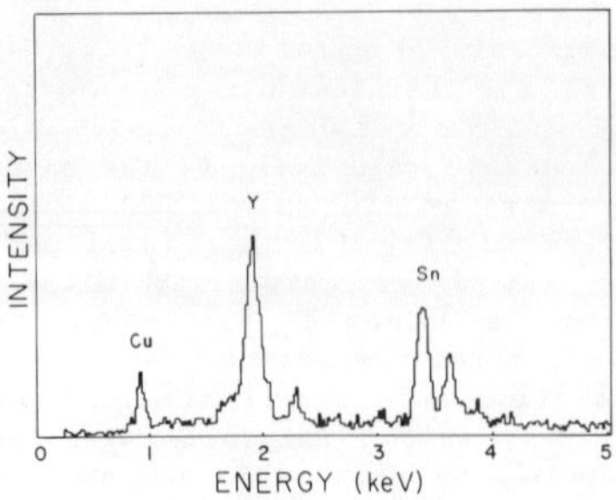


**Fig. 3: EDX spectra of extracted fine precipitates in RQ Ti-5Sn-3Y annealed at 800°C for 16 hr. showing peaks due to Sn and Y in ratio compatible with compound composition $Y_5Sn_3$. The Cu peak is due to Cu grid material.**

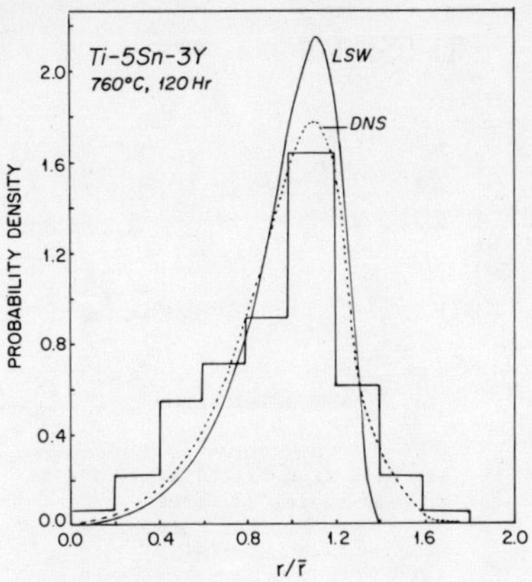


Fig. 4: The distribution of relative particle sizes for $Y_5Sn_3$ precipitates in RQ Ti-5Sn-3Y alloys annealed at 760°C for 120 hr. Fits described by the LSW and DNS models are shown (see text).

diffraction ring pattern agrees with that of the $Mn_5Si_3$ ($D8_8$) type structure of $Y_5Sn_3$. Hence, it is concluded that the particles are, in fact, $Y_5Sn_3$.

b) Particle Coarsening

Fit to MLSW model: The particle size distribution in the Ti-5Sn-3Y alloys annealed at 760°C for 120 hr is shown in Fig. 4, where r and $\bar{r}$ are particle radius and average particle radius, respectively. This particle size distribution deviates significantly from a fit based on the unmodified LSW model [11,12] in which the particle volume is assumed to approach zero. However, the histogram is better represented by a modified LSW (MLSW) particle size distribution predicted by the DNS model [13] where particle coalescence is considered as a mechanism for explaining the volume effect. In Fig. 5, the average particle size $\bar{r}$ for RQ Ti-5Sn-3Y alloys held at three different temperatures is plotted against the annealing time(t), producing smooth, non-linear curves for $\bar{r}$. Evaluation of these data shows that the best linear correlation is obtained by plotting the cube of the average particle size vs. time as shown in Fig. 6. This third power relation indicates that the coarsening of rare earth particles in Ti-5Sn-3Y is governed by a volume diffusion mechanism and may be expressed by the MLSW model for volume diffusion [13-16], yielding

$$\bar{r}^3 - \bar{r}_o{}^3 = f(K,T) \cdot (t - t_o) \qquad (1)$$

with

$$f(K,T) = \frac{8}{9} \cdot \frac{K \cdot C_s(\infty) \cdot D \cdot \gamma \cdot V_m}{R.T} \qquad (1\text{-}1)$$

where

$f(K,T)$ = measured slope of $\bar{r}^3$ in plot of $\bar{r}^3$ vs t

$\bar{r}_o, \bar{r}$ = average particle radii for the onset and final states
K = volume fraction factor
$C_s(\infty)$ = equilibrium solubility of solute in the matrix with particles of infinite size at a given temperature T
D = diffusion coefficient of solute, $cm^2/sec$
$\gamma$ = interfacial free energy of particle, $J/m^2$

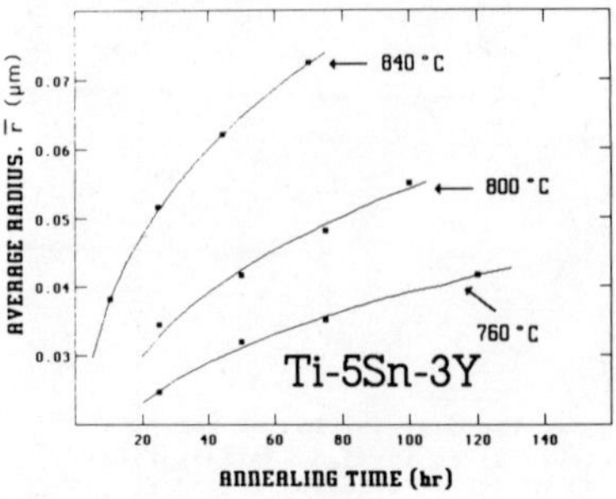


Fig. 5: Dependence of the mean $Y_5Sn_3$ particle radius $\bar{r}$ on time at three aging temperatures, as shown.

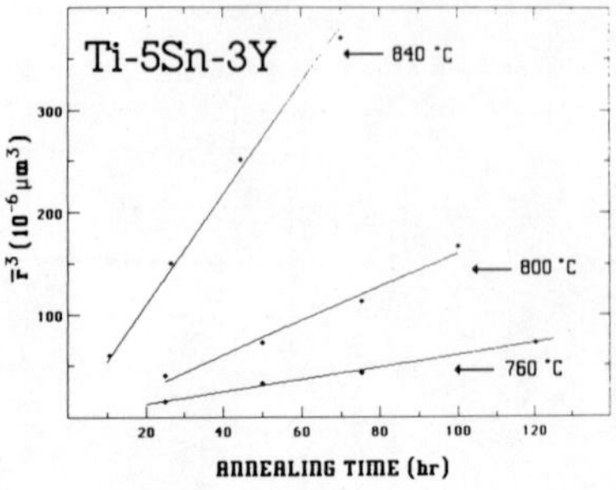


Fig. 6: Data from Fig. 5 replotted as cube of mean radius vs. time.

| | | |
|---|---|---|
| $V_m$ | = | gram-molar volume of precipitate, $cm^3/g$-mole |
| $t_o, t$ | = | onset and final annealing time, sec |
| R | = | gas constant, J/mole-$^oK$ |
| T | = | absolute annealing temperature, $^oK$ |

To evaluate the slopes of the curves taken at the three temperatures and to extract the diffusion coefficient of the solute from them (in order to arrive at a mechanism for this Ostwald ripening) assumptions must be made concerning K and $\gamma$.

Volume Fraction Factor and Interfacial Free Energy: The K value appropriate for the present volume fraction of 3.4 % can be estimated from various models where the K value varies from 1.0 to 1.5 depending upon the model [13]. For the present calculation, we chose the K value 1.18 given by the DNS model since the histogram of particle sizes in the present system agrees well with that given by this model, as discussed above.

Since $\gamma$, the interfacial free energy, is not known for rare earth dispersoids, a value $\gamma = 1$ $J/m^2$ was assumed based on various reported values, such as 0.5 $J/m^2$ for precipitates in the α-Fe - Cu system [17] and 1.5 $J/m^2$ for $ThO_2$ in Ni [18].

Diffusion Coefficient: With the assumptions given, diffusion coefficients D could be calculated from the MLSW relation above. The data and derived D values are summarized in Table I.

Although they are not reported as part of this study, a set of values for coarsened Ti-5Sn-4.5La alloys is included in Table I for purposes of comparison.

We must now assign the observed D values to the diffusion of one of the two solute species involved in the coarsening process, i.e., Sn and Y. We believe that the diffusion coefficients obtained are the diffusion coeffieients for Y in Ti, for two reasons. First, in the temperature range of 880-1600$^o$C the diffusion coefficient of Sn in Ti is nearly identical with the self-diffusion coefficient of Ti [19]. Assuming that this relation also holds at the somewhat lower temperature range studied here, the observed D value is smaller than that predicted for Sn from the above observation by three orders of magnitude in Ti-5Sn-3Y and by four orders of magnitude in Ti-5Sn-4.5La. (It may be observed that any

TABLE I

| Alloy System | Temp. (°C) | f(K,T) ($m^3/s$) | Vol. Fract.(%) | K | $C_s$ (at.%) | $V_m$ ($m^3$/g-mole) | D ($cm^2/s$) |
|---|---|---|---|---|---|---|---|
| Ti-5Sn-3Y | 760 | $1.65x10^{-28}$ | 3.4 | 1.18 | 0.42 | $1.67x10^{-5}$ | $1.9x10^{-13}$ |
| | 800 | $4.66x10^{-28}$ | | | | | $5.6x10^{-13}$ |
| | 840 | $1.52x10^{-27}$ | | | | | $1.9x10^{-12}$ |
| Ti-5Sn-4.5La ($La_2Sn$) | 760 | $1.23x10^{-29}$ | ~4.0 | 1.21 | 0.46 | $2.33x10^{-5}$ | $9.0x10^{-15}$ |
| | 800 | $9.39x10^{-29}$ | | | | | $7.2x10^{-14}$ |

uncertainty in the value of the interfacial energy $\gamma$ assumed for the dispersoid is negligible compared to this difference.) Secondly, if the observed D values were indeed due to the migration of Sn, the values for both Ti-5Sn-3Y and Ti-5Sn-4.5La should be identical or nearly so; however, they differ by as much as an order of magnitude. Hence, it may be concluded that the observed diffusion coefficient in the Ti alloy is that of Y. Regrettably, literature data for the diffusion of Y in Ti which would allow a confirmation of this conclusion are not available.

Activation Energy: Evaluating the observed D values using an Arrhenius plot, an activation energy Q = 272 kJ/mole (65kcal/mole) is found. As for D, literature values for comparison are lacking for Q. While this value is relatively high compared to the activation energy for the diffusion of Ce in Zr in the temperature range of 880-1600°C where Q = 173 kJ/mole was found [20], it must be noted that Zr is BCC at these temperatures while Ti is HCP at the temperatures studied and that the atomic radius of Zr is 8.8% larger than that of Ti. The high activation energy of diffusion is certainly related to the large size difference of Y and Ti and their low chemical interaction energy ($\Delta H_M$). In turn, the high Q values and the low D values are responsible for the slow particle coarsening and hence the potentially interesting high temperature properties of Ti alloys of this family.

## SUMMARY

1. The dispersoids in high purity Ti-5Sn-3Y alloys were identified as the binary $Y_5Sn_3$ intermetallic compound.

2. Particle coarsening of this dispersoid in titanium is controlled primarily by a volume diffusion mechanism of Y in the alloy.

3. The activation energy for diffusion of Y in Ti alloy in the temperature range from 700°C to 900°C is 272 kJ/mole, i.e., it is relatively high.

4. The particle coarsening rate for Y (as well as La) dispersoids in Ti-Sn-RE is very low. This low rate appears to derive from the low diffusion coefficients of rare earth metals in titanium.

## ACKNOWLEDGEMENT

We gratefully acknowledge support by the Office of Naval Research (Contract N00014-85-K-0787) for research on rapidly solidified Ti alloys.

* Communication No. 279 from the Barnett Institute

REFERENCES


1. A.E. Gorum and R. Colton, Proc. Second Int. Conf. on Titanium, edited by R.I. Jaffee and H.M. Burte (Plenum Press, 1973) Vol. 1, p. 11.

2. S.M.L. Sastry, T.C. Peng, P.J. Meschter, and J.E. O'Neal, J. Metals, 35, 21 (1983).

3. S.H. Whang, J. Metals, 36, 34 (1984).

4. S.H. Whang, J. Mat. Sci., (1986) (to be published).

5. F.H. Froes, MRS Sym. Proc., edited by B.C. Giessen, D.E. Polk and A.I. Taub, (Materials Research Society, Pittsburgh, PA, 1986), Vol. 58.

6. Y.Z. Lu, C.S. Chi and S.H. Whang, Proc. 5th Int. Conf. on Rapidly Quenched Metals, edited by S. Steeb and H. S. Warlimont, (Elsevier Science Publishers, New York, 1985), p. 949.

7. S.M.L. Sastry, P.J. Meschter and J.E. O'Neal, Met. Trans. A, 15, 1451, (1984).

8. S.H. Whang, Y.Z. Lu and B.C. Giessen, MRS Sym. Proc., edited by B.H. Kear and B.C. Giessen, (Elsevier Science Publishers, New York, 1984), Vol. 28, p. 367.

9. S.H. Whang and C.S. Chi, MRS Sym. Proc., edited by B.C. Giessen, D.E. Polk and A.I. Taub, (Materials Research Society, Pittsburgh, PA, 1986), Vol. 58.

10. G. Cliff and G.W. Lorimer, Proc. 5th European Congress on Electron Microscopy, (Institute of Physics, Bristol, 1972), p. 140.

11. I.M. Lifshitz and V.V. Slyozov, J. Phys. Chem. Solids, 19, 35 (1961).

12. C. Wagner, Z. Electrochem, 65, 581 (1961).

13. C.K.L. Davies, P. Nash, and R.N. Stevens, Acta Metall., 28, 179 (1980).

14. A.J. Ardell, Acta Metall., 20, 61 (1972).

15. A.D. Brailsford and P. Wynblatt, Acta Metall., 27, 489 (1979).

16. P.W. Voorhees and M.E. Glickman, Met. Trans. A, 15, 1081 (1984).

17. C.S. Smith, Metall. Revs. 9, 1 (1964).

18. P.K. Footer and C.B. Alcock, Met. Trans., 3, 2633 (1972).

19. J. Askill and G.B. Gibbs, Phys. Status Solidi., 11, 557 (1965).

20. A.R. Paul, M.S. Anand, M.C. Naik and R.P. Agarwala, Jül. Conf. 2, (1968), Vol. 1, p. 105.

# SHAPE MEMORY BEHAVIOUR IN RAPIDLY SOLIDIFIED TiNi ALLOYS

M. IGHARO AND J.V. WOOD
The Open University, Dept. of Materials, Walton Hall, Milton Keynes, MK7 6AA, UK.

ABSTRACT

Equiatomic TiNi alloy has been rapidly solidified by melt spinning. The $M_s$ temperature of ribbon is depressed with respect to samples made from elemental powders. This phenomenon is related to grain refinement. Some of the ribbons have been crushed and the consolidation characteristics of the resulting pre-alloyed powder is compared with that of elemental blends. Sintering activity is higher in elemental compacts as a result of the dominant effect of the alloy formation energy in elemental blends.

## Introduction

Although the process of rapid solidification of shape memory alloys has been extensively investigated [e.g 1-3] most of this work has concentrated on copper-base alloys and little has been reported on other alloy systems eg TiNi. The compound TiNi undergoes a thermoelastic martensitic transformation near room temperature. This reaction is a pre-requisite of the shape memory effect (SME). Alloys based on TiNi are widely used for a number of applications although there are considerable problems involved in processing. In a previous attempt to make TiNi by sintering elemental powders, it was observed that as a result of the unequal interdiffusion coefficients of nickel and titanium, extensive porosity remained in the compact [4].

## Experimental Details

TiNi mixtures were produced by compaction and sintering elemental powders. About 10 gram of the pre-sintered compacts were rapidly solidified in an enclosed melt spinning unit under an argon atmosphere. Quartz crucibles were employed with an 0.8 mm orifice diameter. Melt was quenched on to a 230 mm diameter water cooled brass wheel with a peripheral speed of 27 $ms^{-1}$ using an ejection pressure of 100 KPa. Under these conditions, ribbons of about 3 mm width and up to 50 $\mu m$ in thickness were produced. Subsequently some of the ribbons were comminuted to powder followed by die compaction and vacuum sintering. Samples for micro-structural examination were etched in a mixture of 6 parts of HF, 3 parts of $HNO_3$ in water. The transformation temperatures of RS and sintered specimens were determined using a DuPont 1090 differential scanning calorimeter (DSC).

## Results and Discussion

As Solidified Structures: Scanning electron micrographs of rapidly solidified ribbon in both the longitudinal axis and plane cross-section near the bottom surface are shown in Fig. 1. The structure consists predominantly of columnar grains extending from the wheel contact surface through the entire thickness for thin ribbons (1a), although equiaxed grains are found in thicker material near the top surface (1b). Fig 1c demonstrates the degree of grain refinement resulting from melt spinning compared with material made from elemental powders. The grain size ranges from 0.5 to 10 $\mu m$ with a mean value of 4 $\mu m$ which is at least an order of magnitude smaller.

 

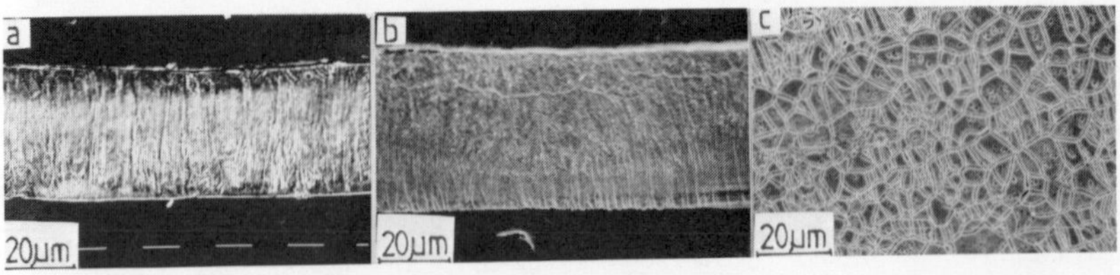


Fig 1 Scanning electron micrographs of RS ribbon
(a,b) longitudinal cross sections
(c) plane section near the bottom surface

Transformation Behaviour: Differential scanning calorimetry (DSC) show that the martensitic start temperature ($M_s$) of this alloy is depressed by rapid quenching. A mean $M_s$ value for spun ribbon is 16°C as compared with 79°C for the larger grained starting material. Typical traces are shown in Fig. 2. Apart from the depression in $M_s$, another notable difference between the transformation behaviour of the two materials is that the exothermic peak associated with the martensitic reaction for the starting material is narrow and clearly defined while that for the RS specimen is broad and appears in two steps.

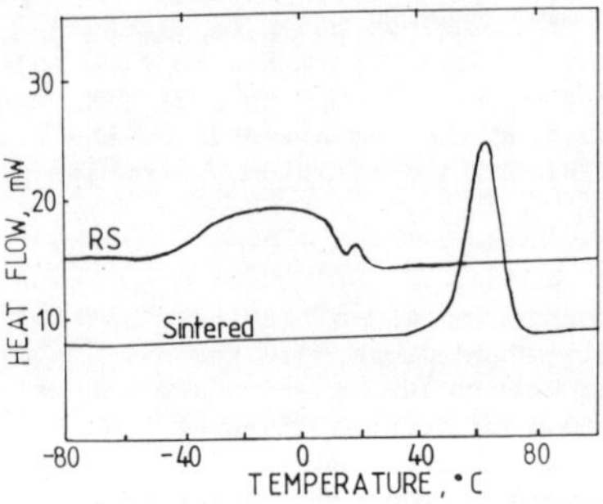


Fig 2 Differential scanning calorimetry cooling curves for spun ribbon and sintered specimen.

It is well known that high-angle grain boundaries act as effective barriers to martensitic transformation [5] and it would be expected that the fine-grained ribbon with a higher grain boundary area would give a lower $M_s$ than the coarse-grained material. This has also been observed in Cu-base SME alloys [2]. Some lattice defects, such as quenched-in vacancies and high dislocation density induced by differential cooling conditions across the ribbon may also affect the $M_s$ but these are known to enhance martensitic transformation and thus tend to raise $M_s$. There is sufficient evidence [e.g. 6, 7] that TiNi specimens quenched from temperatures above the diffusion-controlled order-disorder reaction temperature (600-700°C) have different martensitic structures from those quenched from below. In the present case there is the possibility of partially suppressing the ordering reaction by rapid solidification processing. The effect of ordering on martensitic transformation temperature varies according to the alloy system. For example, while $M_s$ decreases with increasing degree of ordering in shape memory alloys such as Cu-Zn [8] and $Fe_3Pt$ [9], a rise is observed in case of CuZnAl [3,8]. For TiNi alloys, however, the effect of the state of ordering is unclear. Ageing results from treatments on spun ribbons above and below the order-disorder transition range and the possible effect of ordering on the transformation behaviour of this alloy are now being correlated.

Apart from processing variables, the martensitic transformation kinetics of TiNi alloys is highly sensitive to impurities such as oxygen contamination [10]. A small increase in the oxygen content significantly decreases the transition temperature and slows down the rate of reaction and also causes a drastic increase in hardness [11]. In the present case melt spun ribbon has a mean microhardness value of 300 VPN which is not significantly different from 289 VPN for the starting material. This suggests that the change in oxygen content during melt spinning is minimal. Some runs in which the molten material reacted with oxygen from the quartz crucible produced brittle ribbons with hardness values in excess of 450 VPN. A careful combination of fast heating and early ejection of the molten material produced ductile ribbons with low contamination.

Consolidation Response: Since the ribbon was ductile, the yield of the ribbon-powder conversion process was low. This can be enhanced by a hydriding-dehydriding process. The milled particles have a higher hardness than elemental powders and thus gives a lower green density for the same pressing conditions. For example, at a compaction pressure of 774 MPa, the elemental powders achieve a green density of 81% of the theoretical value, while the equivalent value for the pre-alloyed RS powder is 66%. However, unlike the elemental blend compacts which decrease in density during sintering, densification takes place for RS pre-alloyed powders. The fall in density of elemental samples has been related to the difference in interdiffusion rates of titanium and nickel and the presence of a transient liquid phase above the eutectic temperature [4].

The microstructures of pre-alloyed powder compacts pressed at 774 MPa and sintered at various temperatures for a standard time of 5 hours are presented in Fig. 3. The corresponding results for the elemental blends are also given for comparison. At 700° C (Fig. 3a), interdiffusion of titanium and nickel starts in the elemental specimens and compositional homogenisation is complete at 1000° C (Fig. 3b,c). The amount of porosity also increases with sintering temperature. In contrast, RS pre-alloyed powder compacts show no evidence of particle sintering even at 1050°C (Fig. 3d). Only at 1100°C is interparticle neck growth observed leading to densification of the compact (Fig. 3e-f).

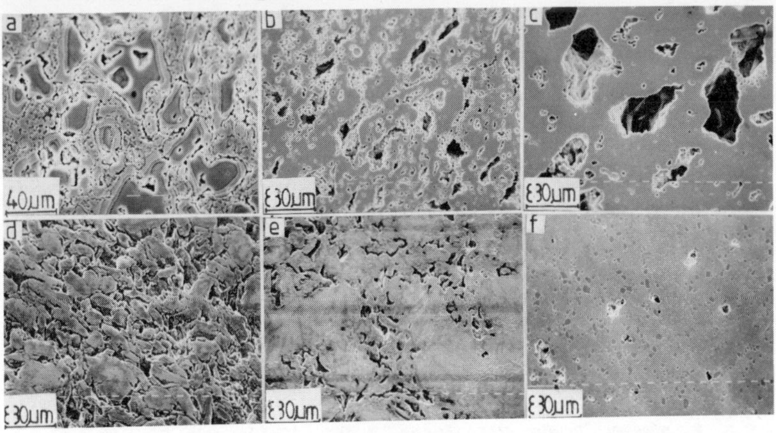


Fig 3 Microstructures of elemental powder compacts sintered at (a) 700°C (b) 900°C (c) 1000°C and pre-alloyed particle compacts sintered at (d) 1050°C, (e) 1100°C and (f) 1150°C.

Although the problem of uneven interdiffusion rates of Ti and Ni in elemental blend compacts has been overcome by using pre-alloyed RS powder, the results show that sintering activity is greater in elemental powder specimens. The main driving force for sintering in pre-alloyed powder is a reduction of surface energy. In elemental blends, there is an additional driving force for alloy formation. The magnitude of these two driving forces has been derived [12] and the alloy formation energy is about four orders of magnitude greater than the total surface energy contained in the pressed compact. This accounts for the higher sintering activity in elemental samples. Despite the need to sinter the RS powder at much higher temperatures the grain size of the resulting compact is still smaller than that found in elemental blends. The grain diameter of the pre-alloyed compact sintered at 1120°C varies from 10 to 20 μm while that of elemental mixture sintered at 1000°C ranges from 20 to 40 μm (Fig. 4). RS pre-alloyed particulate therefore can be employed to process fine grained TiNi compacts with a low Ms temperature.

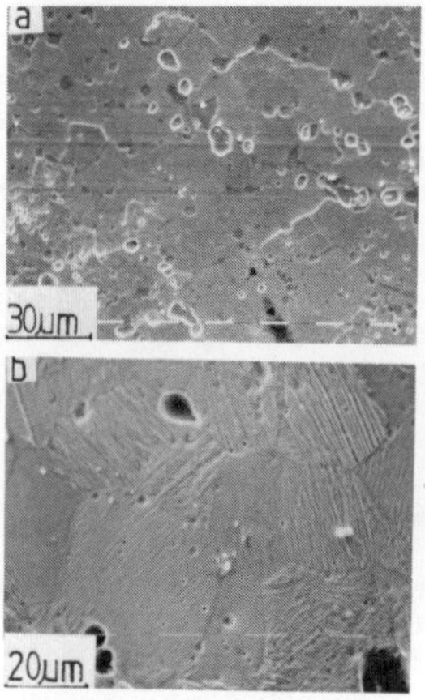


Fig 4 Scanning electron micrographs of (a) pre-alloyed (5 hr at 1120°C) and (b) elemental blends (5 hr at 1000°C compacts showing the difference in grain size.

References

1. J.V. Wood, A. Crossley, and W.M. Stobbs, in Proc 3rd Int. Conf. on Rapidly Quenched Metals, edited by B. Cantor (Metals Society, London, 1978), p.180.
2. J.V. Wood and P.H. Shingu, Metall. Trans. 15A 471 (1984).
3. J. Perkins, Metall, Trans. 13A, 1367 (1982).
4. M. Igharo and J.V. Wood, Powder Metallurgy, 28, 131 (1985).
5. Z. Nishiyama, Martensitic Transformations (Academic Press, London, 1978) pp.238-289.
6. R. G. De Lange and J.A. Zijderveld, J. Appl. Phys., 39 2195 (1968).
7. F.E. Wang, W.J. Buehler, and S. J. Pickart, J. Appl. Phys., 36 3232 (1965).
8. R. Rapacioli and M. Ahlers, Acta Metall., 27 777 (1979).
9. M. Foos, C. Frantz and M. Gantois, in Shape Memory Effects in Alloys, edited by J. Perkins (Plenum Press, New York, 1975). p.407.
10. R. J. Wasilewski, S.R. Butler and J.E. Hanlon, Met. Sci. J. 1 104 (1967).
11. S. E. Rogers, Powder Metallurgy, 7 249 (1961).
12. M. Igharo and J.V. Wood, to be published.

PART V B

# Microstructure and Properties of Crystalline RSP Alloys—Nickel, Iron and Cobalt Alloys

# MECHANICAL PROPERTIES OF RAPIDLY SOLIDIFIED NICKEL-BASE SUPERALLOYS AND INTERMETALLICS

A. I. TAUB AND M. R. JACKSON
General Electric Corporate Research and Development, PO Box 8, Schenectady, New York 12301

## ABSTRACT

The improvements in the mechanical properties of nickel-base alloys that have been made possible by rapid solidification processing are reviewed. The results of processing by powder metallurgy, laser melting, low pressure plasma deposition and spray forming are examined. In general, the increased homogeneity obtained by rapid solidification allows for increased alloying and improved hot workability. The refined grain size improves the low and intermediate temperature strength, but leads to lower strengths at high temperature. For the high temperature applications, post solidification grain growth is required, as for example the directional recrystallization of powder metallurgy preforms. The development of a novel means of producing a fine dispersion from amorphous alloy precursors and the recent work attempting to improve the ductility of the intermetallic phases NiAl and $Ni_3Al$ are also described.

## INTRODUCTION

Over the past three decades nickel-base alloys have been continuously improved in their high temperature strength and creep and fatigue resistance by advances in chemistry and processing. This paper will review those advances which have been made possible by rapid solidification processing. Before proceeding with that discussion however, it is necessary to clarify which processes should in our opinion be classified as rapid solidification. The problem arises because the definition of rapid solidification processing is somewhat arbitrary. We choose to follow the suggestion of Cohen, et.al. [1] that the transition from conventional to rapid solidification occurs at cooling rates of about 100 K/s. More precisely, rapid solidification processing encompasses those techniques which produce microstructures that are either more refined, more homogeneous or completely novel when compared to the structures obtained by solidifying at rates below 100 K/s.

It is convenient to group the advances obtained in nickel-base alloys through rapid solidification processing into two categories. The first category consists of those developments which take advantage primarily of the increased homogeneity provided by rapid solidification and thereby allow for increased alloying to obtain improved mechanical properties [2-5]. Much of the early work in rapid solidification of nickel-base alloys falls into this category which can be described more in terms of process development than alloy development. The most prominent advances have been made in powder metallurgy (P/M) processing. We also include in this category the development of low pressure plasma deposition, layer glazing and spray forming. The second category of interest consists of alloy development studies which have been tailored to take advantage of rapid solidification processing, as for example the introduction of a fine dispersoid into superalloys. More recently, the focus has broadened to include not only complex superalloy-type compositions but also work in single-phase nickel-base intermetallic compounds as well.

In this paper, the developments in both categories of research will be reviewed. We start with a brief discussion of the microstructural aspects of rapid solidification as they apply to the mechanical properties of nickel-base alloys. The various processes designed to produce rapidly solidified nickel-base alloys will then be described and the mechanical properties obtained with these new techniques will be presented. Attention will then turn to the more recent investigations in which new alloys are being developed specifically for rapid solidification processing.

## MICROSTRUCTURAL CONSIDERATIONS FOR MECHANICAL BEHAVIOR

The microstructural aspects of rapid solidification are well established [1,6-8,13]. Among the important attributes of rapid solidification are refined grain size, extended solid solubility, increased homogeneity, uniform dispersion of fine precipitates and metastable phase formation. As will be shown in the following sections, several of these attributes including reduced segregation and the introduction of fine dispersoids are expected to improve the performance and processability of nickel-base superalloys. In addition, the refined grain size will result in improved ambient temperature strength. Figure 1 shows the room temperature yield strength of two rapidly solidified nickel-base superalloys as a function of grain size [9]. The data shows that a Hall-Petch type dependence is obeyed by these alloys. Strength improvements attributed to refined grain size have also been reported for other melt-spun superalloys (see Table 1). For the Nimonic 80A ribbon, an increase of about 10% in hardness was reported through the ribbon thickness indicating that the finer and more homogeneous structure at the chill surface resulted in improved properties compared with the slower-cooled top surface.

While the improvements in ambient temperature strength due to refined grain size are significant, these materials are designed for elevated temperature use and in this regime coarse grains are desired to reduce grain boundary sliding. Figure 2 shows the dependence of tensile strength on temperature for Rene 80 and IN 738 produced by plasma deposition and by conventional casting [12]. The increased strength due to the fine structure of the rapidly solidified product is evident only at low temperatures and the fine grains actually weaken the alloys at high temperature. Therefore, most applications will require some form of post-solidification grain growth.

## IMPROVED PERFORMANCE OF "CONVENTIONAL" SUPERALLOYS BY RAPID SOLIDIFICATION PROCESSING

### P/M Processing

For the most part, the improvements in the elevated temperature capabilities of blade materials have been accomplished by increasing the content and controlling the distribution of $\gamma'$ ($Ni_3Al$, Ti, Ta, Nb); solution hardening by Co, Mo, and W; and introducing carbide formers (Ta, Nb, Hf) [3]. For this engine component, the original interest in P/M processing was not for microstructural considerations but for the introduction of intricate cooling passages and near net shape processing [14]. Only recently has the application of powder processing for blades been considered in terms of the microstructural benefits afforded by rapid solidification processing. The application of interest in this regard is the production of directional recrystallization preforms. This process provides a good example where the benefits of rapid solidification processing

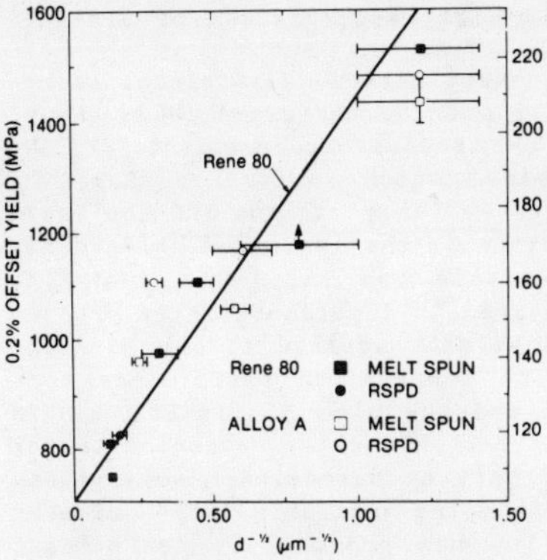


Figure 1. Yield strength at room temperature as a function of inverse square root of the grain size for two nickel-base super alloys processed by melt spinning and plasma deposition (RSPD) [9].

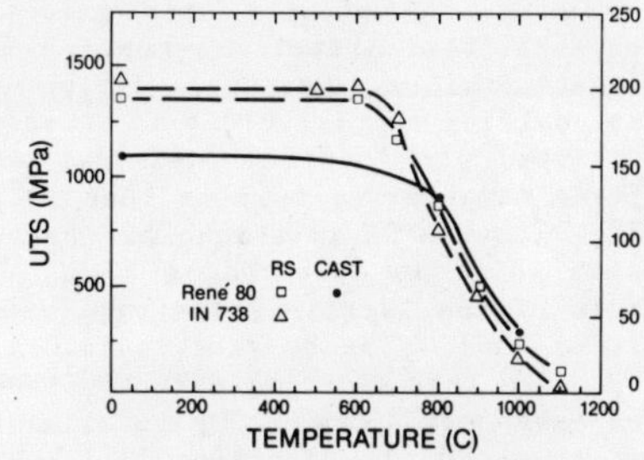


Figure 2. Tensile strength of two commercial nickel-base superalloys processed by plasma deposition (RS) and ingot metallurgy (CAST) [12].

TABLE 1

ROOM-TEMPERATURE MECHANICAL PROPERTIES OF

MELT SPUN VS. CONVENTIONALLY CAST SUPERALLOYS

| Alloy | Process | Tensile Strength (MPa) | Hardness (Vickers) |
|---|---|---|---|
| Rene 80 | Conventional | 1107 ± 5 | - |
| [10] | Melt Spun | 1117 ± 10 | - |
| IN738LC | Conventional | 1107 ± 20 | - |
| [10] | Melt Spun | 1152 ± 24 | - |
| FSX414 | Conventional | 828 ± 5 | - |
| [10] | Melt Spun | 1365 ± 40 | - |
| Nimonic80A | Conventional | - | 180 |
| [11] | Melt Spun | - | 260 |
| Nimonic115 | Conventional | - | 370 |
| [11] | Melt Spun | - | 465 |
| IN 100 | Conventional | - | 340 |
| [11] | Melt Spun | - | 500 |

can be utilized while overcoming the grain size restrictions for elevated temperature.

Since γ' precipitates impede grain boundary motion, directional recrystallization (DR) of nickel-base superalloys must be performed above the γ' - solvus. To optimize this process, the temperature gradient at the recrystallizing interface should be maximized which in turn requires the largest possible temperature difference between the γ' solvus and the incipient melting temperature. As shown in figure 3, the increased homogeneity associated with rapid solidification processing can raise the incipient melting temperature by more than 50°C compared to ingot metallurgy product [15]. The P/M-DR technique has been successfully applied to the Ni-Al-Mo system which is expected to have good high temperature performance as a result of the large γ-γ' lattice mismatch and the high γ' content (70-80 vol. %) [16]. The DR process is desirable for this alloy because attempts to produce this material directly from the melt by directional solidification have been hampered by rotation away from the desirable [112] orientation to the [111] direction [17]. The alloy was produced by centrifugal atomization [18], consolidated by HIP or hot extrusion and subsequently directionally recrystallized. A significant improvement in time to 1% creep was obtained compared to one of the best single crystals presently in use, directionally solidified MAR-M200 + Hf [2].

P/M processing has also been applied to engine disk materials resulting in (a) P/M materials designed specifically to benefit from the RS effect on homogeneity such as APK-1, AF-115, or MERL-76 and (b) materials first developed as cast and wrought materials and later modified chemically to be compatible with the P/M process, such as Rene' 95, Astroloy or IN-100. The increased homogeneity permits greater alloying while still retaining good workability. Space limitations preclude discussion of these materials and the reader is referred elsewhere [2-4,13].

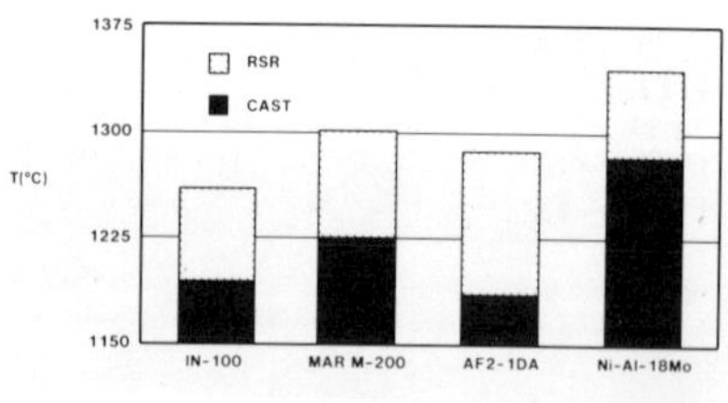


Figure 3. Incipient melting temperature for several nickel-base superalloys processed by centrifugal atomization (RSR) and ingot metallurgy [15].

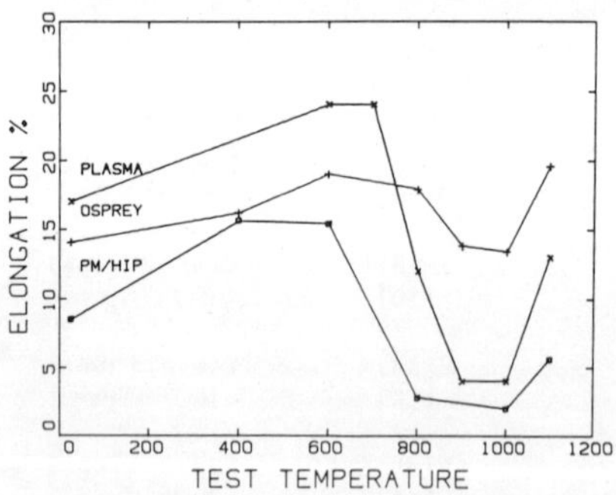


Figure 4. Tensile elongation as a function of test temperature for heat treated ene 80 produced by various rapid solidification techniques: low pressure plasma deposition, spray forming and HIP'ped powder. The respective oxygen contents are about 450, 40 and 250 ppm [25].

## Liquid Dynamic Compaction

Liquid dynamic compaction or equivalently "spray atomization and collection" [20] attempts to eliminate the solid state consolidation step of powder processing by collecting the atomized metal while it is still in the liquid state. Two processes are presently under study which involve deposition of molten droplets on a substrate. In the spray forming process a stream of molten metal is gas atomized much as in powder production. However, before the particles are cooled significantly, they impact on a mandrel and form a deposit. Since there is a spectrum of liquid droplet sizes in the atomized stream, some particles arrive at the mandrel fully liquid, some partially liquid, and some fully solid. Control of heat extraction during atomization is critical to avoid large molten pools (excess heat retained) or gas porosity (insufficient heat retained). The process, as developed by Osprey Metals Ltd. [21-24], produces a product that is low in oxygen content, relatively fine grained (5-10 μm), and reproduces properties of cast and wrought alloys with similar microstructures.

A recent study of Rene' 80 produced by atomization deposition [25] showed the deposit in heat-treated form to be fully dense, clean (50 ppm oxygen), and to be as strong or stronger than conventionally cast Rene' 80 at low and intermediate temperatures. Because of the low oxygen content it showed excellent ductility up to 1100°C (figure 4).

Subsequent detailed characterization of the process [26] has extended the earlier study to include both IN-718 and Rene' 95. The effect of atomizing gas (nitrogen vs argon) on the spray formed product was evaluated. It was found that the nitrogen-atomized deposits were greater than 99% of theoretical density while the argon-atomized deposits had roughly twice the porosity. Oxygen contamination from the spray forming process is quite low (< 30 ppm increase over the remelted ingot) whether nitrogen or argon is used but nitrogen incorporation is quite high for the former (150-300 ppm). In contrast to the earlier study [25] which found improved properties for spray formed Rene 80 compared to the cast product, the spray formed IN-718 and Rene 95 were inferior to their cast and wrought counterparts. Post-solidification deformation processing should improve the properties of the spray formed product.

A second dynamic compaction method is the low pressure plasma deposition process. Powders are fed into the jet of a high velocity inert gas plasma where they are melted, transported and rapidly solidified at the mandrel [12]. Low pressure during processing produces dense structures (>97% theoretical density) with oxygen contents essentially equivalent to that of the powder stock. The process was originally developed to apply coatings to superalloy materials, but the density and cleanliness suggested it is a process to produce the superalloy structure itself. The fine grained structures are much stronger than cast structures of the same alloys, particularly at temperatures below 750°C. At elevated temperatures, strengths are somewhat less than for coarse grained castings (see figure 2). Annealing of the structures just below the melting temperature produces a grain size of 25-50 μm in strong γ' strengthened alloys. The grain size stability makes these materials particularly attractive for intermediate temperature applications, but unattractive for elevated temperatures. In the intermediate temperature regime, ductility problems can be encountered (figure 4) if care is not exercised relative to powder cleanliness. Batch production powders are typically 300 ppm oxygen for the particle sizes normally used for low pressure plasma deposition. This level leads to the loss in ductility noted in the figure. Note that the same problem can be encountered in P/M products if the oxygen level is high.

## Laser Processing

As discussed above, low pressure plasma deposition of structural components evolved from scaling up of a surface coating process. In a similar sense, layerglazing developed as an extension of laser surface processing of materials. During laser surface treatment the surface is melted followed by self-quenching [27-29]. Depending on the laser power density and interaction time, cooling rates as high as $10^{13}$°C/s can be obtained (picosecond pulses to melt 1000 Å layers [30]). More typical applications exhibit rates on the order of $10^{6}$°C/s [27]. These surface treatments result in a rapidly solidified surface layer leaving the bulk of the material unaffected and therefore offer the potential for improving wear and corrosion resistance. For example, laser surface treatment of a high refractory content Ni base alloy (33 wt % Mo, 17 wt % W) more than doubled the hardness at the surface ($H_v$=510 vs 200 for the bulk) [31]. Electron beam surface treatment has also been demonstrated [32,33].

In order to scale up the surface treatment to build structural components, material is continuously fed into the path of the laser beam [27]. In the application of this technique to nickel-base superalloys (AF115, IN100, AF2-1DA), it was found that the complex state of stress produced by post-solidification cooling led to severe cracking. Therefore, this process appears to be amenable only to those alloys which are known to exhibit good resistance to weld cracking. One family of these alloys are based on compositions near the $\gamma$ - $\alpha$ eutectic trough in the NiAlMo system [34]. These alloys have a lower $\gamma'$ volume fraction for ductility during processing, a lower thermal expansion due to the 15-25 vol % of $\alpha$ -Mo and a narrow solidification temperature range [13]. The available data show that at least one alloy developed for this process exhibits tensile properties approaching IN 100 (Table 2). The greater ductility in the radial direction is due to the radially aligned, columnar grain structure.

TABLE 2

TENSILE PROPERTIES OF LAYER GLAZED NiAlMo ALLOYS [27,34]

| Alloy (at.%) | | 25C YS (MPa) | 25C UTS (MPa) | 25C Ductility (%) | 704C YS (MPa) | 704C UTS (MPa) | 704C Ductility (%) |
|---|---|---|---|---|---|---|---|
| 8Al-12Mo-3Ta | (A) | 1055 | 1317 | 20 | 724 | 724 | 2 |
| | (R) | 1027 | 1248 | 40 | 979 | 1124 | 32 |
| 12Al-15Mo | (A) | 1214 | 1393 | <6 | 1179 | 1241 | <6 |
| | (R) | 1207 | 1393 | 26 | 1180 | 1241 | 13 |
| IN 100 | | 1117 | 1558 | 22 | 1089 | 1241 | 18 |

A- axial specimen
R- radial specimen

## ALLOYS DEVELOPED FOR RAPID SOLIDIFICATION PROCESSING

### Fine Dispersions via Amorphous Alloy Precursors

It has been known for some time that fine boride precipitates are capable of strengthening nickel-base superalloys [35,36]. However, the degree of boride alloying in conventional ingot metallurgy has been limited by extensive segregation during solidification and the formation of large, brittle borides at cell and grain boundaries. When precipitates are large, the strengthening potency is reduced and strain localization can occur. The potential of rapid solidification as a means of distributing the precipitates homogeneously by supersaturating the matrix with boron followed by precipitation of a fine distribution of boride phase has generated considerable interest in this area. It has been shown that rapid solidification allows as much as 4 at. % boron to be incorporated in solid solution in nickel base alloys [37] and that subsequent heat treatment results in a fine boride precipitation [38]. However, the more interesting observation is that amorphous phases can be produced at boron levels as low as 5 at. % [38]. The amorphous phase is simply the kinetically frozen liquid state. Therefore, production of an amorphous alloy precursor allows for well controlled devitrification (i.e.- crystallization) in the solid state from a completely homogeneous starting material with the ability to form a fine dispersoid from the metalloid additions. The metalloid serves the dual purpose of promoting glass formation in the precursor material and then as carbide, boride or silicide dispersoids stabilizing the fine matrix grain size that results from devitrification [39]. This approach has led to creation of a new class of nickel base alloys classified as "pyromets" [8,40].

The compositions of the pyromets are formulated with certain critical combinations of transition metals and metalloids which allow the devitrified material to attain its ductility upon heat treatment at high temperatures. Most amorphous alloys have eutectic/hypereutectic compositions with approximately 20 at. pct. boron, phosphorus, carbon or silicon. These new alloys contain these metalloids in ranges of 5-13%, boron being the dominating metalloid [42]. The superior mechanical properties of the new devitrified alloys have their origin in their microstructures. The isotropic glassy phase is crystallized at high temperature (preferably .6-.95 $T_{solidus}$) to produce an aggregate of microcrystalline phases. Annealing below 0.6 $T_{solidus}$ results in sluggish kinetics of transformation and even after long annealing times, the devitrified material remains brittle and weak [43]. TEM studies show that the alloys consist of extremely fine scale, homogeneous microstructures (0.2-0.8 μm grains) that are uniformly dispersed with 0.1 to 0.2 μm borides (20-30 vol %). The result is high hardness (> 500 VHN), high tensile strength (> 200 ksi) and good ductility.

The most promising alloys to date are from the Ni-Mo-B and Ni-Cr-B families. The Cr additions are for increased corrosion resistance [44]. Examination of the ternary Ni-Mo-B system [45] has shown that the hardness monotonically increases as a function of both boron content due to the dispersoid strengthening of $Ni_2MoB_2$ [46] and Mo concentration through solid solution strengthening. In addition, when the Mo/Ni ratio in the matrix exceeds 0.25, the hot hardness is enhanced through the presence of ordered $Ni_4Mo$ or $Ni_3Mo$ depending on composition and heat treatment [46]. Moreover, the strengths are retained at elevated temperature (see figure 5).

Two of these alloys are presently entering the commercial markets: $Ni_{54}Mo_{44}B_2$ (Devitrium 3065) and $Ni_{53}Mo_{35}Fe_9B_2$ (Devitrium 7025) [41]. Devitrium 7025 has been used as a hot extrusion die insert resulting in 2-3 times the life of the Stellite 6 alloy it replaced [47] and as an insert in the gate area of a mold for aluminum die casting resulting in a similar life extension over the H-13 insert it replaced [48]. Devitrium 3065 is designed for high speed cutting and turning operations, and has demonstrated twofold improvement over M42 type tool steel [41]. The powders are produced by Amax Specialty Metals under agreement with Allied Corporation, and are available for about $75/ kg [49].

## Rapid Solidification of Intermetallic Compounds

The early studies in rapid solidification processing of nickel-base alloys for structural components were limited to polyphase materials such as the γ-γ' superalloys and dispersion strengthened alloys discussed above. More recently, attention has focused on rapid solidification processing of single-phase intermetallics, particularly NiAl and $Ni_3Al$. Interest in NiAl stems from the high temperature strength provided by the ordered structure and high melting temperature, and the oxidation resistance and decreased density offered by the high aluminum content. Moreover, the alloy has large solubilities for substitutional third element additions, allowing for substantial strengthening [50,51]. The alloy is limited in application however because of its extreme brittleness at ambient temperature [52,53]. The interest in $Ni_3Al$ is due to its anomalous temperature dependence of strength. $Ni_3Al$ is a member of the family of alloys which exhibit an increasing flow stress with increasing temperature [54-56]. Until recently, this alloy has been considered too brittle for structural applications and use of the unique high temperature behavior has been limited to precipitate strengthening of the γ-γ' superalloys. The discovery of improved ductility through boron-doping [57] is now allowing consideration of its use as a single phase material.

In this section, the attempts to improve the performance of these two compounds by rapid solidification processing is described with emphasis on the attempt to improve ductility and strength. For a more complete discussion of these alloys the reader is referred to the recent review by Stoloff [58].

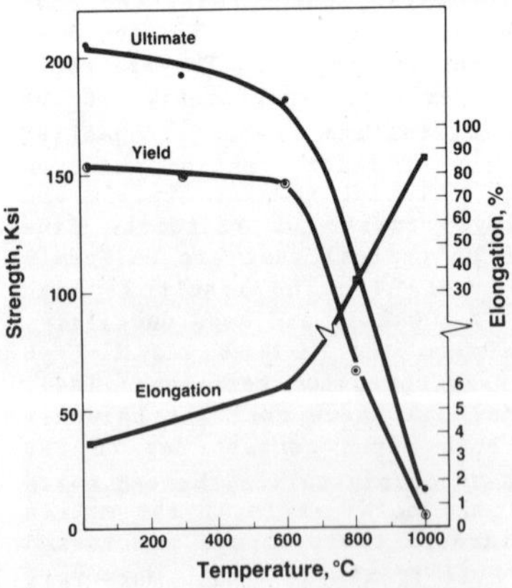


Figure 5. Tensile properties vs temperature for as-HIP'ped and devitrified $Ni_{56.5}Fe_{10}Mo_{23.5}B_{10}$ [45,46].

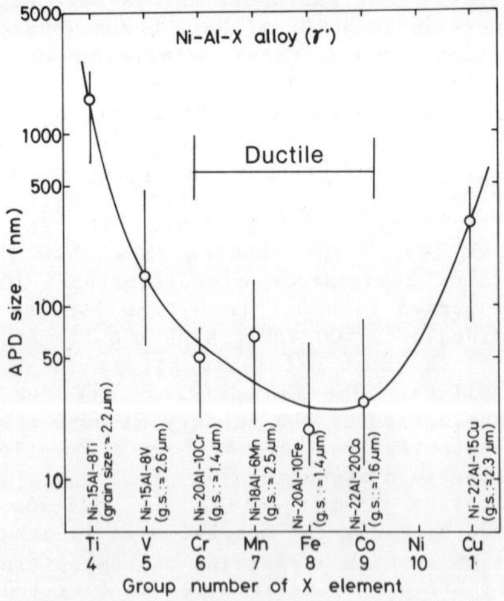


Figure 6. Variation of average anti-phase domain size and average grain size in melt spun ribbons of $Ni_3Al$ alloyed with the indicated elements [60].

## $Ni_3Al$-Base $L1_2$ Alloys

Both conventionally cast and rapidly solidified high purity $Ni_3Al$ are extremely brittle, exhibiting pure intergranular fracture with negligible plastic strain to failure [59]. However, when the compound is alloyed with Cr, Mn, Fe, Co or Si and processed by rapid solidification, high ductilities are obtained [60-63]. Additions of Ti, Zr, V, Nb or Cu do not make the rapidly solidified alloy ductile. These ductility results have been shown to correlate with the density of anti-phase boundaries (APB's) in the material. As can be seen from figure 6, the higher the density of APB's in the as-solidified material, the greater the ductility of the alloy. The effect of post-solidification heat treatment on ductility also appears to correlate with APB density. Annealing the rapidly solidified product results in growth of the anti-phase domains (starting size about 40 nm) eventually leading to complete disappearance of the APB's when each grain is occupied by a single domain (grain size about 2 μm). The annealing also results in a loss of ductility. The ductility loss is accompanied by a change in fracture morphology from dimpled to intergranular and does not appear to be associated with grain growth.

It is worth noting that the introduction of a high APB density is obtainable only by rapid solidification. The ordered $Ni_3Al$ structure remains stable up to melting and the ordering kinetics are very rapid at elevated temperature, making it impossible to achieve disorder by solidification at low rates or by heat treatment in the solid state. Correspondingly, no APB's have been reported for $\gamma'$ solidified at conventional cooling rates. However, a recent investigation has shown that Fe and Mn additions to $Ni_3Al$ result in high ductilities even in conventionally cast material [64]. A study of boron-doped $Ni_3Al$ produced by rapid solidification also showed no correlation of APB structure and ductility [65-67]. Additional work is required to clarify these results in terms of the APB correlation. At this time there is no satisfactory model to explain the empirical relationship between APB density and ductility.

## NiAl-Base $L1_0$ Alloys

Similar results to those just described for the Ni-Al-X $L1_2$ alloys have been reported for $L1_0$ alloys modified with Fe or Co, although the ductilities are considerably lower and the fracture surface is never completely dimpled but contains areas of quasi cleavage [60,63]. In fact, the ductility decreases continuously as the alloying content is varied to change the alloy from pure $\gamma'$ to a mixture of $\gamma'$ and $\beta$ to pure $\beta$. Once again, the ductility does not appear to correlate with grain size.

Although the above discussions have indicated that the refined grain size produced by the rapid solidification is not responsible for the enhanced ductility, a model analysis has suggested that tensile ductility could be imparted to brittle intermetallic polycrystals by refining grains to sizes smaller than a critical value determined from considerations of crack nucleation and propagation [68]. This prediction has been verified for the NiAl system which exhibited dramatic improvements in ductility with decreasing grain size [69,70]. In that study, the grain size was varied by controlling the recrystallization anneal of doubly hot-extruded ingot material. For grain sizes below 20 μm, large elongations ( >10%) were obtained at temperatures as low as 295C. At ambient temperature the ductility was still low. In a later study, free jet melt spinning was employed to produce a fine grained structure directly from the melt [71,72]. The resulting ribbon consisted of columnar grains extending through the ribbon thickness (5 μm diameter by 15 μm long). The room temperature ductility was quite low (1-2% bending strain) but areas of transgranular cleavage were observed particularly after annealing. No tests were performed at the higher temperatures where the thermomechanically processed material exhibited high ductilities, so it is not clear at this time if the fine grained structure produced by the melt spinning will have the same effects as the thermomechanically processed material.

## Boron-Doped $L1_2$ Alloys

At ambient temperature NiAl has only three active, independent slip systems {hk0}⟨001⟩ [53], so limited polycrystalline ductility is expected since extensive deformation requires five operable slip systems. On the other hand, $Ni_3Al$-type compounds slip on the {111}⟨110⟩ system and therefore do not suffer from this restriction. For example, at room temperature single crystals of $Ni_3Al$ have been shown to exhibit high ductilities in all crystal orientations [73]. Nevertheless, the high purity, polycrystalline alloy fractures intergranularly with negligible plastic strain to failure [74,75]. The poor ductility is due to premature fracture along the grain boundaries which are not strong enough to withstand the stresses required for deformation of the matrix grains.

Several years ago it was discovered that boron doping of $Ni_3Al$ suppresses the tendency for intergranular fracture resulting in high ambient temperature ductility [57]. Auger studies have shown that the boron segregates strongly to the grain boundaries of the doped nickel aluminide [76,77]. Grain boundary concentrations of boron greater than 5.0 at. pct. have been observed in alloys doped with less than 1.0 at. pct. boron. The high boron concentration at the boundaries is believed to enhance the grain boundary cohesive strength by strengthening the electronic bonding at the boundary [59,79] and/or changing the free energy of the boundary relative to the free surface [76,80].

In the original study on boron-doping [57], the material was prepared by drop casting in a chill mold. In scaling up of this process, it was found that the ingot required thermomechanical processing to attain high ductility [78,80]. On the other hand, processing of these same alloys by the various rapid solidification techniques results in high ductilities in the as-solidified condition [59,86]. The only reported exception to this result is a study of arc-hammer produced foils [66] which exhibit ductile behavior for thick foils but brittle behavior for very thin specimens. For the as-solidified alloy, slow cooling or very rapid solidification produce brittle material, while intermediate levels of rapid solidification are required to produce high ductilities.

The phenomenology of the boron-induced ductility of $Ni_3Al$ is well established [59,65-67,76-78,80-83]. Referring to figure 7, it is seen that only aluminum-poor alloys (Ni/Al > 3/1) can be made ductile by boron additions. The ductility is also determined by the level of boron doping. Boron concentrations of less than 0.1 at. pct. are capable of producing high ductilities, but at boron doping levels greater than about 1.5 at. pct., $M_{23}B_6$ borides precipitate at the grain boundaries and the alloy once again exhibits brittle intergranular fracture [59]. These chemistry effects on ductility are common to both the thermomechanically processed ingot metallurgy product and the rapidly solidified material [59,76]. However, the effects of chemistry on strength exhibit some differences. The ingot metallurgy product in either the doped or undoped condition exhibits a flow stress minimum at stoichiometry [54,76,84,85] while the rapidly solidified material exhibits a yield strength minimum at 24 at. pct. aluminum [83]. Another difference is manifested at elevated temperature. Figure 8 shows the yield strength and ductility for a boron-doped $Ni_3Al$-base alloy processed by several rapid solidification techniques [86]. In all cases, a severe high temperature ductility loss is observed. As discussed in the section on dynamic liquid consolidation this loss of ductility at elevated temperature is characteristic of many nickel-base superalloys which have been processed by those rapid solidification processes which result in high internal oxygen levels. It has recently been suggested that the oxygen level in the testing environment is more critical than the inherent oxygen level in the material in explaining the loss of ductility in the $Ni_3Al$ alloys [87,88].

The increase in ductility resulting from boron-doping and rapid solidification is not limited to $Ni_3Al$-base alloys [89]. It has been found that $Ni_3Ga$ behaves in a manner very similar to $Ni_3Al$. $Ni_3Si$ can also be made ductile but only if the material is properly heat treated to eliminate the multiphase structure resulting from solidification, producing only the pure $L1_2$ phase. On the other hand, $Ni_3Ge$ was not made ductile by boron doping. These results are consistent with descriptions of grain boundary cohesive strength based on valency and size effects [90] or electronegativity [89].

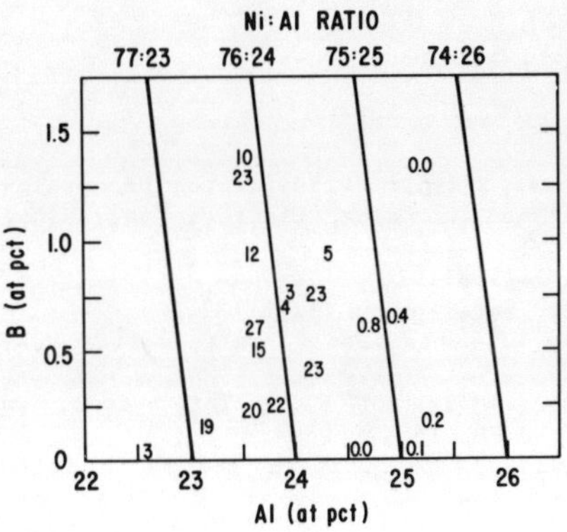


Figure 7. Plastic strain to fracture of melt spun $Ni_3Al$+B as a function of Al and B concentration. The solid lines indicate constant Ni:Al ratio. The ribbons were annealed at 1100C for 2 hours prior to tensile testing [83].

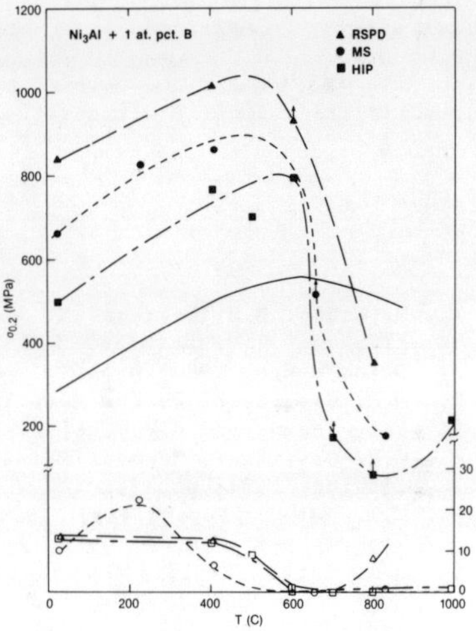


Figure 8. Tensile flow stress and plastic strain to failure of boron-doped γ' processed by three rapid solidification processes: plasma deposition (RSPD), melt spinning (MS) and HIP'ped powder (HIP). The flow stress for thermomechanically processed cast material is included (solid line) [86].

## CONCLUSIONS

The advances in the mechanical properties of nickel-base alloys obtained through rapid solidification processing can be grouped into two categories. The first category consists of those developments based primarily on increased homogeneity and includes the developments in powder metallurgy, low pressure plasma deposition and spray forming. The latter are means of eliminating the solid state consolidation step in powder processing by collecting the atomized metal while it is still molten. The alloys produced by these two processes are stronger than conventionally cast material at low and intermediate temperatures due to the refined grain size but somewhat weaker at high temperature. The buildup of structures by laser melting has not proved to be as generally applicable due to severe cracking on cooling.

The second category encompasses those developments which take advantage of other aspects of rapid solidification including refined grain size, extended solid solubility and metastable phase formation. A novel approach is to alloy with sufficient metalloids so as to form an amorphous phase on solidification. Subsequent devitrification results in a fine dispersion of borides, silicides or carbides. Alloys of this type have recently entered the commercial market. Rapid solidification is also being applied to nickel-base intermetallic phases to produce high ductilities. For NiAl-base alloys, the ductility is enhanced by reduced ordering and refined grain size. For $Ni_3Al$-base alloys, reduced ordering also promotes ductility. In the boron-doped material, rapid solidification produces ductile, as-solidified material without the need for thermomechanical processing.

## REFERENCES


1. M. Cohen, B.H. Kear and R. Mehrabian, "Rapid Solidificaton Processing II", ed. R. Mehrabian, B.H. Kear and M. Cohen, Claitors Publishing, Baton Rouge, 1980, p. 1.
2. G.H. Gessinger, Powder Met. Intl. 13(1981)93.
3. V.A. Tracey and C.P. Cutler, Powder Metallurgy, 1(1981)32.
4. G.H. Gessinger, "Powder Metallurgy of Superalloys", Butterworths Publishing, 1984, pp.348.
5. C.T. Sims, "Superalloys 1984, Proceedings of Fifth Intl. Conf. on Superalloys", ed. M. Gell, et.al., AIME, 1984, p.399.
6. A.I. Taub, "Rapidly Quenched Metals, Proceedings of the Fifth Intl. Conf. on Rapidly Quenched Metals", ed. S. Steeb and H. Warlimont, North Holland, 1985, p.1611.
7. J.C. Williams and J.P. Hirth, "Rapid Solidification Processing III", ed. R. Mehrabian, National Bureau of Standards, 1983, p.135.
8. H. Jones, J. Mat. Sci. 19(1984)1043.
9. A.I. Taub, M.R. Jackson, S.C. Huang and E.L. Hall, "Rapidly Solidified Amorphous and Crystalline Alloys", Mat. Res. Soc. Symp. Proc., 28(1984)389.
10. K. Yasuda, M. Tsuchiyay, T. Kuroda and M. Suwa, "Superalloys 1984, Proceedings of Fifth Intl. Conf. on Superalloys", ed. M. Gell, et. al., AIME, 1984, p. 477.
11. H.A. Davies, N. Shohoji and D.H. Warrington, "Rapid Solidificaton Processing II", ed. R. Mehrabian, B.H. Kear and M. Cohen, Claitors Publishing, Baton Rouge, 1980, p. 153.
12. M.R. Jackson, J.R. Rairden, J.S. Smith and R.W. Smith, J. Metals, 33(1981)23.
13. M.R. Jackson and S.C. Huang, "Mechanical Behavior of Rapidly

Solidified Materials, Proceedings of Symposium, Feb 24-28, 1985", AIME, in press.

14. B. Sugarman and C.T. Jessop, "Symposium on Powder Metallurgy", London, Iron and Steel Institute, 1956, p. 191.
15. R.G. Bourdeau and J.B. Moore, "Rapid Solidification Processing I", ed. R. Mehrabian, B.H. Kear and M. Cohen, Claitors Publishing, Baton Rouge, 1978, p. 334.
16. E.H. Aigeltinger and M. Kersker, Metals Forum, 4(1981)112.
17. E.H. Aigeltinger, A.F. Giamei and D.M. Dimiduk, "Rapid Solidification Processing III", ed. R. Mehrabian, National Bureau of Standards, 1983, p.186.
18. P.R. Holiday and R.J. Patterson, "Apparatus for Producing Metal Powder", U.S. Patent 4,078,873 (1978).
19. R.J. Patterson, A.R. Cox and E.C. VanReuth, J. Metals, 39(1980)34.
20. N.J. Grant, "Rapidly Quenched Metals, Proceedings of the Fifth Intl. Conf. on Rapidly Quenched Metals", ed. S. Steeb and H. Warlimont, North Holland, 1985, p.1611.
21. B. Williams, Metal Powder Report 10(1980)464.
22. R.G. Brookes, A.G. Leatham, G.R. Dunstan and C. Moore, "Powder Metallurgy Superalloys -Aerospace Materials for the 1980's", Metal Powder Report, Shrewsbury, England (1980).
23. R.W. Evans, A.G. Leatham and R.G. Brookes, Proc. Metals Society 1983 Powder Metallurgy Group Meeting, Metals Soc. London (1983).
24. R.G. Brookes, C. Moore, A.G. Leatham and J.S. Coombes, Powder Metallurgy 2(1977)100.
25. R.H. Bricknell, "The Structure and Properties of a Nickel-Based Superalloy Produced by Osprey Atomization-Deposition", Report No. 85CRD043, General Electric Company, Schenectady, NY (1985), also Met. Trans., in press.
26. H.C. Fiedler, T.F. Sawyer and R.W. Kopp, "Spray Forming", Report No. 85CRD073, General Electric Company, Schenectady, NY (1985).
27. D.B. Snow, E.M. Breinan and B.H. Kear, "Superalloys 1980", ed. J.K. Tien, et. al., 1980, ASM, 189.
28. B.L. Mordike and H.W. Bergmann, "Rapidly Solidified Amorphous and Crystalline Alloys", Mat. Res. Soc. Symp. Proc., 28(1984)45.
29. E.M. Breinan and B.H. Kear, "Rapid Solidification Processing I", ed. R. Mehrabian, B.H. Kear and M. Cohen, Claitors Publishing, Baton Rouge, 1978, p. 87.
30. C.J. Lin and F. Spaepen, "Chemistry and Physics of Rapidly Solidified Materials", ed. B.J. Berkowitz and R.O. Scattergood, AIME, 1983, p.273.
31. G. Yunlong, H. Zhuangqi, G. Wei and S. Changhsu, "Rapidly Solidified Amorphous and Crystalline Alloys", Mat. Res. Soc. Symp. Proc., 28(1984)99.
32. P.R. Strutt, J. LeMay and A. Tauqir, "Rapidly Solidified Amorphous and Crystalline Alloys", Mat. Res. Soc. Symp. Proc., 28(1984)87.
33. M. Kurup, A. Tauqir and P.R. Strutt, "Rapidly Solidified Amorphous and Crystalline Alloys", Mat. Res. Soc. Symp. Proc., 28(1984)93.
34. D.B. Snow, "Rapidly Solidified Amorphous and Crystalline Alloys", Mat. Res. Soc. Symp. Proc., 8(1982)523.
35. R.F. Decker and J.W. Freeman, Trans AIME, 218(1960)277.
36. D.H. Maxwell, J.F. Baldwin and J.F. Radavich, Metall. Met. Forming, Oct 1975, 332.
37. V.F. Bashev, I.S. Miroshnichonko and G.A. Sergeev, Russ. J. Inorg. Mat., 17(1981)892.
38. S.C. Huang and K.M. Chang, J. Mat. Sci. 19(1984)1220.
39. B.C. Giessen, D.E. Polk and R. Ray, "High Strength Iron, Nickel and Cobalt Base Crystalline Alloys With Ultrafine Disperion of Borides and Carbides", U.S. Patent 4,297,135, Oct. 27, 1981.

40. R.W. Cahn, Ann. Rev. Mater. Sci., 12(1982)51.
41. D. Raybould, Metal Powder Report, 39(1984)
42. R. Ray, Metal Progress, 125(1982)29.
43. R. Ray, J. Mat. Sci. Letters, 16(1981)2927.
44. S.K. Das and D. Raybould, "Rapidly Quenched Metals, Proceedings of the Fifth Intl. Conf. on Rapidly Quenched Metals", ed. S. Steeb and H. Warlimont, North Holland, 1985, p.1787.
45. C.C. Wan, "Rapidly Solidified Amorphous and Crystalline Alloys", Mat. Res. Soc. Symp. Proc., 8(1982)441.
46. S.K. Das, L.A. Davis, J.R.Y. Wang and D. Kapoor, "Rapid Solidification Processing III", ed. R. Mehrabian, National Bureau of Standards, 1983, p.559.
47. Anon, Metal Progress, 125(1982)38.
48. S. Asheley, Amer. Met. Market Metalworking News, 95(1981)14.
49. Anon, Metal Progress, 128(1985)24.
50. K. Vedula, V. Pathare, I. Aslanidis and R.H. Titran, "High Temperature Ordered Intermetallic Alloys", Mat. Res. Soc. Symp. Proc., 39(1985)411.
51. J.R. Stephens, "High Temperature Ordered Intermetallic Alloys", Mat. Res. Soc. Symp. Proc., 39(1985)381.
52. E.M. Grala, "Mechanical Properties of Intermetallic Compounds", ed. J.H. Westbrook, John Wiley, New York, 1960, 358.
53. A. Ball and R.E. Smallman, Acta Met. 14(1966)1349.
54. R.W. Guard and J.H. Westbrook, Trans TMS-AIME, 215(1959)807.
55. O. Noguchi, Y. Oya and T. Suzuki, Met. Trans. 12A(1981)1647.
56. P.H. Thornton, R.G. Davies and T.L. Johnston, Met. Trans. 1A(1970)207.
57. K. Aoki and O. Izumi, J. Japan Inst. Met., 43(1979)1190.
58. N.S. Stoloff, "High Temperature Ordered Intermetallic Alloys", Mat. Res. Soc. Symp. Proc., 39(1985)3.
59. A.I. Taub, S.C. Huang and K.M. Chang, Met. Trans. 15A(1984)399.
60. A. Inoue, T. Masumoto, H. Tomioka and N. Yano, Int. J. of Rapid Solidification, 1(1984)115.
61. A. Inoue, H. Tomioka and T. Masumoto, J. Mat. Sci. Letters, 1(1982)377.
62. A. Inoue, H. Tomioka and T. Masumoto, Met. Trans. 14A(1983)1367.
63. A. Inoue, H. Tomioka and T. Masumoto, J. Mat. Sci. 19(1984)3097.
64. T. Takasugi and O. Izumi, Acta Met. 33(1985)1259.
65. C.C. Koch, "High Temperature Ordered Intermetallic Alloys", Mat. Res. Soc. Symp. Proc., 39(1985)397.
66. C.C. Koch, J.A. Horton, C.T. Liu, O.B. Cavin and J.O. Scarborough, "Rapid Solidification Processing III", ed. R. Mehrabian, National Bureau of Standards, 1983, p.264.
67. J.A. Horton, C.C. Koch and C.T. Liu, Proceedings 41st Annual Meeting of the Electron Microscopy Society of America, ed. G.W. Bailey, San Francisco Press (1983)248.
68. E.M. Schulson, Res Mech Letters 1(1981)111.
69. E.M. Schulson, "High Temperature Ordered Intermetallic Alloys", Mat. Res. Soc. Symp. Proc., 39(1985)193.
70. E.M. Schulson and D.R. Baker, Scripta Met. 17(1983)519.
71. D.J. Gaydosh, R.W. Jech and R.H. Titran, J. Mat. Sci. Letters 4(1985)138.
72. D.J. Gaydosh and M.A. Crimp, "High Temperature Ordered Intermetallic Alloys", Mat. Res. Soc. Symp. Proc., 39(1985)429.
73. K. Aoki and O. Izumi, J. Mat. Sci., 14 (1979) 1800.
74. T. Takasugi, E.P. George, D.P. Pope and O. Izumi, Scripta Met 19(1985)551.
75. T. Ogura, S. Hanada, T. Masumoto and O. Izumi, Met Trans 16A(1985)441.
76. C.T. Liu, C.L. White and J.A. Horton, Acta. Met. 33 (1985) 213.
77. S.C. Huang, A.I. Taub, K.-M. Chang, C.L. Briant and E.L. Hall,

"Proceedings of the Fifth International Conference on Rapidly Quenched Metals, Wurzburg, Germany, Sept. 1984", ed. S. Steeb and H. Warlimont, North Holland, Amsterdam (1985)1407.
78. C.T. Liu, C.L. White, C.C. Koch and E.H. Lee, Proceedings of the Electrochemical Society on High Temperature Materials, ed. Munir Cubicciotti, Vol. 83-87, Electrochemical Society Inc. (1983), p. 32.
79. R.P. Messmer and C.L. Briant, Acta Met 30(1982)457.
80. C.T. Liu and C.L. White, "High Temperature Ordered Intermetallic Alloys", Mat. Res. Soc. Symp. Proc., 39(1985)365.
81. S.C. Huang, A.I. Taub and K.M. Chang, Acta Met. 32(1984)1703.
82. K.M. Chang, A.I. Taub and S.C. Huang, "High Temperature Ordered Intermetallic Alloys", Mat. Res. Soc. Symp. Proc., 39(1985)335.
83. A.I. Taub, S.C. Huang and K.M. Chang, "Failure Mechanisms in High Performance Materials, Proc. 39th Meeting of Mechanical Failures Prevention Group", ed. J.G. Early, T.R. Shives and J.H. Smith, Cambridge University Press (1985)57.
84. J.A. Lopez and G.F. Hancock, Phys. Stat. Sol. A 2(1970)469.
85. R.D. Rawlings and A.E. Staton-Bevan, J. Mat. Sci. 10(1975)657.
86. A.I. Taub, S.C. Huang and K.M. Chang "High Temperature Ordered Intermetallic Alloys", Mat. Res. Soc. Symp. Proc., 39(1985)221.
87. C.T. Liu, C.L. White and E.H. Lee, Scripta Met, in press.
88. C.T. Liu and C.L. White, Acta Met., in press.
89. A.I. Taub, C.L. Briant, S.C. Huang, K.M. Chang and M.R. Jackson, Scripta Met., in press.
90. T. Takasugi and O. Izumi, Acta Met. 33(1985)1247.

# THE STRENGTHENING MECHANISM IN CONSOLIDATED RAPIDLY SOLIDIFIED ALLOYS

MONDE A. OTOONI*
U.S. Army Research Office, Physics Division, Research Triangle Park, NC 27709-2211

## ABSTRACT

Attempts have been made to study mechanisms of strengthening in the $Ni_{68}W_{22}C_8B_2$ and the $Ni_{45}Co_{20}Cr_{10}Mo_4Fe_5B_{16}$ alloy systems during their amorphous crystalline transitions. In the $Ni_{68}W_{22}C_8B_2$ system where amorphous particulates of varying sizes have been consolidated at high pressure and low temperature (3.6 MPa, and 673°K respectively), the initial stage of crystallization is marked by transformation of the localized regions of the specimen. These crystallized regions contain microcracks and voids. Microhardness measurements from the consolidated specimens indicate an increasing trend in the microhardness with decreasing particulate sizes. Premature failures of the consolidated specimen during tensile stress measurements have been attributed to the presence of microcracks and voids in these specimens. In the $Ni_{45}Co_{20}Cr_{10}Mo_4Fe_5B_{16}$ alloy system isothermal annealing of an initially amorphous alloy has been allowed to produce grains of varying sizes. The tensile stress measurements from these thermally annealed ribbons indicate two distinctly different functional relationships between the strength, $\sigma$ , and the grain size parameter, $\lambda$. In the early stage of transformation where grain reach a maximum growth of up to 400 Å, the functional form of the strength, $\sigma$ , with the grain parameter, $\lambda$ , is $\sigma = \sigma_o + K \log \lambda$ , where $\sigma_o$ and K are constants. During latter stages of transformation, where grains larger than 400 Å have been formed, the strength, $\sigma$ , varies with the inverse square root of the grain sizes. This latter relationship is analogous with the well known Hall-Petch relationships, which describes the strength, $\sigma$ , as a function of the grain sizes in conventionally processed alloys.

## INTRODUCTION

Noncrystalline alloys are produced by rapidly quenching molten alloys at a rate of nearly $10^6$ °K/sec. Although the main reason for this high quench rate is to suppress crystallization during the solidification process, it has the disadvantage of restricting dimensions of the resulting products so that presently only thin ribbons can be formed. This inability to directly produce amorphous specimens of large dimensions has limited the technical application of rapidly solidified materials despite their advanced thermomechanical properties. As a result, much research has been directed towards a technology whereby consolidation of the rapidly solidified processed powder into larger sections is possible [1]. There are several, potentially feasible, techniques of consolidation. These include explosive compaction, dynamic compaction, warm extrusion, and warm die pressing. Despite extensive research, however, thus far the relative merits of each of these techniques has remained questionable. The strength in a well-consolidated mass, having undergone either a purely solid-state compaction (mechanical interlocking of particulates) or a compaction process which is enhanced by diffusional processes, is likely to depend on several variables such as: a) reduced grain size; b) increased solid solubility; c) refined precipitate size; and d) dispersion of oxides introduced through the compaction process. While the improved properties of the amorphous, rapidly-solidified materials may be attributable to the lack of periodic structure, the higher strength attained during high temperature processing is, primarily, a manifestation of the grain boundary pinning by fine precipitates and oxides.

It is not easy, however, to separate these effects routinely. In order to separate the contributions of grain size to the strength, attempts have been made to obtain experimental data as a basis to develop a functional relationship capable of expressing the strength, $\sigma$ , in terms of the grain parameter, $\lambda$ . In the case of the compacted particulates, attempts have been made to demonstrate that the strength of the consolidated matrix is affected by the particulate sizes. In the case of amorphous ribbons which were isothermally annealed to grow grains of various sizes [3] , the intention has been to develop a parametric relationship between the strength and the grains grown during thermal processing and, also, to obtain process parameters useful to evaluate the stability of the amorphous state. It has become plausible to hypothesize the validity of the assumption that one part of the strength is derived from mechanical interlocking of the particulate and the other part is attributable to the grain-size effect developed within each particulate [4], [5].

## EXPERIMENTAL PROCEDURES

Rapidly solidified powders of $Ni_{67}W_{23}C_8B_2$ and $Ni_{45}Co_{20}Cr_{10}Fe_5MO_4B_{16}$ have been obtained from Marko Materials, 144 Rangeway Road, North Billerica, MA 01862 and the Allied Chemical Corporation, Corporate Technology, P.O. Box 1021RL, Morristown, NJ 07960, respectively. Particulates of separate size classes have been prepared by sieving from a main powder stock. Several classes of particulates, ranging from 37μm up to 700μm, have been considered. Specially designed high pressure equipment [4] have been used for isothermal compaction of the specimens at a temperature of $620^{o}$K and a maximum pressure of 3.6 GPa. Macrohardness measurements have been performed by using a microhardness tester made by the Schimitzu Corporation, Seisakusho, LTD, Kyotu, Japan. The same microhardness tester has been employed for the microhardness measurements of isothermally annealed ribbons. High-resolution, electron microscopy has been employed to determine grain sizes of specimens which were isothermally annealed at $623^{o}$K during 5, 10, 15, 20 and 25 minute time intervals. Stress measurements were made by a mini-instron apparatus constructed by Professor R. Weil of the Stevens Institute of Technology. High resolution transmission electron microscopy have been employed in assessing the microstructural characteristics of all specimens studied.

## RESULTS AND DISCUSSION

As the interest in the properties of rapidly solidified alloys has grown, it has become increasingly important to produce large units of these materials by suitable techniques of consolidation. Depending on the technique of consolidation used, however, major mechanical properties of the consolidated specimen alter. For example, samples consolidated during shock impacted experiments contain localized areas which have evidently experienced melting and recrystallization. Similar phenomenon occurs during high pressure compaction of rapidly solidified powders resulting in scattering of the data obtained from these specimens. Figure 1 shows the variation of microhardness plotted as a function of the consolidated particulate sizes. It is seen that the microhardness increases with decreasing particulate size. Attempts, however, to establish a similar graph showing the variation of the tensile stress with grain size remained unattainable because of the occurence of microcracks in the compacted specimens. An electron microphotograph showing detail microstructures of the particulates and their interface boundaries is shown in Figure 2; there is excellent evidence of bonding of "the particulates caused compaction". The same micrograph further reveals that particulates have suffered a transition from the amorphous to the polycrystalline state. Figure 3 shows the variation in tensile strength with grain size for specimens isothermally annealed. Two regimes of grain sizes have been identified. In the early stage of crystallization, when grains have attained a maximum growth of up to 400 Å, the strength increases with an increase in the grain size. For this regime, the strength, $\sigma$, follows as a functional form of the type $\sigma_1 = \sigma_{o1} + K_1 \text{Log } \lambda_1$, where $\sigma_{o1}$ and $K_1$ are constants and $\lambda_1$ is any grain parameter whose value lies between 0 to 400 Å. During the latter states of crystallization, when sufficient amounts of thermal energy have caused grain growth to exceed 400 Å, the parametric relationship which can best describe the strength-grain size relationship is similar to the well-known Hall-Petch formulation. This relationship is as follows:

$$\sigma_2 = \sigma_{o2} + k_2 \lambda_2^{-m}$$

where $\sigma_{o2}$, and $K_2$, and m are constant quantities as shown in the Figure 4. In this investigation it has been found that $m = \frac{1}{2}$ and the parameter $\lambda_2$, the grain size, could attain all values above 400 Å. Figure 5 shows the variation of ductility with grain size. The maximum ductility coincides with the highest strength in specimens annealed for a short time at temperatures below the transition temperature. As the annealing temperature or the annealing time increased, Ni and B - enriched regions, 40 - 60 Å in size, within the parent matrix occur. As it has been previously suggested 3, 4, 6. These effects signal the onset of phase separation from a seemingly homogenous matrix. It is very likely that the loss of the ductility is a direct result of the phase separation [7].

## SUMMARY AND CONCLUSIONS

Samples of rapidly solidified alloys of $Ni_{67}W_{23}C_8B_2$ and $Ni_{45}Co_{20}Cr_{10}Mo_4Fe_5B_{16}$ in ribbon and powder forms have been studied. In the consolidated amorphous particulate of $Ni_{67}W_{23}C_8B_2$ alloy it has been found that microhardness increases with decreasing particulate size. Presence of localized microcracks in these samples, however, excludes the possibility of accurate measurements of the tensile properties, causing premature failure. Samples of rapidly solidified alloys of $Ni_{45}Co_{20}Cr_{10}MO_4Fe_5B_{16}$ have also been isothermally annealed to produce grains of various sizes. Two types of relationships describing the strength, $\sigma$, as a function of the grain size, $\lambda$, have been identified. In the early stage of crystallization, when the grains have grown up to 400 Å, the strength, $\sigma$, obeys a logarithmic function of the grain size as described by the relationship $\sigma_1 = \sigma_{o1} + K_1 \text{Log } \lambda_1$. Here, $\lambda_1$ is the grain size in the range of 0 to 400 Å. During the latter stage of crystallization, when the grains have grown larger than 400 Å, the strength varies with the inverse square root of the grain size, $\sigma_2 = \sigma_{o2} + K_2\lambda_2^{-m}$. This latter relationship is similar to the well known Hall-Petch relationship.

*The author is on temporary assignment as Research Associate in the Physics Division. This work has been completed at the Materials Branch, M&MT Division, ARDC, Dover, NJ 07801.

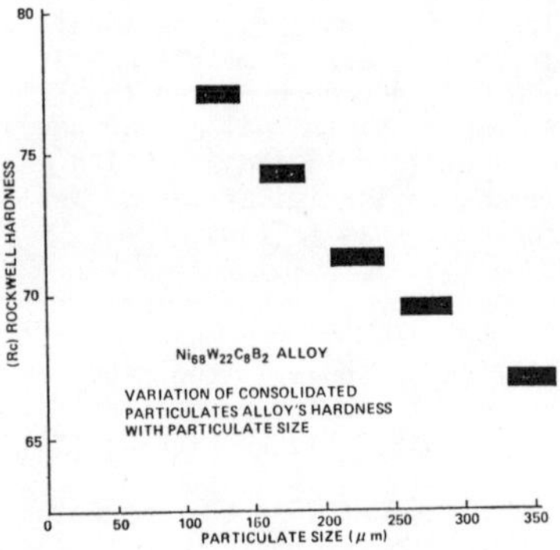


Fig. 1

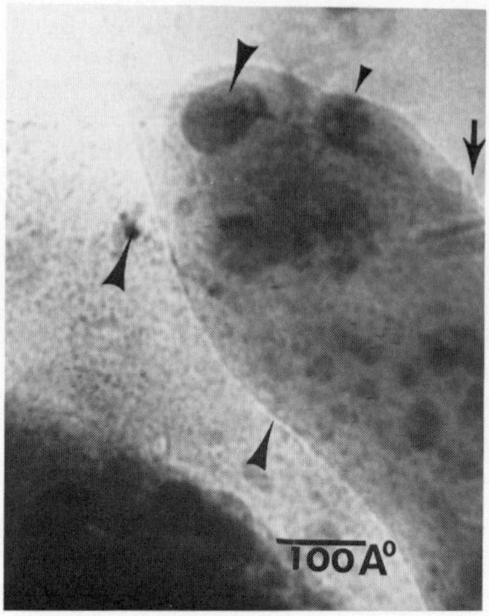


Fig. 2 Electron Photomicrograph showing microstructure of the consolidated $NI_{68}W_{22}C_8B_2$ powder at the particulate boundaries.

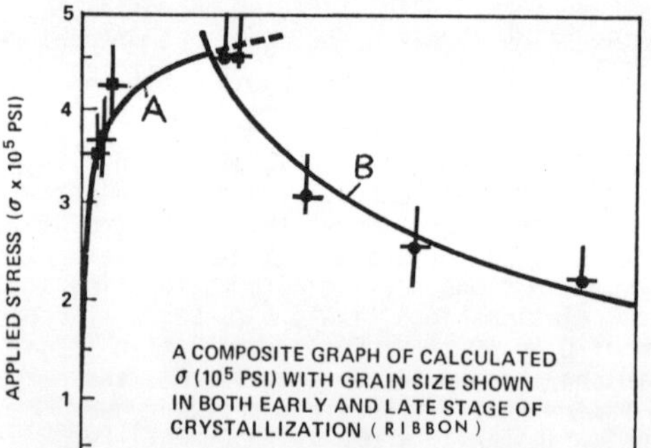


Fig. 3 and 4
(A)= Grains less than 400 Å
(B)= Grains larger than 400 Å

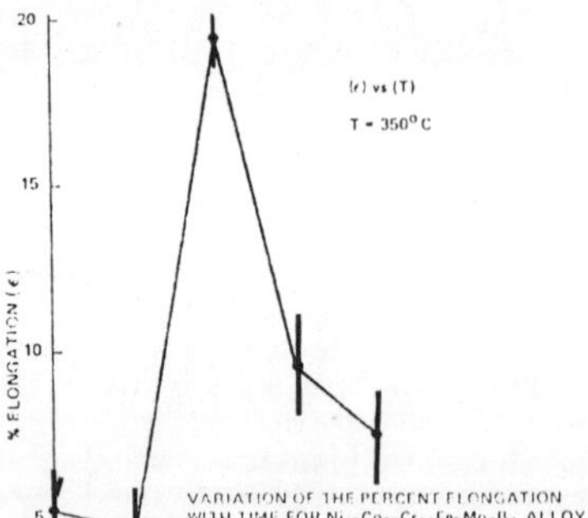


Fig. 5

## REFERENCES

1. S. A. Miller and R. J. Murphy, "The Kinetics of Consolidation of Amorphous $Cu_{60}$-$Zr_{40}$ Powder by Warm Processing, " Proc. 4th Int. Conf. on Rapidly Quenched Materials. Vol. I, (SENDAI) , 1981, pp. 137-140.

2. M. R. Jolly and R. W. K. Honeycombe, "The Properties and Microstructures of Compacted Melt Spun Tin and Tin Based Alloys", Proc. 4th Int. Conf. on Rapidly Quenched Materials, Vol. I, (SENDAI) , 1981, pp. 133-136.

3. M. A. Otooni, "Kinetics of Crystallization in the $Cu_{60}$-$Zr_{40}$ Alloy System", Journal Noncryst.Solids, 65 , pp. 389-402, (1984).

4. M. A. Otooni, "Laser Anneal and the Onset of Amorphous Crystalline Transition in the Amorphous $Cu_{60}$-$Zr_{40}$ Alloy System", Proc. of Mat. Res. Conf., Nov. 1983.

5. A. H. Clauer, et. al. , "Consolidated Studies of Metallic Glass Strips and Microcrystalline Strips and Particulates", from "Amorphous Glassy Metals and Microcrystalline Applications", AFML-TR-78-80, pp. 399-402, Battelle Columbus Laboratories, 1980.

6. M. A. Otooni, "Strengthening Mechanism in the Consolidated Rapidly Solidified Alloy", Tech Report ARSCD-TR-84017, pp. 2-15, 1984.

7. J. Piller and P. Haasen, Acta Met, 1, pp. 30-35, (1982).

# COMPARISON OF STRUCTURES OF GAS ATOMIZED AND OF EMULSIFIED HIGHLY UNDERCOOLED Ni-Sn ALLOY DROPLETS

MICHIHARU YAMAMOTO*+, YANZHONG WU*, YUH SHIOHARA** AND MERTON C. FLEMINGS*
* Dept. of Materials Science and Engineering, MIT, Cambridge, MA 02139
** Materials Processing Center, MIT, Cambridge, MA 02139
\+ now Nippon Mining Co., Ltd., Tokyo, JAPAN

## ABSTRACT

A comparison is made of microstructures of droplets of Ni-Sn alloys rapidly solidified (a) by gas atomization and (b) in a glass emulsifying medium. Cooling rate of the gas atomized particles ranged from $10^3$ to $10^6$K/s depending on droplet diameter (20-230μm). In the hypoeutectic alloy studied (Ni-25wt%Sn), most particles showed a dendritic structure. These same particles, melted and resolidified in a glass medium using DTA (Differential Thermal Analysis), showed undercoolings up to 280K: the structures were dendritic at low undercoolings and non-dendritic at undercoolings above about 220K. It is concluded that (1) the gas atomized particles exhibited little or no undercooling before nucleation, (2) the solidification time of the undercooled emulsified droplets is substantially less than that of gas atomized droplets, and (3) the undercooling required to achieve non-dendritic structure depends on sample size.

## INTRODUCTION

The ultrasonic gas atomization technique is a rapid solidification processing (RSP) method of considerable current interest. It is of interest to understand effects on microstructures of undercooling before nucleation. To understand the effects of high undercooling on the solidification process, it is necessary to precisely measure undercoolings and to gain high undercooling. It is, however, almost impossible to measure the exact undercooling in the gas atomization process because of the very rapid quenching process. Therefore the microstructures of the droplets rapidly solidified by gas atomization were compared with those highly undercooled in a glass emulsifying medium whose undercoolings were directly measured by DTA.

## EXPERIMENTAL PROCEDURE

The Ni-25wt%Sn alloy droplets were produced by the ultrasonic gas atomization technique in an argon atmosphere. The droplets obtained (20-230μm) were sieved to narrow size ranges and the weight for each size range was measured. Then microstructures of as gas atomized particles were observed by SEM (Scanning Electron Microscopy).

In DTA measurements, as gas atomized samples with a mixture of Pyrex* and Vycor* glass were heated to 1620K at a rate of 10 K/min and cooled down at the same rate in the DTA furnace, filled with argon gas. The microstructures of the samples solidified in the glass emulsifying medium were examined.

## RESULTS OF GAS ATOMIZATION

The gas atomized particles were almost spherical, and about 40 wt% of the particles were less than 100 microns diameter. The cross-sectioned microstructures in Figure 1 show fine dendrite structures and primary Ni-rich

* Trademark of Corning Glass Works, New York

 

α phase surrounded by β phase. Fineness of the microstructure increases with decreasing the particle size.

Calculations of the cooling rate for gas atomized droplets were performed to obtain the dependence of the particle size. The Ranz-Marshall equation [1] was employed to calculate the heat transfer coefficient, employing approximate values for particle velocity, etc. Biot numbers for all calculations were less than $10^{-2}$ and so Newtonian cooling approximation was valid in all cases. Results of the calculations show the cooling rate increases from $7x10^3$ to $4x10^5$ K/s with decreasing particle size from 230 to 20 microns.

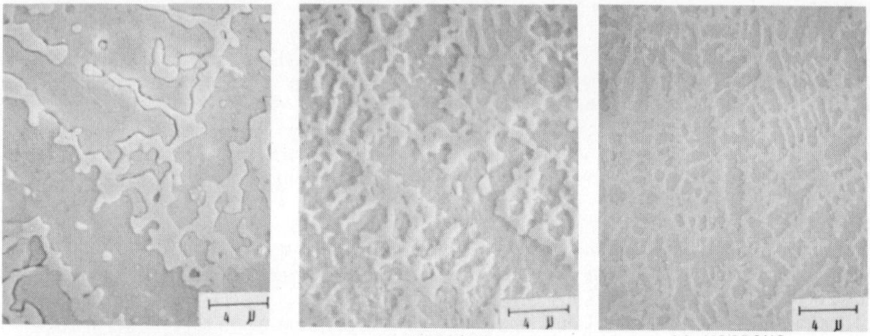


230 MICRONS 90 MICRONS 32 MICRONS

Figure 1 Cross-sectioned microstructures of gas atmized particles.

RESULTS OF DTA MEASUREMENT

Typical heating and cooling curves are shown in Figure 2. The heating curve shows endothermic peaks due to phase transformation, including melting, eutectic and β'-β transformation while the cooling curves show exothermic peaks due to solidification and β-β' transformation. If the primary phase nucleates above the eutectic temperature, two exothermic peaks due to solidification are observed corresponding to primary phase and eutectic solidification.

The proper glass medium, which would prevent particle agglomeration in the emulsion, for the small metal particle was found to be a mixture in which nucleation of the metal particles took place in a narrow temperature range with high undercooling consisted of a weight ratio of metal 2/ Pyrex 3/ Vycor 5.5.

Cross-sectioned microstructures with three different initial undercoolings for the size of 960 μm in diameter were observed and shown in Figure 3. These exhibit the evolution of dendritic morphology with undercooling. Microstructure with initial undercoolings less than 220K shows a degenerate α dendrite surrounded by β phase. Those with initial undercoolings larger than 220K, however, show a non-dendritic structure. These are summarized in Figure 4.

In bulk samples, Walker [2,3] observed a substantial

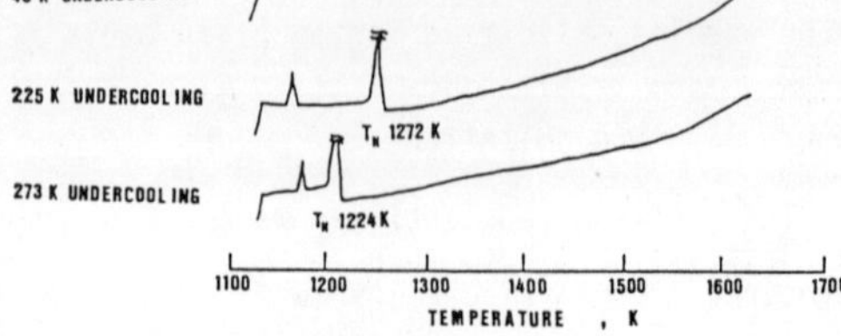


Figure 2 Heating and cooling curves by DTA.

220K UNDERCOOLING 247K UNDERCOOLING 273K UNDERCOOLING

Figure 3 Cross-sectioned microstructures of the same particle size (960 microns in diameter) with various undercoolings.

grain refinement in Ni and Co at a critical initial undercooling of about 175K. Other investigators [4,5] observed a similar transition from dendritic to non-dendritic morphology at almost the same critical undercooling. In current work at MIT on levitation melted Ni-25wt%Sn alloy, the transition was found over a range of undercoolings centering at almost the same undercooling of 175K. However, in these very small diameter droplets, dendritic structure (with primary and secondary arms) was observed up to undercoolings of 220K. The results indicates that the critical undercooling depends on metal droplet size.

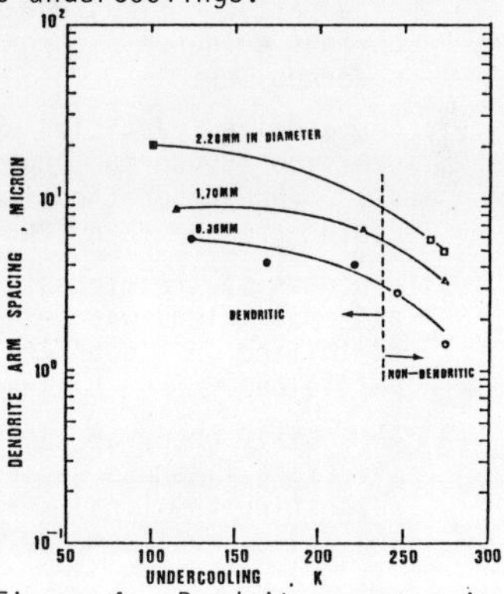


Figure 4 Dendrite arm spacing as a function of undercooling. ( Dotted line is not meant to imply transition is size independent.)

The maximum undercooling achieved in this experiment was about 280K which was independent of particle diameter. Microstructures of samples with the same undercooling but different particle sizes are compared in Figure 5. For the particles larger than 960 μm, dendrite arm spacing is proportional to the cube root of diameter of the droplets, providiing evidence that ripening kinetics determines structure fineness (since soli-

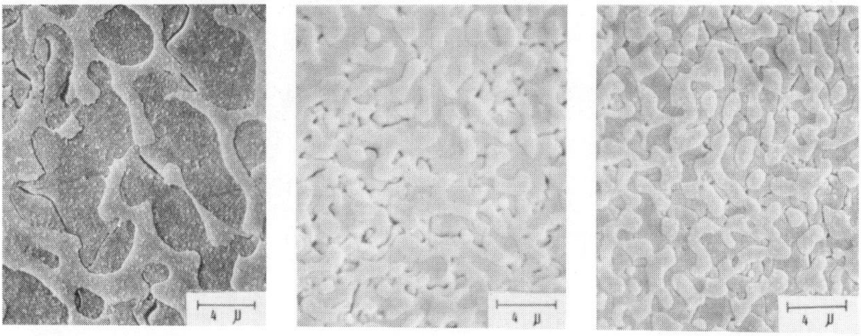


2280 MICRONS 960 MICRONS 90 MICRONS

Figure 5 Cross-sectioned microstructures for different particle sizes with the same undercooling of about 280K.

dification time is proportional to the droplet size for spheres). The arm spacings for the smaller particles are nearly independent of the particle size, perhaps indicating that these spacings are those existing at or soon after the completion of recalescence. Results are summarized in Figure 6 with data of gas atomized powders for comparison.

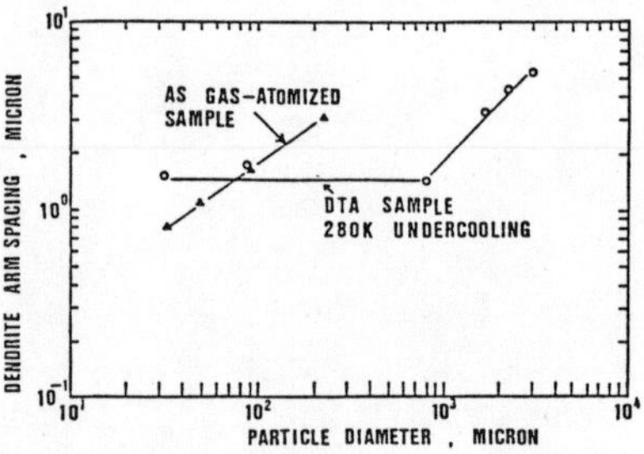


Figure 6 Dendrite arm spacing as a function of metal droplet diameter.

## SUMMARY

(1) Cross-sectioned microstructures of gas atomized particles of Ni-25wt%Sn show fine dendrites which become finer with decreasing particle size, but show no obvious evidence of significant undercooling.

(2) Gas atomized particles were melted and resolidified in a glass emulsion. Undercoolings ranged up to 280K, independent of particle size. Non-dendritic microstructures were found for samples undercooled greater than 220K.

(3) Fineness of the microstructure of most samples (both gas atomized and emulsified) was determined by coarsening. Emulsified samples solidified in a substantially shorter time than the gas atomized particles.

(4) An anomaly observed was that for the smaller emulsified droplets, structure fineness seemed to be independent of particle size, suggesting the final structure observed is that existing at or near the end of recalescence.

## ACKNOWLEDGMENT

The authors express their deep gratitude to NASA for support of this work under Grant No. NSG 7645. They also acknowledge with thanks Professor N.J. Grant and Mr. E. Lavernia for producing the alloy powders by gas atomization. Discussions during the course of the work with Professor T.Z. Kattamis have been particularly helpful.

## REFERENCES

1. G.H. Geiger and D.R. Poirier, Transport Phenomena in Metallurgy, (Addison Wesley, Massachusetts, 1973), p. 249-252.
2. B. Chalmers, Principles of Solidification, (J. Wiley & Sons, New York, 1964), p. 122-125.
3. J.L. Walker, Physical Chemistry of Process Metallurgy, (AIME, New York, 1961), p. 845-853.
4. S. Skoliano, P.S. Liu and T.Z. Kattamis, Proc. Conf. on Grain Refinement in Casting and Welds, (The Metallurgical Society of AIME, Warrendale, PA, 1983), p. 97-116.
5. M.C. Flemings and Y. Shiohara, Materials Science and Engineering, 65, (1984) pp. 157-170.

# IMPROVEMENT OF THERMAL FATIGUE RESISTANCE OF A WROUGHT NICKEL-BASE SUPERALLOY BY LASERGLAZE

Zhao Qi, Ge Yunlong, Hu Zhuangqi, Jiang Ming and Shih Changshu, Institute of Metal Research, Academia Sinica, Shenyang, Liaoning, China

## ABSTRACT

Laserglaze with appropriate post heat treatment has improved the thermal fatigue resistance of a wrought nickel-base superalloy. It was found that laserglaze was able to eliminate the blocky MC phase, refine grains and form a very interesting microstructure of serrated grain boundaries. Careful selection of post heat treatment markedly increased the strength in the laser irradiated region. The initiation and propagation of thermal fatigue cracks were suppressed by this novel microstructure.

## INTRODUCTION

The use of laserglaze for the controlled modification of surface sensitive properties has been a rapidly growing research field in the last decade. It is a very useful metallurgical tool for studying basic mechanisms in material research areas such as rapid solidification, corrosion and oxidation, tribology, etc., as well as a powerful means of beneficial modification of the mechanical and chemical properties of materials for formation of amorphous or microcrystalline metals and alloys. For this reason, laserglazing has attracted vast attention in the world. Some of the advantages have already been reported, for example, refinement of grains, dendrites and phases, elimination of harmful phases, extension of solute solubility and reduction of segregation [1-7]. As a result of the above mentioned surface modification, mechanical, corrosion-resistant and wear-resistant properties can be improved. In this paper we summarize the effect of laserglaze, with appropriate post heat treatment, on thermal fatigue of a wrought nickel-base superalloy.

## EXPERIMENTAL PROCEDURE

The chemical composition of the wrought nickel-base superalloy is listed in Table I. The main strengthening phase is $\gamma'$, about 42wt%. There are also MC, $M_6C$, $M_{23}C_6$ and $M_3B_2$ phases present. The alloy was produced by vacuum induction and vacuum arc duplex melting. The standard heat treatment was: 1220°C, 4hr, AC (air cool) + 1050°C, 4 hr, AC + 950°C, 2hr, AC.

A 2kW CW $CO_2$ laser was used to scan over 25x15x5 mm specimens, notched with a depth of 1.5 mm and a root curvature radius of 0.1 mm. Parameters were chosen as follows: output power 1.3 kW, beam diameter 2.5 mm, scanning speed 15 mm/sec and melt depth about 0.5 mm. Black spray paint was used as an absorbing coating.

We rapidly raised the specimen temperature to 900°C (or 950°C, 1000°C) within 0.5 min., exposed at the upper temperature limit for 2 minutes and water-cooled to 20° ± 5°C. This thermal cycle was repeated many times.

## RESULTS AND DISCUSSION

The key factor controlling the high temperature mechanical properties of this wrought nickel-base superalloy is the $\gamma'$ phase, including its amount, size morphology and distribution. Fig. 1 shows there are two kinds of $\gamma'$ phase, large cuboids and very small globular ones. After laserglaze of this alloy, the microstructure was observed to be in a metastable

Table I The chemical composition of the alloy studied

| Ni | Co | Cr | W | Mo | Al | Ti | V | C | B |
|---|---|---|---|---|---|---|---|---|---|
| bal | 15.16 | 10.17 | 5.45 | 5.51 | 4.21 | 2.35 | 0.26 | 0.05 | 0.017 |

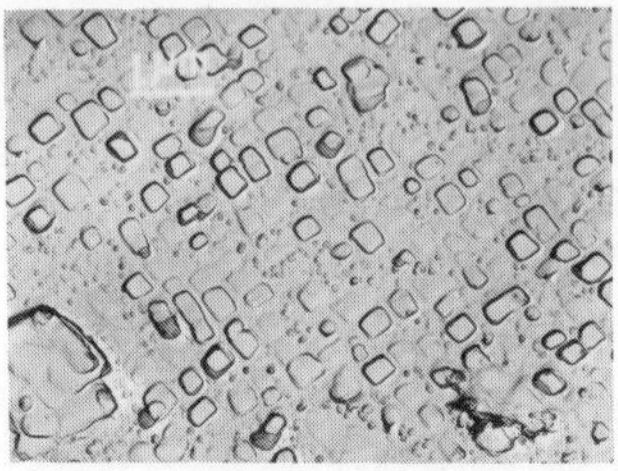

Fig.1 Alloy microstructure after standard heat treatment

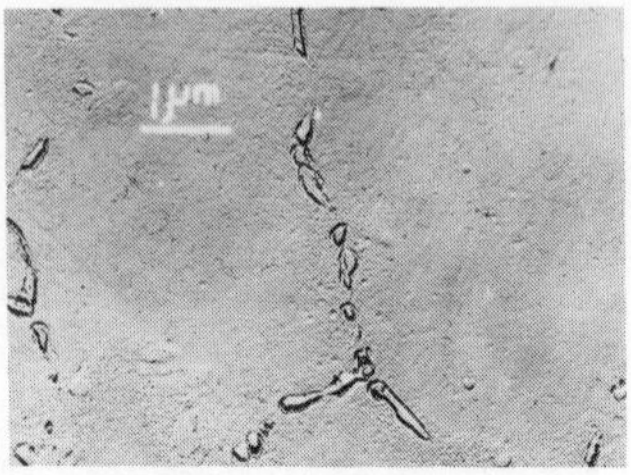


Fig.2 Alloy microstructure after laserglaze

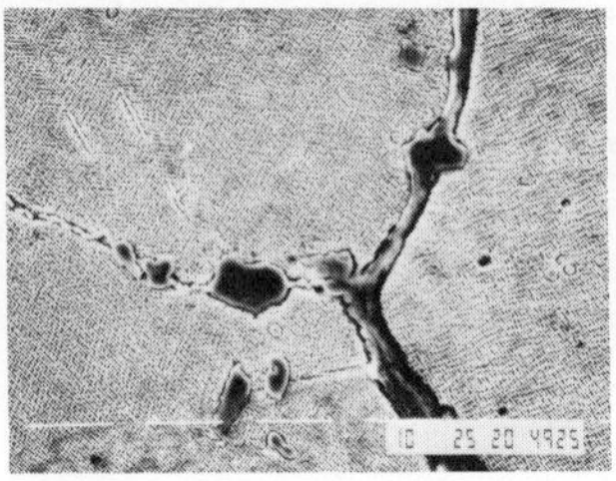

Fig.3 MC in grain boundary after standard heat treatment

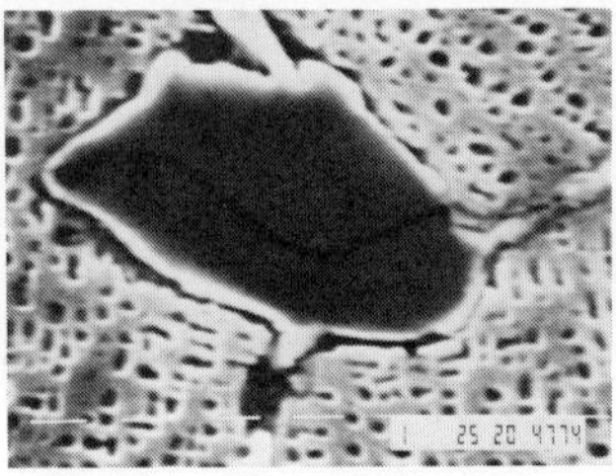

Fig.4 Crack of MC particle under thermal stress

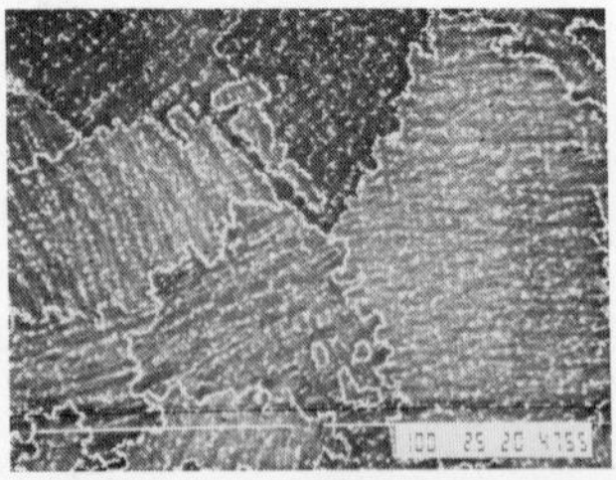

Fig.5 Serrated grain boundary

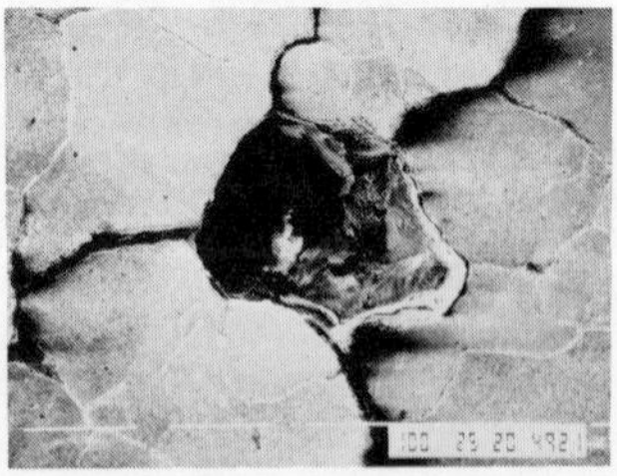

Fig.6 Drop out of loose grain

condition with supersaturation. A large amount of γ' phase as well as carbides at grain boundaries have dissolved as shown in Fig. 2. This may have deleterious effects on the mechanical properties of the alloy. In order to recover the mechanical properties, a post heat treatment after laserglaze was performed to properly reprecipitate γ' phase and carbides both in grains and boundaries. The heat treatment after laserglaze was 1050°C 2hr, AC + 860°C, 6hr, AC, where 860°C and 1050°C were the precipitation peak temperatures of γ' and $M_6$, respectively.

Fig. 3 shows the presence of blocky MC carbide in the grain boundary after standard heat treatment. MC is hard and brittle, and can be easily oxidized at high temperature. During thermal fatigue testing, it became the origin of crack initiation, as shown in Fig. 4. After laserglaze, the blocky MC carbides disappeared and were still not found after post heat treatment. Therefore, the probability of crack initiation was reduced due to the absence of blocky MC carbide, and the number of cycles required to initiate a micro-crack was increased.

If the original alloy was laser treated and properly post heat-treated, a serrated grain boundary was observed in the irradiated region (Fig. 5). Under the same test conditions, micro-cracks were found along the nearly straight grain boundary in the untreated alloy specimen. Even a whole loose grain was dropped out if the maximum testing temperature was controlled at 1000°C, as shown in Fig. 6. On the contrary, if the grain boundary was serrated, the cleavage tendency along grain boundary was reduced, and characteristics of transgranular crack propagation became dominant (Fig. 7). Fig. 8 shows that ductile fracture was found in a laser-irradiated alloy specimen. It can be considered that the serrated grain boundary was beneficial in enhancing the ductility, promoting homogeneous deformation, and suppressing crack initiation and propagation during thermal fatigue.

Due to the rapid solidification, the irradiated zone consisted of refined equiaxed grains. The average grain size was reduced to one-half of the original size. It is interesting that each grain further consisted of several hundred cells, as shown in Fig. 9. This refined microstructure might improve the thermal fatigue life. As a general rule, refined microstructures will enhance the yield strength, which is correlated to the extent of plastic deformation, and in turn to the magnitude of thermal stress. It is another explanation of why laserglaze will lengthen the thermal fatigue life of the material.

Fig. 10a, b, and c are comparisons of thermal fatigue properties between laser irradiated alloy specimens and original ones at 900, 950, and 1000°C, respectively. The larger the number of thermal cycles, the longer the crack length. At all three temperatures laserglaze plus proper post heat treatment resulted in improved thermal fatigue behavior. The influence of maximum temperature on the crack propagation rate is shown in Fig. 11.

## CONCLUSIONS

1. Application of laserglaze coupled with appropriate post heat treatment on a wrought nickel-base superalloy can increase the thermal fatigue life.
2. The main causes of improvement of thermal fatigue properties by laserglaze are elimination of blocky MC phase, formation of serrated grain boundaries and refinement of microstructure.
3. The higher the maximum temperature of a thermal cycle, the faster the thermal fatigue crack propagation rate.

Fig.7 Transgranular crack propagation in laser irradiated specimen

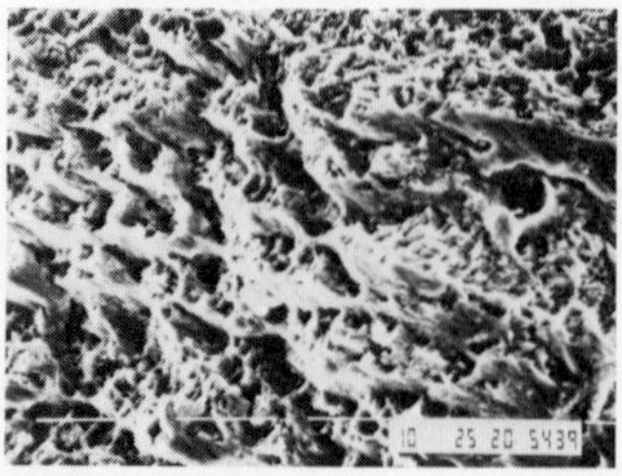

Fig.8 Ductile fracture in laser irradiated specimen

Fig.9 Cell formation due to rapid solidification

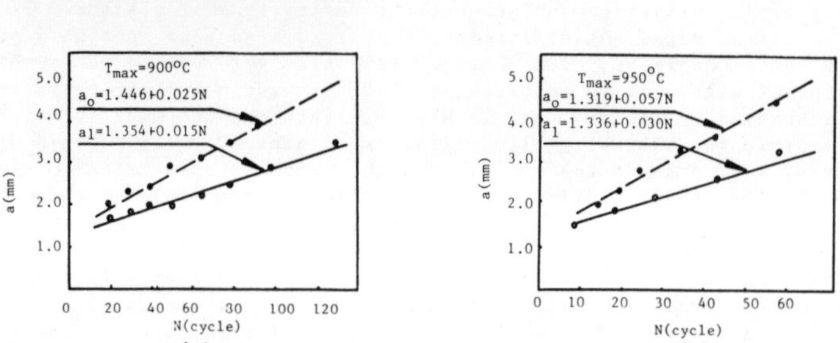


Fig.10 Crack length vs number of cycles $a_1$: laser irradiated $a_o$: original alloy

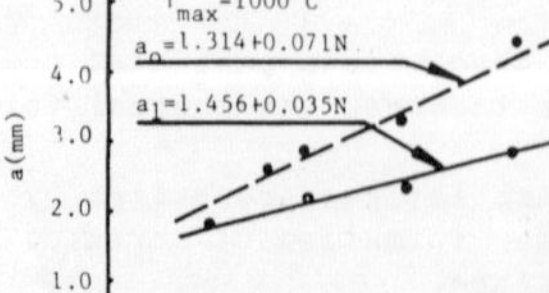


(c)

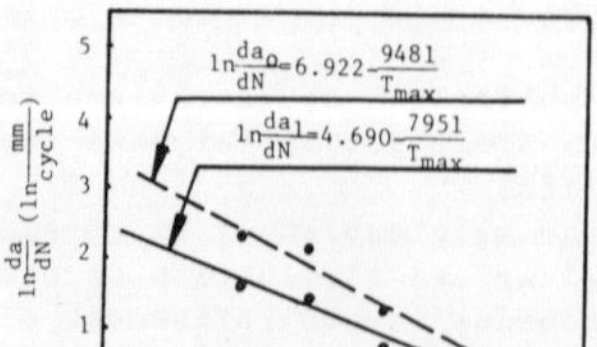


Fig.11 Influence of maximum temperature on crack propagation rate

REFERENCES

[1] E.M. Breinan, Physics Today, No. 11, 29 (1976) 44.
[2] B.H. Kear, E.M. Breinan, L.E. Greenwald, Metals Technology, No. 4, 6 (1979) 121.
[3] E.L. Mordike, H.W. Bergmann, in "Rapidly Solidified Amorphous and Crystalline Alloys" ed. by B.H. Kear, B.C. Giessen, M. Cohen, Boston, MA, USA (1981) 463.
[4] P.R. Strutt, B.C. Lewis, B.H. Kear, in "Rapidly Solidified Amorphous and Crystalline Alloys" ed. by B.H. Kear, B.C. Giessen, M. Cohen, Boston, MA, USA (1981) 485.
[5] E.M. Breinan, in "Proceedings of the International Conference on Rapid Solidification Processing," Reston, VA, USA (1977) 87.
[6] D.B. Snow, in "Superalloy 1980, Proceedings of the Fourth International Symposium of Superalloys), Seven Springs, PA, USA (1980) 189.
[7] F.G. Yunlong, H. Zhuangqi, G. Wei and S. Changhso, Mat. Res. Soc. Symp. Proc. 28 (1984) 99.

# CREEP DEFORMATION OF ULTRAFINE GRAINED $Ni_{75}B_{17}Si_8$

GILLES NUSSBAUM AND DIETER G. AST
Materials Science and Engineering, Bard Hall, Cornell University, Ithaca NY, 14853-1501.

## ABSTRACT

The creep behavior of the fine grained, intermetallic superplastic alloy of overall composition $Ni_{75}B_{17}Si_8$ prepared by crystallizing an amorphous precursor was studied in the temperature range between 500 and 615°C, the stress range between 36 and 800 MPa, and for grain sizes between 6.8 and $1.12*10^{-5}$ mm. The strain rates measured varied between $10^{-7}$ and $6*10^{-5}$ $sec^{-1}$.

Two different creep deformation mechanisms were observed. At low applied stresses, the material deformed in a diffusion controlled mode, with $\dot{\epsilon}$ proportional to $\sigma^1$. No dislocations developed in the grains, and the activation energy for creep was 4 eV. This value is higher than the self-diffusion of Ni in Ni (the principal constituent in the alloy) in agreement with observation in other Ni based alloy systems.

At high applied stresses, a second mechanism became rate limiting. The activation energy of this process is about 0.6 eV and dislocations develop in the interior of grains when this mechanism operates.

The transition between the two regimes is grain size and temperature dependent and occurs at about 400 MPa at a deformation temperature of 550°C and a grain size of 6.8 $10^{-5}$ mm.

An analysis of the grain size dependence of creep strain rate as well as its absolute value indicates that the low stress regime is Coble type creep.

## INTRODUCTION

In a polycrystalline material, different deformation modes can contribute to creep, such as glide and climb of dislocations, subgrain formation, grain boundaries (GB) sliding, point defect diffusion and twinning. If the grain size is very small and the temperature above half the melting temperature, such processes can readily occur and the material may behave in a superplastic manner [1], even when the alloy is an inherently brittle intermetallic alloy.

Fine grained intermetallic alloys can be prepared conveniently by the controlled crystallization of amorphous alloys. This preparation method was chosen by S. Reusswig et al [1], who investigated the deformation of fine grained $Ni_{75}B_{17}Si_8$ with a bimodal grain size distribution in tensile tests at strain rates between $5*10^{-4}$ and $10^{-2}$ $s^{-1}$. They showed that this alloy behaved in a superplastic manner at elevated temperatures but who did not determine activation energies. To explain the observed stress sensitivity of .5, and the presence of dislocations in large grains only, these authors assumed a mixed mode deformation, with diffusional creep dominating in small grains, and dislocation deformation in large grains.

This paper extends this previous study towards much lower strain rates. Special care was taken in crystallizing the material, in order to obtain a single valued grain size distribution. Creep tests were conducted over as wide a temperature range as possible, between $500 \leq T \leq 615$°C, in order to derive a value for the activation energy.

When accommodation occurs by both dislocation climb, diffusion in grains and diffusion along GBs, the processes add and one obtains:

$$\dot{\epsilon} = K_0 \exp(-Q_0/kT)(\sigma)^n + (\sigma/T)[K_1 \exp(-Q_1/kT) + K_2 \exp(-Q_2/kT)] \quad (1)$$

with : $K_0 = f(G)$; $K_1 = 32D_1\Omega/kd^2$; $K_2 = 132wD_2\Omega/kd^3$; $D_c = D_1 \exp(-Q_1/kT)$ and $D_b = D_2 \exp(-Q_2/kT)$ where $D_c$ and $D_b$ are the bulk and GB diffusion coefficients, and $Q_0$, $Q_1$ and $Q_2$ the activation energies for dislocation, Nabarro-Herring and Coble types of creep, respectively. Under conditions where the second term dominates the strain rate is proportional to the stress, a characteristic of superplastic deformation. Eq. 1 can be written as: $\dot{\epsilon} = k\sigma^n + k'\sigma/T$ (2); in this form it will be used below.

One expects that point defect diffusion dominates at low stress and climb controlled diffusion at high stresses, which, depending on the temperature and grain size, can be either Coble creep or Nabarro-Herring creep. Since the two diffusion mechanisms have a different dependence on grain size, a plot of $k'/T = f(\ln d)$ will yield a slope of 2 if bulk diffusion dominates and a slope of 3 if diffusion occurs mostly along GBs.

## EXPERIMENTAL ARRANGEMENTS

$Ni_{75}B_{17}Si_8$, was rapidly quenched by melt spinning into an amorphous ribbon, and then crystallized into a mixture of nickelboride and nickelsilicide [2]:

$Ni_{75}B_{17}Si_8$ (glass) = 8 $Ni_3Si$ (C) + 17 $Ni_3B$ (C).

As shown by U. Koester et al [3,4] nucleation in amorphous metals is heterogeneous and athermal at lower temperatures; i.e. the number of nuclei per unit volume is independent of the annealing time and temperature but depends on the cooling rate of the freezing ribbon when cast. At higher crystallization temperatures, nucleation is thermally activated and the number of nuclei formed is an increasing function of temperature. The transition occurs at the glass transition temperature, $T_g$ [5]. Since $T_g$ is about 500°C for our material [6], nucleation at lower temperatures can only occur at quenched-in nuclei or compositional or topological short range order variations. These nuclei, which initially are too small to be effective nucleation sites, will start growing at the beginning of the annealing treatment. A low heating rate will keep the specimen long enough below $T_g$ to ensure the growth of such nuclei, resulting in a single valued grain size distribution.

To optimize the annealing procedure leading to a single grain size distribution we conducted a series of experiments [7] which led to the adoption of the following annealing treatment: the ribbons were heated at a linear rate of .4°C/s to 600°C, held at this temperature for 10 min, and then cooled down to the temperature at which the creep test was to be performed. TEM of specimens crystallized by this procedure showed that in almost all cases the specimens had a single grain size, of about 0.07 to 0.1 μm, and that the grain size was nearly constant along the ribbon.

The creep tests were carried out in an elliptical radiant furnace. The heating rate and holding temperature was controlled via a H-P 9835 computer to +/- 1°C. No load was applied during heating. Elongation of the sample was measured with a H-P 7 DCDT displacement transducer.

## RESULTS

In all cases we observed a transient state followed by a well defined steady state. At 550°C and 575°C, the data fitted a law represented by (2). Fig. 1, a plot of stress vs strain rate at 550°C shows our data (triangles) together with those extrapolated (closely spaced double line 2) from S. Reusswig et al. [1], and finally lower (line 1) and upper (line 3) estimates for Nabarro-Herring and Cobble creep, as listed in [1]. At strain rates lower than $10^{-5}$ $s^{-1}$, our data fit well the Nabarro-Herring creep in NiSi as calculated in [1] when corrected for the grain size of our samples. At higher strain rates, our data tend smoothly towards those of S. Reusswig et al., with, however, a different slope (.3 vs their .5). At this point the strain rate is found to become independent of grain size.

Inspection of Fig. 1 shows that the measured stress sensitivity, $m = d\ln(\sigma)/d\ln(\dot{\epsilon})$ drops from a value of 1 at low stresses to about 0.3 at high stresses, indicating a change in the rate limiting step from diffusion to dislocation mechanism. TEM observations corroborated this interpretation. Samples tested at low stress (m=1) showed a dislocation free grain interiors, whereas those tested at higher stress (m=.3) did not. Numerically, the creep law at 550 °C is given by:
$\dot{\epsilon} = 8*10^{-4}\sigma^{3.1} + 1.3*10^{-8}\sigma$; with $\epsilon$ in $s^{-1}$, $\sigma$ in $kg/cm^2$; and at 575 °C, by: $\dot{\epsilon} = 7.2*10^{-12}\sigma^{2.5} + 1.6*10^{-8}\sigma$.

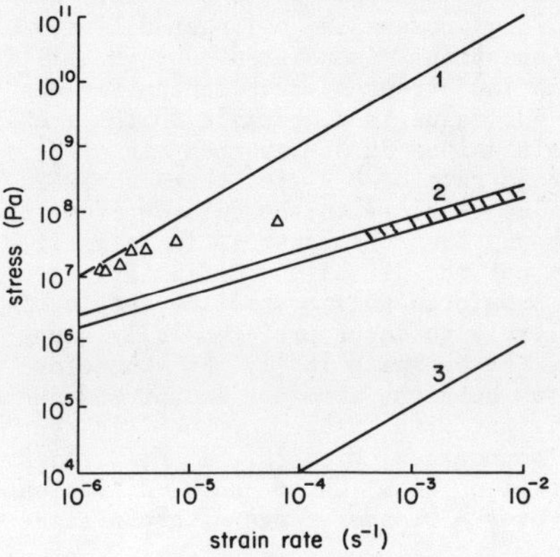


Fig. 1

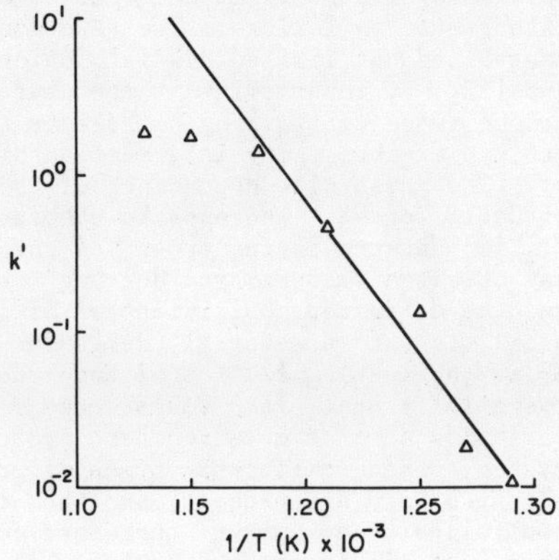


Fig. 2

Fig. 2, a plot of k' vs 1/T, indicates that diffusion accommodated creep prevails for T< 575°C. Above this temperature, another process, with a much lower effective activation energy (about 0.6 eV), becomes rate limiting. A change in the deformation mode is also indicated by TEM observations. Samples tested at 600°C showed evidence of subgrain formation and presence of dislocations. Such effects were not observed in samples tested below 575°C. An analysis of Fig. 2, which shows data for a grain size of $d=8.5*10^{-5}$ mm, indicates an effective activation energy for diffusion controlled creep of 4 eV.

Fig. 3 shows the effect of grain size on the prefactor for diffusional creep, k', for grain sizes between $6.8*10^{-5}$ and $1.12*10^{-4}$ mm. The straight line through the data points shown in Fig. 3 fits the equation: log k' = -3.2 logd - 13.49 when k' is expressed in SI units and d in mm. The grain size exponent, -3.2, indicates that the rate limiting process is Coble creep, at low stress and at temperatures at or below 575°C.

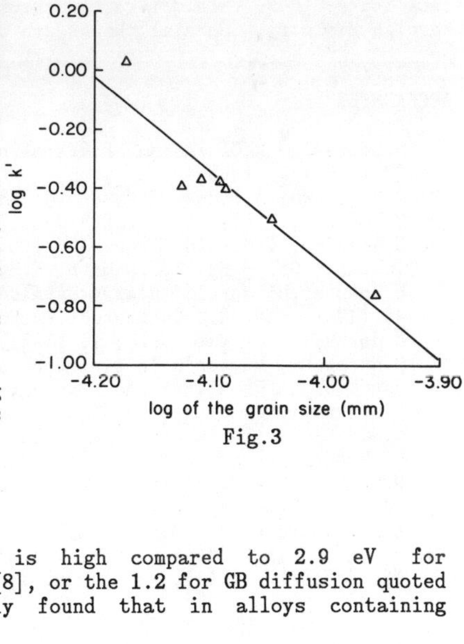


Fig.3

## DISCUSSION

An activation energy of 4 eV is high compared to 2.9 eV for self-diffusion in polycrystalline Ni [8], or the 1.2 for GB diffusion quoted in [1]. However, it is frequently found that in alloys containing intermetallic compounds, such as superalloys based on $Ni_3Al$, the activation energy for creep exceeds that of self diffusion. The activation energy for creep in nickel based superalloys can be as high as twice that of unalloyed nickel and considerably higher than for solid solution alloys of nickel [9]. Furthermore, the value found here is similar to the value of 3.6 eV reported by B. Cantor et al. [10] for diffusion of boron in amorphous $Ni_{40}Fe_{40}B_{20}$ which is believed to have a short range order similar to intermetallic compounds. Thus, the activation energy for creep appears compatible with experiences in related systems.

At a temperature above 575°C, the value of the activation energy of the rate controlling mechanism of creep at low stress drops to 0.6 eV. This was interpreted as follows: The rate controlling process is no longer diffusion controlled but limited by dislocation interaction. As pointed out in [11] and [12], at intermediate temperatures and low stresses, such mechanisms can become rate controlling. We note that our value is comparable to the 1 eV activation energy for intersection of dislocations in Ni reported in [12].

The grain size dependence of the strain rate of 3.2 is characteristic of Coble creep, whereas the absolute value is close to the data derived in [1] for Nabarro-Hering creep. A check showed that the lower estimate in [1] was obtained from Nabarro-Herring formula but the diffusion coefficient used was the diffusion coefficient of Si in a submicron polycrystalline sample of NiSi. In such a material, diffusion is likely to occur preferentially along GB's. Neglecting factors of the order 4, the estimate in [1] is therefore essentially that for Coble creep. Thus both the absolute amount and the variation of $\dot{\epsilon}$ as measured here are consistent with Coble creep. In our system, three different types of grain boundaries, $Ni_3Si/Ni_3Si$, $Ni_3Si/Ni_3B$ and $Ni_3B/Ni_3B$ are present, and the transition from GB to bulk diffusion controlled creep might therefore occur over a broader range of grain sizes than in a single phase material.

## ACKNOWLEDGEMENTS

This research was sponsored by ONR under contract N00014-85-K-0006 and by the Material Science Center at Cornell which operates the central TEM facilities. G.N. would like to acknowledge the assistance of the French Foreign Ministry. Special thanks are due to R. Raj for valuable discussion.

## REFERENCES

1. S.Reusswig, R.Gleichman, P.G.Zielinski, D.G.Ast & R.Raj; Acta Metall., 32 (1984) p. 1553
2. M.Hagiwara, A.Inoue, T.Masumoto; Metal.Trans., 12A (1981) p 1027
3. U.Koester, M.Blanke, Scripta Metall.; 17 (1983), p 495
4. U.Koester, U.Herold, Proc. 4th Int. Conf. on Rapidly Quenched Metals (Sendai 1981), Ed. T.Masumoto, K.Suzuki, p 717
5. U.Koester,U. Herold, Glassy Metals 1,Topics in Applied Physics, 46 (1981), Ed. H.J.Guntherodt, H.Beck, Springer Verlag, p 225
6. H.Jones; J. of Mat. Sci.; 19 (1984), p 1043
7. G.Nussbaum, Master's Thesis, Cornell University (1986)
8. Handbook of Chemistry and Physics, 60th Edition (1980), CRC Press, Ed. R.C.Weast
9. N.S.Stoloff, in Superalloys, Wiley Series on Science and Technology of Materials, ed. C.T.Sims, W.C.Hagel (1972), p 79
10. B.Cantor, R.W.Cahn, in Amorphous Metallic Alloys, Butterworths (1983), Ed. F.E. Luborsky, p 487
11. A.K.Mukherjee, J.E.Bird, J.E.Dorn; Trans. of A.S.M., 62 (1969), p 155
12. H.Conrad, in Mechanical Behavior of Materials at Elevated Temperatures, Ed. J.E.Dorn (1961), Mc Graw Hill, p 218

# Unusual Chemical Composition of Carbides in Rapidly Solidified High Speed Steels

A. Tauqir, H. Nowotny, and P. R. Strutt

Metallurgy Department
University of Connecticut
Storrs, Connecticut 06268

## ABSTRACT

Studies of the elemental composition of carbide phases formed by rapid solidification of a Mo-base high speed steel (M7) have yielded an intriguing finding. Micro-analysis of the extracted carbides ($M_2C$, $M_{23}C_6$ and MC) showed that these were unusually rich in refractory elements with values of Mo in the range 46-53 wt.%, corresponding values for Fe and Cr were low (6-12 wt.% Cr and 1.5-13 wt% Fe). This was in marked contrast with analyses of MC, $M_6C$, $M_7C_3$ and $M_{23}C_6$ in the 'unglazed' material where Fe and Cr were major constituents. A distinctive feature was $M_{23}C_6$ in the rapidly solidified material which contained large amounts of Mo and W and very little Cr. It is proposed that this phase, initially formed as $M_6C$, underwent solid-state transformation.

## INTRODUCTION

Conventional processing of Mo-W based steels, regarded as Fe-X-C (ternary equivalent) where X stands for Mo+W, results in the formation of MC and $M_6C$ carbides [1]. The metallic sites in the carbide can be replaced by other elements, depending on the composition of the alloy; for example in $M_6C$, an otherwise Mo-W rich carbide, Fe can replace the metallic atoms up to 30%, while in MC, vanadium can replace W and Mo [2]. The Cr content of the material does not have much effect on the composition of $M_6C$ [2]; however, it forms Cr-Fe rich $M_{23}C_6$ which has crystallographic similarity to $M_6C$ [3,4]. As for abundance, in M2 tool steel, 90% of the primary carbides are $M_6C$; but $M_2C$ replaces $M_6C$ if cooling rates are increased, specially in the regions of high carbon conenctration [2].

Workers [ 5-7 ] have observed similar carbide phases in the rapidly solidified steels, but the chemical composition of carbide phases in rapid solidification has not been the subject of studies reported to date. In the present work thin foils and carbide extraction replicas from rapidly solidified surface areas are prepared and studied under AEM Philips EM420 to investigate the crystal structure and also the chemical composition of the submicron size carbides present.

## EXPERIMENTAL TECHNIQUE

An oscillated beam of focused electrons was used to produce rapidly solidified surfaces (0.6 cm wide) on M7 high speed steel. Translation velocity of the specimen and the beam power were varied in the range of 0. 2 to 5 $cm.s^{-1}$ and 200 to 1000 kW respectively. Detailed and specific microstructural features were observed and analyzed using (i) single stage extraction replicas and (ii) thin films. Specimens prepared by these techniques were examined under Philips EM420 analytical electron microscope to determine carbide morphology, size, distribution, crystal structure and chemical analysis. Carbide extraction replicas provided particularly useful data since the (i) crystal structure and (ii) chemistry of fine carbide particles could be studied without interference from the iron-base matrix material. For comparative purposes the nature of the

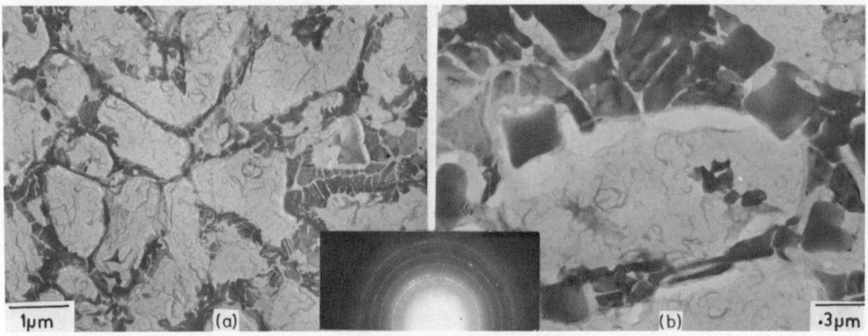


FIG. 1. (a) General view of intercellular carbides; (b) Eutectic $M_2C$, faceted $M_{23}C_6$ and rod-like MC carbides

carbides present prior to electron beam processing was determined by x-ray powder technique on carbides extracted using different chemical solutions.

## OBSERVATIONS AND DISCUSSION

Optical microscopy shows a sharp decrease in the volume % of carbide phases present in the rapidly solidified material as compared to conventionally processed. The detailed nature of carbide network at the boundaries of dendritic cells is well revealed in the single-stage carbide extraction replica in Fig. 1a. Crystallographic study of the carbide phases in the rapidly solidified region using Analytical Electron Microscope confirms the presence of $M_2C$, $M_{23}C_6$, and MC carbides, see Figs. 1-3. Chemical composition of the carbide phases in rapidly solidified zones is summarized in Table I. The most abundant carbide phase present is hexagonal $M_2C$ with the lattice parameter a = $2.92_5$ and c = $4.61_3$. These fan-like eutectic type carbides are smaller than 0.2μm in size. The distribution of $M_2C$ indicates that the carbide is formed eutectically with austenite, which is dissolved during the preparation of extraction replica. $M_{23}C_6$ carbide has cubic morphology with faceted sides and is 3-4 times bigger than $M_2C$. Its morphology, size and distribution suggest that the phase is precipitated out from the liquid entrapped between dendrites of primary phase (see Fig. 1b and 2b). An unusual feature is chemical composition of $M_{23}C_6$ in the rapidly solidified material; it contains only traces of Fe, and is rich in carbide forming elements like V, Mo, and W. Traces of MC cubic carbide with a = $4.35_6$ are also found in the extraction replica and confirmed in thin foils, Figs. 1b and 3; these are rod-like in shape, squeezed in between the dendritic cells. The lattice parameter of MC is quite higher than that of VC (= 4.169), which is probably due to the unusually high concentration (> 50%) of Mo and W; the intriguing feature, however, is that the carbide retains cubic structure and does not transform to the hexagonal structure of WC carbide.

In the 'pre-glazed' material caribes of ≃ 3 μm diameter size are uniformly distributed in a ferrite matrix. Table II summarizes the results obtained by x-ray diffraction on electro-chemically extracted particles. Comparison of this Table with Table I is particularly revealing since it is evident that the phases present differ significantly from those in electron beam processed material. $M_6C$ carbide is abundant in conventionally processed material, while major carbide resulting after rapid solidification is $M_2C$. The chemical analysis shows that $M_{23}C_6$ which is usually rich in Cr and Fe, contains only traces of Fe after electron beam processing, is rich in Mo and V, and as a result has a larger lattice parameter. It is

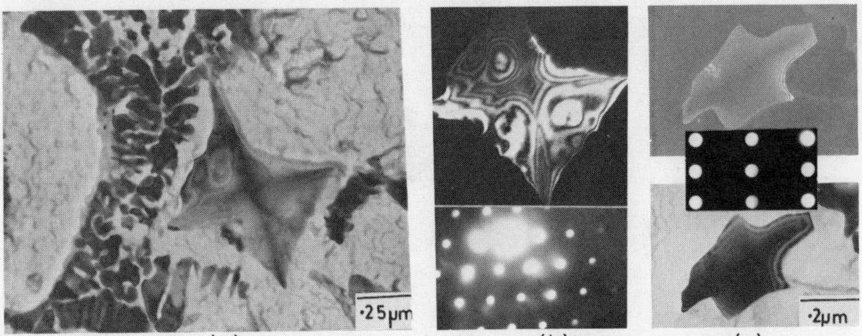


(a) (b) (c)

FIG. 2 (a) Bright field (BF) of carbides near a cell boundary (b) Dark field (DF) and diffraction pattern from $M_{23}C_6$ carbide (c) BF, DF and diffraction pattern from $M_2C$ carbide.

TABLE I

| Carbide Phase | Lattice Parameter | Morphology | Size | Composition (wt%) C | V | Cr | Fe | Mo | W |
|---|---|---|---|---|---|---|---|---|---|
| $M_2C$ | $a=2.92_5$ $c=4.61_3$ | eutectic-like | .1μm | 7.1 | 9.0 | 11.2 | 11.3 | 51.4 | 10.0 |
| MC | a=4.35 | rod-like | .6μm x .1μm | 12.7 | 9.5 | 5.7 | 14.2 | 39.4 | 18.5 |
| $M_{23}C_6$ | $a=10.80_0$ | cubic | .4μm | 4.0 | 30.2 | 5.7 | 1.4 | 45.5 | 13.3 |

TABLE II

| Extraction Soln. | Carbide Phase | Fractional Amount | Lattice Parameter | Density ($gm.cm^{-3}$) | Composition (wt%) C | V | Cr | Fe | Mo | W |
|---|---|---|---|---|---|---|---|---|---|---|
| $10\%HNO_3$ 2volts | $M_7C_3$ | Large | $a=14.00_4$ $c=4.50_2$ | 7 } | 3.6 | 2.9 | 42.8 | 45.4 | 4.4 | 0.9 |
| | $M_{23}C_6$ | Small | $a=10.56_1$ | 10.7 | | | | | | |
| 10%HCl 2volts | $M_6C$ | Large | $a=10.99_1$ | 12 } | 4.3 | 11.9 | 17.6 | 34.7 | 27.6 | 3.8 |
| | MC ($VC_{1-x}$) | Small | $a=4.15_8$ | | | | | | | |
| 2%Br in methanol | $M_6C$ | Large | $a=10.99_3$ | 12 } | 3.6 | 12.1 | 16.1 | 39.8 | 25.0 | 3.5 |
| | MC ($VC_{1-x}$) | Small | $a=4.17_2$ | | | | | | | |

FIG. 3. BF, DF and diffraction pattern from MC carbide in a thin foil.

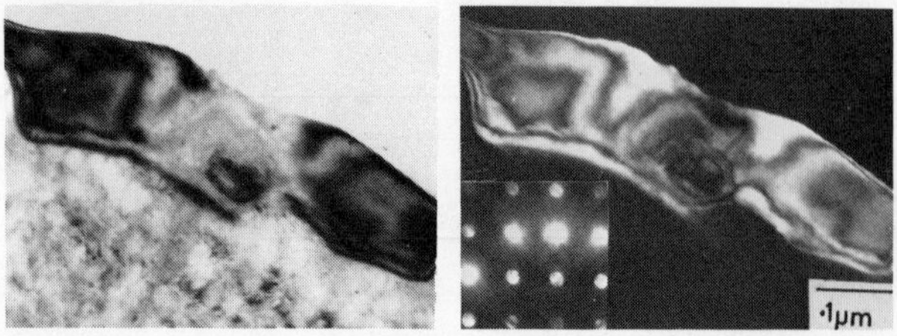


proposed that Mo-W rich $M_6C$ is initially formed and underwent crystallographic transformation during solid state cooling to give $M_{23}C_6$.

## CONCLUSIONS

1. Volume fraction of carbide phases formed during rapid solidification decreases sharply, but concentration of carbide forming elements in them increases.
2. $M_{23}C_6$ is enriched in Mo and W and contains only traces of Cr and Fe in rapid solidification.
3. The concentration of W and Mo in MC exceeds 50 wt%, increasing the lattice parameter (from 4.16$_9$ of VC) to 4.35$_6$. MC carbide remains cubic.

## ACKNOWLEDGEMENT

The authors are pleased to acknowledge support from the Office of Naval Research under Contract N0014-78-C-0580 and continued interest of Dr. B.A. MacDonald.

## REFERENCES

1. E. J. Galda, R. W. Kraft, Met. Trans. 5, 1727 (1974).
2. H. Brandis, E. Haberling, H. H. Weigand, Thyssen Edelstahl Tech. Ber. 7, No. 2, 115 (1981).
3. W. B. Pearson, Handbook of Lattice Spacing and Structures of Metals, (Pergamon Press, Oxford, 1967).
4. E. Staska, R. Bloch and A. Kulmburg, Microchimica Acta (Wien) Suppl. 5, 111-127 (1974).
5. I. R. Sare, R. W. Honeycombe, J. Mat. Sc. 13, 1991 (1978).
6. J. J. Rayment, B. Canter, Met. Sc. 12, 156 (1978).
7. Y. Kim, P. R. Strutt and H. Nowotny, Met. Trans. 10A, 881 (1979).

# PROPERTIES OF SURFACE MELTED SG IRON

B.L. MORDIKE AND H.W. BERGMANN
Institut für Werkstoffkunde und Werkstofftechnik, Technische Universität, Agricolastr. 2, 3392 Clausthal-Zellerfeld, FRG

## ABSTRACT

The surface melting of cast irons is discussed with particular emphasis on composition and effect of impurities. The hardness and wear properties are outstandingly good as shown from tracks under normal working conditions.

## INTRODUCTION

Since it was discovered about the turn of the century that it was possible to use an arc discharge to transform the surface of grey cast iron to white cast iron, many attempts have been made to overcome the variations in normal chill casting [1] by local melting techniques such as electron beam and TIG [2]. Whereas electron beam melting was unsuccessful on an industrial scale due to extensive porosity, TIG melting has proved to be a viable process [3,4]. More recently, laser melting has been attempted with the aim of higher automation, better reproducibility and possibly better properties [5]. The following examples show that a systematic approach can produce excellent results and that it is also a commercially viable proposition.

## EXPERIMENTAL PROCEDURE

The material used for these experiments contained 3.5% C, 2.4% Si, 0.4% Mg, 0.5% Mn and in the case of spheroidal irons $\leq$ 0.001% P and $\leq$ 0.001% S. The surface layer must be crack free, pore free, smooth and have a uniform thickness and microstructure. This requires uniform feed and power to ensure a constant melted depth and rate of quenching. Avoidance of pores requires that no significant evaporation of alloying elements occurs. The formation of cracks is related to the formation of martensite, which in iron depends on the composition, presence of traces of Cu, Mo, Ni, Cr, and on the general melting procedure, eg fixing, sequence of passes, plasma formation. In order to establish the optimum conditions the following parameters were varied:

a) graphite shape - size and distribution
b) inoculants
c) trace elements originating from the mould metal.

The laser used in this work was mainly a Coherent Everlase $CO_2$ laser with a power of 700 W in the continuous and 4 kW in the pulsed mode. The beam was focussed to 0,5 mm and rastered over the surface to be modified. Some experiments were carried out using a Heraeus 5 kW $CO_2$ laser. This is to demonstrate the effect of much thicker remelted layers.

The wear properties were determined by rolling treated rollers against each other (Amsler wear test). The force between them could be varied as could the relative slip S. A vertical load of 800 $N/mm^2$ and 500 $N/mm^2$ was usually employed. Rolling wear is a common form of wear. The results of abrasive wear tests are reported elsewhere [5].

Fatigue behaviour was determined in push-pull on standard Instron machines.

A metallographic investigation was made of the laser melted layers and correlated with the mechanical properties, wear, hardness and fatigue.

## RESULTS AND DISCUSSION

The melting behaviour is affected by the graphite shape, size and distribution. Figs. 1a and b show that melting is only successful for the finer graphite. They also show the unexpected effect that iron with lamellar graphite melts in a completely different manner, depending on their length. Figs. 1a and c show that finely distributed graphite dissolves easily irrespective of its morphology. Fig. 1d shows that large spheres do not dissolve easily causing porosity. A comparison of c and e shows that with high power lasers, thick layers can be melted.

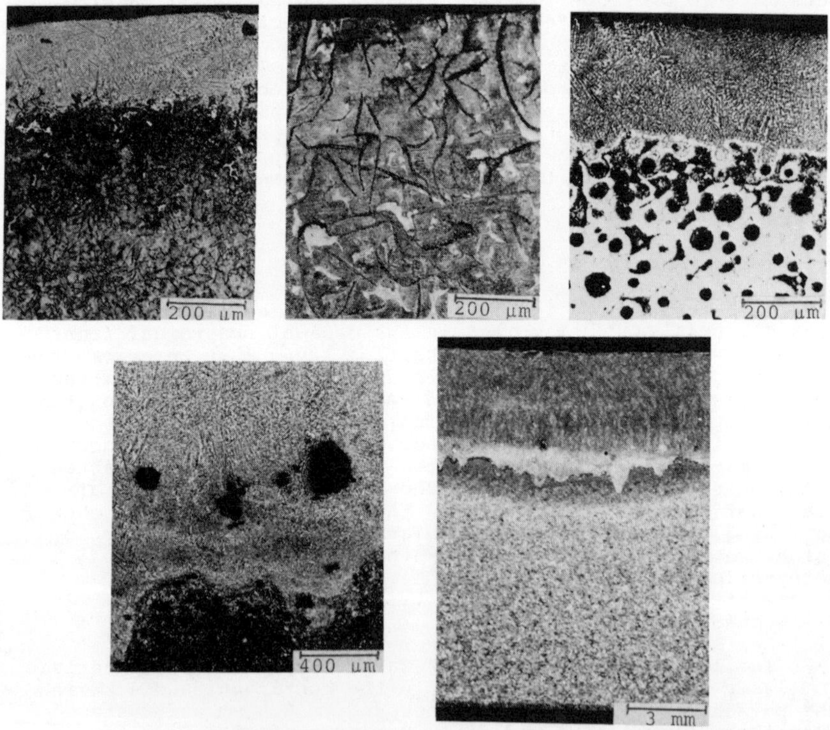


Fig. 1: a,b) Laser treated cast iron with lamellar graphite 1 m/min, 700 W, focussed beam, Argon
c,d) Laser treated cast iron with spheroidal graphite, 1 m/min, 700 W, focussed beam, Argon
e) Laser treated cast iron with spheroidal graphite, 1 m/min, 5 kW, focussed beam, Argon

The hardenability of iron alloys is affected by the amount of Cu, Mo, Ni and Cr. The tendency for the surface layers to crack depends therefore on the amount of these elements present. Cracking may be avoided by preheating the component. This is essential in the case of bainitic irons but may not be necessary in the case of pearlitic or ferritic irons if the alloy content can be maintained sufficiently low. Elements in the surface of the casting which were absorbed from the mould can evaporate and cause porosity problems or ionize and cause a wide plasma, with associated cracking problems 6,7 . The effects of the casting sand and inoculants are shown in the Tables I and II.

The layer, if it is to find application, must be hard, possess a high strength and be of sufficient thickness. A sufficient thickness for cast iron for a specific loading of 1000 N/mm² is 0.2 mm.

Fig. 2 shows the effect of various treatments on the wear. It can be seen that laser surface melting has distinct advantages over other methods. Under extreme loading, the wear for bainitic iron is less than for ferritic or pearlitic as the properties of the substrate are much better.

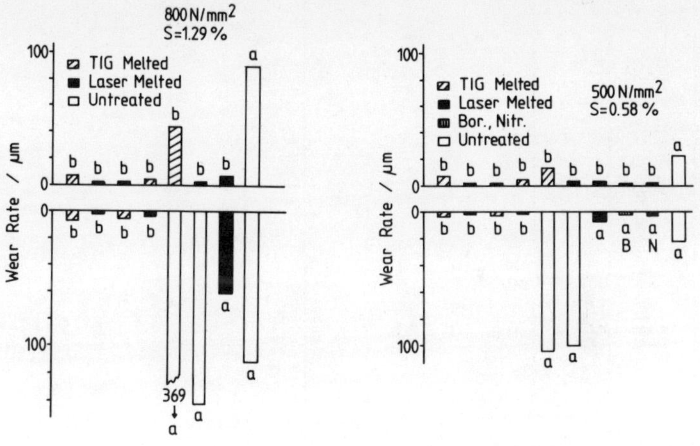


Fig. 2: Wear behaviour after various treatments a represents 0.6%C carbon steel and b surface melted SG iron. The reduction in roller diameter is given for a distance covered of 2000 metres

This improvement in wear may be accompanied by a deterioration in the fatigue properties. Fig. 3 shows the Smith diagrams for ferritic, pearlitic and bainitic cast irons. In each case the surface treatment reduces the permitted stress amplitude.

TABLE I: Influence of Moulding Sand and Mould Wash on the Laser Surface Melting of Grey Cast Iron

| Substrate | Sand | Mould Wash | Pre-heating | Feed Rate m/min | Surface | Roughness | Colour | Cracks | Quantity of Pores |
|---|---|---|---|---|---|---|---|---|---|
| TIG quality | $CO_2$ | - | 400°C | 1 | as cast | medium | bright | no | some |
| " | " | varnish | " | 1 | " | small | bright/ blue | no | medium |
| " | " | alcohol | " | 1 | " | small | " | no | some |
| " | green-sand | - | " | 1 | " | medium | blue | no | medium |
| " | β-set | - | " | 1 | " | rough/ medium | blue | no | many medium |
| " | croming | - | " | 1 | " | medium | blue/ bright | no | medium |
| " | OBB | - | " | 1 | " | small | blue | no | few |
| " | oil | - | " | 1 | " | medium | blue | no | few |

TABLE II: Influence of Inoculants on Laser Melting Behaviour of Cast Irons Cast in Green Sand with no Coating, no Preheating.
(Preheating prevents cracking in all cases and pores for 1, 2, 5, 6)

Analysis of Cast Iron Melts (wt%)

| No. | C | Si | Mn | P | S | Cr | Ni | Mo | Cu | Al | V | Sn | Ti | B |
|---|---|---|---|---|---|---|---|---|---|---|---|---|---|---|
| 1 | 3.21 | 1.91 | .782 | .039 | .029 | .246 | .050 | .010 | .638 | .006 | .005 | .005 | .013 | .001 |
| 2 | 3.19 | 1.92 | .774 | .038 | .029 | .244 | .050 | .009 | .635 | .006 | .005 | .005 | .013 | .001 |
| 3 | 3.21 | 1.92 | .775 | .039 | .031 | .244 | .050 | .010 | .642 | .006 | .005 | .006 | .013 | .001 |
| 4 | 3.23 | 1.95 | .800 | .038 | .026 | .250 | .051 | .010 | .650 | .006 | .005 | .005 | .013 | .001 |
| 5 | 3.27 | 1.89 | .778 | .038 | .028 | .243 | .050 | .009 | .636 | .006 | .005 | .005 | .013 | .001 |
| 6 | 3.30 | 1.90 | .791 | .038 | .025 | .247 | .050 | .010 | .647 | .007 | .005 | .005 | .013 | .001 |
| 7 | 3.26 | 1.82 | .776 | .038 | .030 | .244 | .049 | .010 | .645 | .006 | .060 | .005 | .013 | .001 |

| No. | Inoculant | Analysis of Inoculants (wt%) | | | | | | | | Pores | Cracks | Melted Depth | Depth HAZ |
|---|---|---|---|---|---|---|---|---|---|---|---|---|---|
| | | Si | Al | Ca | Ba | Zr | Sr | Mn | V | | | µm | µm |
| 1 | SB5 | 70 | 1.25 | 1.1 | 2.0 | - | - | - | - | yes | yes | 300 | 200 |
| 2 | ZL 80 | 80 | 2.00 | 2.3 | - | 1.5 | - | - | - | some | some | 300 | 300 |
| 3 | Superceed | 75 | 0.50 | 0.1 | - | - | 0.8 | - | - | yes | no | 300 | 200 |
| 4 | Fe Si 90 | 90 | - | - | - | - | - | - | - | yes | yes | 500 | 200 |
| 5 | SMZ | 65 | 1.00 | 0.8 | - | 6.0 | - | 6.0 | - | some | some | 200 | 200 |
| 6 | SGI | 70 | 3.00 | 0.7 | - | - | - | - | - | no | no | 200 | 250 |
| 7 | Fe V 60 | - | - | - | - | - | - | - | 60 | no | some | 300 | 300 |

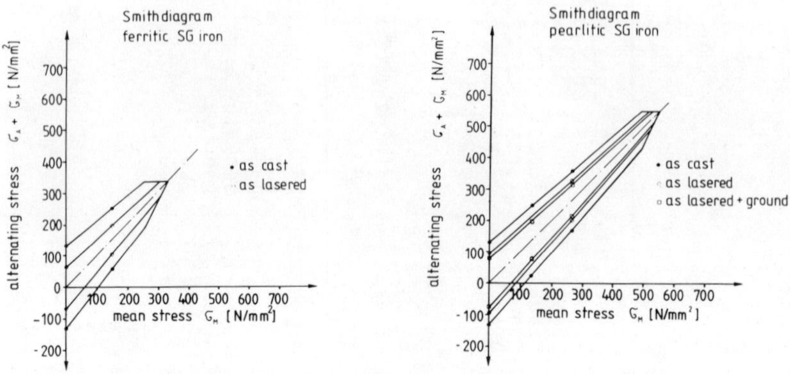


Fig. 3: Smith diagrams
a) ferritic SG iron
b) pearlitic SG iron

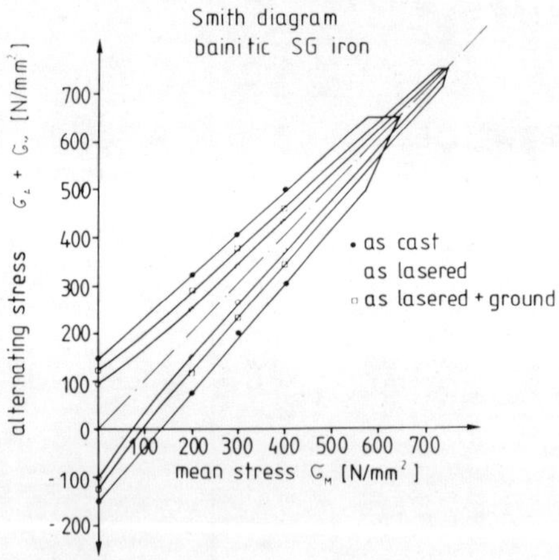


Fig. 3c: Smith diagram of bainitic SG iron

It is clear that bainitic cast iron has a higher fatigue resistance than other grades that is higher permissable amplitudes and mean stresses. Components for automobiles which were made from bainitic cast iron have been tested. They showed significant improvement compared with other treatments. Fig. 4 shows a camfollower after 40.000 km service. The refinement of the microstructure caused by the rapid solidification has increased the ductility of the component surface. Cementite is thus deformed and not fractured, as normally occurs, Fig. 4c.

Fig. 4: Camfollower after 40.000 km service
a,b) general views
c) micrograph

CONCLUSIONS

Laser melting of cast irons can produce a significant improvement in the wear properties.

ACKNOWLEDGEMENT

The work was largely supported by the Volkswagen Foundation. The authors are indebted to Prof. Krause, TU Aachen, for carrying out the wear measurements.

REFERENCES

1. Witzemann, private communication
2. F.H. Reinke, "Local Electric Arc Remelting Process for the Generation of Wear-Resistant White Iron Layers on Workpieces of Grey Cast Iron, especially Camshafts and Camfollowers", AEG Elotherm, Remscheid, FRG, 1985
3. K. Heck, "Local Arc Remelting for the Generation of Wear Resistant White Iron Layers on Grey Cast Iron Camshafts" Wela, Moskau, 1977
4. F. Emde, Elektrowärme International, 37, B3 (1979)
5. H.W. Bergmann, Surface Engineering Vol. 1 No2, 137 (1985)
6. H.W. Bergmann, Proc. Conf. on Laser Surface Treatment of Metals, San Miniato, Italy, 1985, to be published
7. H.W. Bergmann, Int. Conf. on Foundry and Heat Treatment, Johannesburg, 1985, to be published

# MICROSTRUCTURAL CHARACTERIZATION OF RAPIDLY SOLIDIFIED TYPE 304 STAINLESS STEEL

R. N. Wright, J. E. Flinn, and G. E. Korth
Idaho National Engineering Laboratory, EG&G Idaho, Inc., Idaho Falls, ID 83415

## ABSTRACT

The microstructures of rapidly solidified Type 304 stainless steel powders produced by vacuum gas (VGA) and centrifugal atomization (CA) have been examined. The solidification morphology and phase distribution have been characterized using optical and scanning electron microscopy, and the relative amounts of ferrite and austenite have been quantified using x-ray diffraction. Most CA powder particles contain both fcc and bcc phases, with the bcc phase predominating at small particle sizes and the fcc phase at large particle sizes. The VGA powder generally contains less ferrite, with very little dependence on the particle size. The ferrite was metastable and transforms to austenite on annealing at 900°C.

## INTRODUCTION

The potential advantages of rapid solidification processing (RSP), refined microstructures, improved chemical homogeneity, and retention of metastable phases, have been the subject of extensive research since the discovery of metallic glasses in 1960 [1]. To achieve the necessary heat transfer to obtain high cooling rates, the products formed must have very high surface to volume ratios, e.g., ribbons, flakes or powders. Difficulty arises in consolidating the particulates into monolithic forms suitable for engineering applications without losing the benefits of RSP. With the exception of dynamic consolidation [2], the methods currently under development for producing monolithic forms (e.g., hot isostatic pressing or hot extrusion) involve exposing the RSP material to temperatures on the order of half the melting point or above for extended periods of time. The effect of thermal or thermal-mechanical processing on the microstructure of RSP materials is not currently well understood.

This paper presents the results of a study of the microstructure and response to annealing of RSP Type 304 stainless steel powders processed by centrifugal and vacuum gas atomization. The goal of this work is to understand the development of the microstructural features of the rapidly solidified powders and the relationship of the powder characteristics to processing variables. Detailed characterization of the powders, and their response to heat treating, also provides a reference state for consolidation studies and aids in developing processes to produce monolithic forms which retain desirable RSP properties.

## EXPERIMENTAL PROCEDURE

Rapidly solidified Type 304 SS powder made by the CA process was obtained from Pratt and Whitney Government Products Division, West Palm Beach, FL. Powder produced by vacuum gas atomization (VGA) was obtained from Homogeneous Metals Inc., Clayville, NY. Both powders were made from ingots provided by the suppliers; the chemical composition of the powders is given in Table I.

The powders were classified into nine size fractions from -250 μm to -10 μm by screening. The relative volume fractions of ferrite (bcc) and austenite (fcc) phases were determined as a function of particle size using x-ray diffraction. The direct comparison method was used, with graphite crystal monochromated CuKα radiation. It was found that the CA powder could be separated into strongly ferromagnetic and weakly or non-magnetic fractions

 

Table I. Chemical Analysis of RSP Powders

| Alloy | Fe | Ni | Cr | Si | Mn | C | Mo | Cu | S | P |
|---|---|---|---|---|---|---|---|---|---|---|
| VGA | 71. | 9.9 | 19.0 | 0.04 | 0.03 | 0.015 | 0.015 | 0.005 | 0.005 | 0.005 |
| CA | 70.5 | 9.1 | 18.4 | 0.65 | 0.80 | 0.052 | 0.60 | 0.50 | 0.002 | 0.023 |

using a small permanent magnet. The strongly magnetic fraction was classified by screening and the relative amounts of fcc and bcc were determined as a function of particle size. Optical and scanning electron microscopy were used to characterize the morphology and microstructure of the powders.

The annealing response of the powders was studied by measuring the relative amounts of fcc and bcc phases after a series of one hour annealing treatments at temperatures ranging from 300 to 900°C. The powder samples were annealed in evacuated quartz capsules and water quenched.

## RESULTS

The particle size distribution for "grab" samples of the two powders is shown in Fig. 1. The distributions are markedly different in that particle sizes in the 100 to 150 μm range dominate the weight distribution for the CA powder, whereas the VGA appears to be spread equally over the range of 30 to 150 μm. Based on a cumulative % finer than correlation, the average particle sizes are about 40 and 80 μm for the VGA and CA powders, respectively. The size distribution of the magnetically separated CA powder is essentially identical with that shown for the CA "grab" sample in Fig. 1.

Both the VGA and CA powders are primarily spherical. The VGA powder has a large number of satellites, particularly in association with larger particles. Optical microscopy of polished and etched powder samples showed that the CA powder generally had a well-defined dendritic structure, while the VGA powder had a cellular structure. Optical micrographs of the 100 μm CA and VGA powders are shown in Fig. 2.

The volume fraction of fcc phase is shown as a function of the particle size in Fig. 3 for CA and VGA powders and also for the magnetic fraction of the CA powder. The volume fraction of fcc phase in the CA powder shows a strong dependence on the particle size, with the amount of fcc increasing with increasing particle size. In the VGA powder there is a trend for the larger particles to have slightly less fcc. The magnetic fraction of the CA powder exhibits the same trend as the "grab" sample, however, the magnetic material has less fcc at any given particle size.

It was found that the bcc phase in both powders is essentially stable up to annealing temperatures of 500°C, and the volume fraction decreases rapidly between 600 and 800°C, as can be seen in Fig. 4. In the VGA powder the ferrite is almost entirely transformed to austenite after annealing for one hour at 800°C, while the magnetic fraction of the CA powder retains about 25% of the original ferrite at this temperature.

## DISCUSSION

Under equilibrium conditions Type 304 SS solidifies as primary ferrite and upon cooling transforms to a fully austenitic structure at room temperature. It has been observed in other austenitic stainless steels that during rapid solidification primary crystallization can occur either as austenite or ferrite depending on the cooling rate and degree of supercooling [3,4]. Type 303 SS powder produced by centrifugal atomization has been studied in detail [3]. It was observed that small particles solidified as primary ferrite, while larger particles solidified as primary austenite. It was concluded from this work that small particles were unlikely to contain potent nucle-

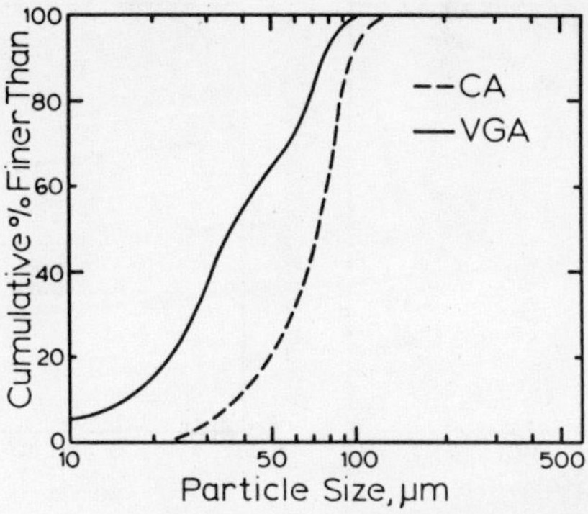


Fig. 1. Particle size distribution for RSP powders.

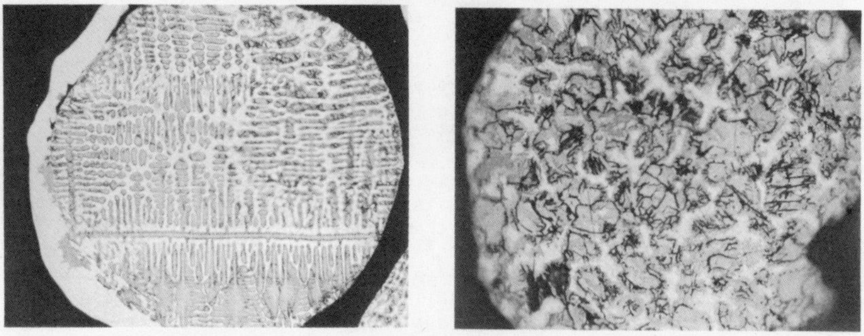

Fig. 2. Optical micrographs of 100 μm (a) CA and (b) VGA RSP powders.

ation sites for crystallization of austenite and therefore the liquid experienced deep supercooling which favored crystallization of ferrite. The large particles, which statistically are more likely to contain heterogeneous nucleation sites, solidified with much less supercooling as primary austenite. A study of Type 308 SS [4] which was solidified by laser melting and also by splat quenching supports the conclusions of Kelly et al. [3]. Type 308 SS solidifies under equilibrium conditions as primary ferrite, however in the laser melting and splat quenching experiments austenite was found to be the primary crystallization phase. In laser melting experiments, solidification begins at the melt pool/solid substrate interface, and in splat quenching solidification begins at the surface of the anvils. These solid surfaces would provide many nucleation sites for solidification, which therefore occurs with little supercooling, and austenite is the primary solidification phase.

It seems likely that the VGA powder, which is atomized by the expansion of gas evolving from the melt into a vacuum where cooling is only by radiation, solidifies principally as primary austenite. The CA powder, which is cooled by a combination of radiation and forced convection with a high velocity gas, apparently solidifies as primary ferrite as a result of deep supercooling of the small particles and as primary austenite in the larger particles. The dendritic structure and finer scale of the microstructure in the CA powder indicate that it probably had a higher overall solidification rate than the VGA which shows a cellular structure [5].

One interesting difference between the results of this study and the work on Type 303 SS [3] is the observation that the Type 304 CA powder particles

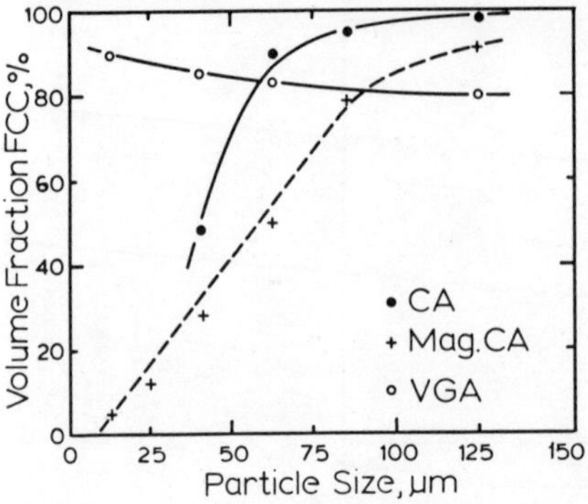


Fig. 3. Volume fraction of fcc phase as a function of particle size.

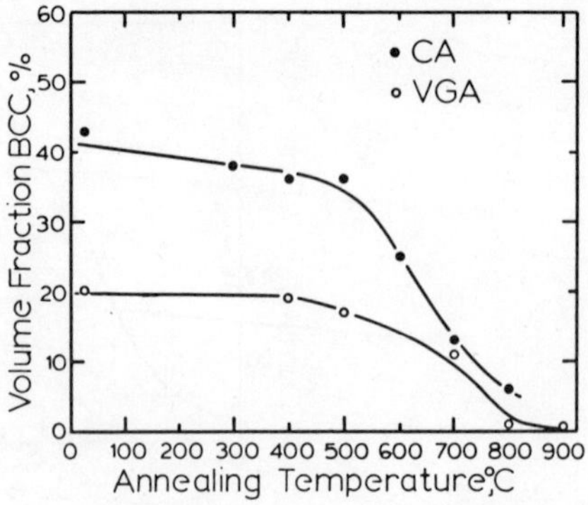


Fig. 4 Volume fraction of bcc phase as a function of temperature for 1 h annealing time.

are a mixture of fcc and bcc phases at all particle sizes, whereas the Type 303 was found to be only bcc at small particle sizes and only fcc at large particle sizes. The reasons for this difference in behavior are not currently understood; detailed STEM analysis is underway to help resolve this question.

The annealing data in Fig. 4 clearly show that the bcc phase identified by x-ray diffraction is metastable and is not retained in the microstructure after annealing above about 900°C. A consolidation study using the CA powder has shown that the minimum temperature necessary for hot isostatic pressing or hot extrusion of this material is also about 900°C, and therefore the bcc phase is not retained in monoliths produced by these processes [6]. Optical metallography has shown, however, that the dendritic structure is still retained in material which has been hot isostatically pressed to full density at 900°C even after the bcc phase has transformed to fcc. The effects of the metastable bcc phase on the mechanical properties of the material are hard to predict, however the retention of the fine scale dendritic structure and improved chemical homogeneity should still result in improvements over conventional ingot metallurgy products.

## ACKNOWLEDGMENTS

This research was performed as part of the Idaho National Engineering Laboratory's Strategic and Critical Materials Research Program sponsored by the U.S. Bureau of Mines and conducted under Department of Energy Contract No. DE-AC07-76ID01570. The authors also gratefully acknowledge the assistance of D.V. Miley, V.L. Smith, G.L. Fletcher, M.D. Harper and M.W. Ellingford in conducting these experiments.

## REFERENCES

1. P. Duwez, R.H. Willens, and W. Klement, J. Appl. Phys., 31, 1136 (1960).
2. J.E. Flinn, Rapid Solidification Technology for Reduced Consumption of Strategic Materials, Noyes Pub., Park Ridge, N.J. (1984).
3. T.F. Kelly, M. Cohen and J.B. Vander Sande, Metall. Trans., 15A, 819 (1984).
4. J.M. Vitek, A. Dasgupta and S.A. David, Metall. Trans., 14A, 1833 (1983).
5. R. Mehrabian, Int. Met. Rev., 27, 185 (1982).
6. J.E. Flinn, G.E. Korth and R.N. Wright, to be published.

# MECHANICAL PROPERTIES OF RAPIDLY SOLIDIFIED HIGH CARBON IRON BASE ALLOYS

D. BURCHARDS, K.U. KAINER, B.L. MORDIKE
Institut für Werkstoffkunde und Werkstofftechnik, Technische Universität, Agricolastr. 2, 3392 Clausthal-Zellerfeld, FRG

## ABSTRACT

High carbon content iron base alloys (2-4 wt%) with various alloying additions ,eg., Cr, Ni, Si or Te were produced in powder form by water atomisation. Depending on the alloying addition the powder particles solidify, microcrystalline with either dendritic or cellular microstructures with supersaturated solid solution and eutectic. The proportion of carbide is dependent on the composition and atomizing conditions. The powder, after suitable preparation, was consolidated by extrusion or hot isostatic pressing. The consolidation behaviour of the powders is influenced by the composition and particle size distribution. The microstructures and hardness values of the powder in the as quenched state are compared with those of the consolidated state with particular emphasis on the thermal stability. The mechanical properties of extruded powders are compared with those for hot isostatically pressed.

## INTRODUCTION

During atomization alloys solidify in metastable states [1]. They can be characterized as follows: finely dispersed microstructures, high degree of supersaturation of matrix, metastable phases and homogeneous distribution of phases and elements. The as-quenched powder is then extruded or hot isostatically pressed to semi finished products and this can lead to changes in the microstructural state. Alloying elements can be chosen so that the fine microstructure is retained producing materials with good properties [2]. Such materials possess high strength and at the same are very ductile. An alloy with 3% C and 1.5% Cr exhibited superplastic behaviour at 700°C [3,4].

## EXPERIMENTAL DETAILS

The powders were prepared using a water atomizer from Davy McKee, Sheffield, Table I. The liquid alloy stream was atomized with a 20 MPa water jet producing powders $d_{50}$ = 38 µm. After drying in vacuum the powder was classified by sieving and air classifying into the particle sizes 10 µm, 20 µm, 35 µm, 70 µm. Billets for extrusion were prepared by filling steel containers and cold compressing. The containers were evacuated and sealed and extruded at temperatures between 860 and 920°C at a pressure of 780 MPa.
For comparative purposes some alloys were hot isostatically pressed at Vacuum Generators, Hastings, England. The powder was again put into containers, sealed under vacuum and consolidated under 200 MPa at 675°C for 5 h. Metallographic and mechanical testing was carried out on the compacted rods.

 

TABLE I Chemical Analysis of the Alloys Prepared

| No. | C | Si | Mn | P | S | Cr | Ni | Ti | Mo |
|---|---|---|---|---|---|---|---|---|---|
| 5 | 2.60 | 2.03 | 0.12 | 0.011 | 0.008 | - | 0.030 | - | - |
| 6 | 3.40 | 1.98 | 0.08 | 0.016 | 0.007 | 0.003 | 0.040 | - | - |
| 8 | 2.76 | 0.64 | 0.48 | 0.017 | 0.100 | 6.840 | 0.130 | 0.005 | - |
| 10 | 2.65 | 0.60 | 0.57 | 0.014 | 0.163 | 2.410 | 6.900 | 0.128 | - |
| 11 | 3.20 | 0.89 | 0.59 | 0.035 | 0.035 | 0.260 | 3.440 | 2.040 | - |
| 12 | 3.80 | 0.17 | 0.03 | 0.025 | 0.013 | 0.040 | 0.030 | - | 0.02 |
| 13 | 3.70 | 0.18 | 0.03 | 0.026 | 0.010 | <0.04 | 2.220 | - | 0.06 |
| 14 | 3.70 | 0.20 | 0.04 | 0.022 | 0.011 | 1.900 | 0.030 | - | 0.05 |
| 15* | 3.85 | 0.19 | 0.03 | 0.025 | 0.008 | <0.04 | 0.030 | - | 0.06 |

* inoculated with 0.01% Te

RESULTS AND DISCUSSION

The rapidly quenched powders exhibited microstructures which ranged from dendritic to cellular matrix with a eutectic, see Fig. 1, the properties of which depended on the particular alloy.

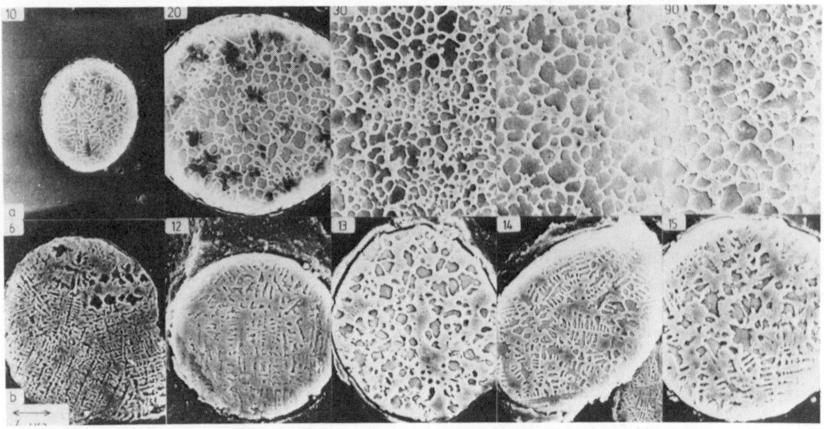


Fig. 1 a) Microstructure as a function of particle size from 10 μm to 90 μm from alloy 10
b) Microstructure of 20 μm particle for different alloys (No 6, 12, 13, 14, 15)

The solidification velocity, as measured by the secondary dendrite arm separation, increases with decreasing particle size [5]. A value of $2x10^5$ K/s was obtained from Fig. 1b, alloy 14. The alloy composition also plays a role in determining the microstructure formed (Fig. 1b). For constant particle size (d = 20 μm) and almost constant carbon content C ≅ 3.8 % the effect of the various elements Si, Ni, Cr, Te (nucleation addition) on the plain Fe-C alloy (alloy 12) can be seen. The plain iron-carbon alloy (alloy 12) solidifies dendritically, as does the chromium containing alloy (alloy 14). Chromium reduces the proportion of eutectic. 0.01% Te (alloy 15) grain refiner leads to a cellular/dendritic structure. Ni containing alloys (alloy 13) solidify cellularly. 2% Si increases the proportion of eutectic (alloy 6) as silicon is dissolved in the matrix. The influence of the alloy elements on the microhardness, as compared with the plain iron-carbon alloy, is apparent in Fig. 2.

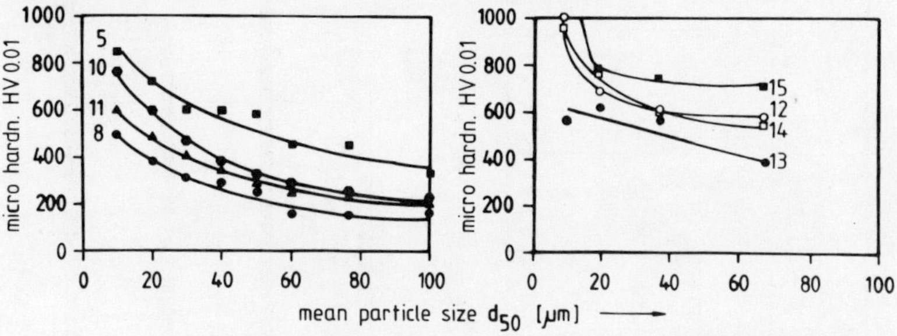


Fig. 2 Microhardness as a function of the particle size and chemical analysis. The numbers refer to the alloy I.D.

The hardness values of the consolidated powder (Fig. 3) are not dependent on the particle size.
The extrusion at the temperature 860-920 °C leads to a change in the microstructure of the alloys.

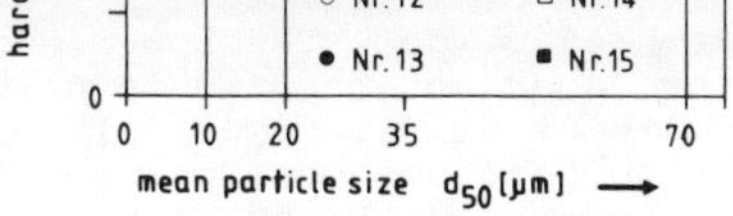


Fig. 3 Hardness HV10 for the consolidated powder as a function of particle size

It is particularly important especially for the alloy 13 to restrict the time the billet is held at the extrusion temperature. By preheating for only 15 min at 900 °C hardness and tensile stress values equal to those for the alloys 12, 14 and 15 could be produced on extrusion. The microstructure consisted of a continuous carbide network. A longer annealing time, 25 min, produced the values of about 200 HV10 shown in Fig. 3. The carbide network has coalesced. Nickel does not stabilize the carbon size as it increases the tendency to graphitization. The alloys 12, 14 and 15 are less prone to carbide growth and hence retain their strength. Fig. 4 shows the ultimate strength plotted against particle size and composition. Fig. 5 shows the ductility.

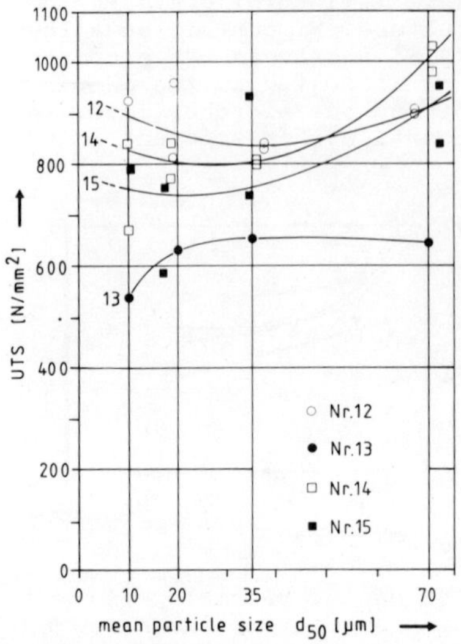


Fig. 4 Tensile strength as a function of particle size and chemical composition

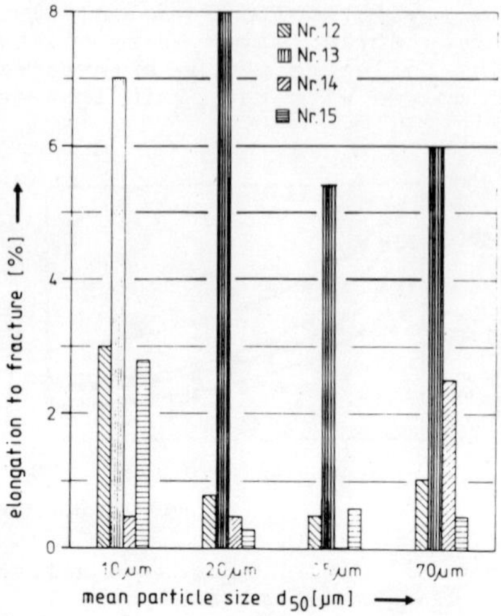


Fig. 5 Ductility in tension as a function of particle size and chemical composition

The ductility is greatest for Ni containing specimens because the carbon has precipitated to some extent as graphite and the carbide network has coalesced. The alloy 13 with a mean particle size of $d_{50}$ = 20 µm is particularly ductile (100% elongation to fracture) at high temperature T = 0.6 $T_m$ = 670°C. A strain rate of 5.9 x $10^{-6}$ $s^{-1}$ was used. Elongations of 30-50% were observed for other compositions and particle sizes.

Hot isostatically pressed powder with 2% Cr (specimen 14) shows a somewhat higher hardness than extruded powder. The lower consolidation temperature in hipping produced an increase in hardness. On the other hand the alloy with 2% Ni (specimen 13) does not tend to graphitize at the lower temperature of hipping and a higher hardness (HV10 = 471) is retained. With the 2% Si alloy graphitization even at the lower hipping temperature is complete and the high hardness of the initial powder is lost. The optimum hipping parameters have not yet been determined and thus it is not yet possible to state definitely which consolidation process is the best in each case.

REFERENCES

1. H.Jones, Rapid Solidification of Metals and Alloys, (Inst. of Metallurgists, London, 1980)
2. B.Walser, O.D.Sherby, Metall. Trans. A 10, 1461 (1979)
3. J.Wadsworth, O.D.Sherby, J. Materials Science 13 (1978)
4. O.A.Ruano, L.E.Eiselstein, O.D.Sherby, Metall. Trans. A 13, (1982)
5. C.G.Levi, R.Mehrabian, Metall. Trans. A 13, (1982)

# FRACTURE PROCESSES IN FULLY CRYSTALLIZED $Co_{84}Nb_{10}B_6$ METALLIC GLASS

JANEZ MEGUSAR, TINA VARGAS AND NICHOLAS J. GRANT
Department of Materials Science and Engineering, Massachusetts Institute of Technology, Cambridge, MA 02139, USA

## ABSTRACT

Fracture processes have been studied in a fully crystallized $Co_{84}Nb_{10}B_6$ glass, with the mean grain size ranging from 43 nm to 0.65 μm. Tensile tested specimens with a mean grain size $\gtrsim$ 0.3 μm showed a ductile (dimple) fracture with microvoids initiated at the boride-cobalt matrix interfaces. As the mean grain size (more correctly, the corresponding mean free path of the deforming cobalt phase) was $\leq$ 0.1 μm, the fracture process may be similarly initiated by the microvoid formation at the boride-cobalt interface. However, one-to-one correspondence between dimples and boride particles may not be preserved. As a result, dimples become more shallow and may include several boride particles instead. Although the fracture surface of the brittle fine grained $Co_{84}Nb_{10}B_6$ alloy gives an appearance of a cleavage fracture on a macroscale, it is distinctly different, by allowing for the microvoid formation on a microscale.

## INTRODUCTION

Most metallic glasses are brittle when fully crystallized. This has been associated with a high volume fraction of brittle phases in the crystallized state. In order to develop a ductile crystalline material from the glassy state, it is therefore necessary to reduce the metalloid content. This was demonstrated for Fe-based metallic glasses where the boron content has been reduced to about 10 at% [1,2]. Recently, it has been shown [3] that $Co_{84}Nb_{10}B_6$ glass with a metalloid content of only 6 at% can be similarly devitrified to develop a ductile crystalline material. As shown in Fig. 1, the bend diameter of the melt spun ribbon reaches its maximum when the $Co_{84}Nb_{10}B_6$ glass is annealed at 973 and 1073 K; it then becomes progressively more ductile with increasing temperature, becoming fully ductile and

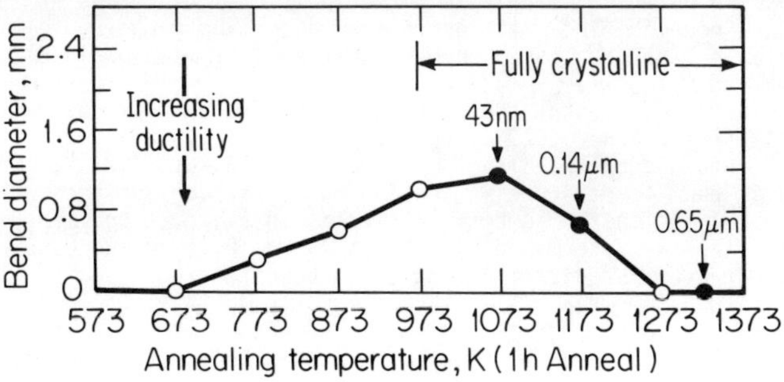


Fig. 1: Bend ductility of a crystallized $Co_{84}Nb_{10}B_6$ glass as a function of annealing temperature (1 h anneal). $T_x$ (onset) is 717 K for a heating rate of 20 K $min^{-1}$.

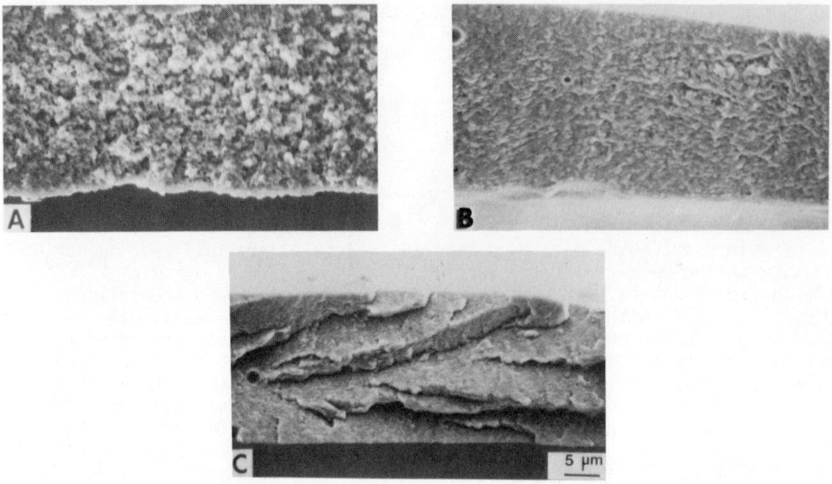

Fig. 2: Fracture surfaces (SEM) of fully crystallized $Co_{84}Nb_{10}B_6$ glass annealed for 1 h at temperatures of 1323 K (A), 1173 K (B) and 1073 K (C). Also see Figs. 3B and 4B, which are higher magnifications of Figs. 2A and 2C, respectively.

bendable 180 degrees without fracturing after an anneal for 1h at 1273 K.

Crystallization of $Co_{84}Nb_{10}B_6$ glass has been systematically studied [3] in order to explain the process of development of the ductile crystalline material from the amorphous state. X-ray diffraction showed that the $Co_{84}Nb_{10}B_6$ glass crystallized into a cobalt-based solid solution phase and an unidentified boride phase, presumably also with cubic symmetry. There is no further phase transformation in the crystalline state on further heating. STEM and high resolution TEM showed no evidence of a fine scale precipitation at the cobalt-boride interface. Isochronal and isothermal annealing indicated the same ductility versus grain size relationship, namely, that ribbons could be bent 180 degrees without fracturing when the mean grain size coarsened to at least 0.3 μm. TEM and SEM studies indicated that the spacing between the brittle boride particles decreases as the microstructure becomes finer with decreasing annealing temperature. This leads to brittle behavior, based on bend testing, for mean grain sizes finer than or equal to 0.1 μm.

In this paper we present the results of the fracture processes in fully crystallized, fine grained $Co_{84}Nb_{10}B_6$ alloy. Fully crystallized melt spun ribbons were obtained by annealing for 1h from 973 K to 1323 K. Specimens were tested at room temperature in tension and fracture surfaces were examined by SEM. As indicated by full circles in Fig. 1, specimens with a mean grain size (as determined by TEM) of 43 nm, 0.14 μm and 0.65 μm were selected for fracture studies. According to bend testing shown in Fig. 1, specimens with 0.65 μm mean grain size were fully ductile, while specimens with 43 nm grain size showed maximum embrittlement.

## EXPERIMENTAL PROCEDURES

The alloy was prepared by melting pure metals (99.9%) using a non-consumable water-cooled tungsten arc melting technique under an argon atmosphere. The nominal composition is that calculated from the initial weights

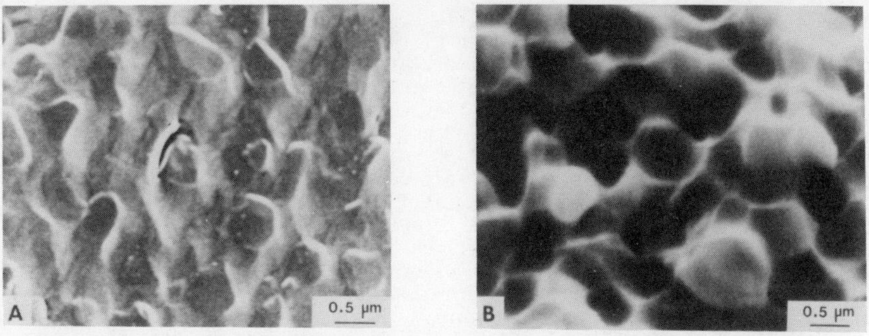


Fig. 3: SEM fractures of fully crystallized $Co_{84}Nb_{10}B_6$ glass, annealed for 1 h at 1323 K: (A) showing initiation of fracture at the boride-cobalt interface at the side of the specimen below the fracture; (B) the resulting ductile fracture.

of the components. Rapid solidification was accomplished by melt spinning (using a fused silica crucible) and quenching against the outside surface of a rotating copper drum under a helium atmosphere. The continuous ribbons were typically 30-40 μm thick and 2-3 mm wide. Room temperature tensile testing of ribbons was performed on an Instron testing machine at a strain rate of $2.7 \times 10^{-3}$ $s^{-1}$. Fracture surfaces were examined in the AMR SEM and the JEOL 200CX TEM operated in a scanning mode.

## EXPERIMENTAL RESULTS AND DISCUSSION

Fig. 2 shows SEM fracture surfaces of tensile specimens of the fully crystallized $Co_{84}Nb_{10}B_6$ glass. Annealing for 1h at 1323 K produced a fully ductile material (Fig. 1) and the corresponding ductile fracture is shown in Fig. 2A. Annealing for 1 h at 1073 K resulted in maximum embrittlement (Fig. 1) and the corresponding fracture surface in Fig. 2C has an appearance of a cleavage fracture. Fracture characteristics in Fig. 2B appear to be intermediate between those of the ductile fracture (Fig. 2A)

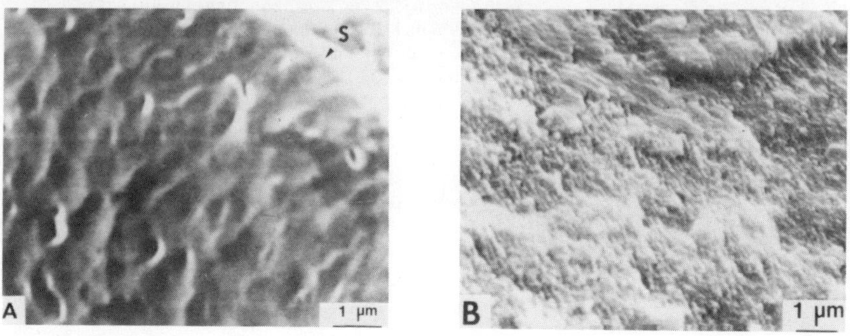


Fig. 4: (A) SEM view of the side of the ribbon immediately below the cleavage fracture surface, marked S, showing crack initiation sites for material annealed 1 h at 1173 K. See Fig. 2B for the fracture surface; (B) enlarged view of fracture surface shown in Fig. 2C.

and the apparent cleavage fracture (Fig. 2C).

In order to study the initiation of the fracture process in the fully crystallized $Co_{84}Nb_{10}B_6$ glass, the tensile tested ribbons (approximately 35μm thick and 0.7 mm wide) were mounted on copper grids and prepared metallographically for SEM observation in the JEOL 200CX transmission electron microscope. Specimens were ion milled and etched with Villela's reagent in order to relate the fracture initiation with the microstructure consisting of the brittle boride phase and the soft cobalt-based solid solution phase. As shown in Fig. 3A for 1 h annealing at 1323 K microvoids are initiated during plastic flow at the interface between the boride particle and the cobalt matrix at the side of the specimen below the fracture. As plastic flow increases, the existing microvoids grow and new microvoids are initiated; eventually the enlarged microvoids grow into close enough proximity so that the thin ridges separating them rupture. The resultant fracture surface in Fig. 3B shows a typical dimple fracture. It is worth noting that the size and shape of dimples in Fig. 3B are very uniform and this is due to the uniform size of boride particles and the cobalt matrix.

Since the plastic flow is confined to the soft cobalt phase, the mean free path of the deforming phase may be a better parameter than the mean grain size in studying deformation and fracture behavior of fully crystallized $Co_{84}Nb_{10}B_6$ alloy [4]. It appears that the fracture process may be similarly initiated by the microvoid formation at the boride-cobalt interface when the mean free path is decreased at lower annealing temperatures of 1173 and 1073 K and resulting in progressively brittle behavior. Several voids may be observed in Fig. 4A, at the side of the specimen below the fracture, in the specimen annealed for 1 h at 1173 K. However, one-to-one correspondence between dimples and boride particles may not be preserved with decreased mean free path. As a result, dimples become more shallow and may include several boride particles instead. Close examination of the fracture surface in Fig. 4B of the specimen annealed for 1 h at 1073 K shows that it is distinctly different from the cleavage fracture observed in conventionally processed and brittle bcc alloys. Although the fracture surface of the brittle fine grained $Co_{84}Nb_{10}B_6$ alloy gives an appearance of a cleavage fracture on a macroscale (Fig. 2C), it is distinctly different, by allowing for the microvoid formation on a microscale.

## ACKNOWLEDGEMENT

This research was sponsored by the National Science Foundation through the Center for Material Science and Engineering, Massachusetts Institute of Technology, under Contract No. 8119295-A04-DMR. One of the authors (T.V.) was partially supported by the M.I.T. Undergraduate Research Opportunities Program.

## REFERENCES

1. R. Ray, J. Mater. Sci., 16, 2924 (1981).
2. C.P. Ashdown, Ph. D. Thesis, M.I.T. (1984).
3. J. Megusar, G. Franks and N.J. Grant, Int. J. Rapid Solid., 1, 291 (1984-85).
4. W.R. Cribb, Scripta Metall., 12, 893 (1978).

PART V C

# Microstructure and Properties of Crystalline RSP Alloys— Other Systems

# COPPER-BASE ALLOYS PROCESSED BY RAPID SOLIDIFICATION AND ION IMPLANTATION

J.V. Wood and C.J. Elvidge, Department of Materials, The Open University, Milton Keynes, UK.

E. Johnson, A. Johansen, L. Sarholt-Kristensen and O. Henriksen, Physics Laboratory II, University of Copenhagen, Denmark.

## ABSTRACT

Alloys of Cu-Sn and Cu-B have been processed by both melt spinning and ion implantation. In some instances (eg Cu-Sn alloys) rapidly solidified ribbons have been subjected to further implantation. This paper describes the similarities and differences in structure of materials subjected to a dynamic and contained process. For example in Cu-B alloys (up to 2wt% Boron) extended solubility is found in implanted alloys which is not present to the same degree in rapidly solidified alloys of the same composition. Likewise the range and nature of the reversible martensitic transformation is different in both cases as examined by electron microscopy and differential scanning calorimetry.

## INTRODUCTION

The production of extended solubilities and non-equilibrium phases (including amorphous structures) are characteristics of both rapid solidification and ion implantation. This paper presents some preliminary data for copper alloys processed by both techniques. Previous work on copper alloys has concentrated either on their 'model' properties for analysing various rapid solidification techniques (1) or has been confined to compositions found to display a shape memory effect (2-4). In the latter instance the temperature of the martensitic transformation associated with the shape memory behaviour, is lowered in comparison with cast and wrought products. This depression is linked with retained defect concentrations and the refinement of grain size.

In this investigation we report on two simple binary copper alloys: Cu-B and Cu-Sn which have been processed by both techniques. The first represents an investigation into a system where boron can act substitutionally and interstially and extensions in solid solubility would be expected. Cu-Sn by contrast is a system which has a number of complex phase boundaries and for compositions around 14 at.% there is a potential for formation of metastable martensite.

## ION IMPLANTATION

All experiments for Cu-Sn were undertaken on pure Cu single crystal spark machined into discs with <110> surface normal. Implantations at random incidence were performed at room temperature in a heavy ion isotope separator below $2\cdot10^{-5}$ Pa. A beam flux of $3\cdot10^{17}m^{-2}sec^{-1}$ ensured that the samples were not heated during implantation. Analyses were made by combining Rutherford backscattering (RBS) and channeling techniques.

For Cu-B, single crystal and polycrystalline samples of Cu were implanted with 40 keV boron ions to fluences of $5\cdot10^{20}m^{-2}$ and $2\cdot10^{21}m^{-2}$. Nuclear reaction analyses were performed using the reaction $^{11}_{5}B(p,\alpha)^{8}_{4}Be$ which has two resonance peaks. The broad peak at 660 keV was used for lattice location analysis of the implanted boron atoms. The lower peak at 163 keV, which is very narrow, was used to determine boron implant concentrations by depth profiling (5).

## RAPID SOLIDIFICATION PROCESSING

Approximately 10 g of homogenized sample was injected from a quartz cru-

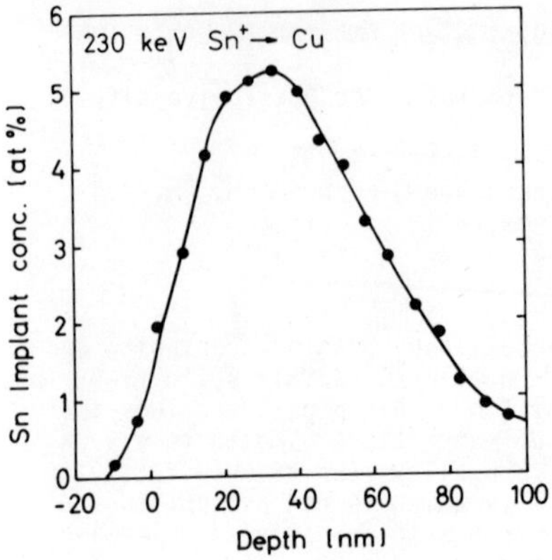


Fig 1. The depth of implanted tin in a copper single crystal.

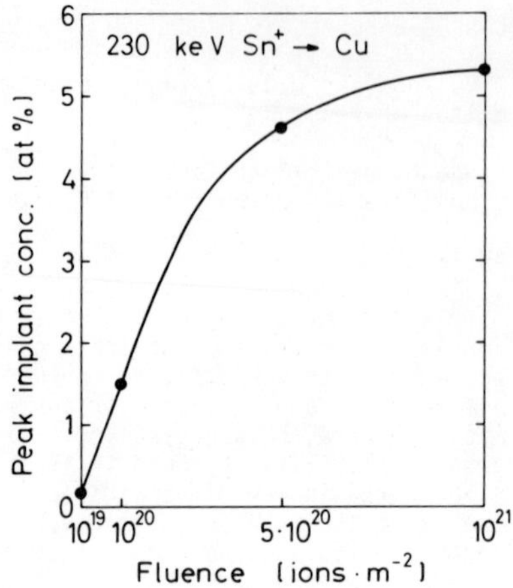


Fig 2. The peak concentration of tin atoms implanted into copper as a function of fluence.

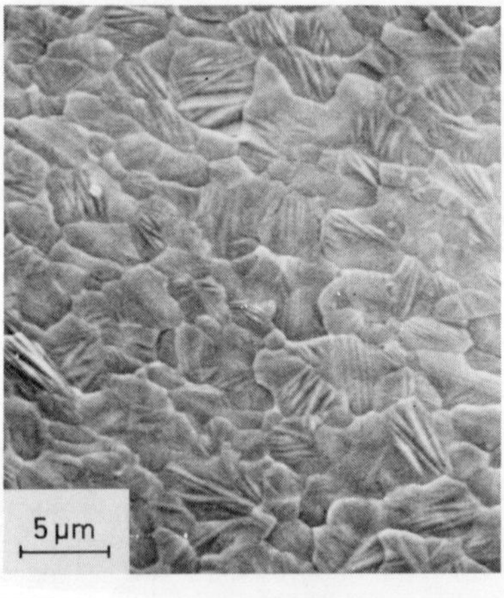


Fig. 3. Fully martensitic surface of melt spun Cu-23·9%Sn ribbon.

cible through 0.8 mm diameter nozzle by applying a top pressure of argon. The crucible to wheel distance was kept at 5 mm and the peripheral wheel speed was 24 $m \cdot sec^{-1}$.

RESULTS AND ANALYSIS

Cu-Sn alloys

The depth distribution of tin in a copper single crystal implanted at random incidence with 230 keV $Sn^+$ ions to a fluence of $10^{21} m^{-2}$ is shown in fig. 1. Theoretical values of 32nm for the range and 18 nm for the straggling were obtained from Winterbon et al. (6) Fig. 2. shows the peak tin concentrations obtained from depth distributions on crystals implanted with different fluences. A calculation of the saturation of tin in copper may be obtained from the following relation:

$$C_s = \frac{1}{Y + 1}$$

and using a value of 15 for the sputtering yield Y, (7) the saturation concentration limit is calculated to be 6 at.% in agreement with fig.2.

Lattice location of the implanted atoms can be derived by investigation of the channeling behaviour combined with RBS analysis. This shows a nearly full substitution of tin in copper after implantation. This result has been confirmed by transmission electron microscopy (8), where a single phase structure saturated

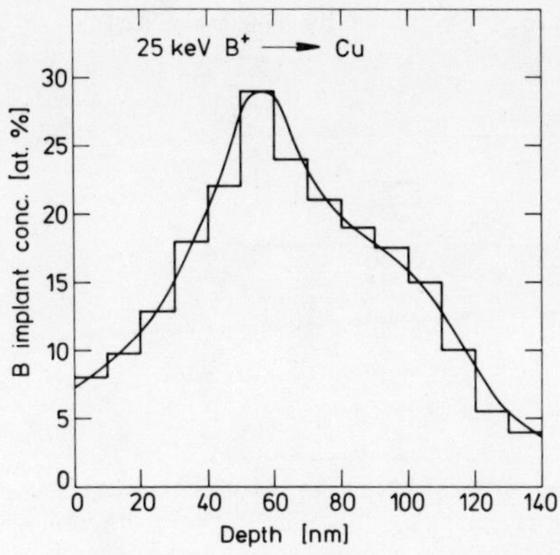


Fig. 4. Concentration profile of boron implanted copper.

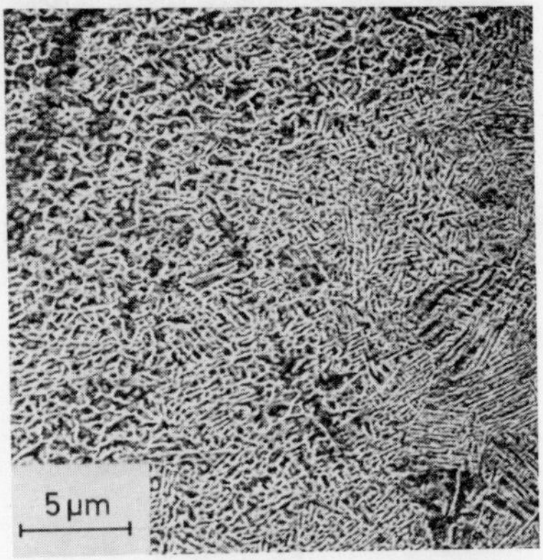


Fig. 5. Structure of melt spun Cu-2%B.

with dense dislocation tangles is observed. Reference to the phase diagram (9) shows that at the implantation temperature the structure of the alloys should be a two phase mixture consisting of $\alpha$ and $\varepsilon$ phases.

Rapid solidification studies on $\alpha$ phase alloys (6.4 at. %) give a single phase structure with a low dislocation density in comparison with the implanted alloys. Tin atoms are again substitutional although there is some evidence of tin clustering. At higher concentrations containing the $\beta$ phase, the structures are complex, since several solid state reactions are possible during the cooling process. For alloys with compositions on either side of the $\beta$ phase field the microstructure is dendritic which is typical of many melt spun alloys. For $\beta$ phase alloys, the $M_s$ temperature is dependant upon composition and segregation. Fig. 3 shows a fully martensitic structure in Cu-23·9% Sn alloy which does not display segregation effects. According to the work of Oshima et al. (4), the $M_s$ temperature is expected to be below room temperature although this is not observed in the present experiments. The reason for this difference in $M_s$ temperature is not clear at present.

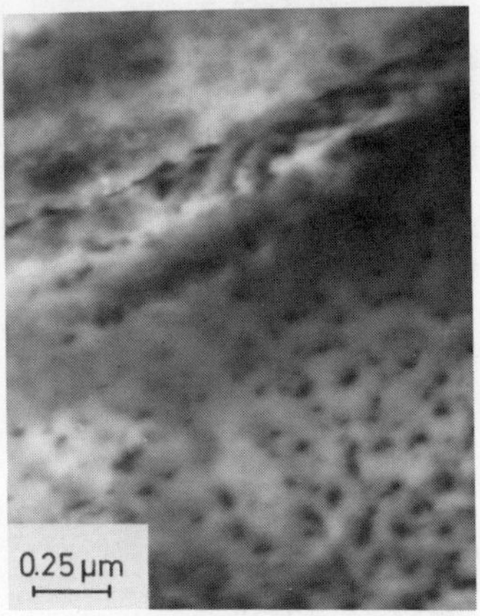


Fig. 6. Boride precipitation in melt spun Cu-1%B.

## Cu-B alloys

The concentration profile of a boron implanted copper sample is shown in fig. 4. Throughout the sample there was no evidence of boride precipitation, and it can be seen that the boron concentration greatly exceeds the solubility limit at 0.5 at.% as well as the eutectic point at10.7 at.%. Channeling studies indicate that the samples are still crystalline (8). The implanted boron atoms

Table 1 shows the nominal compositions of alloys prepared by the melt spinning technique.

Table 1

| Cu | Sn | | Cu | B | |
|---|---|---|---|---|---|
| Wt.% | Wt.% | At.% | Wt.% | Wt.% | At.% |
| 90.20 | 9.80 | 6.38 | 99.50 | 0.50 | 2.20 |
| 82.81 | 17.19 | 11.20 | 99.00 | 1.00 | 5.30 |
| 79.70 | 20.30 | 13.22 | 98.00 | 2.00 | 10.70 |
| 76.98 | 23.02 | 14.99 | | | |
| 76.10 | 23.90 | 15.57 | | | |
| 75.21 | 24.79 | 16.15 | | | |

are both situated on both substitutional and interstitial lattice sites within the implantation zone. Even after annealing for 30 minutes at 600°C there is no evidence of boron redistribution.

By contrast, rapidly solidified copper-boron alloys are not capable of matching this large extension in boron solubility. Scanning Electron microscopy (fig 5) of a Cu 1wt%B alloy shows a fine cell and dendrite boundary phase out of solution. TEM (fig 6) indicates that the matrix contains clusters and precipitates. These and the boundary phase have as yet not been indentified. These results indicate that only a negligible increase in solid solubility of boron can be achieved by rapid solidification.

## DISCUSSION AND CONCLUSION

Data from Cu-B and Cu-Sn alloys produced by the two non-equilibrium techniques show apparent similarities in structures for the $\alpha$ copper phase. Since ion implantation is a low temperature process there is insufficient atomic mobility to respond to the driving force for phase separation for any of these alloys where the composition exceeds the equilibrium terminal solubility limit. Rapid solidification on the other hand does allow some long range atomic movements during solidification and subsequent solid state quenching. The resultant structures in this case are modifications of the equilibrium structures with small grain size, minimal segregation and some extended solubility. Work is now in progress to examine the structures obtained after heavy ion implantation of rapidly solidified alloys.

## ACKNOWLEDGEMENTS

This work is partly sponsored by an EEC contract under the 'stimulation' programme.

## REFERENCES

1) J.V. Wood, Journal de Physique 43 (1982) C4-755.
2) S. Euken and E. Hornbogen, in Rapidly Quenched Metals, edited by S. Steeb and H Warlimont (Elsevier Science Publishers B.V., 1985) pp 1429-1434.
3) Dong-Sheng Li, Qi-Quang Rong and Yun Zhu, ibid, pp 1425-1428.
4) R. Oshima, M. Tanimoto, T. Oka, F.E. Fujika, Y. Hanadate, T. Hamada and M. Miyagi, Journal de Physique 43 (1982) C4-749.
5) E. Ligeon, A. Boutemps, J. Fontenille and G. Guernet, Communication Collogne de Microelectronique Avence, 1970, p.50.
6) K.B. Winterbon, Ion Implantation Range and Energy Distributions, Vol.2 (IFI/Plenum Data Company, New York 1975).
7) H.H. Andersen and H.L. Bay, in Physical Sputtering of Single-element Solids, ed. R. Behrisch (Springer Verlag, 1981), P.145.
8) O. Henriksen, J.V. Wood, E. Johnson, Y. Zysin, A. Johansen and L. Sarholt-Kristensen in Rapidly Quenched Metals (Elsevier Science Publishers B.V., 1985) pp. 827-830.
9) M. Hansen, Constitution of Binary Alloys (McGraw Hill Book Company, New York, 1958, p.633.

# RAPID SOLIDIFICATION OF AN $Al_2O_3$-25WT.PCT.$ZrO_2$ OXIDE ALLOY

TEIICHI ANDO*, YUH SHIOHARA** and M.C. FLEMINGS***
* Toyo Kohan Co. Ltd., Visiting scientist, Materials Processing Center,
** Materials Processing Center, Massachusetts Institute of Technology,
*** Dept. of Materials Science and Engineering, Massachusetts Institute of Technology, Cambridge, MA 02139

## ABSTRACT

TEM observation of the morphology of alumina dendrites in a rapidly solidified $Al_2O_3$-25w/o$ZrO_2$ oxide alloy revealed three distinct dendrite morphologies corresponding to $\alpha$-, $\gamma$- and $\delta$- aluminas. STEM analysis showed the center of the $\alpha$ and $\gamma$ dendrites to be solute poor, but the $\delta$ dendrite center to be solute rich, indicating possible solute trapping. Theoretical calculations show interface growth velocities in excess of 3m/sec.

## INTRODUCTION

The $Al_2O_3$-$ZrO_2$ psuedobinary phase diagram shows only a limited solubility of $ZrO_2$ in $\alpha$-$Al_2O_3$[1]. Relatively few data have been reported on the solubility extension by rapid solidification(RS) of oxide alloys[2]. Among them, the $MgAl_2O_4$(spinel) phase in the $Al_2O_3$-$ZrO_2$ system has been reported to exist in compositions extending to pure alumina[2].

Polymorphism has been reported to exist in RS pure alumina[3]. The reported non-equilibrium modifications, $\gamma$-(cubic) and $\delta$-(tetragonal) aluminas have a spinel structure with varying degrees of lattice disorder[4]. Such polymorphism may be directly linked with the formation of non-equilibrium phases with high solute solubilities in the solidification of a highly undercooled liquid oxide alloy.

The purpose of this investigation was to examine the interrelationship between solidification conditions and phase formation in a rapidly solidified $Al_2O_3$-25 w/o $ZrO_2$ oxide alloy. Of particular interest was to examine possible metastable structures.

## EXPERIMENTAL

Rapidly solidified(RS) samples of an $Al_2O_3$-25 w/o $Zr_O2$ alloy were produced by melt extraction. Sintered pellets 12mm in diameter were prepared from fine grain reagent grade powders of $Al_2O_3$ and $ZrO_2$. The top surface of the pellets was melted in air with an oxidizing oxy-acetylene flame and a copper extraction wheel was pressed against the liquid ceramics which produced thin (50-150 $\mu$m) flakes of RS $Al_2O_3$-$ZrO_2$ alloy samples.
The RS flakes were ion-milled for examination by transmission electron microscopy (TEM) and scanning transmission electron microscopy (STEM).

## RESULTS AND DISCUSSION

The overall microstructure of the RS oxide alloy appears dendritic. Furthermore, the dendrites were found to have three distinct morphologies as shown in Fig.1. The coarsest dendrites in Fig.1(a) show the expected hexagonal symmetry of $\alpha$-$Al_2O_3$, whereas the finer dendrites with well developed secondary and tertiary arms, Fig.1(a), indicate an orthogonal symmetry. The third, and the finest morphology shown in Fig.1(b) also has an orthogonal symmetry although no tertiary arms are clearly observed. The

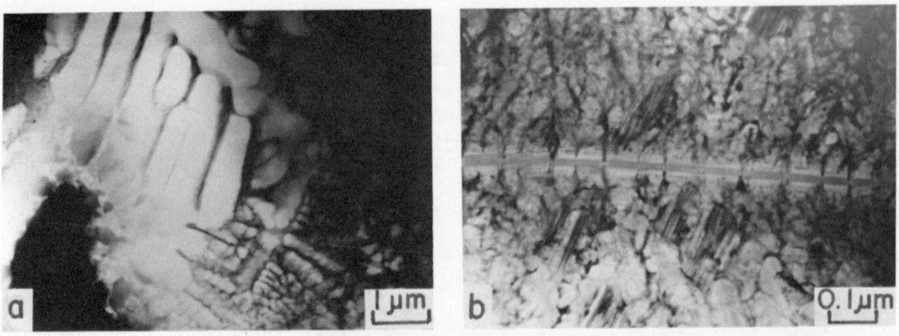


Fig.1 Annular dark field images of dendrites in RS $Al_2O_3$-25wt.pct.$ZrO_2$ alloy.

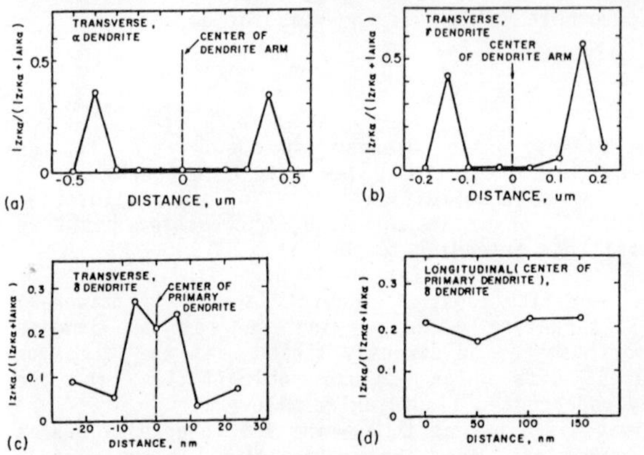


Fig.2 Profiles of normalized ZrKα intensity across (a) coarse, (b) medium and (c) fine dendrites. (a) and (b) were obtained from the area in Fig.1(a). (c) and (d) were obtained from the dendrite in Fig.1(b).

Table 1 Secondary dendrite arm spacings (SDAS) of coarse, medium and fine dendrites in $Al_2O_3$-25$ZrO_2$.

| | SDAS |
|---|---|
| Coarse | 0.57 - 1.1 μm |
| Medium | 0.16 - 0.60 |
| Fine | 0.04 - 0.09 |

values of secondary dendrite arm spacings determined on several dendrites for each dendrite type are listed in Table I.

Fig.2 shows the profiles of the normalized Zr Kα intensity across the three types of dendrites shown in Fig.1. The I ZrKα profiles across the coarse and medium dendrites Fig.2(a) and (b), indicate only very low concentrations of Zr. The fine dendrite, however, has a high Zr content in the 'mid rib' region, Fig.2(c) and (d). Whether the solute enrichment in the fine dendrites indeed occurs as a result of solute entrapment has yet to be proven since solute enrichment beyond the equilibrium solubility is still possible if the solid/liquid interface assumes the local equilibrium defined by the metastable extension of the liquidus and the solidus.

Since the solubility of $ZrO_2$ in α-$Al_2O_3$ is limited, the existence of a non-equilibrium modification with a higher solubility, such as a spinel-type phase, is necessitated to facilitate high supersaturations. The orthogonal symmetry implied by the dendrite morphology appears to support the above hypothesis. In fact TEM does reveal the existence of the γ- and δ- symmetries for the medium and fine dendrites, respectively, Figs.3 and 4. The indexed diffraction patterns unambiguously show that the dendritic growth occurred in the (100) γ and (110) δ directions. The fine

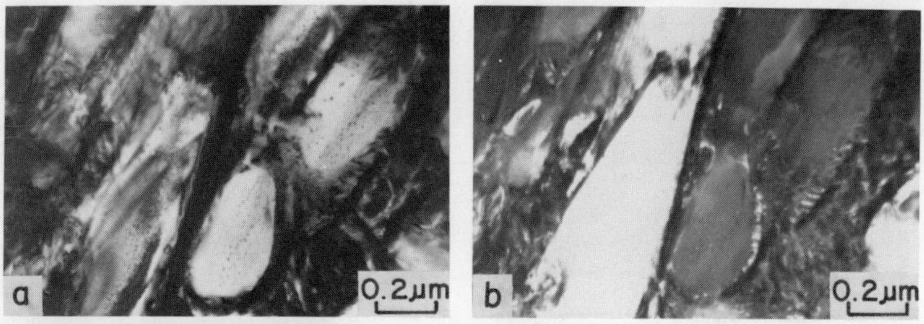


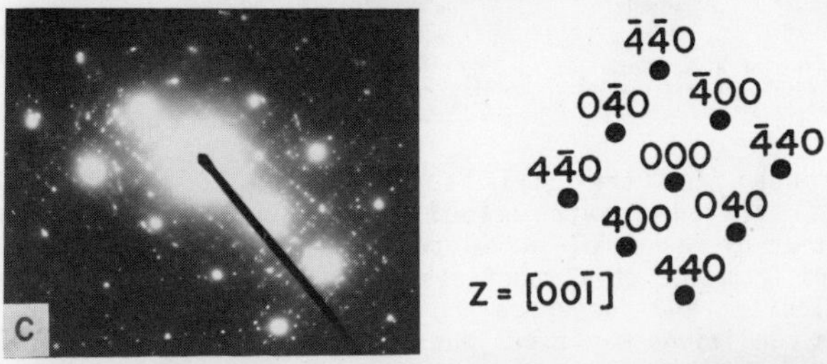


Fig.3 γ-alumina dendrites. Note the growth direction is approximately parallel to $[100]_\gamma Al_2O_3$ (a) bright field image, (b) dark field image produced by $(040)_\gamma$ reflection, (c) selected area diffraction.

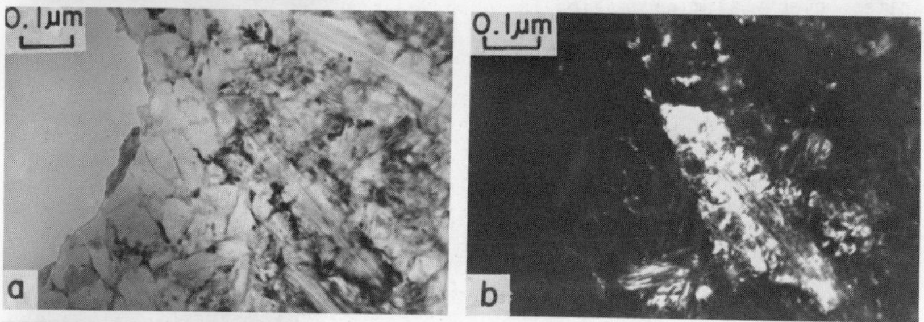


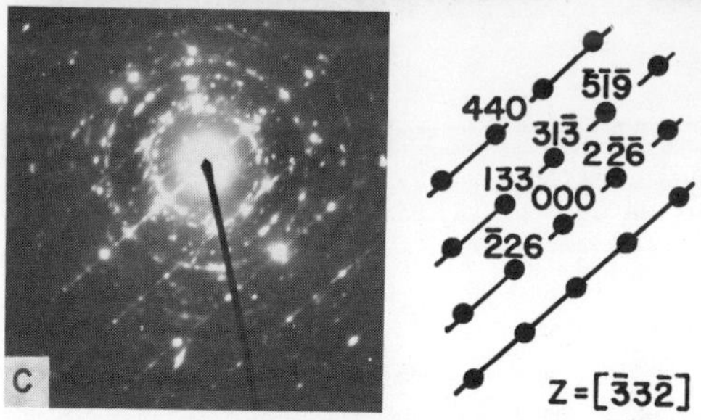


Fig.4 δ-alumina dendrites with growth directions parallel to $[110]_\delta Al_2O_3$. (a) bright field image, (b) dark field image produced by $(4\cdot\bar{4}\cdot\bar{1}\bar{2})_\delta$ reflection (c) selected area diffraction.

striations in the delta alumina dendrites seem to indicate fine scale faulting within the crystal structure and are manifested by the streaks around the diffracted beam spots in the direction perpendicular to the striations.

To achieve the metastable structures found, and certainly to achieve solute trapping, would require high interface velocities with attendant high tip undercoolings. Calculations based on one dimensional heat flow and no mold-casting resistance, Fig.5, show growth rates and cooling rates to vary, respectively, from 0.4 to 0.002 m/s and $2x10^6$ to $3x10^3$ K/s with

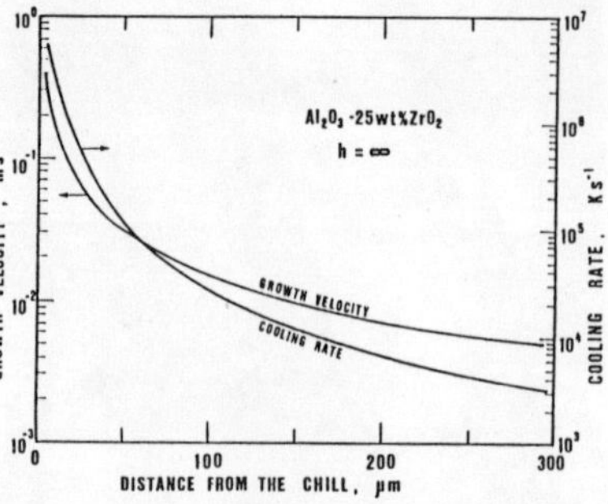


Fig.5 Heat flow and solidification of $Al_2O_3$-25wt%$ZrO_2$ alloy against a copper chill, no mold-casting resistance to heat transfer.

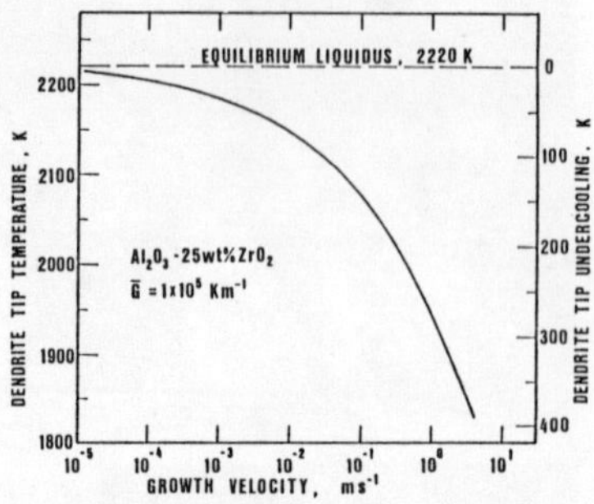


Fig.6 Dendrite tip temprature versus tip velocity, columnar growth of alpha alumina in $Al_2O_3$-25wt%$ZrO_2$ alloy.

increasing the distance from the chill surface from 50 to 300 μm. Comparison of the calculated growth velocities and the expected tip undercooling calculated by a model developed recently (5,6), Fig.6, suggests that at 50 μm from the chill surfaces, an α-$Al_2O_3$ dendrite would grow at a tip velocity and undercooling of 0.03 m/s and 100K, respectively[7]. Such conditions may favor the formation of δ-alumina dendrites, over α-alumina, making the observed high $ZrO_2$ content possible.

## CONCLUSIONS

1. Rapid solidification of an $Al_2O_3$-25wt.Pct.$ZrO_2$ alloy produces a fine microstructure consisting of α-, γ- and δ- alumina dendrites, with the δ-alumina dendrites being the finest.
2. The fine δ-alumina dendrites are characterized by enrichment of Zr in the core region of the dendrites, indicating the possibility of solute entrapment.
3. Heat flow and dendrite tip growth calculations indicate sufficiently high dendrite tip velocities to cause high tip undercoolings and possible solute entrapment.

## ACKNOWLEDGEMENTS

The authors gratefully acknowledge support by Norton Co. Thanks are also due to Dr. M. Simpson of Norton Co. for valuable discussions.

## REFERENCES


1. A.M. Alper, R.N. McNally and R.C. Doman, Amer.Ceram.Soc.Bull. 43,643(1964).
2. P.T. Sarjeant and R. Roy, J.Appl.Phys. 38,4540(1967).
3. P.T. Sarjeant and R. Roy, J.Amer.Ceram.Soc. 50,500(1967).
4. B.C. Lippens and J.H. deBoer, Acte Cryst. 17,1312(1964).
5. R. Trivedi, J.Cryst.Growth. 49, 219(1980).
6. R. Trivedi, and K. Somboonsuk, Mater.Sci.Eng. 65,65(1984).
7. T. Ando, Y. Shiohara, and M.C. Flemings to be published.

# Author Index

# Subject Index